The Euphorbiales

Chemistry, Taxonomy and Economic Botany

Edited by S. L. Jury, T. Reynolds,
D. F. Cutler and F. J. Evans

Proceedings of a joint symposium organized by the Linnean
Society of London and the Phytochemical Society of Europe
in celebration of the Linnean Society Bicentenary

Reprinted from the Botanical Journal of the Linnean Society
Volume 94, Numbers 1 & 2, 1987

Published for the Linnean Society of London

ACADEMIC PRESS

Harcourt Brace Jovanovich, Publishers

London   Orlando   San Diego   New York
Austin   Boston   Sydney   Tokyo   Toronto

ACADEMIC PRESS INC. (LONDON) LIMITED
24/28 Oval Road
London NW1
(Registered Office)

US edition published by
ACADEMIC PRESS INC.
Orlando
Florida 32887

All rights reserved
No part of this book may be reproduced in any form by photostat, microfilm or any other means
without written permission from the publishers

© 1987 The Linnean Society of London
ISBN 0 12 392480 4

Printed in Great Britain by
The Whitefriars Press Ltd., Tonbridge

Contents

Preface 1–2

The saga of the spurges: a review of classification and relationships in the Euphorbiales
GRADY L. WEBSTER 3–46

Segregate families from the Euphorbiaceae
ALAN RADCLIFFE-SMITH. 47–66

Problems of distinction among succulent *Euphorbia* species from eastern tropical Africa
SUSAN CARTER 67–78

Members of the Euphorbiaceae in primitive and advanced societies
RICHARD E. SCHULTES 79–95

Fuel oils from euphorbs and other plants
MELVIN CALVIN 97–110

Wood anatomy of the Euphorbiaceae, in particular of the subfamily Phyllanthoideae
ALBERTA M. W. MENNEGA 111–126

A survey of pollen morphology in Euphorbiaceae with special reference to *Phyllanthus*
WILLEM PUNT. 127–142

Laticifers in Euphorbiaceae—a conspectus
PAULA J. RUDALL. 143–163

Laticifers and the classification of *Euphorbia*: the chemotaxonomy of *Euphorbia esula* L.
PAUL G. MAHLBERG, D. G. DAVIS, D. S. GALITZ AND G. D. MANNERS 165–180

New aspects of rubber biosynthesis
BERNARD L. ARCHER AND B. G. AUDLEY 181–196

Tumour promoters of the irritant deterpene ester type as risk factors of cancer in man
ERICH HECKER 197–219

The biosynthesis of tigliane and related diterpenoids; an intriguing problem
RICHARD J. SCHMIDT 221–230

Activity correlations in the phorbol ester series
    FRED J. EVANS AND MARY C. EDWARDS 231–246

The activation of protein kinase C by daphnane, ingenane and tigliane diterpenoid esters
    ALASTAIR AITKEN 247–263

A review of the evidence from *in vitro* and *in vivo* studies for a role for phorbol ester tumour promoters from the Euphorbiales in the selection and clonal expansion of specific cell populations
    ANNE R. KINSELLA 265–282

Phorbol esters as probes of the modulatory site on protein kinase C—an overview
    PETER M. BLUMBERG, T. NAKADATE, B. S. WARREN, MARIE DELL'AQUILA, T. SAKO, GABRIELLA PASTI AND NANCY A. SHARKEY. 283–292

The chemical constituents and economic plants of the Euphorbiaceae
    ABDEL-FATTAH M. RIZK 293–326

Preface

This symposium was the latest of a number held over the years by the Phytochemical Society, usually in conjunction with the Linnean Society, on a number of important plant groups which have included the families Cruciferae, Umbelliferae, Solanaceae, Compositae and Rutales. It was held in April 1986 at the Jodrell Laboratory, Royal Botanic Gardens, Kew, and attracted an international audience of over 80 scientists who took part in the often controversial discussions which accompanied most papers. This was the first of a series of joint meetings arranged with a number of societies to celebrate the bicentenary of the Linnean Society (1988).

The Euphorbiales is a large Order whose delimitation has been disputed by taxonomists for many years, still with no agreement. It is strange that *Euphorbia*, the genus which gave its name to the family and the order, is atypical in that its flowers are much reduced and surrounded by a cup of bracts with glands, giving a distinctive inflorescence, the cyathium. Therefore, it is not surprising that this volume contains several, cometimes conflicting, ideas on interrelationships in the order, seen from a variety of points of view. These differences of opinion and approach have vindicated one aim of the symposium by highlighting the present research problems in the Euphorbiales, as well as indicating promising pathways for new advances.

The order contains many species of economic or potentially economic importance. One of these, rubber, is a major world crop. This aspect was not neglected during the symposium, and it was the bringing together of so many diverse subjects which aroused so much interest: it is hoped that this will stimulate a multidisciplinary approach in the future. Economic uses of a plant are almost always reflected in its chemistry, and the Euphorbiales provides a seemingly endless vista of chemical problems. One particular and rather obvious interest relates to the diterpenes which show tumour-promoting or anti-tumour activity, and the importance of this subject is reflected in the number of papers on the topic.

Many chemical aspects of the Euphorbiales involve the laticifer system, and it is here that we again encounter a frontier between diverse disciplines. In plant anatomy, Dr. C. R. Metcalfe, former Keeper of the Jodrell Laboratory, made a study of laticifers a quarter of a century ago. Even before that, at the turn of the century, D. H. Scott and co-workers did much of the original work on these structures, at Kew.

The organizers are grateful to the Royal Botanic Gardens, Kew, for providing conference facilities and to the following organizations for help in assembling an international cast of speakers: the Royal Society, the British Council, the Cancer Research Campaign, Ciba-Geigy Ltd., Dr Madeus and Co. (Cologne), the Pirelli Group, Shell Chemicals (UK) Ltd. and the Wellcome Foundation. Special thanks are due to Miss J. Dring (Kew) and Miss S. Darell-Brown (Linnean Society) for help with organization, and to Mr B. Halliwell (Kew) for mounting an exhibition of living examples of Euphorbiales. Several other

PREFACE

members of Kew staff gave invaluable help, and we are especially grateful to the Director, Professor E. A. Bell, for his support and interest.

T. REYNOLDS
Phytochemical Society

D. F. CUTLER
Linnean Society

The saga of the spurges: a review of classification and relationships in the Euphorbiales

GRADY L. WEBSTER

Department of Botany, University of California, Davis, California, U.S.A.

Received June 1986, accepted for publication September 1986

WEBSTER, G. L., 1987. **The saga of the spurges: a review of classification and relationships in the Euphorbiales.** The order Euphorbiales, defined by Lindley in 1836, has undergone many vicissitudes to the present day. Over 30 families have been referred to the Euphorbiales by various authors, but most of them no longer appear closely related to the Euphorbiaceae. Several families commonly referred to Euphorbiales (or Tricoccae) in the 19th century now appear better located in other orders: Buxaceae in Hamamelidales (*sensu lato*), Empetraceae in Ericales, and Callitrichaceae in Lamiales. Several putatively related families, including the Aextoxicaceae, Stackhousiaceae, and Dichapetalaceae, appear to be of Celastralean affinity. The Simmondsiaceae, a problematical family once referred to the Buxaceae, have been included in the Euphorbiales in recent classifications, but the evidence for this is still inconclusive. Thymelaeaceae show a number of similarities to Euphorbiaceae, but share a greater number of distinctive characters with Malvales. The Euphorbiales show affinities to both the Malvales and Geraniales (*sensu lato*), and separation of these three orders between two major dicotyledon subclasses (Dilleniidae and Rosidae) appears questionable. The evidence does not support recognizing segregate families from the Euphorbiaceae such as the Bischofiaceae, Picrodendraceae and Stilaginaceae; the Euphorbiales are therefore best construed as containing a single major family, the Euphorbiaceae, as suggested by Hutchinson; the small family Pandaceae is also included in the Euphorbiales but may not be separable from the Euphorbiaceae. The recent classifications of Airy Shaw and Webster agree in many respects, and areas of disagreement indicate the need for additional data gathering and analysis.

ADDITIONAL KEY WORDS:—Aextoxicaceae – Buxaceae – Dichapetalaceae – Euphorbiaceae – Malvales – Simmondsiaceae – Thymelaeaceae.

CONTENTS

Introduction	3
Taxonomic History of the Euphorbiales	7
Relationships of the Euphorbiales	17
Classification of the taxa of Euphorbiaceae	23
Conclusions	38
Acknowledgements	39
References	39

INTRODUCTION

The first detailed treatment of the Euphorbiales was by Lindley (1836), and it is somewhat sobering to consider that after a century and a half we are still debating many of the same taxonomic questions.

Before reviewing the taxonomic history of the Euphorbiales, we have to face the fact that the order has been given very different circumscriptions by different systematists. To go back no further than the most recent treatments, we find that Croizat (1973) and Hutchinson (1973) include only one family, the Euphorbiaceae, in the Euphorbiales, whereas the order includes four to six families in the classifications of Stebbins (1974), Takhtajan (1980), Cronquist (1981), Dahlgren (1983) and Thorne (1983); and the family composition is different in all five of the multi-family classifications! This disagreement among both contemporary and past authors is indeed a major justification for holding this symposium.

To approach the problem of the history of the Euphorbiales, perhaps the most pragmatic solution is to begin with the uni-familial concept of Hutchinson, as the Euphorbiaceae is clearly the 'core' family in the circumscriptions of all workers. Let us then briefly review the salient characteristics of the Euphorbiaceae as defined by Webster (1967, 1975), Hutchinson (1969b, 1973), and Cronquist (1981). This will establish a central mode, or standard, for the Euphorbiales, and the problem then becomes how to decide which other families show sufficient similarity to the Euphorbiaceae that they should be included in the same order.

The Euphorbiaceae, with about 8000 species in 300 genera, are one of the largest and most diversified families of angiosperms. Although all taxa have unisexual flowers, pollination is achieved by a wide variety of agents, including wind, insects, birds, bats and non-flying mammals. There is great diversity in growth form, from tall rain forest trees to lianas, shrubs, perennial and annual herbs, geophytes, succulents, and floating aquatics; only the epiphytic habit is lacking among the major 'niches' of vegetative adaptation. However, because of the primarily tropical distribution of the family, and the often minute flowers, the Euphorbiaceae have been much less thoroughly studied than other large angiosperm families, and there have previously not been any symposia on Euphorbiaceae such as those treating the Compositae, Cruciferae, Gramineae, Leguminosae, Solanaceae and Umbelliferae.

Nevertheless, despite this relative neglect, the classification of the Euphorbiaceae has been studied during the past 150 years by a number of outstanding systematists, including Adrien de Jussieu (1823, 1824), Baillon (1858), Boissier (1862), Jean Mueller (1866a), Bentham (1878, 1880), Pax (1890–1931), Hurusawa (1954), Hutchinson (1969b), and Airy Shaw (1971, 1975, 1980a). However, the last complete monographic treatment of the family is that of Boissier (1862) and Mueller (1866a) in A. P. de Candolle's *Prodromus Systematis Naturalis Regni Vegetabilis*, and only one large genus, *Manihot*, has been recently monographed (Rogers & Appan, 1973). Since the time of Pax & Hoffman (1931), considerable outstanding systematic work has been accomplished on individual groups of Euphorbiaceae, especially the tribe Euphorbieae (Allem & Irgang, 1975; Anton, 1974; Bally, 1961; Carter, 1982; Croizat, 1938, 1939, 1940, 1972; Dressler, 1957, 1962; Leach, 1976; Mahlberg & Pleszczynska, 1984; Wheeler, 1941, 1943; White, Dyer & Sloane, 1941). Important regional contributions to our knowledge of Euphorbiaceae have been provided for Asia and Australia (Airy Shaw, 1971, 1975, 1980a, b, 1981, 1982), Africa (Leonard, 1962; Vindt, 1953), Madagascar (Leandri, 1957); North America (Webster, 1967; Webster & Burch, 1968; Webster & Huft, 1987) and

South America (Jablonski, 1967; Lanjouw, 1931; Lourteig & O'Donell, 1943). Many other important contributions (partly cited in Webster, 1967) to the systematics of the family are not mentioned here because they fall outside the scope of this review.

In technical terms, the outstanding diagnostic characteristics of the Euphorbiaceae include the stipulate simple or palmately compound leaves; basically cymose inflorescences (often highly modified and raceme-like or capitate); hypogynous radially symmetrical unisexual flowers; a receptacular nectary disk (sometimes secondarily lost); petals showy to absent; stamens 1 to many and often connate; pollen grains 2- or 3-nucleate and very diverse in exine sculpturing; gynoecium of mostly 3–5 connate carpels with styles undivided to multifid; the ovules solitary or paired in each carpel, anatropous (or less commonly hemitropous) and epitropous, crassinucellate, with two integuments; fruits mostly capsular (but sometimes drupaceous or baccate), primarily septicidal, the carpels (cocci) separating from a central columella on dehiscence; seeds 1 or 2 in each coccus, with exotegmic seed coat development, with or without endosperm, often carunculate; embryo large, often with broad cotyledons (Webster, 1967).

This characterization of the Euphorbiaceae understates the considerable morphological variation within the family. Relationships among the 300 genera are still poorly understood, but I have proposed to group them in five subfamilies (Webster, 1975); salient characters of the subfamilies are given in Table 1. Biochemically, the family is very diverse as indicated in the summaries of Hegnauer (1966) and Gibbs (1974); alkaloids, cyanogenic glycosides, fatty acids, glucosinolates and terpenoid compounds have provided a large body of biochemical information of great systematic interest.

Considering its size, the family is rather poorly known with respect to morphological and biochemical characters, although there are summaries of the work on anatomy by Metcalfe & Chalk (1950), laticifers by Mahlberg (1975), cytology by Hans (1973), floral morphology by Michaelis (1924), pollen by Punt (1962) and Köhler (1965), seed coat anatomy by Netolitzski (1926) and Corner (1976), and embryology by Davies (1966) and Rao (1970). Our knowledge in all these areas is very uneven and is much more satisfactory for a small minority of genera of economic importance (*Hevea*, *Manihot*, and *Ricinus*) or horticultural interest (succulent *Euphorbia*).

I now consider those families that have been added to make up the membership of the Euphorbiales. As indicated in Table 2, at least 30 families have been proposed for membership. It is especially interesting to see how fashions have changed; of those families placed in the Euphorbiales by 19th century systematists, only one, the Buxaceae, is still considered as possibly Euphorbialean in the late 20th century (and that is a minority opinion). In the following survey of the taxonomic history of the Euphorbiales, the reasons for this striking shift in opinion are given.

TAXONOMIC HISTORY OF THE EUPHORBIALES

Although the order Euphorbiales technically dates from the publication of the name in the *Nixus Plantarum* of Lindley (1833), groupings of plants that included the spurges (Euphorbiaceae) and supposedly related taxa were recognized much

Table 1. Summary of characteristics of the taxa of Euphorbiales; the five subfamilies of Euphorbiaceae are given as circumscribed by Webster (1975); characters of the Pandaceae are mostly from Forman (1966, 1968, 1971) and Villiers (1973, 1975)

Euphorbiaceae A. L. de Jussieu, *Genera Plantarum:* 384 (1789)
   Petals imbricate or valvate but rarely hooded or concave; disk present or absent; fruits capsular, baccate, or drupaceous; ovules anatropous to hemitropous, obturator present
   Subfamily 1. Phyllanthoideae Ascherson, *Flora der Provinz Brandenburg, 1:* 59 (1864). Laticifers absent; leaves stipulate, simple (except *Bischofia*), usually alternate, and entire; petals and disk present or absent; pollen grains 2-nucleate, exine not spinulose; ovules 2 per locule; seeds rarely carunculate. 13 tribes with 50 genera, *c.* 2000 species; common in wet to seasonal habitats, but rare in deserts
   Subfamily 2. Oldfieldioideae Köhler & Webster, *Journal of the Arnold Arboretum, 48:* 308 (1967). Laticifers absent; leaves exstipulate, simple or compound, alternate to opposite or whorled, entire or dentate; petals and disk absent; pollen grains 2-nucleate, exine usually distinctly spinulose; ovules 2 per locule; seed often carunculate. 4 tribes with 25 genera and *c.* 100 species; in wet to desert habitats
   Subfamily 3. Acalyphoideae Ascherson, *Flora der Provinz Brandenburg, 1:* 58 (1864). Laticifers generally absent; leaves stipulate, simple to lobed (rarely compound), alternate (rarely opposite), entire to dentate; petals and disk present or absent; pollen grains 2-nucleate, exine neither spinulose nor with 'croton' pattern; ovules 1 per locule; seeds carunculate or not. 19 tribes (some of which perhaps should be combined) with 110 genera, *c.* 1500 species; in many habitats
   Subfamily 4. Crotonoideae Pax, *Natürlichen Pflanzenfamilien, 3(5):* 14 (1890) Articulated and/or non-articulated laticifers usually present; leaves stipulate, simple to compound, alternate (rarely opposite), entire to dentate; sepals mostly imbricate; petals and disk usually present; pollen grains 2-nucleate or 3-nucleate, exine mostly with 'croton' pattern; ovules 1 per locule; seeds often carunculate. 11 tribes with *c.* 70 genera, *c.* 2000 species; well represented from rain forest to desert
   Subfamily 5. Euphorbioideae. Non-articulated laticifers usually present; leaves mostly stipulate (except in *Euphorbia*), simple to lobed, alternate (rarely opposite), entire to denate; calyx mostly valvate to obsolete (rarely imbricate); petals and usually disk absent; pollen grains 2-nucleate or 3-nucleate, exine reticulate or tectate; ovules 1 per locule; seeds carunculate or not. 5 tribes with *c.* 40 genera, *c.* 2500 species; commonest in seasonal to arid regions.

Pandaceae Pierre, *Bulletin de la Société Linnéenne de Paris,* 1255 (1896); Engler & Gilg, *Syllabus der Pflanzenfamilien, Ed. 7:* 223 (1913)
   Petals slightly imbricate to valvate, often hooded; disk absent; fruits drupaceous, endocarp often with processes or canals; ovules anatropous to orthotropous, obturator absent
   Tribe 1. Dicoelieae Hurusawa, *Journal of the Faculty of Science, University of Tokyo, III. Botany, 6:* 322 (1954). 1 genus, *Dicoelia* Bentham, with 1 or possibly 2 species. Monoecious; flowers in axillary thyrses; pistillode with slender branches; styles entire, elongated; ovules 2 in each locule; fruit capsular
   Tribe 2. Galearieae Bentham, *Genera Plantarum, 3:* 247, 287 (1880) Bennettieae R. Brown ex Schnizlein, *Iconographia Familiarum Naturalium Regni Vegetabilis, 3:* pl. 172 (1860?). 3 genera, *Galearia, Microdesmis,* and *Panda,* with *c.* 30 species. Dioecious; flowers in terminal thyrses or axillary clusters; pistillode unlobed; styles bifid; ovules 1 in each locule; fruit drupaceous

earlier. Morison (1680) appears to have been the earliest botanist to publish a distinct group, under the name *Plantae tricoccae purgatrices*. Linnaeus, in collaboration with van Royen (1740), established a class Tricoccae that included several Euphorbiaceous genera as well as a number of extraneous ones (*Cneorum, Montia, Phylica, Osyris*). He proposed a similar treatment in the *Philosophia Botanica* (1751) under the name Tricocca, and (as Tricoccae) in an appendix to the sixth edition of the *Genera Plantarum* (1764). Of the 26 genera listed by Linnaeus under Tricoccae, only 16 are now considered Euphorbiaceous, and only five of the ten extraneous genera (*Buxus, Carica, Cneorum, Sterculia* and *Thryallis*) belong to families considered by later writers as of Euphorbialean affinity.

The name Tricoccae, for the group including the Euphorbiaceae and related families, continued to be used for more than two centuries; it was adopted by

Table 2. Families that have been included in the Euphorbiales (or Tricoccae) by major botanical authors. Segregate families (see Table 3) are not listed, except for Pandaceae. Only selected citations are included for some families. Incidental and provisional placements have mostly been omitted.

Family	Reference
Aextoxicaceae	Barkley, 1948; Dahlgren, 1980; Takhtajan, 1980; Thorne, 1980
Aquifoliaceae	Bartling, 1830
Batidaceae	Lindley, 1853
Buxaceae	Eichler, 1878; Gundersen, 1950; Cronquist, 1981
Callitrichaceae	Horaninov, 1847; Lindley, 1853; Wettstein, 1935
Celastraceae	Bartling, 1830; Lindley, 1833, 1836
Daphniphyllaceae	Wettstein, 1935; Emberger, 1960; Novák, 1961; Soó, 1967; Stebbins, 1974
Dichapetalaceae (Chailletiaceae)	Baillon, 1873; Pulle, 1938; Emberger, 1960; Novák, 1961; Thorne, 1983
Didymelaceae	Novák, 1961
Empetraceae	Lindley, 1836; Endlicher, 1841; Eichler, 1878; Warming, 1904
Forestieraceae	Brongniart, 1850
Fouquieriaceae	Lindley, 1836
Geissolomataceae	Hooker, 1876
Gyrostemonaceae	Lindley, 1853
Haloraghaceae	Horaninov, 1847
Lacistemaceae	Hooker, 1876
Lophopyxidaceae	Emberger, 1960
Malpighiaceae	Lindley, 1833
Nepenthaceae	Lindley, 1847
Pandaceae	Webster, 1967; Cronquist, 1968, 1981; Thorne, 1983
Penaeaceae	Hooker, 1876
Pittosporaceae	Bartling, 1830
Polygalaceae	Grisebach, 1854
Rhamnaceae	Bartling, 1830
Simmondsiaceae	Takhtajan, 1980; Cronquist, 1981; Dahlgren, 1983; Thorne, 1983
Stackhousiaceae	Bartling, 1830; Martius, 1835; Lindley, 1836
Thymelaeaceae	Thorne, 1968, 1983
Tremandraceae	Grisebach, 1854
Trigoniaceae	Grisebach, 1854

Endlicher (1841), Klotzsch (1859, 1860), Eichler (1878), Rendle (1925), Wettstein (1935), Copeland (1957) and Lanjouw et al. (1968). However, although its use still appears to be legal in the International Code Of Botanical Nomenclature (Voss et al., 1983), the overwhelming number of 20th century systematists have used the name Euphorbiales. However, this nomenclatural concensus is rather misleading since, as mentioned earlier, the circumscription of the order Euphorbiales has been extremely variable.

The only other ordinal name that gained any currency during the 19th century is Cocciferae, which appears to have been introduced by Batsch (1802), who included the family Tricoccae. Martius (1835) treated Cocciferae as a "Cohors" apparently equivalent to the order Euphorbiales of Lindley. Grisebach (1854), the last major systematist to use the name Cocciferae, treated it as a Nexus with four families: Polygalaceae, Tremandraceae, Trigoniaceae, and Euphorbiaceae in an inclusive sense (including Putranjiveae, Pseudantheae, Scepaceae, Batideae, and Antidesmeae).

In reviewing the literature, I find that over 30 families have been included in the Euphorbiales by leading botanists in the 19th and 20th centuries (Table 2);

and the number would be considerably higher if all of the families segregated from the Euphorbiaceae were included. Redefinitions of the boundaries of the Euphorbiales have steadily gone on since the initial listing of the constituent genera of Tricoccae by Linnaeus (1751, 1764). During the critical period 1825–1840 when the natural system was replacing the Linnean system, the publications of Lindley (1833, 1836) were most influential in establishing the order as the standard hierarchical group that includes one or more families (despite the fact that Lindley used the term "alliance" in place of order, and "order" for family, his use of the endings -ales and -aceae established the precedent for later usage). Lindley (1836) included five families within Euphorbiales: Euphorbiaceae, Empetraceae, Stackhousiaceae, Fouquieriaceae, and Celastraceae. Although the latter three families were not regarded as Euphorbialean by subsequent workers, the Empetraceae continued to be assigned to the Euphorbiales by Endlicher (1841), Eichler (1878), and Warming (1904). In the third edition of his *Vegetable Kingdom*, Lindley (1853) offered his matured judgement about the membership of the Euphorbiales, and, as sometimes happens, it was retrogressive; Stackhousiaceae, Fouquieriaceae, and Celastraceae were excluded, but their replacements were families that on the whole would be regarded as even less closely related to the Euphorbiaceae: Batidaceae, Callitrichaceae, Gyrostemonaceae, and Nepenthaceae.

A particular role in the taxonomic history of the Euphorbiales has been played by the family Buxaceae. The boxwoods were included in Linnaeus's Tricoccae, and in the Euphorbiaceae by Adanson (1763; as Tithymali), A. L. Jussieu (1789), and Adrien Jussieu (1824). Since Lindley (1836) also included Buxaceae within Euphorbiaceae, the Euphorbiaceous core of the order was tainted with Buxacean characters. Baillon (1856, 1858, 1859) deserves the credit for clearly demonstrating that *Buxus* does not belong to the Euphorbiaceae, although he appears to have been anticipated by Plée (1854). Baillon (1859), in his monograph of the Buxaceae, reaffirmed his earlier statements by providing a detailed description and excellent illustrations of floral morphology in *Buxus* and related genera. He showed clearly that the Buxaceae differ significantly from the Euphorbiaceae in characters of the gynoecium, especially in the peripheral position of the styles on the carpels and the apotropous ovules. Agardh (1857) had recently described and illustrated the differences in ovule position that involve the relative placement of the funiculus (raphe) with respect to the placentation. Although Baillon did not use Agardh's terminology, it is clear from his description that the ovules of *Buxus*, with the raphe turned away from the placenta (outside the micropyle), are strikingly different in orientation from the ovules of typical Euphorbiaceae, where the raphe is next to the placenta (and the micropyle outside) and the ovules would be described as apotropous. The systematic value of this character has been disputed, for example by Asa Gray (1879); and Bentham (1878, 1880), after finding exceptions to most of Baillon's diagnostic characters for Buxaceae, did not consider epitropous ovules of sufficient weight to merit placing *Buxus* in a family separate from Euphorbiaceae. However, Mueller (1866a) and Pax (1884, 1890) were convinced by Baillon's arguments and accepted Buxaceae as a distinct family. Although Eichler (1878) retained Buxaceae in the Euphorbiales (Tricoccae), Engler (1892, 1897) transferred the Buxaceae to the Celastrales. During the 20th century, the influence of Bentham and Eichler is apparent in the retention of the Buxaceae in the Euphorbiales by Rendle (1925), Wettstein (1935) and Cronquist (1981).

Although Dang-Van-Liem (1962) affirmed an embryological similarity between Buxaceae and Celastrales, the preponderance of evidence pointed toward a position in or near the Hamamelidales; this is well summarized by Takhtajan (1980). Some floral features of Buxaceae, such as the tardily closing carpels with peculiar development of placentation, interpreted as parietal by Baillon, and elongated outer integuments suggest features of the Hamamelidales as discussed by Endress (1977). Also, the presence of laterocytic stomata in Buxaceae (Baranova, 1983) supports a connection with Hamamelidales. Embryologically, the Buxaceae diverge sharply from the Euphorbiaceae in the presence of cellular endosperm and a different pattern of differentiation of the integuments (Davis, 1966; Wiger, 1935; Wunderlich, 1968). At the same time, it must be noted that the Buxaceae differ from other Hamamelidalean taxa in embryological characters and other features. Behnke (1982) has recently shown that the sieve-tube plastids of Buxaceae (*sensu stricto*) are of a unique type different from *Daphniphyllum* and other Hamamelidales, as well as *Simmondsia* and the Euphorbiaceae.

A recurrent problem with the Buxaceae is the controversial circumscription of the family, due mainly to questions regarding the aberrant genus *Simmondsia*, which was described by Nuttall (1844) as resembling Garryaceae and Euphorbiaceae, and placed in the Euphorbiaceae by Lindley (1853). Baillon (1858) followed Nuttall in assigning it to the Garryaceae, a disposition rejected by Mueller Argoviensis (1866a, 1869), who referred *Simmondsia* to the Buxaceae. van Tieghem (1897a, 1898) removed *Simmondsia* to a separate family, which he suggested belonged to the Centrospermae in the neighborhood of the Tetragoniaceae. Until recently, most botanists followed Mueller in treating *Simmondsia* as a genus of Buxaceae, although Melikian (1968) supported van Tieghem on the basis of seed coat structure. Wagenitz (1975) notes that the ovular structure in *Simmondsia* is incompatible with membership in the Centrospermae, but doubts that Simmondsiaceae, Buxaceae, and Daphniphyllaceae are closely related to the Hamamelidales. Scogin (1980) has claimed support for Lindley's Euphorbiaceous assignment on the grounds of similarities in serological experiments. Thorne (1981, 1983), on the basis of Scogin's evidence, now classifies Simmondsiaceae as a family of Euphorbiales closely related to Euphorbiaceae and far removed from Buxaceae (*sensu stricto*), which Thorne refers to the Pittosporales.

The decision of Scogin and Thorne to place Simmondsiaceae in the Euphorbiales is contested by Köhler & Bruckner (1983), who find that the pollen of *Simmondsia*, as originally indicated by Erdtman (1952), is quite different from other taxa assigned to Buxaceae. In its large rather poorly defined operculate apertures it is also different from any known Euphorbiaceae (Köhler & Bruckner, 1983; Webster, unpubl.). Corner (1976) regards the seed coat structure in *Simmondsia* as typical for Buxaceae, which belongs to his mesotestal class, in contrast to the exotegmic Euphorbiaceae. In view of the ambiguity that surrounds interpretation of many serological studies, and the contrary evidence from pollen and seed structure, I think that the case for assigning Simmondsiaceae to Euphorbiales is weak. Furthermore, the recent study of sieve-element plastids in Buxaceae by Behnke (1982) shows that *Simmondsia*, with S-type plastids, strongly differs from the Buxaceae (*sensu stricto*), which have a unique type of plastid with a central globular protein crystal. This would seem to leave the Simmondsiaceae as a very isolated group without close relationship to either Buxaceae or Euphorbiaceae. Nevertheless, as Köhler & Bruckner (1983) point out, the resemblances between *Simmondsia* and the Buxaceae in wood anatomy, stomatal type, chromosome number, pollen exine stratification and seed coat development do not preclude a taxonomic association, even if not very close, between the two families. On the basis of present evidence, I believe that there is support for the decision of Novák (1961) to place Buxaceae in a separate order Buxales; and perhaps the Simmondsiaceae should be referred to the same order.

A genus that has at times accompanied the Buxaceae in these labyrinthine taxonomic paths is *Daphniphyllum*. Lindley (1853) originally followed the suggestion of Blume (1826) that *Daphniphyllum* might belong to the Rhamnaceae, but Planchon (1854) referred it to the Scepaceae (i.e. Euphorbiaceae tribe Aporuseae), and Baillon (1858) accepted it as a genus of Euphorbiaceae. Mueller (1866a) disagreed and suggested that it might belong to the "Ilicineae"; later (1869) he proposed the family Daphniphyllaceae and noted resemblances to Buxaceae, Euphorbiaceae and Ilicineae. Bentham (1880) returned to Baillon's opinion that *Daphniphyllum* belongs in the Euphorbiaceae, and due perhaps to his influence the Daphniphyllaceae, although usually treated as a family distinct from the Euphorbiaceae, was often juxtaposed in taxonomic treatments (e.g. by Rosenthal, 1919, 1931). Hallier (1904) broke radically with this tradition by transferring *Daphniphyllum* into a tribe of the Hamamelidaceae, an opinion that displayed great taxonomic insight. Support has been given to Hallier's position on the basis of wood anatomy by Janssonius (1950), and on grounds of general morphology by Hutchinson (1959) and Airy Shaw (1966). Although some recent workers (Novák, 1961; Scholz, 1964; Soó, 1967; Stebbins, 1974) have continued to place Daphniphyllaceae in the Euphorbiales, it is excluded by others. Bhatnagar & Garg (1977) note that Daphniphyllaceae differ from Euphorbiaceae in their cellular endosperm development, as well as in pollen characters. Takhtajan (1980) has the Daphniphyllaceae, associated with the Simmondsiaceae, in suborder Buxineae in the Hamamelidales; Cronquist (1981, 1983) recognizes a separate order Daphniphyllales (Hurusawa, 1954) in the Hamamelidae; Dahlgren (1983) places it in the Buxales along with Buxaceae and Didymelaceae; and Thorne (1983) puts it in suborder Buxineae of Pittosporales. Based on the characters discussed above, it appears to me that the classifications of Takhtajan and Cronquist best agree with available evidence.

A recent phenetic study of families in the Hamamelidae by Barabé *et al.* (1982) found the Daphniphyllaceae, Buxaceae, Simmondsiaceae and Didymelaceae to be most closely related not to the 'classic' Hamamelidalean families but rather to fringe families such as the Leitneriaceae, Barbeyaceae, Balanopaceae and Eucommiaceae. The authors note a closer phenetic resemblance of Daphniphyllaceae to Balanopaceae rather than with Buxaceae, and suggest that the Buxaceae and Simmondsiaceae should probably be excluded from the Hamamelidales. These results are not necessarily incompatible with the classifications of Takhtajan, Dahlgren, and Thorne, but seem inconclusive; a thorough cladistic analysis of the complex of families might well produce a more significant insight into relationships.

The small Madagascar genus *Didymeles* has been regarded as allied to the Leitneriaceae by some workers (Baillon, 1873b; Leandri, 1937) and with Euphorbiaceae by others (Novák, 1961; Thorne, 1968, 1976). However, the suggestion of a Hamamelidalean affinity made by Hallier (1912) has been supported on anatomical grounds by Melikian (1973) and Cronquist (1981), who both accept a separate order Didymelales, adjacent to Hamamelidales. In contrast, the palynological study of *Didymeles* by Köhler (1980) indicates a resemblance to Buxaceae; and Carlquist (1982) detects anatomical similarities to Buxaceae. Thorne (1983), therefore, places the Didymelaceae along with Buxaceae, Daphniphyllaceae and Balanopaceae in a suborder of his Pittosporales.

Whether the Hamamelidalean or the Pittosporalean affinity of Buxaceae and Daphniphyllaceae is preferable remains an open question, but at least there is something of a concensus (not entirely shared by Cronquist, 1981) that the Buxaceae and Daphniphyllaceae are not closely related to Euphorbiales. Another family whose Euphorbialean connections are now generally discredited is the Callitrichaceae. Lindley (1853), approving an earlier suggestion by Reichenbach (1837), placed the Callitrichaceae in the Euphorbiales and questioned whether it should even be considered distinct from the Euphorbiaceae. Baillon (1957, 1858) concurred and actually demoted the Callitrichaceae to a tribe of Euphorbiaceae. This was rejected by Mueller (1866a), who accepted the older view of Robert Brown that *Callitriche* is related to the Hippuridaceae. Notwithstanding, the prevailing opinion in the century after Baillon was that Callitrichaceae were close to Euphorbiaceae; Pax (1890, 1931) associated *Callitriche* with Euphorbiaceae in a common suborder; and Rendle (1925) and Wettstein (1935) had both families in the Euphorbiales (Tricoccae). However, Clarke (1865) noted a gynoecial resemblance between Callitrichaceae and Boraginaceae, and much later Jorgensen (1923, 1925) pointed out that the 4-lobed gynoecium and seeds with cellular, haustorial endosperm development are characteristic of taxa in the Tubiflorae (Asteridae). There is still some disagreement about the position of the Callitrichaceae, since Cronquist (1981) associates it with the Hippuridaceae in an order Callitrichales, whereas Takhtajan (1980) and Thorne (1983) put it into the Lamiales; but since Cronquist's Callitrichales immediately follows Lamiales, the difference is not very significant.

A more difficult case than that of the Callitrichaceae is presented by the Dichapetalaceae, which Lindley (1853) placed in the Rhamnales (under the old family name Chailletiaceae). Baillon (1874) included the Dichapetalaceae as a tribe within the Euphorbiaceae. Engler (1897), perhaps influenced by Baillon, placed the Dichapetalaceae in a separate suborder of the Geraniales adjacent to the suborder Tricoccae. Van Tieghem (1903), on the other hand, related the Dichapetalaceae to the Convolvulaceae because of the sympetalous corolla of some taxa. The Dichapetalaceae have more recently been classified in the Euphorbiales by Wettstein (1935), Novák (1961), Takhtajan (1980), Dahlgren (1983) and Thorne (1983).

However, the resemblances of Dichapetalaceae to Euphorbiaceae are not very convincing, and alternative suggestions of relationship have been made, especially with the Celastrales. Lobreau (1969) indeed suggested a relationship between Dichapetalaceae and Euphorbiaceae on the basis of pollen morphology, but later (Lobreau-Callen, 1976) she decided that Euphorbiaceous pollen itself shows similarities to Celastralean families such as the Aquifoliaceae and Icacinaceae. Punt (1975), on the basis of a detailed survey of the pollen of Dichapetalaceae, disagrees that there is any significant similarity with the pollen of Euphorbiaceae, studied by him earlier (Punt, 1962). Punt suggests that there are more impressive resemblances between the pollen of Dichapetalaceae and Malpighiaceae. Prance (1972), in a revision of the neotropical Dichapetalaceae, also supports a Celastralean affinity, with particular ties to the Celastraceae and Icacinaceae. In this instance, it appears that the concensus of our phylogeneticists is not very well-founded, and the best classification appears to be that of Cronquist (1981), who places the Dichapetalaceae in the Celastrales (but admittedly hedges his bets by having it immediately precede the Euphorbiales).

A small family that has long been associated in the literature with the Euphorbiales is the Aextoxicaceae, a monogeneric family endemic to Chile. W. J. Hooker (1836), when describing the genus *Aextoxicon*, ascribed it with some doubt to the Euphorbiaceae; similar opinions were indicated by Endlicher (1841) and Lindley (1853). Baillon (1858) regarded it as forming a group intermediate between the Celastraceae and "Ilicinées"; Miers (1862) related *Aextoxicon* to Icacinaceae (as part of Aquifoliaceae). Decaisne (1858), possibly influenced by the prominence of Monimiaceae in Chile, suggested a comparison with Monimiaceae, which was viewed with approval by Mueller (1866a). Bentham (1878, 1880), who was impressed by the resemblance of the lepidote indumentum of *Aextoxicon* to that of *Hyeronima*, as well as other resemblances between the Chilean plant and various Euphorbiaceae, firmly placed *Aextoxicon* in the tribe Phyllantheae of the Euphorbiaceae. Pax (1890) at first followed Bentham, but later (Pax & Hoffmann, 1917) reconsidered and created a separate family, the Aextoxicaceae, located adjacent to the Icacinaceae. In the latest edition of Engler's *Syllabus*, Scholz (1964) assigns the Aextoxicaceae to the Sapindales. Both 19th century alternative traditions have persisted into recent times, with Takhtajan (1980) and Thorne (1983) referring the Aextoxicaceae to the Euphorbiales, while Hutchinson (1973) and Cronquist (1981) follow the tradition of Baillon, Miers and Pax in treating them as a family of Celastrales. In my opinion, the preponderance of evidence favours the Celastralean view, but *Aextoxicon* is still a very poorly known plant and needs much more detailed study before its relationships can be clarified.

Another small family that has been referred to the Euphorbiales is the Lophopyxidaceae. Although only relatively recently given family status by Pfeiffer (1951), *Lophopyxis* was originally described by J. D Hooker (1887) as a genus of Euphorbiaceae, but of doubtful affinity. Pax (1890) expressed even greater doubt about relationship with the Euphorbiaceae, and Engler (1893) created a subfamily Lophopyxidoideae within the Icacinaceae. Van Tieghem (1897b) recognized it as a distinct family, but with the invalid name Lophopyxidacées. Although Hallier (1910) and Emberger (1960) have reaffirmed the affinity with Euphorbiaceae, most recent workers would disagree. Airy Shaw (1966) indicated a relationship with the Rhamnaceae (tribe Gouanieae), but this also has received little support. Hutchinson (1959) referred *Lophopyxis* to the Celastraceae, and Scholz (1964) placed it in the subfamily Tripterygioideae of Celastraceae. Sleumer (1968), on the other hand, argues that although related to Celastraceae, Lophopyxidaceae should be recognized as a distinct family. Sleumer's argument is reinforced by the pollen study of Lobreau (1969), who found the pollen structure of *Lophopyxis* to be more similar to Tiliales than to Celastrales. Takhtajan (1980) accepts the Lophopyxidaceae as a distinct family of Celastrales, but Cronquist (1981) regards it as only an aberrant genus of Celastraceae. Here we seem to have a recent concensus, at least at the ordinal level; Lophopyxidaceae is Celastralean in affinity and not related to the Euphorbiales.

The family Caricaceae has not often been mentioned as a candidate for admission into the Euphorbiales, although Adanson (1763) included *Carica* in the Euphorbiaceae. However, Hallier (1896, 1903) argued for a close relationship between Papayaceae (=Caricaceae) and Euphorbiaceae on the basis of resemblance in seed coat structure. As indicated by Corner (1976), the seed coat in Caricaceae is indeed exotegmic in development, but differs in a number of ways that are more like Violales than Euphorbiales. The rather striking vegetative and floral resemblances between *Carica* and the Euphorbiaceous genus *Cnidoscolus* appear to be only a striking example of convergence (it is notable that the flowers in both are primarily lepidopteran-pollinated). There is a rather tenuous biochemical similarity in the presence of glucosinolates in both Caricaceae and Euphorbiaceae (Kjaer, 1973), but so far glucosinolates are reported only from the isolated tribe Drypeteae in the Euphorbiaceae, and the resemblance seems more likely to be a convergence. Spencer & Seigler (1984) show that *Carica papaya* is the only known species to produce both glucosinolates and cyanogenic glycosides. These glycosides are of two types: prunasin, a phenylalanine-derived glucoside; and tetraphyllin B, with a cyclopentene ring. The latter compound is characteristic of the Violalean families Flacourtiaceae, Turneraceae and Passifloraceae; so the biochemical evidence at least in part agrees with morphological similarities in suggesting a Violalean position for Caricaceae; and there is indeed a concensus on this point in the classifications of Takhtajan (1980), Cronquist (1981), Dahlgren (1983) and Thorne (1983).

The last, but definitely not the most insignificant, family to be discussed with regard to possible Euphorbialean affinities is the Thymelaeaceae. Although traditionally placed in the Myrtales (e.g. by Cronquist, 1981; Eichler, 1878; Grisebach, 1854; and Wettstein, 1935) or in a separate order Thymelaeales (Gundersen, 1950; Hutchinson, 1973; Pfitzer, 1894; Wagenitz, 1964), the alternative of a relationship with the Euphorbiaceae has surfaced in recent years. Domke (1934), on the basis of floral morphological characters, suggested an overall Malvalean affinity, with particular floral resemblances between Thymelaeaceae and Euphorbiaceae/Dichapetalaceae. Erdtman (1952) emphasized the palynological similarities ("crotonoid" exine) between genera of Euphorbiaceae and Thymelaeaceae, which apparently influenced Thorne (1968) in his placement of the Thymelaeaceae in the Euphorbiales. There are also some biochemical similarities. Paris (1963) notes that the coumarin glycoside daphnetin has been reported only from Thymelaeaceae and Euphorbiaceae. Ourisson (1973) indicates that Thymelaeaceae and Euphorbiaceae also share common diterpenes (phorbol esters). In opposition to these tantalizing suggestions of phylogenetic affinity must be set some considerable morphological differences: Thymelaeaceae have exstipulate leaves, polytelic inflorescences, strongly perigynous flowers with anthers inserted on the hypanthial tube, and often a pseudomonomerous gynoecium. The study of floral anatomy by Heinig (1951) does not indicate any special resemblance to the Euphorbiales. The most impressive recent evidence for a relationship comes from studies of seed coat anatomy by Wunderlich (1968) and Corner (1976), who find that most genera in both families have an exotegmic seed coat with a palisade layer derived from the outer epidermis of the inner integument. As pointed out by a number of writers, this also furnishes evidence in favour of a placement of both Thymelaeaceae and Euphorbiaceae near the Malvales. In my opinion, the many floral differences make it undesirable to locate the Thymelaeaceae within the Euphorbiales, as is done by Thorne (1983). The classifications of Takhtajan (1980) and Dahlgren et al. (1981), in which the Euphorbiales and Thymelaeales are juxtaposed within a superorder Malviflorae, would better accord with the facts. My own inclination is to tentatively associate the Euphorbiales and Thymelaeales together, but in a superorder separate from the Malviflorae. However, it is not yet certain that this is the best solution, and there is still the alternative of a Myrtalean affinity for the Thymelaeales, which we must now consider.

Dahlgren & Thorne (1985) argue that the Thymelaeaceae cannot be regarded as related to typical Myrtales because of a number of biochemical and embryological characters (e.g. lack of ellagic acid, presence of coumarins, different pollen grains, ovules with obturator, and exotegmic seed coat development in Thymelaeaceae). They therefore reaffirm the Malvalean/Euphorbialean position of Thymelaeaceae proposed by Thorne (1968, 1983) and adopted by Takhtajan (1980) and Dahlgren (1983). On the other hand, characters of Thymelaeales more or less concordant with Myrtales are reported in leaf histology (Keating, 1985), wood anatomy (van Vliet & Baas, 1985), sieve-element plastids (Behnke, 1985), and a number of reproductive features (Schmid, 1985). Cronquist (1985) provides a vigorous rebuttal to the Dahlgren/Thorne classification, and notes that most of the resemblances between Thymelaeales and Euphorbiales cited by those authors are either primitive (plesiomorphic) or else, e.g. the common occurrence of daphnetin in *Daphne* and in *Euphorbia*, may represent convergence. It is clear that the jury is still out, and more evidence and analysis is needed to break this deadlock between the experts. However, whatever the position of the Thymelaeaceae may prove to be, I am in sympathy with Cronquist's opinion that it should be assigned to a separate order, and not be included within the Euphorbiales. Perhaps it is worth reconsidering the earlier classification of Dahlgren (1975), in which the Thymelaeales are placed in a superorder Thymelaeanae immediately following the Euphorbiales.

Looking back over the families we have reviewed as candidates for admission to the Euphorbiales, it is apparent that the credentials of most of them are seriously defective. The Buxaceae, although long included within the Euphorbiaceae, now appear to lie outside the Euphorbiales because of both reproductive characters (carpel morphology, apotropous ovules, cellular endosperm and mesotestal seed coat development). The Simmondsiaceae, although approaching the Euphorbiaceae in some respects, differ fundamentally in pollen structure, and may be aberrant relatives of the Buxaceae. Daphniphyllaceae, although not closely related to Buxaceae as has sometimes been suggested, may also be of general Hamamelidalean affinity, and this is probably also true of the Didymelaceae. Certain small families, notably the Aextoxicaceae, Lophopyxidaceae, Stackhousiaceae, and perhaps the Dichapetalaceae, appear to lie closer to the Celastrales. The Callitrichaceae are now generally accepted as belonging in the Asteridae in or near the Lamiales.

After excluding all of these families, there is only one remaining to consider for inclusion in the Euphorbiales: the Pandaceae. This small Old World group was first recognized as possibly meriting family status by Robert Brown (Bennett & Brown, 1852) when he published the genus *Bennettia* from Java, based on the plant now treated as *Galearia filiformis* (Bl.) Boerl. (Forman, 1971). Brown's family name Bennettiaceae is invalid because he proposed it only in a provisional sense, but the name was validated (with the spelling Bennettieae) by Schnizlein (*c*. 1860). Schnizlein agreed with Brown that *Bennettia* showed resemblances to Euphorbiaceae, and especially with Antidesmaceae, then generally considered a distinct family. Schnizlein also suggested a possible affinity between *Bennettia* (i.e., *Galearia*) and *Phytocrene* (Icacinaceae).

Mueller (1866a) demoted the Bennettieae to a subtribe of his tribe Hippomaneae, noting that it was atypical in having the sepals scarcely imbricate. Bentham (1878, 1880) clearly considered *Galearia* and allied genera as occupying a distinctive position in the Euphorbiaceae, since he created Galearieae as one of the six major groups (tribes) in the family. Bentham specifically noted an affinity between *Pentabrachion*, a genus of subfamily Phyllanthoideae, and *Microdesmis*, a genus described by Hooker (1848). Bentham's Galearieae was heterogeneous, as he also included in it *Pogonophora* and *Tetrorchidium*, two genera of subfamily Acalyphoideae and Crotonoideae, respectively. Pax (1890) at first improved the definition of the group, treated by him as subtribe Galeariinae of tribe Clutieae, by removing *Tetrorchidium*; however, he later (Pax, 1911) added the genus *Syndyophyllum* Schum. & Lauterbach, which differs from the other Galearieae in opposite leaves, monoecious flower production and capsular fruit. Airy Shaw (1960) has shown that the branching pattern in *Syndyophyllum* is quite different from that in *Galearia* or *Microdesmis*, and that it has much more in common with *Erismanthus* and *Moultonianthus*. In the system of Webster (1975), *Syndyophyllum*, *Erismanthus* and *Moultanianthus* are associated in tribe Erismantheae of subfamily Acalyphoideae.

After it had been relegated to obscurity for the better part of a century due to the influence of Pax, the Bennettiaceae of Robert Brown and Schizlein was revived by Forman (1966) as the family Pandaceae. Forman provides a detailed history of the family, beginning with the recognition of the Pandaceae by Pierre (1896) concomitantly with his description of the new genus *Panda*. Pierre, in the same publication, recognized the affinity of *Microdesmis* with *Panda*; and the following year (Pierre, 1897) he added *Galearia* to the family. Engler (1912a, b) disagreed with Pierre's conclusions, mainly because the orthotropous ovules of *Panda* contrasted with the more-or-less anatropous ovules of *Galearia* and *Microdesmis*; he restricted the Pandaceae to include only *Panda*, and erected a new order, the Pandales. Forman (1966) confirmed the disposition of Pierre with a detailed study of fruit structure and branching pattern, with contributions on anatomical characters by Metcalfe and Parameswaran.

In a review of the problems of circumscription of the family Euphorbiaceae (Webster, 1967), I agreed that it appeared expedient to recognize the Pandaceae as a distinct family because of a constellation of distinctive features (unusual configuration of petals and stamens in male flowers, divergent fruit type, orthotropous ovules in *Panda*, and the unusual chromosome number of $n = 15$ in *Microdesmis*). In my synopsis of the subfamilies and tribes of Euphorbiaceae (Webster, 1975), I omitted mention of the genera of Pandaceae, thus implicitly continuing recognition of Forman's family concept.

Now, two decades after Forman's proposal to reinstate the Pandaceae, my opinion is still equivocal. The three genera of Pandaceae *sensu* Forman appear to form a natural group, but they show many similarities to the Euphorbiaceae. It seems probable that the ancestry of the Pandaceae is to be sought within the subfamily Phyllanthoideae, and a significant clue to phylogenetic relationships was given by Bentham (1878), who noted the resemblance in floral structure between *Dicoelia* and *Galearia*, and in his arrangement of genera (1880) placed *Dicoelia* as the final genus of Phyllantheae, immediately preceding *Galearia*. Admittedly, *Dicoelia* differs from *Galearia* in its capsular fruit, monoecious flower production (in axillary thyrses), and undivided styles, but it nevertheless is similar in its habit, in the conformation of the staminate flowers (due to position of anthers in concavities in the petals), and in the lack of a floral disk. Although *Dicoelia* is still very poorly known, my own impression is that Bentham's intuition is probably well founded, and that *Dicoelia* stands in an ancestral or collateral relationship to *Galearia* and the other Pandaceae. The separation of *Dicoelia* from these other genera because of its biovulate locules has been overemphasized; there is no reason, after all, why reduction to uniovulate locules could not have occurred, along with other specialized characters, during the evolution of the Pandaceae. In the tabulation of the taxa of Euphorbiales in this paper, I have therefore transferred tribe Dicoelieae from the Euphorbiaceae to the Pandaceae.

However, this does not really solve the problem of family delimitation. Adding a monoecious genus with biovulate carpels to the Pandaceae blurs the distinctions, already difficult to articulate, from the Euphorbiaceae. As Backhuizen van den Brink (1968) remarked, it is a weak argument to recognize the Pandaceae as distinct in order to make the Euphorbiaceae more homogeneous. I believe that the addition of *Dicoelia* to the Pandaceae is an improved reflection of phylogenetic relationships, but it admittedly raises at least as many questions as it answers. Although *Dicoelia* is certainly an isolated genus when placed in the Euphorbiaceae subfamily Phyllanthoideae, it does not appear totally alien either. One could argue that the recognition of the relationship between *Dicoelia* and *Galearia* just as logically suggests incorporating the Pandaceae into the Euphorbiaceae, perhaps as a separate subfamily. The phylogeny of subfamily Acalyphoideae is still very poorly understood, but it is possible that the genera of the Pandaceae represent the sister group to other tribes of Acalyphoideae, such as Erismantheae and Cheiloseae, and perhaps lie near the stem of the Acalyphoideae as a whole. If that proved to be so, the Pandaceae might shrink to the status of a tribe within the Acalyphoideae. The recognition of the Pandaceae as a family distinct from Euphorbiaceae in this paper is therefore done provisionally and for reasons of expedience, not because I am convinced that these two families represent the optimal suprageneric grouping for the taxa of Euphorbiales. In effect, I agree with Croizat (1973) and Hutchinson (1973) in accepting the Euphorbiaceae as the only major family within the Euphorbiales.

RELATIONSHIPS OF THE EUPHORBIALES

To a considerable extent, the history just reviewed of the varying concepts of membership in the Euphorbiales gives an impression of the way systematists have interpreted the relationships of the order to other suprafamilial groups of dicotyledons. The early treatments by Linnaeus (1751, 1764), Adanson (1763) and A. L. de Jussieu (1789) included genera that would now be referred to Malvales, Rhamnales, Geraniales, Sapindales and Violales (in the broad Englerian sense of these orders). However, from the beginning systematists also offered speculations about possible relationships of the Euphorbiaceae; Adanson (1763), for example, noted that his Tithymali approached the Pistaciae (a mixture of mostly Geraniales and Sapindales) in characters of stamens, ovules, and stem exudates.

Ventenat (1799) remarked that "les Tithymaloides se rapprochent, par plusiers caractères, des Rhamnoides". Adrien de Jussieu (1824), the first monographer of the Euphorbiaceae, did not offer any speculations about its affinities with other families (perhaps reflecting his greater knowledge of the magnitude of the problem!). Lindley (1836) listed the Euphorbiales immediately after the Rhamnales, and commented on similarities of the Euphorbiales with both Malvales and Rhamnales; all three orders were included in a common group Syncarposae. However, later he placed Euphorbiales following Urticales (Lindley, 1853), but with lateral affinities to the Malvales and Rhamnales. Lindley in 1853 appears to have had a working model of a progression from a simpler to more complex floral form, as he says: "Euphorbiaceae may be regarded then as a higher form of Urticads, and accordingly we find their lateral affinities also pointing to groups with a more complicated structure; as for example to the Rhamnads in the perigynous, and Malvads in the hypogynous Sub-class".

The first really detailed consideration of the affinities of the Euphorbiaceae was provided by Baillon (1858). Baillon, who always expressed his systematic conclusions in an idiosyncratic form, did not explicitly define an order Euphorbiales, but his discussion implies that the Euphorbiaceae represented an independent order. He found the greatest affinity of the Euphorbiaceae to be with the Malvales, and detected a parallel set of taxa, with different tribes of Euphorbiaceae corresponding to various families of Malvales. In a striking geometrical model (reconstructed in Fig. 1A), he imagined the Euphorbiaceae and Malvales arranged in parallel on the triangular face of a pyramid, with *Euphorbia* and *Malva* (an odd couple!) at the base, ascending through the families Byttneriaceae, Bombacaceae, and Sterculiaceae, with the "Helicterées" opposite the Scepaceae, and at the apex of the triangle (and pyramid) the Urticaceae, reflecting the double relation of that family (in Baillon's mind) to both Malvales and Euphorbiaceae. On the second face of this pyramid he placed the Rhamnales, with emphasis on the Chailletiaceae (Dichapetalaceae), at the margin opposite the Euphorbiaceae, and the Rutales on the more distant margin; the third face to be occupied by the Geraniales, especially the Linaceae. This model can reasonably be converted into a dendrogram of putative relationships between the orders (Fig. 1B). It should be noted that Baillon's pyramidal design was probably copied from that of Weddell (1857), who used a very similar model (with many of the same families) to illustrate the affinities of the Urticaceae.

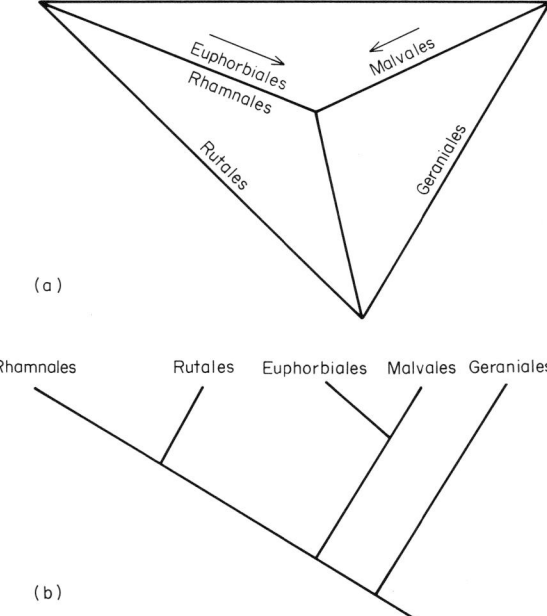

Figure 1. Baillon's model of relationships of Euphorbiaceae with other taxa. A. Pyramidal model described by Baillon (1858). B. Cladogram inferred from Baillon's description of relationships.

This model of Baillon epitomizes relationships observed by Lindley earlier, and other botanists who published later. Mueller (1866a) noted for the Euphorbiaceae that "affinitas ordinis vasti est multiplex", and gave priority of position to ties with Malvaceae and Urticaceae, but also mentioned possible affinities with Rhamnaceae, Dichapetalaceae, Sapindaceae, Burseraceae and Menispermaceae. Bentham (1878, 1880) was very non-committal about ordinal relationships, but he was impressed by the resemblances between tribe Phyllantheae of Euphorbiaceae with Rhamnaceae and Celastraceae, and between the Hippomaneae and Urticaceae. Eichler (1878) emphasized the taxonomic isolation of the Euphorbiales more than any previous systematist by creating for them a Reihe Tricoccae, one of his nine major groups of the dicotyledons (equivalent to the subclasses or superorders of contemporary botanists). In recent times, a similar degree of isolation has been proposed only by Emberger (1960), whose Phylum (=superorder) Tricoques has a single order Tricoques including Euphorbiaceae, Daphniphyllaceae, Dichapetalaceae and Lophopyxidaceae. Pax (1890) assigned the Euphorbiaceae to a suborder of Geraniales, a position probably dictated for him by Engler. It is rather surprising, incidentally, that Engler's system, which seems to have evolved from Eichler's (Stafleu, 1965), demoted the Euphorbiales from a superorder to a suborder of Geraniales. In an explication of his system, Engler (1897) does not offer any indication why this change was made.

An original but uninfluential hypothesis about the phylogenetic placement of the Euphorbiales was offered by Wettstein (1901, 1935). He regarded the apetaly of Euphorbiaceae as primitive, apparently due to his inability to imagine the biological reasons for reversion of bisexual flowers to unisexual ones, or the poor development of a corolla in insect-pollinated plants. In Wettstein's phyletic diagram (Fig. 2A) the Tricoccae are indicated as a connecting link between the Amentiferae and the Malvalean line, with the "Columniferae" derived from an Euphorbialean ancestry. In the "Stammbaum" of Wettstein's system published posthumously by Janchen (1932), the Euphorbiaceae are shown as a nodal family leading to the origin of both the Malvalean and Geranialean families; this resembles the model of Pax, discussed below, but with the direction of evolution reversed. Wettstein's theory that the Euphorbiaceae link the Amentiferae and Malvales has found little support among subsequent authors, although it appears to have influenced Däniker (1946), who found similarities between the Euphorbiaceae and Balanopaceae, and sought to derive the Juglandaceae from the Euphorbiaceae via the Balanopaceae.

When given the opportunity to express his own opinions, Pax (1924) did so in a discussion that summed up his conclusions after 40 years of work on the Euphorbiaceae. His solution to the problem of multiple affinities suggested by previous workers was to propose a frankly polyphyletic origin from both the Malvales and Geraniales (Fig. 2C), although he did not consider this to be true polyphylesis, since the Euphorbiaceae were considered to have branched off early in the evolution of both Malvales and Geraniales (". . . die Euphorbiaceen sich sehr früh von den Urtypen der Geraniales und Malvales lösten und zu einer gewissen Selbständigkeit entwickelten"). Pax suggested that the euphorbiaceous taxa with basically imbricate sepals evolved from near the Geraniales, and those with valvate sepals from the Malvales. As an evolutionary model, this is defective because the 'valvate' taxa of Euphorbiaceae belong mainly to the derived subfamilies Acalyphoideae and Crotonoideae, and appear to be derived from 'imbricate' ancestors within the Euphorbiaceae (Webster, 1967, 1975).

The Paxian concept of a polyphyletic origin of the Euphorbiaceae apparently influenced Hutchinson (1926), who in his phyletic diagram indicated an origin of the Euphorbiales from the Malvales and Sapindales, an opinion concurred with by Croizat (1940). Later (1969a), Hutchinson explicitly stated (but without producing any new evidence) that the Euphorbiaceae are a composite group derived from a number of other families, including Tiliaceae, Sterculiaceae, Malvaceae, Celastraceae, Rhamnaceae, and Sapindaceae. In his last phyletic diagram (Hutchinson, 1973), the Euphorbiales are indicated as derived from the Malpighiales, Celastrales and Malvales; and they are the only order of dicotyledons shown to have a composite origin!

At about the same time as Pax's essay on the phylogeny of the Euphorbiaceae, serological workers also placed the Euphorbiaceae into juxtaposition with both Geraniales and Malvales, as is shown in the phylograms of Hoeffgen (1922) and Mez & Ziegenspeck (1926; Fig. 3). Similar placement of Euphorbiales in the vicinity of both Malvales and Geraniales has been made by a number of subsequent workers, such as Novák (1961, Fig. 2E) and Soó (1967, Fig. 2D).

We have already discussed some of the suggestions of affinity between the Euphorbiales and Thymelaeaceae, most of which seem to have been inspired by the similarity in pollen grains pointed out by Erdtman (1952). Croizat (1960), although most strongly favouring a Sterculiacean affinity for Euphorbiaceae, was also impressed by the similarity to the Thymelaeaceae. As I have noted,

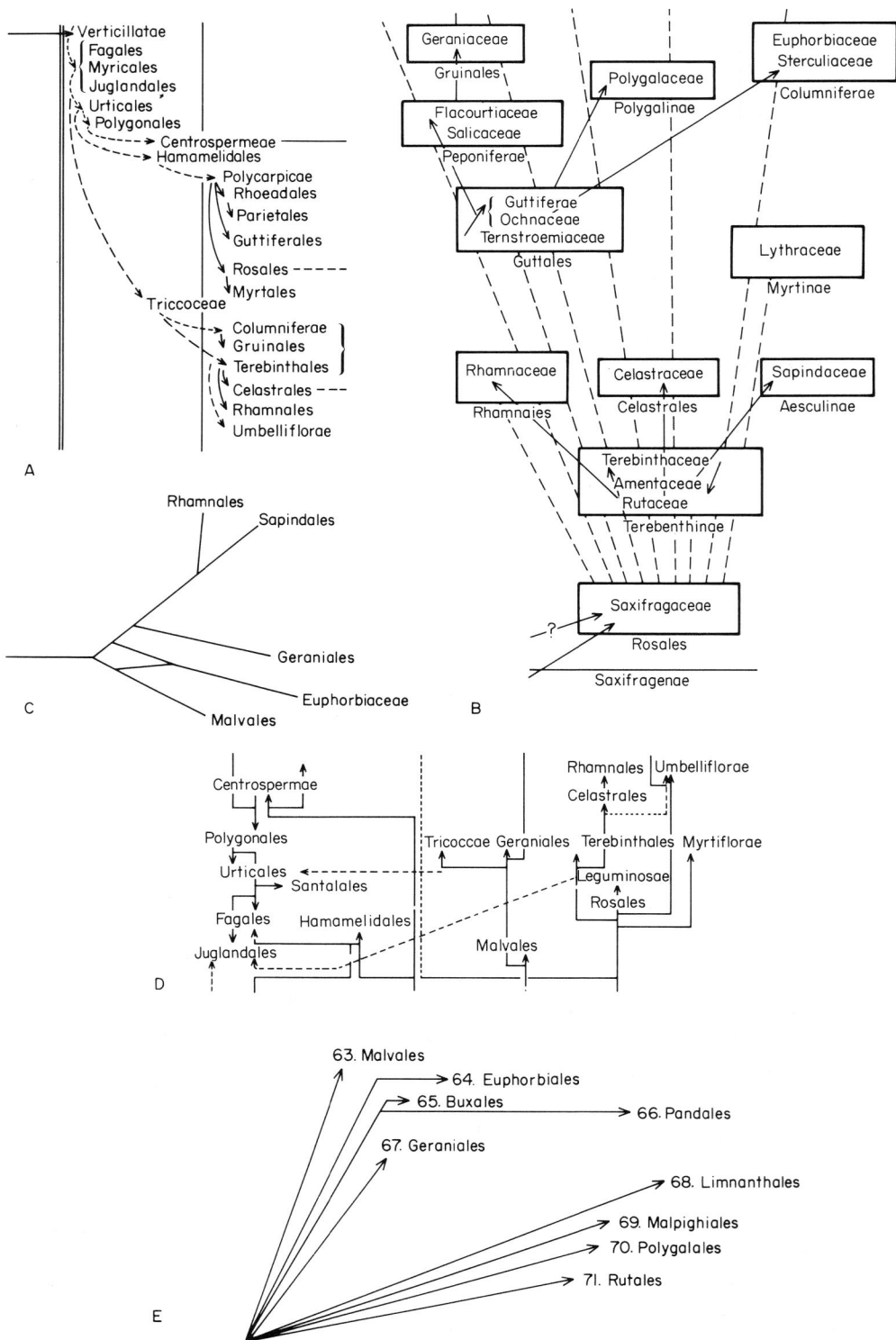

EUPHORBIALES CLASSIFICATION: A REVIEW

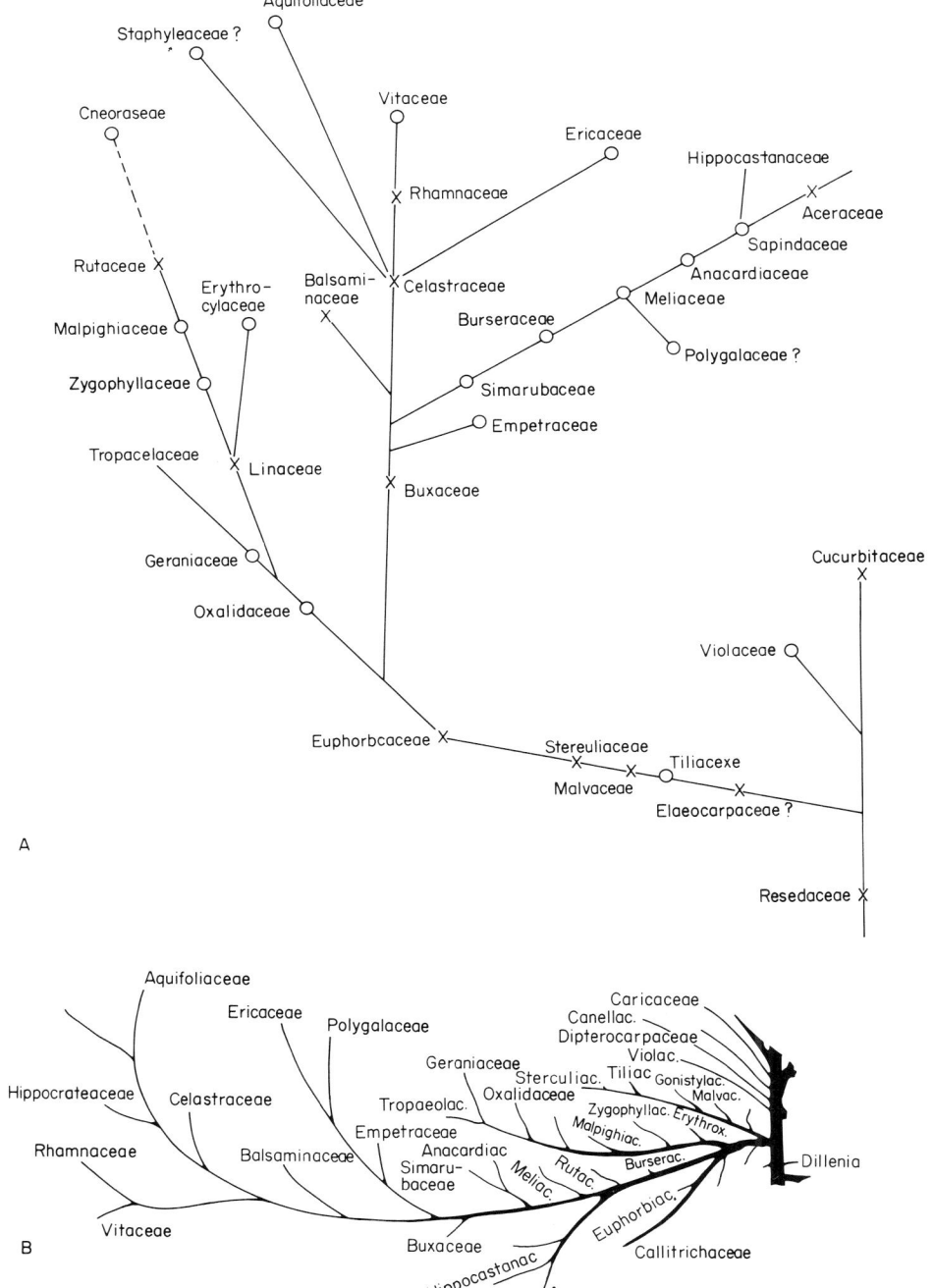

Figure 3. Phylograms based on serological studies. A. Hoeffgen (1922). B. Mez & Ziegenspeck (1926).

Figure 2. Phylograms proposed by various authors to illustrate the putative geneological relationships of the Euphorbiaceae (partly adapted from Grossheim & Sakhokia, 1966). A. Wettstein (1905). B. Hallier (1912). C. Pax (1924). D. Soó (1967). E. Novák (1961).

there is still uncertainty about the relative merits of a Myrtalean or Euphorbialean affinity for the Thymelaeaceae.

Another order that has been mentioned as having Euphorbialean connections is the Violales. The possible affinity of Caricaceae to Euphorbiaceae has already been mentioned, but in addition to that suggestion, Hallier (1903, 1912) also emphasized a possible relationship between Flacourtiaceae and Euphorbiaceae (Fig. 2B). This seems not to have been suggested by any prominent 19th century systematist, perhaps because the difference in placentation seemed too great. However, Janssonius (1950) commented on the resemblance in wood anatomy between certain genera in the two families, and Miller (1975) also sees some similarities. Croizat (1960) regarded a link between Flacourtiaceae and Euphorbiaceae as patently obvious, and Hutchinson (1967) pointed out similarities in gross floral morphology. Takhtajan (1966, 1969) stated that the Euphorbiaceae showed affinities to both Malvales and Flacourtiales. Keating (1973) has demonstrated palynological similarities between the Flacourtiaceae and Euphorbiaceae subfamily Phyllanthoideae. However, it is notable that in their recent treatments, Takhtajan (1980), Cronquist (1981) and Thorne (1983) do not mention any Violalean affinities of the Euphorbiales. I believe that this reflects a concensus (however subconscious) among our phyleticians that the similarities between Flacourtiaceae and Euphorbiaceae mentioned above are mainly unspecialized (plesiomorphous) characters that do not provide any compelling evidence for a common origin.

The most emphatic recent claim of Violalean ancestry for the Euphorbiales is made by Hickey & Wolfe (1975), on the basis of the occurrence of "Violoid" teeth in both groups. In the Euphorbiaceae subfamily Phyllanthoideae, Levin (1986) reports "Theoid" teeth in the genera *Drypetes*, *Putranjiva* and *Bischofia*; presumably the "Violoid" teeth of Hickey & Wolfe were reported from subfamilies Acalyphoideae and Crotonoideae. The presence of "Violoid" or "Theoid" teeth in both Euphorbiaceae and Flacourtiaceae could be regarded as singificant if indeed this represents a shared advanced character due to common ancestry (synapomorphy). However, within the Euphorbiaceae subfamily Phyllanthoideae, the genera *Drypetes*, *Putranjiva* and *Bischofia* have advanced floral characters and cannot be regarded as primitive in the family; the genera that appear basal, such as *Heywoodia*, *Savia* and *Wielandia*, have entire leaves. It appears that the "Theoid" teeth in subfamily Phyllanthoideae, and the "Violoid" teeth in subfamilys Acalyphoideae and Crotonoideae, were "reinvented" from ancestral forms with entire leaves, so it can be argued that the similarity in the leaf tooth character does not represent a significant homology useful in determining the phyletic relationships between Euphorbiaceae and Flacourtiaceae. In order to justify use of this character, one would have to invoke the principle of "apomorphous tendencies" advanced by Cantino (1982, 1985). This idea that shared tendencies to change in morphological characters can be used to define phylogenetic relationship has been criticized as unrigorous by Rasmussen (1983), although it looks as if such a principle indeed reflects the thinking of phylogeneticists such as Cronquist, Takhtajan and Thorne. There is some evidence for relating Flacourtiaceae to Euphorbiaceae because of similarities in seed coat development (Corner, 1976), but these are also shared by taxa of Celastrales. In my opinion, the failure so far to demonstrate additional synapomorphies between Euphorbiales and

Flacourtiales or Celastrales puts the burden of proof on those who follow either the claim of Hickey & Wolfe (1975) that the Euphorbiales came from the Violalean branch of the "palmate Dilleniidae", or the assertion of Cronquist (1981) that they should be associated with the Celastrales.

We have seen that with the exception of the Violalean connection initiated by Hallier and by Hickey and Wolfe, most suggestions of the affinity of Euphorbiales by 20th century systematists fall into two classes with regard to a putative sister group: Celastralean, supported by Stebbins (1974) and Cronquist (1981); and Malvalean, supported by Croizat (1973), Takhtajan (1980), Dahlgren (1983) and Thorne (1983). Nevertheless, I still believe that my own suggestion of Geranialean affinity (Webster, 1967) also merits consideration. The flowers in families such as Ixonanthaceae (Forman, 1965; Noteboom, 1967) appear closer to those of primitive Euphorbiaceae (subfamily Phyllanthoideae) than do those of corresponding genera in Celastrales or Malvales. I suggest that the Linales (*sensu* Cronquist, 1981) must be taken into account in searching for the closest phyletic lines to Euphorbiaceae.

CLASSIFICATION OF THE TAXA OF EUPHORBIACEAE

Although there have been many changes in the circumscription of the Euphorbiales during the past 150 years, these have generally not involved questions of delimitation of the family Euphorbiaceae, with the notable exception of the position of Buxaceae. The outstanding problems of interest, both in the 19th century and now, involve intergeneric relationships and delimitation of the infrafamilial taxa of Euphorbiaceae.

In the first family treatment, that of Adanson (1763), 32 genera were enumerated, of which nine are extraneous (*Buxus, Carica, Cissampelos, Clusia, Cupania, Hernandia*, and three genera of Polygalaceae). Adanson also attempted the first infrafamilial division of the family, but his two sections (founded on whether the stamens are free or united) are wholly unnatural. The treatment by A. L. de Jussieu (1789) is more impressive in terms of generic placement, since only five of 86 genera admitted into the Euphorbiaceae are not Euphorbiaceous: three Buxaceae, *Lacistema* and *Xylosma*. However, Jussieu's attempt to subdivide the family on the basis of stylar, rather than stamen union, was no more successful than that of Adanson.

The first great advance in taxonomic insight was provided by Adrien de Jussieu (1824), who subdivided the Euphorbiaceae into six well-defined sections based on several characters: ovule number, stamen insertion, presence of petals and type of inflorescence. Although Jussieu unfortunately did not provide names for his sections, they were recognized and given formal status as tribes by Dumortier (1829), Bartling (1830) and Spach (1834). The classification based on these six tribes (Buxeae, Phyllantheae, Ricineae, Acalypheae, Hippomaneae and Euphorbieae) has provided the paradigm for all later work on classification of the groups of genera within the family.

The most important insight of Adrien de Jussieu was to recognize the taxonomic importance of ovule number; the distinction between taxa with uniovulate and biovulate ovary loculi has been accepted as fundamental by subsequent workers. Zollinger (1845) was the first to use this character to divide the family into two subfamilies: Monosporae and Disporae. Pax (1884, 1890) assigned most of the genera in the Euphorbiaceae into two subfamilies, Phyllanthoideae and Crotonoideae, that correspond to Zollinger's Monosporae and Disporae. Mueller (1866a) and Bentham (1880) did not recognize subfamilies, but the ovule number character was still important in their definition of tribes.

In addition to providing the first monograph of the Euphorbiaceae, Adrien de Jussieu (1823) also wrote an interesting essay in which he analysed character variation and described his systematic methodology. Here he noted the primary importance of ovule number in establishing taxa, and he showed a pragmatic approach to choice of characters: for example, he observed that the pistillode character used to establish two sections in the biovulate taxa was of no use for the uniovulate taxa, where it had to be replaced with inflorescence characters.

Although he stated that his selection of characters to define major taxa was based on a modification of the Aristotelian method of logical division, A. Jussieu went on to explain that his actual working method was to select a few large and clearly delimited genera (*Buxus*, *Phyllanthus*, *Croton*, *Acalypha*, *Sapium* and *Euphorbia*) and then to build up his tribes by accretion, adding genera to the core of each group according to progressively decreasing strength of resemblance. The result of this typological procedure, which in fact is still one of the basic approaches used by taxonomists today, is that the groups are modally well-defined but not always sharply circumscribed. In Jussieu's words: ". . . le limites de ces sections ne sont pas bien tranchées, et certains genres etablissent le passage de l'une à l'autre. Ce qui me semble important, c'est que les genres s'y trouvent placés toujours près de ceux avec lesquels ils ont les affinités les plus fortes et les plus nombreuses; c'est que la série soit le plus naturelle possible". It is clear that although Jussieu was playing by the rules of De Candolle's *Theorie Elementaire* (1813), his actual practice in constructing a classification was largely empirical and inductive (as indeed that of De Candolle and most other leading contemporaries was in large part).

Immediately following the monograph of A. Jussieu (1824), these followed what we might call the 'romantic period' in Euphorbiaceous taxonomy, in which attempts were made to recognize a number of small families segregated from the Euphorbiaceae (Table 3). These included Stilaginaceae for *Antidesma* (Agardh, 1825), Scepaceae for *Aporusa* and *Hymenocardia* (Lindley, 1836), Putranjiveae for *Putranjiva* (Endlicher, 1837) and Prosopidoclineae (later Peraceae) for *Pera* (Klotzsch, 1841, 1859, 1860). Most of these segregations involved genera with reduced flowers, and the proposed new families were associated with wind-pollinated groups; thus, Lindley (1836) juxtaposed the Stilaginaceae with Urticaceae, and Scepaceae with Betulaceae.

Reichenbach (1828, 1837, 1841) was the first major systematist after A. Jussieu to propose an original classification of the Euphorbiaceae. He treated it as a subordinate group of the Rutaceae, a proposal that has not been followed by anyone else, and created the most complex classification before that of Mueller (1866a). There are three tribes, each with three subtribes: Euphorbieae genuinae (Callitrichineae, Tithymaleae, Ricineae); Crotoneae (Micrantheae, Argythamneae, Crozophoreae); and Buxeae (Phyllantheae, Cluytieae, Buxeae genuinae). Reichenbach's Crotoneae is on the whole a natural taxon, roughly corresponding to Pax's Crotonoideae. Reichenbach did not provide detailed

Table 3. Enumeration of family names referable to the Euphorbiales (the family Euphorbiaceae as defined by Webster, 1975); family references are given by Bullock (1958a, 1958b). Names are given in their original form of publication, except where noted. The names of some taxa that have been included in the Euphorbiales are here omitted because of typification they must be excluded from the order (e.g. Forestieraceae Endl., based on a genus of Oleaceae)

Acalyphaceae J. G. Agardh, *Theoria Systematis Plantarum:* 258 (1858) = Euphorbiaceae subfamily Acalyphoideae Ascherson (*pro. parte*).

Antidesmeae Endlicher, *Genera Plantarum secundum Ordines Naturales Disposita:* 287 (1837); Antidesmaceae Walpers, *Annales Botanices Systematicae, 3:* 391 (1852) = Euphorbiaceae tribe Antidesmeae (Endl.) Hurusawa.

Aporoseae Lindl. ex Planchon, *Annales des Sciences Naturelles, Botanique, IV, 1:* 265 (1854); ex Miquel, *Flora van Nederlandsch Indie, 1(2):* 430 (1859) = Euphorbiaceae tribe Aporuseae Airy Shaw.

Bennettiaceae R. Brown, in J. J. Bennett & R. Brown, *Plantae Javanicae Rariores:* 250 (1852; *nom. prov.*); Bennettieae Schnizlein, *Iconographia Familarum Naturalium Regni Vegetabilis, 3:* 172 (c. 1860; *nom. illegit.*, based on *Bennettia* R. Brown, a later homonym of *Bennettia* S. F. Gray) = Pandaceae.

Bertyeae J. G. Agardh, *Thoeria Systematis Plantarum:* 190 (1858) = Euphorbiaceae tribe Ricinocarpeae Muell. Arg.

Bischofiaceae (Muell. Arg.) Airy Shaw, *Kew Bulletin, 18:* 261 (1965) = Euphorbiaceae tribe Bischofieae (Muell. Arg.) Hurusawa.

Crotoneae J. G. Agardh, *Theoria Systematis Plantarum:* 258 (1858) = Euphorbiaceae subfamily Crotonoideae Pax.

Euphorbiaceae A. L. Jussieu, *Genera Plantarum:* 384 (1789; as Euphorbieae).

Galeariaceae Pierre, *Bulletin de la Société Linnéenne de Paris*, 1327 (1897; *nom. illegit.*) = Pandaceae.

Hippomaneae J. G. Agardh, *Theoria Systematis Plantarum:* 244 (1858) = Euphorbiaceae tribe Hippomaneae A Juss. ex Spach.

Hymenocardiaceae Airy Shaw, *Kew Bulletin, 18:* 261 (1965) = Euphorbiaceae tribe Hymenocardieae (Muell. Arg.) Hutchinson.

Linozosteae Sprengel, *Anleitung zur Kenntnis der Gewaechse, Zweite... Ausgabe, 2(1):* 371 (1817; rank of "Ordnung" probably equivalent to subfamily).

Micrantheae J. G. Agardh, *Theoria Systematis Plantarum:* 182 (1858) = Euphorbiaceae tribe Caletieae Muell. Arg.

Pandaceae Pierre, *Bulletin de la Société Linnéenne de Paris, 2:* 1255 (1896).

Peracées A. Jussieu in C. V. D. D'Orbigny, *Dictionnaire Universel d'Histoire Naturelle, 5:* 773 (1844; *nom. illegit.*); Peraceae Klotzsch, *Monatsberichte der Königlich Preussischen Akademie der Wissenschaften zu Berlin, 1859:* 246 (1859) = Euphorbiaceae tribe Pereae (Kl.) Pax & Hoffm.

Phyllantheae J. G. Agardh, *Theoria Systematis Plantarum:* 249 (1858); Phyllanthaceae Klotzsch, *Monatsberichte der Königlich Preussischen Akademie der Wissenschaften zu Berlin, 1859:* 246 (1859) = Euphorbiaceae subfamily Phyllanthoideae Ascherson (*pro parte*).

Picrodendraceae Small, *Journal of the New York Botanical Garden, 18:* 184 (1917) = Euphorbiaceae tribe Picrodendreae (Small) Webster.

Porantheraceae (Muell. Arg.) Hurusawa, *Journal of the Faculty of Science, University of Tokyo, Section III Botany, 6(6):* 224 (1954) = Euphorbiaceae tribe Poranthereae (Muell. Arg.) Grüning.

Pseudantheae Endlicher, *Genera Plantarum Secundum Ordines Naturales Disposita:* 328 (1838; probably of subfamilial rank) = Euphorbiaceae tribe Caletieae Muell. Arg.

Putranjiveae Endlicher, *Genera Plantarum Secundum Ordines Naturales Disposita:* 287 (1837) = Euphorbiaceae tribe Drypeteae (Griseb.) Hurusawa.

Ricinocarpaceae (Muell. Arg.) Hurusawa, *Journal of the Faculty of Science, University of Tokyo, Section III Botany, 6(6):* 224 (1954) = Euphorbiaceae tribe Ricinocarpeae Muell. Arg.

Scepaceae Lindley, *A Natural System of Botany, Second Edition:* 171 (1836) = Euphorbiaceae tribe Aporuseae Airy Shaw.

Stilagineae C. A. Agardh, *Aphorismi Botanici, 14:* 199 (1824); Stilaginaceae Lindley, *The Vegtable Kingdom, Third Edition:* 259 (1853) = Euphorbiaceae tribe Antidesmeae (Endl.) Hurusawa.

Tithymali Adanson, *Familles des Plantes:* 346 (1763; *nom. illegit.*) = Euphorbiaceae A. L. Jussieu (*nom. cons.*).

Trewiaceae Lindley, *A Natural System of Botany, Second Edition:* 174 (1836) = Euphorbiaceae tribe Acalypheae Dumort. (subtribe Rottlerinae Meisn.).

Uapacaceae (Muell. Arg.) Airy Shaw, *Kew Bulletin, 18:* 270 (1965) = Euphorbiaceae tribe Uapaceae (Muell. Arg.) Hutchinson.

descriptions or documentation, and his treatment appears to have had little influence on later taxonomic work.

More in the mainstream was the classification of Meisner (1841), which was clearly influenced by that of A. Jussieu. Meisner's tribe Crotoneae included a subtribe, Jatropheae, in which *Aleurites*, *Jatropha*, *Manihot*, *Ricinocarpus*, and *Trigonostemon* are correctly associated; this appears to be the first published system in which there is a clearly defined taxon for genera with 'crotonoid' pollen.

With the publication of *Étude Générale du Groupe des Euphorbiacées* by Baillon (1858), a significant milestone in Euphorbiaceous classification was attained. This great work is Baillon's major effort at a monograph of a plant family, even though he went on in the *Histoire des Plantes* to accomplish the only extensive multi-volume monograph of the angiosperms attained by a single individual. In the *Étude*, Baillon demonstrated a dazzling virtuosity in morphological observation (especially as seen with the microscope), and his superb analytical drawings (which make the plates of A. Jussieu seem extremely crude by comparison) remain unsurpassed. Baillon created many new genera and sections, redefined the circumscription of the family by excluding the Buxaceae, and provided many interesting discussions of relationships between genera, often with great insight. However, as remarked by both Mueller and Bentham, Baillon's work is regrettably marred by its failure to provide a classification appropriate to reflect degrees of affinity. Instead of using a proper hierarchical system, Baillon created 14 "series", within which each genus was usually characterized by its differences from a previously cited genus. This method of creating a classification by a chain of archetypes was contrary to better systematic practice, even in Baillon's day, and it fortunately did not have any appreciable influence on the next generation of workers on Euphorbiaceae.

Baillon's monograph appeared almost simultaneously with two other systems and shortly was followed by that of Jean Mueller, so that the decade following 1858 can well be considere the 'golden age' of Euphorbiaceous classifications. The treatment by Klotzsch (1859, 1860) initiated a new tradition of splitting and inflation of rank in the Euphorbiaceae. Klotzsch treated the Euphorbiaceae as an order ("Klasse") divided into six families ("Ordnungen"): Euphorbiaceae (*sensu stricto*), Peraceae, Acalyphaceae, Buxaceae, Phyllanthaceae and Antidesmaceae. Although this classification has six major taxa, as did that of A. Jussieu, the two are not very similar. Only the Euphorbiaceae (*sensu stricto*), Buxaceae and Phyllanthaceae correspond fairly closely to the comparable sections of Jussieu; the Peraceae and Antidesmaceae are segregate groups, while the Acalyphaceae combine the Jussieu/Spach tribes Ricineae, Acalypheae and Hippomaneae in a rather infelicitous juxtaposition. Subsequent work has not dealt very kindly with this classification, since not a single one of Klotzsch's major taxa is currently accepted with the same circumscription, and only a few recent workers (e.g Barkley, 1948; Hurusawa, 1954) have accepted such narrowly circumscribed families.

Grisebach (1859, 1860), in a contribution that has been neglected, discussed the criteria for infrafamilial subdivision, and proposed a new classification of the Euphorbiaceae that in some respects was the best published up until that time. Although he unfortunately retained the tribe Buxeae in the family (despite Baillon's well-founded assertion that they should be excluded), Grisebach placed all of the uniovulate taxa into a single tribe Crotoneae with seven subtribes: Jatropheae, Ricineae, Eucrotoneae, Ditaxideae, Acalypheae, Hippomaneae and Euphorbieae. This may be criticized as placing too much emphasis on the pecularities of *Ricinus*; but the subtribes Jatropheae and Crotoneae, characterized largely on inflorescence characters, contain mostly genera now referred to subfamily Crotonoideae; the Ditaxideae and Acalypheae mostly belong to subfamily Acalyphoideae; and the Hippomaneae and Euphorbieae to subfamily Euphorbioideae. Grisebach based his classification mostly on Caribbean taxa, and it was soon eclipsed by that of Mueller.

The first, and only, monograph of the Euphorbiaceae to the species level appeared in de Candolle's *Prodromus* in two sections, the tribe Euphorbieae by Boissier (1862), and the other tribes by Jean Mueller (Mueller Argoviensis) in 1866. Boissier's revision of *Euphorbia*, the last complete review of the genus, is justly famous for its judicious use of characters in defining species groups; Boissier's sections have survived remarkably intact (although recently challenged; Gilbert, in press), considering the fact that the number of species of *Euphorbia* has at least tripled since his work.

Mueller's infrafamilial classification of the Euphorbiaceae (1864, 1866a) was, if anything, more revolutionary than that of A. Jussieu four decades earlier; it is much more complex and sophisticated than its predecessor. The hierarchy of taxa in Euphorbiaceae as developed by Mueller was probably the most elaborate devised for any plant family up to that time. The major infrafamilial division is the novel recognition of two "tribal series" Stenolobeae and Platylobeae, defined on cotyledon shape; although accepted by Bentham (1880) and Pax (1890, 1931), this division appears now to have been an unfortunate choice, since it results in unnatural associations of genera.

Within each "series", Mueller's system has tribes characterized on the basis of a sequential hierarchy of three characters: ovule number, aestivation of calyx, and orientation of anthers in the bud. Within most of the tribes, subtribes are defined on the basis of the presence or absence of an involucre or of petals; an additional character of subordinate value is the position of insertion of the stamens. An outstanding peculiarity of Mueller's system arises from his use of disk-segments and petals as characters. Some apetalous taxa are placed among those with "flores petalis praediti", because the disk-segments are opposite the sepals; taxa with disk-segments alternate to the sepals are assigned to groups with "flores absolute apetali." However, since the disk-segments are often united, partially fused, or absent, this character is in practice difficult, or impossible, to use. Furthermore, it is most doubtful that the theoretical distinction intended by Mueller is valid; the disk is more plastic and variable within the basic floral plan than Mueller's scheme would allow.

Mueller's classification, with 10 tribes instead of the six recognized earlier by A. Jussieu, was actually not as different at the tribal level as might appear at first sight. With exclusion of the Buxaceae, it continued use of the five other Jussieuian tribes and added three Stenolobeae (Caletieae, Ricinocarpeae and Ampereae) plus two new Platylobeae (Bridelieae and Dalechampieae). The recognition of these five new tribes depended on Mueller's systematic philosophy and use of characters. Fortunately, we have his own testimony as to his working methods published shortly after the treatment in the *Prodromus* (Mueller, 1866b). Reflecting the philosophy of De Candolle in his *Théorie Élémentaire de*

Botanique (de Candolle, 1813), Mueller states as self-evident that the reproductive parts of the flower should receive highest value because of their functional importance; this is the familiar Aristotelian precept inherited from Caesalpino and Linnaeus. However, in the selection of characters, Mueller claimed that he had tested them as to their generality of occurrence and hence had used an inductive method. This is true to a certain extent, but Mueller seems not to have realized that he had pre-selected the characters to be tested on the basis of *a priori* philosophical biases.

As noted by Briquet (1896, 1940), Mueller showed a very strong mathematical bent as a student, and it would appear that he had an inherently formalistic cast of mind. It is perhaps not surprising that he erected a rigid scaffolding for his systematic work, or that he followed strict Aristotelian practice in selecting an embryonic character (cotyledons) as his *fundamentum divisionis*. Here he was unlucky, for it happens that the narrow cotyledons of the Australian Stenolobeae represent a striking instance of morphological convergence between unrelated taxa now referred to four different subfamilies. The likelihood of such convergence happening only in Australia (and not in other climatic areas with ericoid-leaved Euphorbiaceae) would indeed seem rather small, and it still has not been satisfactorily explained; however, there are some non-Australian taxa with narrow cotyledons, such as *Reverchonia* in the Phyllantheae (Webster & Miller, 1963). Bentham (1878, 1880), evidently inmpressed by the geographical correlation of the cotyledon character, recognized the Stenolobeae as a natural tribe, and Pax (1931) retained the group for 40 years after he had already decided that it was polyphyletic.

However, while he was simply unfortunate in selecting an embryo character for his primary division, Mueller was even less justified in his use of floral aestivation for the major tribal divisions. Adrien de Jussieu (1823) had already pointed out the differences in "préfloraison" of the perianth, but had not placed great weight on the character. The peculiarity of Mueller's usage resides in his apparent assumption that this character, once tested and found to show strong correlations with generic differences, could be applied across the board to separate nearly all of the groups at tribal rank. This assumption that a character found to yield correlations could be assigned an absolute weight or value in assigning rank to taxa occurs in the work of other 19th century systematists, but is nowhere as consistently and uncompromisingly pursued as in the work of Mueller.

Mueller's insistence on the hierarchical value of perianth aestivation had the most pernicious consequences in his tribe Hippomaneae, which, in his circumscription, included most of the uniovulate taxa with imbricate sepals. Baillon (1858) had already shown a number of instances in which this character was variable or inconstant, and in the Hippomaneae (*sensu stricto*), which have a more-or-less reduced calyx it is particularly difficult to interpret. Furthermore, recent studies, especially of pollen (Erdtman, 1952; Punt, 1962; Webster, unpublished) clearly show that the valvate/imbricate dichotomy separates many natural lines of affinity and makes both the Acalypheae and Hippomaneae of Mueller highly unnatural groups. The case of the Hippomaneae is particularly striking, since all of the 13 genera assigned to the tribe (as section V) by A. de Jussieu (1824) are retained in it by Pax & Hoffmann (1931), except for the Buxaceous genus *Styloceras*. In contrast, Mueller's Hippomaneae also included a large number of genera now assigned to subfamilies Acalyphoideae or Crotonoideae, such as *Chaetocarpus*, *Clutia*, *Jatropha* and *Manihot*.

The unanalysed assumption by Mueller that biased his approach was specified by himself (Mueller, 1866b) as a dogma: "In jeder systematischen Einheit muss der innere Werth des Characters dem hierarchischen Höhengrad der Einheit entsprechen". This can only mean that the classification is intended to reflect a descending series of Platonic archetypes. For Mueller, this principle of the near-absolute equivalence between character and rank extended down to the level of genera, sections, and species. However, in his actual work at the level of genera and species, Mueller followed a more empirical procedure. He reported that he carefully analysed three or four well-separated species of a genus in order to arrive at the most useful characters, and that he then applied these character differentials in placing all of the other taxa. Nevertheless, it is notable that in descriptions of genera, sections, and species only, reproductive characters are given as diagnostic; vegetative characters are always relegated to a second section of less important supplementary characters.

Mueller's realignments in classification were accompanied by many nomenclatural changes. He adopted a strict rule of priority, including the priority of species epithets over species names that was common practice in the mid-19th century, but in addition cited himself as the authority for the names of genera to which he had given a different circumscription. These innovations in nomenclature occasioned more discussion from other botanists (as noted below) than the changes in classification itself.

Mueller's monograph of the Euphorbiaceae was generally well received, although there were various complaints and reservations by his peers. Jaubert (1866) was dissatisfied with the use of the embryo character to separate the primary groups Stenolobeae and Platylobeae, mainly because he felt that it was too difficult to use in practice. Seemann (1866) applauded the competence of Mueller while deploring his nomenclatural innovations. Gray (1867) was unfavourably disposed towards Mueller's citations of authority as well as his practice of adopting herbarium or manuscript names without crediting them to the original 'author'; this went against the prevailing standards of gentlemanly practice. Alphonse de Candolle (1867), in reply to Seemann, disavowed responsibility for Mueller's innovations with the candid statement that he did not edit any of the manuscripts submitted for the *Prodromus*; but he did defend Mueller's work for its great precision and characterized it as "un véritable travail monographique".

Mueller's work could hardly have escaped some critical review from the acidulous pen of Henri Baillon, since many of his dispositions contradicted those in Baillon's classification of 1850. Even before the *Prodromus* volume appeared, Baillon (1865) had already criticized Mueller's generic circumscriptions in the Hippomaneae. Later (1867) he adversely commented on Mueller's placement of *Petalostigma* near *Phyllanthus*, arguing that it belonged closer to *Drypetes* and *Longetia*. Finally, in a long article titled *Nouvelles Observations sur les Euphorbiacées* (Baillon, 1873a), he launched a full-scale attack on Mueller's classification. After briefly complimenting Mueller on his reductions of a number of genera, Baillon stated: "Mais je cesse de partager la manière de voir de l'auteur du *Prodromus* quant à la valeur générique qu'il accorde à la forme des anthères, au degré de développement de la caroncule, et quant aux caractères de tribus qu'il fonde sur le préfloraison. Je pense qu'en l'imitant, on briserait à chaque pas les liens les plus naturels". Baillon decisively rejected the cotyledon character used by Mueller for the primary division of the family, partly for the practical reasons already stated by Jaubert. With considerable asperity, he also contested the circumscription and placement of many taxa. A number of his strictures, such as his disapproval of the calyx aestivation character to separate Acalypheae from Hippomaneae, today seem well targeted. However, in several instances Baillon went to almost grotesque lengths of amalgamating genera in opposition to Mueller's concepts; for example, he proposed to unite *Pentabrachium*, *Bridelia* and *Cleistanthus* with *Amanoa*.

Mueller replied with an angry counter-attack (Mueller, 1875) which is full of sarcasm and in places shows a surprising lapse in decorum. In replying to Baillon's major criticisms, Mueller made few concessions, and in turn pointed out some supposed errors committed by Baillon (e.g. in calling the corolla of *Jatropha* subgenus *Curcas* gamopetalous). After reviewing the characters that had been contested by Baillon, he concluded sarcastically that following Baillon would logically lead to amalgamating all of the Euphorbiaceae into a single genus. Finally, Mueller framed a rhetorical question: how could two outstanding botanists, after working for years, come to such diametrically opposed conclusions? His explanation is: "Dr. Baillon ist Adansonianer, er zählt die Differenzen, er perhorrescirt mehr oder weniger die Subordination der Charactere; ich dagegen gehöre dem systematischen Bekenntniss der Jussieu, R. Brown, de Candolle, etc., an, ich lege der Charactere auf die Wagschale, ich taxire die Differenzen, und die Subordination der Charactere ist für mich elementares und absolutes Bedürfniss".

This exchange between Mueller and Baillon, which seems to have been the first full-scale confrontation between two outstanding taxonomists of Adansonian and non-Adansonian persuasions, was not to be equalled for three-quarters of a century. In terms of our present-day knowledge, the verdict is curiously mixed. Some of Baillon's criticisms, especially the use of cotyledon shape and calyx aestivation as major characters, have been fully substantiated. On the other hand, Mueller's system of tribes and subtribes, despite his problems with the Hippomaneae, has held up much better than that of Baillon. Mueller's polemic against Baillon would have been much more effective if he had had in hand volume five of the *Histoire des Plantes*, which appeared in 1874 but which Mueller obviously had not seen when he sent his reply to the *Botanische Zeitung* on 29 December 1874. In volume five, Baillon presented his own mature (and final) system of the Euphorbiaceae, in which the genera (reduced to 146) are grouped in six series (not counting the Dichapetalaceae and Callitrichaceae, which were erroneously included). These six series are of extremely unequal size, with Euphorbieae, Ricineae, and Crotoneae each containing only two, three and four genera, respectively, while the Jatropheae have 88. In terms of circumscription (exclusion of extraneous elements) and information content (explication of relationships between taxa), it is probably the worst major classification of the Euphorbiaceae ever published. Mueller's accusation that Baillon merely enumerated the differences underscores the major weakness of Baillon as a systematist: he excelled in analysis but seems to have lacked the capacity (or interest) for synthesis. Baillon could also be just as dogmatic as Mueller, as is indicated, for example, by his stubborn anachronistic insistence that the cyathium of *Euphorbia* is a hermaphroditic flower. He believed as strongly as did Mueller that genera should be sharply defined, and an excessively rigorous logic led him to reject even slightly unreliable characters and to combine genera separated by such characters. In this practice he was curiously non-Adansonian.

Stevens (1984) has cited Baillon (in a later work, Baillon, 1891) as a promoter of the "chaining technique" of producing higher taxa by stringing related taxa together. It is not surprising that Baillon and Mueller, writing during the decade immediately following the publication of Darwin's *Origin of Species*, do not express themselves in phylogenetic terms. However, it is clear that Mueller's elaborate hierarchical system produced nested arrays of taxa and hence came far closer to being a representation of evolutionary affinities than did the classification of Baillon.

As pointed out by Leandri (1962), Baillon had little interest in the historical (phylogenetic) aspects of systematics; and judging from his published works, the same could be said of Mueller. At about the time of their exchange of polemics, a third botanist of very different temperament, George Bentham, was taking up a study of the family for the *Genera Plantarum*. Bentham's essay on Euphorbiaceae (1878) remains today as readable and interesting as his better-known review of the Compositae. It provides a judicious history of earlier taxonomic work, with major emphasis on Baillon and Mueller, both of whom he chided for their arbitrariness in use of characters.

Since Bentham appeared to be less constrained by preconceptions and dogmatic attitudes than other 19th century students of the Euphorbiaceae, it is of some interest to see how this was reflected in his classification. Bentham frankly acknowledged that his own classification was based on that of Mueller, but he was closer to Jussieu in recognizing only six tribes: Euphorbieae, Stenolobeae, Buxeae, Phylleae, Galearieae and Crotoneae. By present standards, his treatment was retrograde in including Buxaceae within the family after both Baillon and Mueller had excluded it. His unfortunate adoption of the Stenolobeae was perhaps influenced by the striking resemblances in habit that he must have noticed while preparing the Euphorbiaceae for the *Flora Australiensis* (1873). On the other hand, his recognition of the Galearieae as a distinct tribe was an improvement on the treatments of Baillon and Mueller, since *Galearia* and *Microdesmis* are indeed divergent from typical Euphorbiaceae and have recently been removed to the family Pandaceae by Forman (1966, 1968, 1971).

Within his large tribe Crotoneae, which includes most of the uniovulate Euphorbiaceae, Bentham created subtribes comparable to some of Mueller's tribes, and made a number of perceptive indications of affinity. Particularly felicitious were his demotion of Mueller's tribe Crotoneae to subtribal rank, and his creation of a subtribe Plukenetiinae. Bentham's demotion of the Dalechampieae to an element of the Plukenetieae gave a much better expression of affinity for *Dalechampia* than indicated in most other classifications. Despite various defects of Bentham's system, it was the most nearly natural that had been produced up until that time. However, today perhaps the major interest in his essay is in the final section on geographical distribution, which appears very modern because of its patently phylogenetic reasoning, even though Bentham did not use evolutionary phraseology.

Shortly after Bentham's treatment in the *Genera Plantarum* (1880), a young German botanist, Ferdinand Pax, began working on the systematic anatomy of the Euphorbiaceae. His first paper (Pax, 1884) initiated a series of publications that extended over more than 50 years. In his essay of 1884 Pax reviewed the previous systems of Klotzsch, Baillon, Mueller and Bentham, and then offered his own findings based on anatomical observations (mainly on cross-sections of twigs on herbarium specimens, many of which were furnished by Mueller from the Geneva herbarium). In one of the most striking cases of a missed opportunity in 19th century plant systematics, he considered using pollen characters but rejected them as taxonomically useless!

Pax in 1884 offered his own original classification, based on anatomical characters, and in some respects it was superior to those he produced later. He presented for the first time since Zollinger in 1845 a bipartite classification with all of the biovulate taxa in the subfamily Phyllanthoideae and the uniovulate in Crotonoideae. Perhaps his most important innovation was the use of laticifers as a primary character; this was the first time that non-floral characters had been given such emphasis, and reflects a further decline in the use of Linnaean essentialistic criteria. His other major innovation in classification was to divide the Crotonoideae into two taxa of supertribal rank, the Acalyphoineae and Hippomanoineae, defined on the presence of articulated and non-articulated laticifers, respectively. This was not a very natural division, since the Ricinocarpeae and Joannesieae, with 'crotonoid' pollen, were placed in the Acalyphoineae, while the Crotoneae, with similar pollen, went into the Hippomanoineae. As various authors pointed out later, Pax's observations on laticifers were inexact, and he did not always distinguish articulated laticifers from rows of tannin cells. Nevertheless, description and definition of laticifers still plagues workers in the 20th century, and I agree with Dr Mahlberg (pers. comm.) that Pax's classification of 1884 deserves re-examination.

With a display of that curious inconsistency that he was to show to the end of his career, Pax concluded that his findings supported the system of Mueller, despite the fact that they clearly contradicted it in some important aspects. For example, his anatomical work showed the heterogeneity of the Stenolobeae of Mueller and demonstrated that Baillon (1873a) was correct in dismembering it and referring the genera to appropriate positions among the biovulate and uniovulate taxa. In his dendrogram illustrating phylogenetic relationships among the tribes (Fig. 4A), the first published for the family, Pax showed the Stenolobeae as biphyletic, and in his systematic table of tribes, he referred the Stenolobean Caletieae to the Phyllanthoideae and the Ricinocarpeae to the Crotonoideae; yet in his concluding discussion he appears to have accepted the Stenolobeae as a legitimate taxon ("systematisch jedenfalls gut begründete").

As a result of his studies, Pax arrived at several conclusions about the phylogeny of the Euphorbiaceae that had been partly anticipated by Bentham and that have been accepted by most later workers: greater antiquity of the Phyllanthoideae, early separation of the two subfamilies, origin of the 'acalyphoid' line mainly in the Old World and the 'hippomanoid' mainly in the New World, as well as the biphyletic origin of the Stenolobeae. In his treatment in the first edition of *Die Natürlichen Pflanzenfamilien* (1890), he contradicted some of his conclusions of 1884 and reverted to Mueller's treatment in retaining Platylobeae and Stenolobeae as the major divisions of the family. Otherwise, in

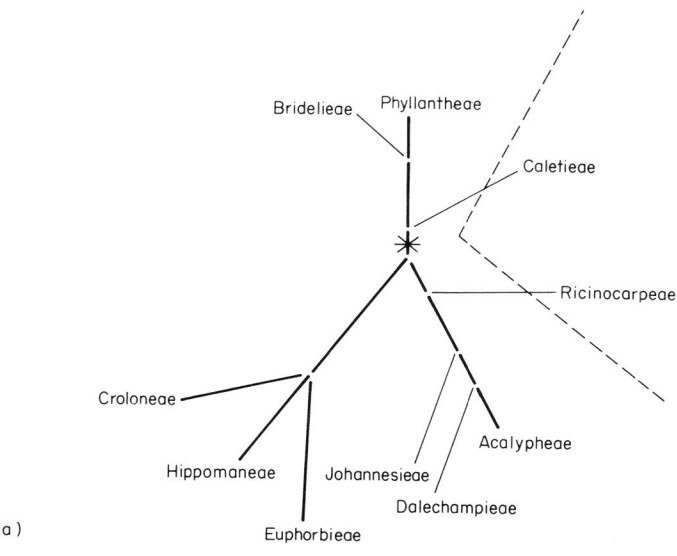
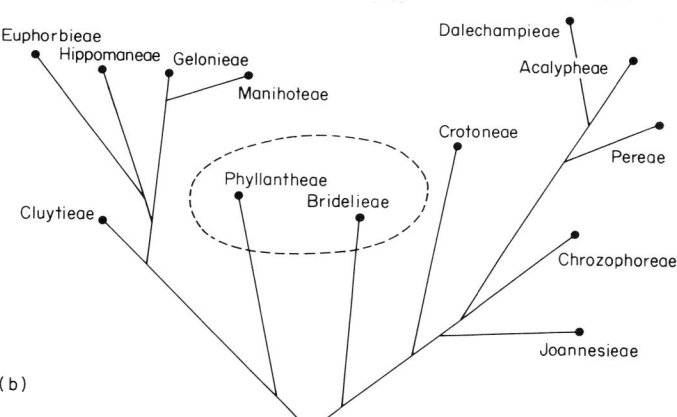

Figure 4. Phylograms proposed by Pax to illustrate putative relationships among the tribes of Euphorbiaceae. A. Version of 1884. B. Version of 1924.

his divisions into tribes and subtribes Pax usually followed the dispositions of Mueller, but did accept many of Bentham's suggestions. However, one suggestion of Bentham that he should have ignored was the re-inclusion of *Daphniphyllum* in the Euphorbiaceae after its earlier exclusion by Mueller (1869). Some of Pax's changes were distinct improvements; for example, he brought together most of the Oldfieldioideae into two adjacent subtribes, Toxicodendrinae and Bischofinae.

Pax's treatment of the Crotonoideae as comprising eight tribes was heavily dependent on Bentham, with Bentham's subtribes raised to tribal rank. His tribes Acalypheae and Jatropheae were largely natural, but in following Bentham in the circumscription of the tribe Manihoteae (based on Bentham's

Adrianeae), he seconded one of Bentham's worst mistakes. Pax degraded Bentham's tribe Galearieae by renaming it Cluytieae and including many extraneous elements. On the other hand, by associating the Plukenetiinae with the Acalypheae and making the Hippomaneae (*sensu* A. de Jussieu) coordinate in rank with the Euphorbieae, Pax provided a classification that reflected relationships better than previous systems.

Beginning with the treatment of the Jatropheae in *Das Pflanzenreich*, Pax (1910) initiated a series of monographic treatments of the Euphorbiaceae (assisted after 1910 by Käthe Hoffmann) that ran for 17 instalments with over 2200 pages (summarized in Webster, 1967), thus establishing a quantitative superiority over Mueller. With the completion of the treatment of the Acalyphinae in 1924, most of the family had been covered, except for the subtribes with the three largest genera: *Phyllanthus*, *Croton*, and *Euphorbia*.

Pax did not introduce any important modifications of his system of 1890 until the publication of Additamentum VI in *Das Pflanzenreich* (Pax & Hoffmann, 1919). Here he remarked that in 1890 he had based his tribes Jatropheae and Manihoteae on Bentham's corresponding subtribes of Crotoneae, but that he had gradually come to see the Benthamian taxa as unnatural. In effect, Pax in 1919 largely returned to the system of Mueller and readopted calyx aestivation as a primary diagnostic character. With regard to the Manihoteae, Pax was on firm ground when he restricted the tribe to *Manihot* and expelled the other genera included by Bentham. However, in realigning the genera of Jatropheae Pax was disastrously mistaken, for this was one of the most natural of all Bentham's taxa. Now *Aleurites*, *Jatropha*, *Hevea*, and *Micrandra* were widely separated in four different tribes; and 11 genera of the Jatropheae of 1890 were referred to eight different subtribes. In his final general review of the phylogeny of the family, Pax (1924) presented a phylogram (Fig. 4B) in which the Phyllanthoideae is straightforwardly shown as an unnatural group. In this essay he did associate *Hevea* and *Joannesia* together in the tribe Joannesieae, but otherwise this fragmentation of the original Jatropheae was allowed to stand in the final version of Pax's classification (Pax & Hoffmann, 1931). On the whole, the classification of 1931 now appears only marginally better than that of 1890, and was definitely inferior in its treatment of genera with 'crotonoid' pollen.

In the 55 years since the last detailed classification of Pax & Hoffmann (1931), three complete classifications of the family down to the tribal level have been produced: Hurusawa (1954), Hutchinson (1969) and Webster (1975). Because of the shorter historical perspective, it is difficult to evaluate these more recent classifications, but a brief discussion seems warranted. Hurusawa's classification is the only 20th century system to return to the tradition of Klotzsch, in which the Euphorbiaceae are dismembered into several families. However, Hurusawa's four families, Antidesmataceae, Euphorbiaceae, Porantheraceae and Ricinocarpaceae, really represent the four subfamilies of Pax raised to family rank, and hence do not indicate any novelty of classification. On the other hand, Hurusawa provides a more original treatment at the subfamily level, with three biovulate subfamilies (Bridelioideae, Antidesmatioideae and Phyllanthoideae) and four uniovulate (Euphorbioideae, Acalyphoideae, Crotonoideae, and Sapioideae). The value of Hurusawa's paper would appear to rest mainly on his detailed discussions of Oriental taxa and delimitation of the new tribes Epiprineae, Aleuritideae, Hureae, Drypeteae and Bischofieae. To the best of my knowledge, none of Hurusawa's proposed segregate families have been accepted.

Hutchinson (1969b), in contrast, produced a very original classification in which subfamilies are abandoned and the genera assigned to 40 tribes, many of them new. Hutchinson had begun work on the family over 50 years earlier with his treatment of part of the Euphorbiaceae for the *Flora of Tropical Africa* (Hutchinson, 1911–1912). In his first classification of the families of dicotyledons (1926), he enumerated no subfamilial taxa, but did characterize the family as polyphyletic. His classification of 1969 is featured by a synoptic key to tribes, illustrations, tribal descriptions and enumerations of genera.

How well does Hutchinson's system reflect affinities? If it is measured against the treatment of Webster (1975), it is notable that 11 of the 40 Hutchinsonian tribes have genera belonging to at least three different tribes (*sensu* Webster). Among the taxa that were placed in Acalypheae subtribe Mercurialinae by Pax & Hoffmann (1931), Hutchinson's treatment appears especially unfortunate. His tribe Malloteae includes genera belonging to five different tribes, while he refers to seven different tribes genera that in my opinion should be brought together in the single tribe Acalypheae. On the other hand, his classification does have the virtue of awarding tribal status to phylogenetically isolated genera such as *Uapaca*, and largely natural groups of genera, such as the Codiaeeae and Alchornieae.

Overall, Hutchinson's system does not appear greatly superior to that of Pax in terms of elucidating phylogenetic affinities, and in some respects it is retrograde. Partly, this disappointing result seems due to Hutchinson's reliance on Gestalt of habit and gross floral morphology without seeking correlations among anatomical, cytological, or palynological characters. It is apparent that in the Euphorbiaceae Hutchinson's intuition was not very successful in identifying lines of natural affinity. His insistence on using only a single rank between family and genus would in itself result in considerable distortion of relationships compared to a more hierarchical system.

Some of the most significant contributions to improving our understanding of the classification of the Euphorbiaceae have been made by the late Mr H. K. Airy Shaw. Although he dealt primarily with taxa of tropical Asia and Australia, he also had occasion to consider the problem of delimitation of the family (Airy Shaw, 1965; Willis, 1966). He concluded that in addition to the Euphorbiaceae *sensu stricto*, a number of small segregate families should be recognized: Androstachydaceae, Bischofiaceae, Hymenocardiaceae, Pandaceae, Peraceae, Picrodendraceae, Stilaginaceae and Uapacaceae. Earlier (Webster, 1967) I expressed scepticism about some of these segregations, and, on the basis of studies done since then, I am not persuaded that any of the splinter-families should be recognized, except possibly the Pandaceae.

Within the scope of the present review, it is not practicable to discuss in detail the arguments for and against recognizing these small families. However, it seems worthwhile to note recent studies that provide evidence for evaluating Airy Shaw's proposals. Evidence from pollen (Punt, 1962; Köhler, 1965) and wood structure (Metcalfe & Chalk, 1950; Mennega, this volume) does not support recognition of the Stilaginaceae and Uapacaceae (Airy Shaw in his revision of Willis, 1966) as distinct families. My suggestion (Webster, 1967, 1975) that *Picrodendron* should be located in the Euphorbiaceae subfamily

Oldfieldioideae is supported on anatomical and palynological grounds by Hayden (1977) and Hayden *et al.* (1984). Inclusion of *Androstachys* in subfamily Oldfieldioideae is also supported by anatomical studies of Hayden (1980).

There is no doubt that *Bischofia* is an isolated genus in the Euphorbiaceae, without any morphologically similar neighbours in the subfamily Phyllanthoideae. However, Airy Shaw's claim (Airy Shaw, 1965; Willis, 1966) that it is not only a distinct family but is more closely related to the Staphyleaceae is contradicted by the study of Bhatnager & Kapil (1974, 1979), who demonstrate that *Bischofia* is fully compatible with Euphorbiaceae in embryological characters and quite different from Staphyleaceae.

The proposal by Airy Shaw (1965) to create a segregate family Hymenocardiaceae remains controversial. Leonard & Mosango (1985) accept family status for *Hymenocardia*, citing floral and pollen differences. It is true that both Punt (1962) and Köhler (1965) have remarked on the palynological distinctiveness of *Hymenocardia*; however, this could possibly reflect the shift to wind pollination in the genus, rather than a wide phylogenetic divergence. Against the pollen evidence, one may cite the ovule structure, typical for Euphorbiaceae (Baillon, 1858); chromosome number of $n=13$ as in other Phyllanthoideae (Hans, 1973); and wood anatomy similar to Antidesmeae (Metcalfe & Chalk, 1950; Mennega, this volume). A very interesting contribution towards resolving the problem is the discovery by Pais *et al.* (1967) of a new alkaloid, hymenocardine, in *Hymenocardia acida* Tul.; this belongs to an unusual class ("alcaloides peptidiques") that has also been isolated from Rhamnaceae and Pandaceae. Whether or not this furnishes evidence for placing *Hymenocardia* in a separate family, it provides a very suggestive indication of its relationships. The conflict between evidence from pollen and possibly alkaloids versus that from other organs clearly requires explanation, but at present I see no compelling need to remove *Hymenocardia* from the Euphorbiaceae, unless indeed it is shown to be more closely related to the Pandaceae.

It is unfortunate that Airy Shaw never found time to complete his system of suprageneric taxa in the Euphorbiaceae. However, in his revision of Willis's dictionary and in various floristic treatments (Airy Shaw, 1972, 1975, 1980; Willis, 1966), he has offered a fragmentary draft of his revised system, which is based mainly on taxa of the Oriental and Australasian regions; a brief modified version of his system is given by Radcliffe-Smith (1978). For comparison with the system of Webster (1975), his taxa are listed in Table 4. If one ignores differences in rank, it can be seen that the two systems show many correspondences overall. Airy Shaw's tribes Crotoneae, Jatropheae, and Euphorbieae are more-or-less comparable to the subfamilies Acalyphoideae, Crotonoideae and Euphorbioideae *sensu* Webster. The main differences are due to the recognition (by creation of suprageneric taxa) that Airy Shaw has given to certain individual genera, such as *Acalypha*, *Baloghia*, *Borneodendron* and *Ricinus*. His suggestions for placement of genera in his enlarged tribe Jatropheae are based on considerable personal experience and merit serious consideration where they differ from those of other workers.

For obvious reasons of self-reference, an objective appraisal of the system of Webster (1975) cannot be offered in this review. Eventually, it will no doubt be at least partly superseded by a different classification, as the Euphorbiaceae is still a very poorly understood family and much work at the alpha taxonomic

Table 4. Comparison of the classifications of Euphorbiaceae by Airy Shaw 1972, 1975, 1980; Willis 1966) and Webster (1975); segregate families recognized by Airy Shaw are given in capitals; taxa not treated by Airy Shaw are indicated by ——. The infrafamilial taxa of Airy Shaw, as explicitly noted by him, are mostly 'informal' taxa that have not been validly published

	Webster	Airy Shaw
Subfamily	PHYLLANTHOIDEAE	
Tribe	Wielandieae	——
	Amanoeae	(Phylantheae)
	Bridelieae	Bridelieae
	Poranthereae	——
	Spondiantheae	——
	Antidesmeae	STILAGINACEAE
	Aporuseae	Aporuseae, Baccaureeae
	Drypeteae	Drypeteae
	Phyllantheae	Phyllantheae
	Hymenocardieae	HYMENOCARDIACEAE
	Uapaceae	UAPACACEAE
	Bischofieae	BISCHOFIACEAE
Subfamily	OLDFIELDIOIDEAE	
	Hyaenancheae	Dissiliarieae, Austrobuxeae
	Petalostigmateae	Petalostigmateae
	Caletieae	Drypeteae (ex parte)
	Picrodendreae	——
Subfamily	ACALYPHOIDEAE	
	Clutieae	——
	Pogonophoreae	——
	Chaetocarpeae	Chaetocarpeae
	Cheiloseae	Cheiloseae
	Erismantheae	Crotoneae–Erismanthinae
	Ampereae	——
	Chrozophoreae	Crotoneae–Chrozophorinae
	Agrostistachyideae	Crotoneae–Agrostistachydinae
	Caryodendreae	——
	Pycnocomeae	Crotoneae–Blumeodendrinae, Mallotinae
	Bernardieae	——
	Epiprineae	Crotoneae–Epiprininae
	Adelieae	——
	Alchornieae	Crotoneae–Alchorneinae
	Acalypheae	Acalypheae; Crotoneae–Mallotinae, Homonoiinae, Mercurialinae; Ricineae
	Plukenetieae	Crotoneae–Plukenetiinae
	Dalechampieae	Crotoneae–Dalechampiinae
	Omphaleae	Jatropheae–Jatrophinae
	Pereae	PERACEAE
Subfamily	CROTONOIDEAE	
	Micrandreae	——
	Manihoteae	Jatropheae–Manihotinae
	Adenoclineae	Crotoneae–Endosperminae
	Gelonieae	Suregadeae
	Elateriospermeae	Jatropheae–Jatrophinae
	Joannesieae	Jatropheae–Jatrophinae
	Codiaeeae	Baloghieae; Crotoneae–Codiaeinae, Ostodinae
	Ricinocarpeae	Borneodendreae; Crotoneae–Alphandiinae
	Trigonostemoneae	Jatropheae–Ostodinae
	Aleuritideae	Jatropheae–Jatrophinae
	Crotoneae	Crotoneae–Crotoninae
Subfamily	EUPHORBIOIDEAE	
	Stomatocalyceae	Euphorbieae–Pimelodendrinae
	Hippomaneae	Euphorbieae–Hippomaninae
	Pachystromateae	——
	Hureae	Euphorbieae–Hippomaninae
	Euphorbieae	Euphorbieae–Euphorbiinae

level still remains to be done, to say nothing of what will be possible with more sophisticated techniques. I would predict that if currently espoused practices in cladistic taxonomy (Wiley, 1981) are followed, the boundaries of the subfamilies and many tribes would have to be drastically modified; it is evident, for example, that the subfamilies Phyllanthoideae, Crotonoideae, and Acalyphoideae, as presently constituted, are paraphyletic taxa.

CONCLUSIONS

The complexity of relationships within the Euphorbiales has challenged the best efforts of systematists for two centuries, and the resolution of problems will probably continue for many more years. The greatest uncertainty remaining at present is the position of the entire order on the phylogenetic map of the angiosperms. The competing claims of Malvales and Geraniales for the status of sister group makes it impossible at present to confidently assign the Euphorbiales to the appropriate superorder.

A striking result of researches during the 20th century has been to remove, one after another, extraneous families that do not belong in the Euphorbiales. The most recent classifications by well-known students of phylogeny all list families that I believe should be excluded from the Euphorbiales: Aextoxicaceae (Dahlgren, 1983; Takhtajan, 1980; Thorne, 1983); Buxaceae (Stebbins, 1974; Cronquist, 1981); Simmondsiaceae (Cronquist, 1981; Dahlgren, 1983; Takhtajan, 1980; Thorne, 1983); and Thymelaeaceae (Thorne, 1983). Only the Pandaceae survives a close scrutiny of characters to emerge as the only other family of the order besides Euphorbiaceae. However, despite the arguments of Forman (1966, 1968, 1971), the evidence to support status of the Pandaceae as a separate family is very weak, and rests more on our ignorance than on our knowledge. The Pandaceae is very tentatively accepted in this paper as the only other family of Euphorbiales, with the understanding that future studies may demonstrate that it is better treated as a group within the Euphorbiaceae.

The question of the position of the Euphorbiales in the phylogenetic system of the dicotyledons remains unresolved. During the years since my earlier statement favouring a placement of the Euphorbiaceae in the Rosidae (Webster, 1967), I have come to be less certain about rejecting a Malvalean affinity. The morphological and biochemical similarities between Euphorbiales and various taxa of the superorders Malvanae, Rutanae and Celastranae (*sensu* Takhtajan, 1980) appear to me to contradict the assignment of these groups to two different dicotyledon subclasses, the Dilleniidae and Rosidae (*sensu* Cronquist, 1968, 1981). It is notable in this regard that Kubitzki (1969, 1977) and Merxmüller (1972), on the basis of biochemical and floral characters, have pointed out the arbitrariness of the boundary between Dilleniidae and Rosidae. There are other indications, besides problems in tracing affinities of the Euphorbiales, that there is a serious problem with the delimitation of the two dicotyledon subclasses by Cronquist. Lobreau-Callen (1977), for example, considers that the pollen of Celastrales *sensu stricto* agrees with that in the Rosidae, whereas other families that have been placed in the Celastrales (e.g. Aquifoliaceae, Icacinaceae) are palynologically closer to Dilleniidae. Therefore, I am in sympathy with the criticisms of Kubitzki, Merxmüller, and the recent statement by Ehrendorfer (1983) expressing an agnostic point of view towards recognizing the currently defined limits of Dilleniidae and Rosidae. At present, it appears that the best reading of the evidence is to associate the Euphorbiales with the Malvales and Geraniales, and possibly with the Celastrales, in a superorder that straddles the arbitrary boundary between Dilleniidae and Rosidae suggested by Cronquist (1981). Among recently published classifications, phylogenetic affinities seem to be best expressed in the system of Soó (1967), who links the Malvales, Geraniales and Euphorbiales into a single phyletic line (Fig. 2D), even though he does not create a formal supraordinal taxon.

Despite over 150 years of study of infrafamilial relationships in the Euphorbiaceae, following the pioneering work of A. de Jussieu (1824), the subfamilies and tribes of Euphorbiaceae are still not satisfactorily defined. The system of Webster (1975), with five subfamilies and over 50 tribes, is broadly compatible with data from the fields of wood anatomy, cytology, and biochemistry, but detailed morphological and anatomical studies are still needed for many genera before the scaffolding of the classification of Euphorbiaceae can be considered secure.

ACKNOWLEDGEMENTS

The conclusions in this paper have been strongly influenced by research done over the past 30 years with support from the National Science Foundation. I am indebted to Mr Radcliffe-Smith and the staff of the library at Kew for assistance in completing the bibliographic work.

REFERENCES

ADANSON, M., 1763. *Familles des Plantes*. Paris: Vincent.
AGARDH, C. A., 1824. Stilagineae, in *Aphorismi Botanici:* 199–200. Lund: Berling.
AGARDH, C. A., 1825. *Classes plantarum*. Lund: Berling.
AGARDH, J. G., 1857. Om äggets läge inom ovariet hos de Phanerogama vexterna. *Kongelige Svenska Vetenskaps Academiens Handlingar II, 1(1):* 1–12.
AGARDH, J. G., 1858. *Theoria Systematis Plantarum*. Lund: Gleerup.
AIRY SHAW, H. K., 1960. The genus *Syndyophyllum* K. Schum. & Lauterb. in Sumatra and Borneo. *Kew Bulletin, 14:* 392–394.
AIRY SHAW, H. K., 1965. Diagnoses of new families, new names, etc., for the seventh edition of Willis's "Dictionary". *Kew Bulletin, 18:* 249–273.
AIRY SHAW, H. K., 1966. *A Dictionary of the Flowering Plants and Ferns, by J. C. Willis*, 7th edition. Cambridge: University Press.
AIRY SHAW, H. K., 1972. The Euphorbiaceae of Siam. *Kew Bulletin, 26:* 191–363.
AIRY SHAW, H. K., 1975. The Euphorbiaceae of Borneo. *Kew Bulletin Additional Series, 4:* 1–245.
AIRY SHAW, H. K., 1980a. The Euphorbiaceae of New Guinea. *Kew Bulletin Additional Series, 8:* 1–253.
AIRY SHAW, H. K., 1980b. The Euphorbiaceae–Platylobeae of Australia. *Kew Bulletin, 35:* 577–700.
AIRY SHAW, H. K., 1982. The Euphorbiaceae of Sumatra. *Kew Bulletin, 36:* 239–374.
ALLEM, A. C., & IRGANG, B. E., 1975. Euphorbiaceae tribo Euphorbieae. *Flora Ilustrada do Rio Grande do Sul, 11:* 1–97.
ANTON, R., 1974. *Étude chimiotaxonomique sur le genre* Euphorbia *(Euphorbiacées)*. Thesis, Université Lothis Pasteur, Strasbourg.
BAILLON, H., 1856. Sur la véritable organisation du buis. *Bulletin de la Société Botanique de France, 3:* 285–287.
BAILLON, H., 1857. Recherches sur l'organogénie du *Callitriche* et sur ses rapports naturels. *Bulletin de la Société Botanique de France, 5:* 337–341.
BAILLON, H., 1858. *Étude Générale du Groupe des Euphorbiacées*. Paris: Victor Masson.
BAILLON, H., 1859. *Monographie des Buxacées et de Stylocerées*. Paris: Victor Masson.
BAILLON, H., 1865. Remarques générales sur les Stillingiidées du Brésil. *Adansonia, I., 5:* 336–344.
BAILLON, H., 1867. Recherches complémentaires sur les Euphorbiacées australiennes. *Adansonia, I., 7:* 352–360.
BAILLON, H., 1873a. Nouvelles observations sur les Euphorbiacées. *Adansonia I., 11:* 72–138.
BAILLON, H., 1873b. Ochnacées, Rutacées. *Histoire des Plantes, 4:* 357–520. Paris: Hachette.

BAILLON, H., 1874. Euphorbiacées. *Histoire des Plantes, 5:* 105–256. Paris: Hachette.
BAILLON, H., 1891. *Dictionnaire de Botanique, 4:* 160. Paris: Hachette.
BALLY, P. R. O., 1961. *The genus Monadenium.* Berne: Bentelli.
BARABÉ, D., BERGERON, Y. & VINCENT, G, A., 1982. La position des Daphniphyllaceae, Buxaceae, Simmondsiaceae et Cecropiaceae dans la sous-classe des Hamamelididae. Étude numérique. *Comptes Rendus de l'Académie des Sciences, Paris, 294:* 891–896.
BARANOVA, M., 1983. On the laterocytic stomatotype in angiosperms. *Brittonia, 35:* 93–102.
BARKLEY, F. A., 1948. Lista de los ordenes y familias de las Anthophyta con ejemplos genericos seleccionados. *Revista Facultad Nacional de Agronomía (Medellín), 8(31):* 153–372.
BARTLING, F. G., 1830. *Ordines Naturales Plantarum Eorumque Characteres et Affinitates Adjecta Generum Enumeratione.* Goettingen: Dietrich.
BATSCH, A. J. G. C., 1802. *Tabula Affinitatum Regni Vegetabilis.* Weimar: Landes-Industrie-Comptoir.
BEHNKE, H.-D., 1982. Sieve-element plastids, exine sculpturing and the systematic affinities of the Buxaceae. *Plant Systematics and Evolution, 139:* 257–266.
BEHNKE, H.-D., 1985. Ultrastructure of sieve-element plastids of Myrtales and allied groups. *Annals of the Missouri Botanical Garden, 71:* 824–831.
BENNETT, J. J. & BROWN, R., 1852. *Plantae Javanicae Rariores,* part 4. London: W. H. Allen.
BENTHAM, G., 1873. Euphorbiaceae, in *Flora Australiensis, 6:* 41–153.
BENTHAM, G., 1878. Notes on Euphorbiaceae. *Journal of the Linnean Society of London, Botany, 17:* 185–267.
BENTHAM, G., 1880. Euphorbiaceae. In G. Bentham & J. D. Hooker (Eds), *Genera Plantarum ad Exemplaria Imprimis in Herbariis Kewensibus Servata Definita, 3:* 239–340. London: Lovell Reeve & Co.
BHATNAGAR, A. K. & GARG, M., 1977. Affinities of *Daphniphyllum*—palynological approach. *Phytomorphology, 27:* 92–97.
BHATNAGAR, A. K. & KAPIL, R. N., 1974. *Bischofia javanica*—its relationship with Euphorbiaceae. *Phytomorphology, 23:* 264–267.
BHATNAGAR, A. K. & KAPIL, R. N., 1979. Ontogeny and taxonomic significance of anther in *Bischofia javanica. Phytomorphology, 29:* 298–306.
BLUME, C. L., 1826. *Bijdragen tot de Flora van Nederlandsch Indie,* stukke 12. Batavia: Lands Drukkerij.
BOISSIER, E., 1862. Euphorbieae. In A. P. de Candolle (Ed), *Prodromus Systematis Naturalis Regni Vegetabilis, 15(2):* 3–188. Paris: Treuttel & Würtz.
BRIQUET, J., 1896. Notice sur la vie et les oeuvres de Jean Müller. *Bulletin de l'Herbier Boissier, I., 4:* 111–133.
BRIQUET, J., 1940. Biographies des botanistes à Genève. *Berichte der Schweizerischen Botanischen Gesellschaft, 50a:* 1–494.
BRONGNIART, A. T., 1850. *Énumération des Genres des Plantes cultivés au Muséum d'Histoire Naturelle de Paris,* ed. 2. Paris: Baillière.
BULLOCK, A. A., 1958a. Indicis nominum familiarum angiospermarum prodromus. *Taxon, 7:* 1–35.
BULLOCK, A. A., 1958b. Indicis nominum familiarum angiospermarum prodromus. Additamenta et corrigenda I. *Taxon, 7:* 158–163.
CANTINO, P. D., 1982. Affinities of the Lamiales: a cladistic analysis. *Systematic Botany, 7:* 237–248.
CANTINO, P. D., 1985. Phylogenetic inference from nonuniversal character states. *Systematic Botany, 10:* 119–122.
CARLQUIST, S., 1982. Wood anatomy of Buxaceae: correlation with ecology and phylogeny. *Flora, 172:* 463–491.
CARTER, S., 1982. New succulent spiny Euphorbias from East Africa. *Hooker's Icones Plantarum, 39(3):* 1–118, pls. 3851–3875.
CLARKE, C. B., 1865. On the structure and affinities of the Callitrichaceae. *Transactions of the Linnaean Society of London, 22:* 411–414.
COPELAND, H. F., 1957. Forecast of a system of the Dicotyledons. *Madroño, 14:* 1–9.
CORNER, E. J. H., 1976. *The Seeds of Dicotyledons.* 2 vols. Cambridge: Cambridge University Press.
CROIZAT, L., 1938. Notes on the Euphorbiaceae, with a new genus of the Euphorbieae. *Philippine Journal of Science, 64:* 397–412.
CROIZAT, L., 1939. Euphorbiae species et subgenera nova ex Americana Latina. *Revista Sudamerica Botanica, 6:* 10–14.
CROIZAT, L., 1940. On the phylogeny of the Euphorbiaceae and some of their presumed allies. *Revista Universitaria (Universidad Católica de Chile), 25:* 205–220.
CROIZAT, L., 1960. *Principia Botanica,* 2 vols. Caracas: the author.
CROIZAT, L., 1972. An introduction to the subgeneric classification of "Euphorbia" L., with stress on the South African and Malagasy species. III. *Webbia, 27:* 1–221.
CROIZAT, L., 1973. Les Euphorbiacées vues en elles-mêmes, et dans leurs rapports envers l'angiospermie en général. *Memórias de Sociedade Broteriana, 23:* 5–206.
CRONQUIST, A., 1968. *The Evolution and Classification of Flowering Plants.* Boston: Houghton Mifflin.
CRONQUIST, A., 1981. *An Integrated System of Classification of Flowering Plants.* New York: Columbia University Press.
CRONQUIST, A., 1983. Some realignments in the dicotyledons. *Nordic Journal of Botany, 3:* 75–83.
CRONQUIST, A., 1985. A commentary on the definition of the order Myrtales. *Annals of the Missouri Botanical Garden, 71:* 780–782.

DAHLGREN, R., 1975. A system of classification of the angiosperms to be used to demonstrate the distribution of characters. *Botaniska Notiser, 128:* 119–147.
DAHLGREN, R., 1980. A revised system of classification of the angiosperms. *Botanical Journal of the Linnean Society, 80:* 91–124.
DAHLGREN, R., 1983. General aspects of angiosperm evolution and macrosystematics. *Nordic Journal of Botany, 3:* 119–149.
DAHLGREN, R., JENSEN, S. R. & NIELSEN, B. J. 1981. A revised classification of the angiosperms with comments on correlations between chemical and other characters. In D. A. Young & D. S. Seigler (Eds), *Phytochemistry and Angiosperm Phylogeny:* 149–199. New York: Praeger.
DAHLGREN, R. & THORNE, R. F., 1985. The order Myrtales: circumscription, variation, and relationships. *Annals of the Missouri Botanical Garden, 71:* 633–699.
DANG-VAN-LIEM, 1962. 'Recherches sur l'embryogénie des Tricoques'. Thèse, Paris.
DÄNIKER, A. U., 1946. Über die Euphorbiaceen und die Entwicklung der Monochlamydeae. *Archiv der Julius Klaus-Stifting für Vererbungsforschung Sozialanthropologie und Rassenhygiene, 21:* 465–469.
DAVIS, G., 1966. *Systematic Embryology of the Angiosperms.* New York: Wiley & Sons.
DECAISNE, J., 1842. *Bischoffia.* In C. d'Orbigny, (Ed.), *Dictionnaire Universel d'Histoire Naturelle.* Paris: Bureau Principal de l'Editeur.
DECAISNE, J., 1858. Sur le genre *Aegotoxicum. Bulletin de la Société Botanique de France, 5:* 214–215.
DE CANDOLLE, A. L. P. P., 1867. Dr. Mueller's monograph of the Euphorbiaceae. *Journal of Botany, 5:* 151–153.
DE CANDOLLE, A. P., 1813. *Théorie élémentaire de la Botanique.* Paris: Deterville.
DE JUSSIEU, A., 1823. Considérations sur la famille des Euphorbiacées. *Mémoires du Muséum National d'Histoire Naturelle, Paris, 10:* 317–355.
DE JUSSIEU, A., 1824. *De Euphorbiacearum Generibus Medicisque Earumdem Viribus Tentamen.* Paris: Didot.
DE JUSSIEU, A. 1845., Euphorbiacées. In A. D. d'Orbigny, *Dictionnaire Universel d'Histoire Naturelle, 5:* 773–776.
DE JUSSIEU, A. L., 1789. *Genera Plantarum Secundum Ordines Naturales Disposita.* Paris: Herissant.
DOMKE, W., 1934. Untersuchungen über die systematische und geographische Gliederung der Thymelaeaceen nebst einer Neubeschreibung ihrer Gattungen. *Bibliotheca Botanica, 111:* 1–151.
DRESSLER, R. L., 1957. The genus *Pedilanthus* (Euphorbiaceae). *Contributions from the Gray Herbarium of Harvard University, 182:* 1–188.
DRESSLER, R. L., 1962. A synopsis of *Poinsettia* (Euphorbiaceae). *Annals of the Missouri Botanical Garden, 48:* 329–341.
DUMORTIER, B. C. J., 1829. *Analyse des Familles des Plantes.* Tournai: Casterman.
EHRENDORFER, F., 1983. Summary statement. *Nordic Journal of Botany, 3:* 151–155.
EICHLER, A. W., 1878. *Blütendiagramme construirt und erläutert,* 2. Leipzig: W. Engelmann.
EMBERGER, L., 1960. *Traité de Botanique (Systématique), Tome II. Les Végétaux Vasculaires.* Paris: Masson.
ENDLICHER, S., 1841. *Enchiridion Botanicum Exhibens Classes et Ordines Plantarum.* Leipzig: W. Engelmann.
ENDRESS, P. K., 1977. Evolutionary trends in the Hamamelidales-Fagales Group. *Plant Systematics and Evolution, Supplement 1:* 321–347.
ENGLER, A., 1892. *Syllabus der Vorlesungen über specielle und Medicinisch-Pharmaceutische Botanik.* Berlin: Borntraeger.
ENGLER, A., 1893. Icacinaceae. In A. Engler & K. Prantl (Eds), *Die Natürlichen Pflanzenfamilien, 3(5):* 233–257. Leipzig: W. Engelmann.
ENGLER, A., 1897. Übersicht über die Unterabteilungen, Klassen, Reihen, Unterreihen und Familien der Embryophyta Siphonogama. In A. Engler & K. Prantl, (Eds), *Die Natürlichen Pflanzenfamilien, Nachträge zum II-IV Teil.* Leipzig: W. Engelmann.
ENGLER, A., 1912a. *Panda oleosa* Pierre, ein ölsamenbaum Westafrikas. *Notizblatt des Königlichen Botanischen Gartens und Musems zu Berlin, 5:* 274–276.
ENGLER, A., 1912b. *Syllabus der Pflanzenfamilien,* Ed. 7. Berlin: Borntraeger.
ERDTMAN, G., 1952. *Pollen Morphology and Plant Taxonomy: Angiosperms.* Stockholm: Almqvist & Wicksell.
FORMAN, L. L., 1965. A new genus of Ixonanthaceae with notes on the family. *Kew Bulletin, 19:* 517–526.
FORMAN, L. L., 1966. The reinstatement of *Galearia* Zoll. et Mor. and *Microdesmis* Hook. f. in the Pandaceae. *Kew Bulletin, 20:* 309–321.
FORMAN, L. L., 1968. The systematic position of *Panda* Pierre. *Proceedings of the Linnean Society of London, 179:* 269–270.
FORMAN, L. L., 1971. A synopsis of *Galearia* Zoll. and Mor. (Pandaceae). *Kew Bulletin, 26:* 153–165.
GIBBS, R. D., 1974. *Chemotaxonomy of Flowering Plants.* 4 vols. Montreal: McGill-Queen's University Press.
GILBERT, M. G., in press. The new geophytic species of *Euphorbia,* with comments on the subgeneric grouping of the African members of the genus. *Kew Bulletin, 42.*
GRAY, A., 1879. *Gray's Botanical Text-Book, 1. Structural Botany.* New York: American Book Co.
GRISEBACH, A., 1854. *Grundriss der systematischen Botanik.* Göttingen: Dieterichschen Buchhandlung.
GRISEBACH, A., 1859. Euphorbiaceae, in *Flora of the British West Indian Islands:* 31–54. London: Lovell Reeve.
GRISEBACH, A., 1860. Erlauterungen ausgewählter Pflanzen des Tropischen Amerikas. *Abhandlungen der Königlichen Gesellschaft der Wissenschaften zu Göttingen, 9:* 3–58.

GROSSHEIM, A. A. & SAKHOKIA, M. F. 1966. *Review of Modern Systems of Flowering Plants*. Tbilisi: Metsniereba.
GUNDERSEN, A., 1950. *Families of dicotyledons*. Waltham: Chronica Botanica.
HALLIER, H., 1896. Betrachtungen über die Verwandtschaftsbeziehungen der Ampelideen und anderer Pflanzenfamilien. *Natuurkundig Tijdschrift voor Nederlandsch-Indië, 56:* 300–331.
HALLIER, H., 1903. Über die Verwandtschaftsverhältnisse dei Engler's Rosalen, Parietalen, Myrtifloren und in anderen Ordnungen der Dikotylen. *Abhandlungen aus dem Gebiete der Naturwissenschaften, 18:* 1–98.
HALLIER, H., 1904. Über die Gattung *Daphniphyllum*, ein übergangsglied von den Magnoliaceen und Hamamelidaceen zu den Katzchenblütlern. *Botanical Magazine (Tokyo), 18:* 55–69.
HALLIER, H., 1910. Ueber Phanerogamen von unischerer oder unrichtiger Stellung. *Mededeelingen van's Rijks-herbarium, 1:* 1–41.
HALLIER, H., 1912. L'origine et le système phylétique des Angiosperms exposés à l'aide de leur arbre généalogique. *Archives Néerlandaises des Sciences Exactes et Naturelles, III, 1:* 146–234.
HANS, A. S., 1973. Chromosomal conspectus of the Euphorbiaceae. *Taxon, 22:* 591–636.
HAYDEN, W. J., 1977. Comparative anatomy and systematics of *Picrodendron*, genus *Incertae sedis*. *Journal of the Arnold Arboretun, 58:* 257–279.
HAYDEN, W. J., 1980. 'Systematic Anatomy of Oldfieldioideae (Euphorbiaceae)'. Unpublished Ph.D. dissertation, University of Maryland, College Park.
HAYDEN, W. J., GILLIS, W. T., STONE, D. E., BROOME, C. R. & WEBSTER, G. L., 1984. Systematics and palynology of *Picrodendron:* further evidence for relationship with the Oldfieldioideae (Euphorbiaceae). *Journal of the Arnold Arboretum, 65:* 105–127.
HEGNAUER, R., 1963–1973. *Chemotaxonomie der Pflanzen*. 6 vols. Basel: Birkhauser.
HEGNAUER, R., 1977. Cyanogenic compounds as systematic markers in Tracheophyta. *Plant Systematics and Evolution, Supplement 1:* 191–209.
HEINIG, K. H., 1951. Studies in the floral morphology of the Thymelaeaceae. *American Journal of Botany, 38:* 113–132.
HICKEY, L. J. & WOLFE, J. A., 1975. The bases of angiosperm phylogeny: vegetative morphology. *Annals of the Missouri Botanical Garden, 62:* 538–589.
HOEFFGEN, F., 1922. Sero-diagnostische Untersuchungen über die Verwandtschaftsverhältnisse innerhalb des Columniferen-astes der Dicotylen. *Botanisches Archiv, 1:* 81–99.
HOOKER, J. D., 1848. *Microdesmis puberula. Hooker's Icones Plantarum, 8:* t. 758.
HOOKER, J. D., 1876. On the classification of plants by the natural method with an analysis of their classes, cohorts, and orders, as arranged in this work. In E. Le Maout & J. Decaisne, *A General System of Botany*. London: Longmans, Green & Co.
HOOKER, J. D., 1887. *Lophopyxis. Hooker's Icones Plantarum, 18:* t. 1714.
HOOKER, W. J., 1836. *Aextoxicum punctatunm. Hooker's Icones Plantarum, 1:* t. 12.
HORANINOV, P. F., 1847. *Characteres Essentiales Familiarum*. St Petersburg: Wienhoberianis.
HURUSAWA, I., 1954. Eine nochmalige Durchsicht des herkömmlichen systems der Euphorbiaceen im weiteren Sinne. *Journal of the Faculty of Science, University of Tokyo, Botany, 6:* 209–342.
HUTCHINSON, J., 1911–12. Euphorbiaceae [in part]. In W. T. Thistleton-Dyer (Ed.), *Flora of Tropical Africa, 6(1):* 441–1059.
HUTCHINSON, J., 1926. *The Families of Flowering Plants I. Dicotyledons*. London: Macmillan.
HUTCHINSON, J., 1959. *The Families of Flowering Plants I. Dicotyledons*, 2nd edition. Oxford: Clarendon Press.
HUTCHINSON, J., 1967. *The Genera of Flowering Plants (Angiospermae), 1*. Oxford: Clarendon Press.
HUTCHINSON, J., 1969a. *Evolution and Phylogeny of Flowering Plants*. London: Academic Press.
HUTCHINSON, J., 1969b. Tribalism in the family Euphorbiaceae. *American Journal of Botany, 56:* 738–758.
HUTCHINSON, J., 1973. *The Families of Flowering Plants*, 3rd edition. Oxford: Clarendon Press.
JABLONSKI, E., 1967. Euphorbiaceae, in Botany of the Guayana Highland—Part VII. *Memoirs of the New York Botanical Garden, 17(1):* 80–190.
JANCHEN, E., 1932. Entwurf eines Stammbaumes der Blütenpflanzen nach Richard Wettstein. *Oesterreichische Botanische Zeitschrift, 81:* 161–166.
JANSSONIUS, H. H., 1950. Wood anatomy and relationship. *Blumea, 6:* 407–461.
JAUBERT, H.-F., 1866. Sur les Euphorbiacées et sur un genre nouveau de Buxacées de Zanzibar. *Bulletin de la Société Botanique de France, 13:* 461–468.
JORGENSEN, C. A., 1923. Studies on Callitrichaceae. *Botanisk Tidsskrift, 38:* 81–122.
JORGENSEN, C. A., 1925. Frage der systematischen Stellung der Callitrichaceae. *Jahrbuch für Wissenschaftliche Botanik, 64:* 440–442.
KEATING, R. C., 1973. Pollen morphology and relationships of the Flacourtiaceae. *Annals of the Missouri Botanical Garden, 60:* 273–305.
KEATING, R. C., 1985. Leaf histology and its contribution to relationships in the Myrtales. *Annals of the Missouri Botanical Garden, 71:* 801–823.
KJAER, A., 1973. The natural distribution of glucosinolates: a uniform group of sulfur-containing glucosides. In G. Bendz & J. Santesson (Eds), *Chemistry in Botanical Classification*. Nobel 25. New York: Academic Press.
KLOTZSCH, J. F., 1841. Neue und weniger gekannte südamerikanische Euphorbiaceen-Gattungen. *Archiv für Naturgeschichte, 7(1):* 175–204.

KLOTZSCH, J. F., 1859. Linne's natürliche Pflanzenklasse Tricoccae des Berliner Herbarium's im Allgemeinen und die natürliche Ordnung Euphorbiaceae insbesondere. *Monatsberichte der Königlichen Preussichen Akademie der Wissenschaften zu Berlin, 1859:* 236–254. [Reprinted and amplified with same title as Klotzsch (1860).]

KLOTZSCH, J. F., 1860. *Abhandlungen der Königlichen Akademie der Wissenschaften zu Berlin, 1859:* 1–108.

KÖHLER, E., 1965. Die Pollenmorphologie der biovulaten Euphorbiaceae und ihre Bedeutung für die Taxonomie. *Grana Palynologica, 6:* 26–120.

KÖHLER, E., 1980. Zur Pollenmorphologie und systematischen Stellung der Didymelaceae Leandri. *Feddes Repertorium 91:* 581–591.

KÖHLER, E. & BRÜCKNER, P., 1983. Zur Pollenmorphologie und systematischen Stellung der Gattung *Simmondsia* Nutt. *Wissenschaftliche Zeitschrift der Friedrich-Schiller-Universität Jena, Mathematisch-Naturwissenschaftliche Reihe, 32:* 945–955.

KUBITZKI, K., 1972. Probleme der Grossgliederung der Blütenpflanzen. *Berichte der Deutschen Botanischen Gesellschaft, 85:* 259–277.

KUBITZKI, K., 1977. Classification and evolution of higher plants. *Plant Systematics and Evolution, Supplement 1:* 21–31.

LANJOUW, J., 1931. *The Euphorbiaceae of Surinam.* Thesis. Amsterdam: de Bussy.

LANJOUW, J. et al. (Eds), 1968. *Compendium van de Pteridophyta en Spermatophyta.* Amsterdam: A. Oosthoek's Uitgeversmaatschappij.

LEACH, L. C., 1976. Distributional and morphological studies of the tribe Euphorbieae (Euphorbiaceae) and their relevance to its classification and possible evolution. *Excelsa, 6:* 3–19.

LEANDRI, J., 1937. Sur l'aire et la position systématique du genre malgache *Didymeles* Thouars. *Annales des Sciences Naturelles, Botanique, Série 10, 19:* 309–317.

LEANDRI, J., 1957. Euphorbiacées; tome I. In H. Humbert (Ed.), *Flora de Madagascar et des Comores, 111:* 1–209.

LEANDRI, J., 1962. Un grand systématicien français émule d'Adanson: Ernest-Henri Baillon 1827–1895. *Adansonia, Série 2, 2:* 3–14.

LEONARD, J., 1962. Euphorbiaceae, part I. *Flore du Congo et du Rwanda-Burundi, 8(1):* 1–214.

LEONARD, J. & MOSANGO M., 1985. Hymenocardiaceae. *Flore d'Afrique Centrale: Spermatophytes.* Bruxelles: Jardin Botanique National de Belgique.

LEVIN, G. A., 1986. Systematic foliar morphology of Phyllanthoideae (Euphorbiaceae). I. Conspectus. *Annals of the Missouri Botanical Garden, 73:* 29–85.

LINDLEY, J., 1833. *Nixus Plantarum.* London: Ridgway & Sons.

LINDLEY, J., 1836. *A Natural System of Botany.* London: Longman, Rees, Orme, Brown, Green, and Longman.

LINDLEY, J., 1853. *The Vegetable Kingdom,* 3rd edition. London: Bradbury & Evans.

LINNAEUS, C., 1751. *Philosophia Botanica.* Stockholm: Kiesewetter.

LINNAEUS, C., 1764. *Genera Plantarum: Editio Sexta.* Stockholm: Salvius.

LOBREAU, D., 1969. Les limites de l'"ordre" de Celastrales d'après le pollen. *Pollen et Spores, 11:* 499–555.

LOBREAU-CALLEN, D., 1976. Ultrastructure de l'exine de quèlques pollens de Celastrales et des groupes voisins. *Adansonia, Série 2, 16:* 83–92.

LOBREAU-CALLEN, D., 1977. Nouvelle interprétation de l'"ordre" des Celastrales à l'aide de la Palynologie. *Comptes Rendus Hebdomadaires des Séances de l'Académie des Sciences, Paris, Série D, 284:* 915–918.

LOURTEIG, A. & O'DONELL, C. A., 1943. Euphorbiaceae tribus I, IV, V, VII (III, IV by C. A. O'Donell & A. Lourteig). In H. R. Descole (Ed.), *Genera et Species Plantarum Argentinarum 1:* 143–317. Buenos Aires: G. Kraft.

MAHLBERG, P. G., 1975. Evolution of the laticifer in *Euphorbia* as interpreted from starch grain morphology. *American Journal of Botany, 62:* 577–583.

MAHLBERG, P. G. & PLESZCZYNSKA, W. J., 1984. Evolution of succulent *Euphorbia* as interpreted from latex composition. In W. Rauh (Ed.), *Tropische und Subtropische Pflanzenwelt, 45:* 94–108. Wiesbaden: Steiner.

MEISNER, C. F., 1841. Euphorbiaceae, in *Plantarum Vascularium Genera, 1:* 337–344; 2: 249–255. Leipzig: Weidmann.

MELIKIAN, A. P., 1968. Position of the families Buxaceae and Simmondsiaceae in the system. *Botanicheskij Zhurnal, 53:* 1043–1047.

MELIKIAN, A. P., 1973. Seed coat types of Hamamelidaceae and allied families in relation to their systematics. *Botanicheskij Zhurnal, 58:* 350–359.

MERXMÜLLER, H., 1972. Systematic botany—an unachieved synthesis. *Biological Journal of the Linnean Society, 4:* 311–321.

METCALFE, C. R. & CHALK, L., 1950. *Anatomy of the Dicotyledons.* Oxford: Clarendon Press.

MEZ, C. & ZIEGENSPECK, H. 1926. Der Königsberger serodiagnostische Stammbaum. *Botanisches Archiv, 13:* 483–485.

MICHAELIS, P., 1924. Blütenmorphologische Untersuchungen an den Euphorbiaceen. *Botanische Abhandlungen (Jena), 3:* 1–150.

MIERS, J., 1862. On *Aextoxicum* and *Bursinopetalum. The Annals and Magazine of Natural History, Third Series, 9:* 214–223.

MILLER, R. B., 1975. Systematic anatomy of the xylem and comments on the relationshops of Flacourtiaceae. *Journal of the Arnold Arboretum, 56:* 20–102.
MORISON, R., 1680. *Plantarum Historiae Universalis, 2.* Oxford: e Theathro Sheldoniano.
MUELLER, J., 1864. Systemen der Euphorbiaceen. *Botanische Zeitung, 22:* 324.
MUELLER, J., 1866a. Euphorbiaceae. In A. P. de Candolle (Ed.), *Prodromus Systematis Naturalis Regni Vegetabilis, 15(2):* 189–1261. Paris: Masson.
MUELLER, J., 1866b. Nachschrift zu meiner systematischen Arbeit über Euphorbiaceen. *Botanische Zeitung, 24:* 333–345.
MUELLER, J., 1869. Daphniphyllaceae, Buxaceae. In A. P. de Candolle (Ed.), *Prodromus Systematis Naturalis Regni Vegetabilis, 16(1):* 1–6, 7–23. Paris: Masson.
MUELLER, J., 1875. Replik auf Dr. Baillon's "Nouvelles observations sur les Euphorbiacées". *Botanische Zeitung, 33:* 223–240, 254–256.
NETOLITZKY, F., 1926. *Anatomie der Angiospermen-Samen.* Berlin: Gebrüder Borntraeger.
NOTEBOOM, H. P., 1967. The taxonomic position of Irvingioideae, *Allantospermum* Forman and *Cyrillopsis* Kuhlm. *Adansonia, Serie 2, 7:* 161–168.
NOVÁK, F. A., 1961. *Vyšší Rostliny. Tracheophyta.* Praha: Ceskoslovenska Academie.
NUTTALL, T., 1844. On *Simmondsia*, a new genus of plants from California. *London Journal of Botany, 3:* 400–401.
OURISSON, G., 1973. Some aspects of the distribution of diterpenes in plants. In G. Bendz & J. Santesson (Eds), *Chemistry in Botanical Classification:* 129–134. New York: Academic Press.
PAIS, M., MARCHAND, J., MONSEUR, X., JARREAU, F.-X. & GOUTREL, R., 1967. Alcaloïdes peptidiques. Structure de l'hymenocardine, alcaloïde de *l'Hymenocardia acida* Tul. (Euphorbiacées). *Comptes Rendus Hebdomadaires des Séances de l'Academie des Sciences, Paris, Série C, 264:* 1409–1411.
PARIS, R., 1963. The distribution of plant glycosides. In T. Swain (Ed.), *Chemical Plant Taxonomy:* 337–358. New York: Academic Press.
PAX, F., 1884. Die Anatomie der Euphorbiaceen in ihrer Beziehung zum System derselben. *Botanische Jahrbücher für Systematik, Pflanzengeschichte und Pflanzengeographie, 5:* 384–421.
PAX, F., 1890. Euphorbiaceae. In A. Engler & K. Prantl (Eds), *Die Natürlichen Pflanzenfamilien, Erste Auflage, III. 3(5):* 1–119. Leipzig: W. Engelmann.
PAX, F., 1910. Euphorbiaceae-Jatropheae. In Engler, A. (Ed.), *Das Pflanzenreich, IV. 147. I (Heft 42):* 1–148.
PAX, F., 1911. Euphorbiaceae-Cluytieae, Additamentum II. In A. Engler (Ed.), *Das Pflanzenreich, IV, 147. IV (Heft 47):* 1–124.
PAX, F., 1924. Die Phylogenic der Euphorbiaceen. *Botanische Jahrbücher für Systematik, Pflanzengeschichte und Pflanzengeographie, 59:* 129–182.
PAX, F. & HOFFMANN, K., 1917. Systematische Stellung der Gattung *Aextoxicon*. *Jahresbericht der Schlesischen Gesellschaft für Vaterländische Cultur, Zoologisch-Botanische Sektion (Band I, Abteilung IIb), 1916:* 17–21.
PAX, F. & HOFFMANN, K., 1919. Euphorbiaceae-Additamentum VI. In A. Engler (Ed), *Das Pflanzenreich, IV. 147. XIV (Heft 68):* 1–63.
PAX, F. & HOFFMANN, K., 1931. Euphorbiaceae. In A. Engler & K. Prantl (Eds), *Die Natürlichen Pflanzenfamilien, Zweite Auflage, 19c:* 11–233. Leipzig: W. Engelmann.
PIERRE, L., 1896. Plantes du Gabon. *Bulletin Mensuel de la Société Linnéenne de Paris, 1896:* 1250–1256.
PIERRE, L., 1897. Sur le genre, *Plagiostylis*. *Bulletin Mensuel de la Société Linnéenne de Paris, 1897:* 1326–1328.
PFEIFFER, H. H., 1951. *Lophopyxis* als typus einer Eigenen familie. *Revista Sudamerica Botanica, 10:* 3–6.
PFITZER, E. H. H., 1894. *Übersicht des natürlichen Systems der Pflanzen.* Heidelberg: Winter's Universitätsbuchhandlung.
PLANCHON, J.-E., 1854. Affinités et synonymie de qualques genres nouveaux ou peu connus. *Annales des Sciences Naturelles, Botanique, Série 4, 1:* 256–266.
PLÉE, F., 1854. *Types de Chaques Familles et des Principaux Genres des Plantes Croissant Spontanément en France.* Paris: Baillière.
PRANCE, G. T., 1972. A monograph of the neotropical Dichapetalaceae. *Flora Neotropica, 10:* 1–84.
PULLE, A., 1938. *Compendium van de Terminologie, Nomenclatur en Systematiek der Zaadplanten.* Utrecht: Oosthoek.
PUNT, W., 1962. Pollen morphology of the Euphorbiaceae with special reference to taxonomy. *Wentia, 7:* 1–116.
PUNT, W., 1975. Pollen morphology of the Dichapetalaceae with special reference to evolutionary trends and mutual relationships of pollen types. *Review of Palaeobotany and Palynology, 19:* 1–97.
RAO, P. N., 1970. Euphorbiaceae. In B. M. Johri, (Ed.), *Symposium on Comparative Embryology of Angiosperms. Bulletin of the Indian National Science Academy, 41:* 136–141.
RADCLIFFE-SMITH, A., 1978. Euphorbiaceae. In V. H. Heywood (Ed.), *Flowering Plants of the World.* New York: Mayflower Books.
RASMUSSEN, F. N., 1983. On "apomorphic tendencies" and phylogenetic inference. *Systematic Botany, 8:* 334–337.
REICHENBACH, H, G. L., 1828. *Conspectus Regni Vegetabilis.* Leipzig: Cnobloch.
REICHENBACH, H. G. L., 1837. *Handbuch des Natürlichen Pflanzensystems.* Dresden: Arnoldischen Buchhandlung.
REICHENBACH, H. G. L., 1841. *Repertorium Herbarii sive Nomanclator Generum Plantarum.* Dresden: Arnold.
RENDLE, A. B., 1925. *The Classification of Flowering Plants, 2. Dicotyledons.* Cambridge: The University Press.

ROGERS, D. J. & APPAN, S. G., 1973. *Manihot, Manihotoides* (Euphorbiaceae). *Flora Neotropica, 13:* 1–272.
ROSENTHAL, K., 1919. Daphniphyllaceae. In A. Engler (Ed.), *Das Pflanzenreich, IV. 147a (Heft 68):* 1–6.
ROSENTHAL, K., 1931. Daphniphyllaceae. In A, Engler & H. Harms (Eds), *Die Natürlichen Pflanzenfamilien, Zweite Auflage, 19c:* 233–235.
SCHMID, R., 1985. Reproductive anatomy and morphology of Myrtales in relation to systematics. *Annals of the Missouri Botanical Garden, 71:* 832–835.
SCHNIZLEIN, A., c. 1860. Bennettieae R. Brown. *Iconographia Familiarum Naturalium Regni Vegetabilis, 3:* pl. 172. Bonn: Cohen & Sohn.
SCHOLZ, H., 1964. Geraniales, Rutales, Sapindales, Celastrales. In H. Melchior (Ed), *A. Engler's Syllabus der Pflanzenfamilien*, ed. 12, 2: 246–300. Berlin: Borntraeger.
SCOGIN, R., 1980. Serotaxonomy of *Simmondsia chinensis* (Simmondsiaceae). *Aliso, 9:* 555–559.
SEEMANN, B., 1866. Prodromus Systematis Naturalis Regni Vegetabilis . . . Pars XV, Sectio Posterior, Fasc. II, sistens Euphorbiaeas [review]. *Journal of Botany, 4:* 387–388.
SEIGLER, D. S., 1977. The naturally-occurring cyanogenic glycosides. In L. Reinhold, J. Harborne & T. Swain (Eds), *Progress in Phytochemistry, 4:* 83–120. Oxford: Pergamon Press.
SLEUMER, H., 1968. The genus *Lophopyxis* Hook. f. (Lophopyxidaceae). *Blumea, 16:* 321–323.
SOÓ, R., 1967. Die modernen Systeme der Angiospermen. *Acta Botanica Academiae Scientiarum Hungaricae, 13:* 201–233.
SPACH, R., 1834. *Historie Naturelle des Végétaux, 2*. Paris: Roret.
SPENCER, K. C. & SEIGLER, D. S., 1984. Cyanogenic glycosides of *Carica papaya* and its phylogenetic position with respect to the Violales and Capparales. *American Journal of Botany, 71:* 1444–1447.
STAFLEU, F., 1965. Engler's Syllabus [review]. *Taxon, 14:* 23–25.
STEBBINS, G. L., 1974. *Flowering Plants: Evolution above the Species Level.* Cambridge, Massachusetts: Belknap Press of Harvard University Press.
STEVENS, P. F., 1984. Metaphors and typology in the development of botanical systematics 1690–1960, or the art of putting new wine in old bottles. *Taxon, 33:* 169–211.
TAKHTAJAN, A., 1966. *Systema et Phylogenia Magnoliophytorum*. Moscow: Nauka.
TAKHTAJAN, A., 1969. *Flowering Plants, Origin and Dispersal.* Edinburgh: Oliver & Boyd.
TAKHTAJAN, A., 1980. Outline of the classification of flowering plants (Magnoliophyta). *The Botanical Review, 46:* 225–359.
THORNE, R. F., 1968. Synopsis of a putatively phylogenetic classification of the flowering plants. *Aliso, 6:* 57–66.
THORNE, R. F., 1976. A phylogenetic classification of the Angiospermae. *Evolutionary Biology, 9:* 35–106.
THORNE, R. F., 1981. Phytochemistry and angiosperm phylogeny: a summary statement. In D. A. Young & D. S. Seigler (Eds), *Phytochemistry and Angiosperm Phylogeny:* 233–295. New York: Praeger.
THORNE, R. F., 1983. Proposed new realignments in the angiosperms. *Nordic Journal of Botany, 3:* 85–117.
VAN DEN BRINK, R. C. Backhuizen, 1968. Addenda et corrigenda. In C. A. Backer & R. C. Backhuizen van den Brink, *Flora of Java, 3:* 642–660. Groningen: Walters–Noordhoff.
VAN ROYEN, A., 1740. *Florae Leydensis Prodromus.* Leyden: Luchtmans.
VAN TIEGHEM, P., 1897a. Sur les Buxacées. *Annales des Sciences Naturelles, Série 8, Botanique, 5:* 289–338.
VAN TIEGHEM, P., 1897b. Sur les phanerogames sans graines, formant le division des Inseminées. *Bulletin de la Société Botanique de France, 44:* 99–139.
VAN TIEGHEM, P., 1898. Sur le genre Simmóndsie considéré comme type d'une famille distincte, les Simmondsiacées. *Journal de Botanique, 12:* 103–112.
VAN TIEGHEM, P., 1903. Structure de l'ovule des Dichapetalacées et place de cette famille dans la classification. *Journal de Botanique, 17:* 229–233.
VAN VLIET, C. J. C. M. & BASS, P., 1985. Wood anatomy and classification of the Myrtales. *Annals of the Missouri Botanical Garden, 71:* 783–800.
VENTENAT, E. P., 1799. *Tableau du Regne Végétale, selon la Méthode de Jussieu.* Tome Troisième. Paris: Drissonier.
VILLIERS, J.-F., 1973. Pandaceae. In A. Aubréville & J.-F. Leroy (Eds), *Flore du Gabon, 22:* 14–30. Paris: Muséum National d'Histoire Naturelle.
VILLIERS, J.-F., 1975. Pandaceae. In A. Aubréville & J.-F. Leroy (Eds), *Flore du Cameroun, 19:* 42–58. Paris: Muséum National d'Histoire Naturelle.
VINDT, J., 1953. Monographie des Euphorbiacées du Maroc. Première Partie: Revision et Systématique. *Travaux de l'Institut Scientifique Chérifien, 6:* 1–217.
von MARTIUS, C. F. P., 1835. *Conspectus Regni Vegetabilis.* Nürnberg: Schrag.
VOSS, E. G., et al., 1983. *International Code of Botanical Nomenclature.* Utrecht & Antwerpen: Bohn, Scheltema & Holkema.
WAGENITZ, G., 1964. Thymelaeales. In H. Melchior (Ed.), *A. Engler's Syllabus der Pflanzenfamilien*, ed. 12, 2: 316–322. Berlin: Borntraeger.
WAGENITZ, G., 1975. Blütenreduktion als ein zentrales Problem der Angiospermen-Systematik. *Botanische Jahrbücher für Systematik, Pflanzengeschichte und Pflanzengeographie, 96:* 448–470.
WARMING, E., 1904. *A Handbook of Systematic Botany*, 2nd edition. London: Allen & Unwin.
WEBSTER, G. L., 1967. The genera of Euphorbiaceae in the southeastern United States. *Journal of the Arnold Arboretum, 48:* 303–430.

WEBSTER, G. L., 1975. Conspectus of a new classification of the Euphorbiaceae. *Taxon, 24:* 593–601.
WEBSTER, G. L. & BURCH, D., 1968. Euphorbiaceae, in Flora of Panama, Part VI, Family 97. *Annals of the Missouri Botanical Garden, 54:* 211–350.
WEBSTER, G. L. & HUFT, M., 1986. Revised synopsis of Panamanian Euphorbiaceae. *Annals of the Missouri Botanical Garden, 72:* in press.
WEBSTER, G. L. & MILLER, K. I., 1963. The genus *Reverchonia* (Euphorbiaceae). *Rhodora, 65:* 193–207.
WEDDELL, H.-A., 1857. Considérations générales sur la famille des Urticées, suivie de la description des tribus et des genres. *Annales des Sciences Naturelles, Série 4, Botanique, 7:* 307–398.
WETTSTEIN, R., 1901. *Handbuch der Systematischen Botanik.* Leipzig: Deuticke.
WETTSTEIN, R., 1935. *Handbuch der Systematischen Botanik. Vierte, umgearbetete Auflage.* Leipzig: Deuticke.
WHEELER, L. C., 1941. *Euphorbia* subgenus *Chamaesyce* in Canada and the United States exclusive of southern Florida. *Rhodora, 43:* 97–154, 168–205, 223–286.
WHEELER, L. C., 1943. The genera of living Euphorbieae. *American Midland Naturalist, 30:* 456–503.
WHITE, A., DYER, R. A. & SLOANE, B. L., 1941. *The Succelent Euphorbieae (Southern Africa).* 2 vols. Pasadena: Abbey Garden Press.
WIGER, J., 1935. Embryological Studies on the Families Buxaceae, Meliaceae, Simarubaceae and Burseraceae. Thesis, Lund.
WILEY, E. O., 1981. *Phylogenetics: The Theory and Practice of Phylogenetic Systematics.* New York: Wiley.
WILLIS, J. C., 1966. *A Dictionary of the Flowering Plants and Ferns.* 7th ed. revised by H. K. Airy Shaw. Cambridge: Cambridge University Press.
WUNDERLICH, R., 1968. Some remarks on the taxonomic significance of the seed coat. *Phytomorphology, 17:* 301–311.
ZOLLINGER, H., 1845. Observationes phytographicae, praecipue genera et species nova nonnulla respicientes. *Natur-en Geneeskundig Archief voor Neerland's-Indie, 2:* 1–19.

… With 11 Figures

Segregate families from the Euphorbiaceae

A. RADCLIFFE-SMITH

Herbarium, Royal Botanic Gardens, Kew, Richmond, Surrey

Received May 1986, accepted for publication August 1986

RADCLIFFE-SMITH, A. 1987. **Segregate families from the Euphorbiaceae.** Several families have been segregated from the Euphorbiaceae. Some of these are of long-standing, and are now generally accepted. Others are more recent, and somewhat controversial. Five such have recently been recognized at Kew, and are discussed here, as are two others which have not been so recognized.

ADDITIONAL KEYWORDS:—Aextoxicaceae – Androstachydaceae – *Androstachys* – *Antidesma* – *Aristogeitonia* – *Bischofia* – Bischofiaceae – Buxaceae – *Celaenodendron* – Daphniphyllaceae – Didymelaceae – *Hyaenanche* – *Hymenocardia* – Hymenocardiaceae – *Mischodon* – *Oldfieldia* – *Paivaeusa* – Paivaeusinae – Palo Prieto – Pandaceae – *Pera* – Peraceae – Picrodendraceae – *Picrodendron* – Piraiera Preta – *Piranhea* – Piranheira – *Simmondsia* – Stilaginaceae – *Uapaca* – Uapacaceae.

CONTENTS

Introduction	47
Segregate families	48
Buxaceae	48
Aextoxicaceae	48
Didymelaceae	49
Daphniphyllaceae	49
Pandaceae	49
Controversial segregates	49
Stilaginaceae	50
Hymenocardiaceae	52
Uapacaceae	54
Bischofiaceae	54
Peraceae	57
Androstachydaceae	58
Paivaeusinae	59
Conclusion	64
Dedication	64
Acknowledgements	64
References	65

INTRODUCTION

In his paper on the floral morphology of *Picrodendron*, Hakki (1985) says that those who have studied the Euphorbiaceae in detail emphasize the difficulty encountered in the characterization of such a huge group with so many diverse members. He comments that this makes the recognition of an atypical euphorbiaceous plant difficult, and for that reason, perhaps, *Picrodendron* was incorrectly assigned systematically until recently, even though it had been known for some 280 years. This confusion was added to by Pax & Hoffmann (1931) who erroneously described it as having an inferior ovary. However, it now seems to be firmly ensconced in the Euphorbiaceae, and Webster (1975) places it in his subfamily Oldfieldioideae.

Bentham (1878), referring to the work of Baillon (1858) and Mueller of Aargau (1866) remarked: "Two men, indeed, both of high standing in the science, and with comparatively ample materials at their command, have recently worked up the Order with great care and attention independently of each other, and I would have readily followed the lead of either of them, but that the two have so frequently come to conclusions diametrically opposed to each other, that I have been compelled to steer a course of my own through a labyrinth of tribes, subtribes, genera, sections or vaguely indicated affinities". This problem has not gone away!

The recognition or rejection of families segregated from Euphorbiaceae over the years has, unfortunately, been based largely on opinions formed from the same data. Little new information has been assimilated, except for some chromosome counts, pollen and wood anatomical features. It is inevitable that little improvement can be expected until a concerted attempt is made to rationalize all available facts, and fill the gaps in knowledge.

Some of the segregate families do have wide general acceptance, and these will be discussed first. The status of the more controversial 'families' will then be considered.

Recently Kew staff have been reviewing plant families recognized in the Herbarium; reference will be made to those that relate to Euphorbiaceae.

SEGREGATE FAMILIES

Buxaceae

The Buxaceae have had a long history as a distinct family, but Bentham & Hooker (1880) regarded them as deserving of only tribal rank among the Euphorbiaceae, and for a long time they were so regarded at Kew. They are generally still included in the Euphorbiales, but differ from the Euphorbiaceae in having apotropous ovules with a dorsal raphe. Furthermore, they are without stipules or disc, although those deficiencies alone would not debar them. The Buxaceae itself suffered dismemberment when Van Tieghem (1897) separated the monotypic genus *Simmondsia* into a family of its own. *Simmondsia chinensis* (Link) Schneider, the Jojoba, differs in a number of macro- and micromorphological, anatomical, palynological and metabolic features from the Euphorbiales, and is considered by some to have a closer affinity with the Monimiaceae in the Laurales, although Cronquist (1981) still keeps it in the Euphorbiales.

Aextoxicaceae

Another family removed from the Euphorbiales, by some to the vicinity of the Monimiaceae, by others, e.g. Cronquist (1981), to the Celastrales, is the Aextoxicaceae. It has apotropous anatropous ovules borne in only one of the two loculi, the style is strongly deflexed to one side and the seeds have a ruminate endosperm. It was undoubtedly kept in the Euphorbiaceae partly on account of its very striking superficial resemblance to certain *Croton* species, enhanced by the lepidote indumentum. Anatomically, it would not appear to be out of place in the Celastrales.

Didymelaceae

This is another monogeneric family which has been associated with the Euphorbiales through the Buxaceae. However, it is now generally considered to belong to the Hamamelidales (as were the Buxaceae and the Daphniphyllaceae by Hutchinson, 1959), or, as Cronquist (1981) now has it, as a distinct order, the Didymelales, in the subclass Hamamelidae.

Daphniphyllaceae

The monogeneric family Daphniphyllaceae, some species of which bear a superficial resemblance in habit to rather dull rhododendrons although without the showy flowers, was considered a distinct family by Mueller of Aargau (1869), although Bentham & Hooker (1880) later placed *Daphniphyllum* in the Euphorbiaceae in their tribe Phyllantheae between the two closely related genera *Aporosa* and *Baccaurea*. The family is exstipulate, and the minute apical embryo is located in a copious endosperm. They are considered by Cronquist (1981) as constituting an order of their own in the Hamamelidae. They produce the unique alkaloid daphniphylline, and have a basic chromosome number of $x = 8$, whereas *Aporosa*, *Baccaurea*, etc. have $x = 7$.

Pandaceae

This family originally comprised the sole genus *Panda*, described by Pierre (1896), and was the only representative of the Order Pandales. It is a West African tree with a large fruit having a massive stony endocarp. Forman (1966) showed, on account of the shoot structure, the nature of the fruit and anatomical features, that the family should also include the genera *Galearia* and *Microdesmis*, hitherto placed in the Euphorbiaceae, and for which Bentham & Hooker (1880) had established their artificial tribe Galeariae. Pax & Hoffmann (1931) later reduced this to subtribal status under their tribe Clutieae. Pierre (1896, 1897) himself had indeed drawn attention to the similarity between the three genera, among other things on account of the orthotropous or anatropous ovules which lack an obturator. Airy Shaw (1966) included the genus *Centroplacus* here also. The Pandaceae is generally kept within the pale of the Euphorbiales.

CONTROVERSIAL SEGREGATES

There is now general agreement that the above families should be excluded from the Euphorbiaceae, and in some cases even from Euphorbiales. There are however more controversial groups.

Comprehensive multidisciplinary studies have been carried out in the last decade on *Picrodendron*, resulting in its having been brought in to the

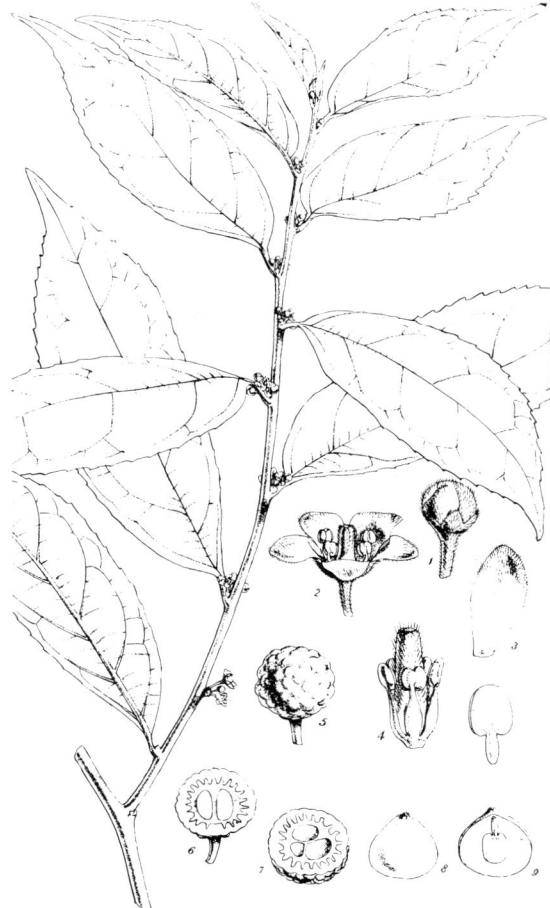

Figure 1. *Microdesmis puberula* Hook. fil. ex Planch. Drawing made for *Hooker's Icones Plantarum, VIII*, t. DCCLVIII (1848). Flowering shoot with enlargements & dissections: 1. Male flower-bud. 2. Male flower. 3. Petal. 4. Stamens surrounding the pistillode. 5. Fruit, dried. 6. Vertical section of fruit. 7. Transverse section of fruit. 8. Seed. 9. Vertical section of seed. (Enlargements not enlarged upon). Based on material collected by Vogel in W. Trop. Africa.

Euphorbiaceae from an uneasy affinity with the Sapindales. Similar work now needs to be carried out on certain anomalous genera which have long resided in the Euphorbiaceae but about which there is no general consensus as to whether they should be kept therein.

Stilaginaceae

Antidesma is a genus of about 170 species from the Old World Tropics with a few species in Tropical Africa and Madagascar but the majority in Tropical Asia. It was so named by Burmann (1737) on account of its having been used to prepare an antidote to certain snake poisons. Agardh (1825) regarded this genus as constituting a separate family which he named the Stilaginaceae (from the superfluous generic name *Stilago* L., meaning wandering style, from the often

Figures 2 & 3. *Antidesma madagascariense* Lam. Photographed by M. J. E. Coode in the Crater on the Mascarine Island of Mauritius. Fig. 2. Male inflorescences. Fig. 3. Fruiting branch.

excentric subterminal or laterally-disposed style in this genus). Endlicher (1837) renamed the family Antidesmataceae. However, *Antidesma* was regarded as euphorbiaceous by most authors (Mueller (1866), for example, placed it near *Thecacoris*, Pax & Hoffmann (1931) near *Aporosa* and Hutchinson (1969) near *Hieronyma*) until recently when Airy Shaw (1966) expressed the opinion that it appears to have no close relatives in the family, stating that the "small drupaceous fruits with their characteristic foveolate–reticulate often flattened and somewhat obliquely-orientated endocarps differ widely from anything found in the Euphorbiaceae, but show much similarity to those of several genera of Icacinaceae—Celastrales—e.g. *Rhyticaryum*, *Natsiatum*, *Iodes* etc.; the leaf-venation is also very similar to that of *Rhyticaryum* etc. There are, however, also obvious and profound differences from the Icacinaceae". One of these is that in the latter the ovules have a dorsal instead of a ventral raphe. Cronquist (1981), however, considers that the Euphorbiales and the Sapindales are both closely allied to the Celastrales, and the shifting of families from one to another of the three orders does not reflect any great difference in the concepts of relationship. Broadly speaking, the relationships between the three orders may be briefly summarized as follows:

1 Flowers ± reduced and unisexual with the perianth poorly developed or lacking . . . . Euphorbiales
1' Flowers not markedly reduced . . . .
   2 Leaves mostly compound . . . . . . Sapindales
   2' Leaves simple, seldom compound . . . . . Celastrales

Webster (1975) preferred to keep *Antidesma* within the subfamily Phyllanthioideae. At Kew, we recently decided to recognize the Stilaginaceae, but have now had second thoughts.

The pollen of *Antidesma* is not distinctive, but is in fact similar to that of *Hieronyma*, *Thecacoris*, *Cyathogyne*, *Apodiscus*, *Protomegabaria*, *Spondiathus* and *Maesobotrya*, as Punt (1962) showed when he grouped these together with *Antidesma* in his *Antidesma*-subtype of the *Antidesma* pollen-type. Pax & Hoffmann (1931) had put all these genera together in their subtribe Antidesminae of the tribe Phyllantheae on the basis of inflorescence characters. According to Hans (1970), *Antidesma* should not be segregated from the Euphorbiaceae on karyological grounds. Endlicher (1837) incidentally had long ago included in the family Scepaceae most of the genera later referred by Pax & Hoffmann to their Antidesminae except for *Antidesma* itself. This family is not now considered worth upholding.

Anatomically, the wood of *Antidesma* shows a similarity to that of certain Flacourtiaceae—Violales— for example, *Caloncoba* and *Kiggerlaria*; but then, so apparently does that of certain woody *Phyllanthus* species.

Another segregation made by Agardh (1825) was that of *Drypetes* and its allies as the family Putranjivaceae, but this has never been generally accepted.

Hymenocardiaceae

A second controversial segregate family which we have decided to recognize here at Kew is the Hymenocardiaceae, established by Airy Shaw (1965). This

Figure 4. *Hymenocardia acida* Tul. Drawing made in January 1926, by William Edward Trevithick, published in the *Flora of West Tropical Africa*, by J. Hutchinson & J. M. Dalziel. In Vol. 1(2) of the 2nd edition (1958), revised by R. W. J. Keay, it appeared as Fig. 132 on p. 376. A. Male shoot. B. Male flower with rudimentary ovary. C, D. Stamen, back and front. E. Female shoot. F. Female flower. G. Fruiting branch.

again has but one component genus, *Hymenocardia*, a small genus with a few species in Tropical Africa and one species in Tropical Asia. *Hymenocardia acida* Tulasne is the most widespread of the African species. The large extrorse anthers, with a terminal gland, fold hard back after dehiscence in a manner reminiscent of some Ulmaceae. The di- or tri-porate pollen differs completely from that of all other Euphorbiaceae investigated, and Punt (1962) designated a special pollen-type to accommodate it. No disc is present in either the male or the female flowers. The ovary with its long undivided styles is also somewhat reminiscent of the Ulmaceae, e.g. *Celtis* spp., and furthermore, the flattened fruit is winged. However, unlike the Ulmaceae, the fruit is perfectly bilaterally symmetrical and dehiscent. The flattening of the fruit is at right-angles to the septum, producing a superficial resemblance to the samara of *Acer* spp., but upon dehiscence the central axis or columella is left on the plant in much the same way as one is left by the dehiscing fruits in many Euphorbiaceae, except that in the latter they are generally triquetrous instead of bilateral.

Anatomically, the wood of *Hymenocardia* also has some point in common with that of certain flacourtiaceous genera, more specifically that of *Homalium*, *Casearia* and *Samyda*; but then so apparently does that of *Aporosella* and *Glochidion*. Commenting on the two different base-numbers (13,14) reported in *Hymenocardia*, Hans (1973) says, 'If $x = 14$ is confirmed in the genus, Willis' (i.e. Airy Shaw's) view may be endorsed'.

Mueller (1866) had *Hymenocardia* as constituting a subtribe of his Tribe Phyllantheae, Webster (1975) as constituting a Tribe of his subfamily Phyllanthoideae; Hutchinson (1969) also had a Tribe Hymenocardieae but included in it *Aporosa*, *Aporosella*, *Baccaurea*, *Didymocistus* and *Martretia* as well, whereas Bentham & Hooker (1880) and Pax & Hoffmann (1931) accorded it no special recognition at all. Léonard & Mosango (1985) recognize the Hymenocardiaceae.

Uapacaceae

Kew recognized this monogeneric family for a time, but has now revoked that decision. *Uapaca* (pronounced 'Wa-pa-ka') derives its strange name from the Malagasy vernacular name for two Madagascan species. It is a genus of about fifty species restricted to Tropical Africa and Madagascar. Figure 5 shows the stilt-roots of an *Uapaca* species taken in Nigeria, and Fig. 6 is of *Uapaca kirkiana* Muell. Arg. in the Ntchisi Forest Reserve, Malawi. The genus is, incidentally, often a subdominant in *Brachystegia* association. Uapacas are dioecious pachycaul trees or shrubs, with the leaves crowded towards the ends of the branches. The small male flowers are crowded into dense globose capitate pedunculate inflorescences, and are surrounded by a quasicalycine whorl of 5–10 imbricate bracts. No disc is present in either the male or the female flowers. The female flowers are solitary, but are surrounded by a similar quasicalycine whorl of bracts as are the heads of male flowers. This particular type of involucre would be unique in the Euphorbiaceae were *Uapaca* to be left in it. The styles are strap-like, much-laciniate and recurved over the ovary. The fruit is drupaceous, with a tough, fleshy exocarp, and encloses three pyrenes which are dorsally bisulcate. On the basis of the pachycaul habit and the texture and venation of the leaves, Airy Shaw (1965), who also established this family, suggested an affinity with the Anacardiaceae.

Anatomically, the wood-structure of *Uapaca* is highly aberrant, and Metcalfe & Chalk (1950) could not readily fit it into their main Euphorbiaceous groups.

Palynologically, *Uapaca* does not appear to be out of the ordinary, for it has a pollen-type very like that of a number of phyllanthoid genera such as *Savia*, *Cleistanthus* and *Meineckia*. Furthermore, Hans (1973) states that it "deserves its place in the tribe Phyllantheae of Bentham and Hooker on the chromosomal evidence".

As with *Hymenocardia*, Mueller (1866) had *Uapaca* as constituting a subtribe of his Tribe Phyllantheae. Pax & Hoffmann (1931) also kept *Uapaca* as the sole representative of their subtribe Uapacinae. Hutchinson (1969) upgraded this to tribal status, but this time with no other genera added, and Webster (1975) also kept *Uapaca* at the tribal level, in the subfamily Phyllanthoideae. Bentham & Hooker (1880) had accorded no special status to *Uapaca*, placing it within an unnamed subdivision of their Tribe Phyllantheae along with *Aporosa*, *Daphniphyllum*, *Baccaurea*, *Cometia*, *Antidesma*, *Hieronyma*, *Maesobotrya* and *Aextoxicon*.

Bischofiaceae

A fourth controversial family briefly recognized at Kew is the Bischofiaceae, with the sole genus *Bischofia*, which has two species, *Bischofia javanica* Bl. and *B. polycarpa* (Lévl.) Airy Shaw (*B. racemosa* Cheng & Chu). The first is a timber-tree ranging from China south through Southeast Asia to southern India and Polynesia. The second is found only in China. From the euphorbialean

Figures 5 & 6. Fig. 5 *Uapaca* sp., showing stilt-roots. Photographed by J. Lowe, 3 February 1968, at Sapele in Nigeria. Fig. 6. *Uapaca kirkiana* Muell. Arg. A male tree photographed by Dr R. K. Brummitt in the Ntchisi Forest Reserve, Malawi.

standpoint, *Bischofia* is unique in that it has pinnate leaves, and superficially appears to be more at home in the Sapindales than the Euphorbiales. Figure 7 shows the female plant. The small, generally dioecious flowers are borne in large thyrses. The calyx is valvate in the male flower, but imbricate in the female flower. No disc is present in either sex. The stamens are opposite to and enclosed by the strongly adaxially-concave and cucullate sepals. The style is short, but the stigmas are elongated and spreading. The fruit is a small, globose, fleshy drupe. The spherical pollen-grains are not as aberrant for the family

Figure 7. *Bischofia javanica* Bl. Drawn & lithographed by Walter Hood Fitch for *Hooker's Icones Plantarum*, IX, t. DCCCXLIV (1852). Female flowering, and fruiting, shoots with magnifications & dissections: 1. Female flower. 2. Transverse section of ovary. 3. Transverse section of fruit, showing two loculi aborting. 4. Seed. 5. Vertical section of seed. 6. Embryo abstracted from albumen. Based on material from the Bombay Ghauts at 16°N, *c*. 600 m above sea level.

Euphorbiaceae as are those of *Hymenocardia*, but Punt (1962) nevertheless regards them as constituting a distinct type which is not readily to be associated with those of his phyllanthoid configurations. Anatomically, the wood of *Bischofia* is not as aberrant as that of *Uapaca*, but in many ways resembles that of *Antidesma* and *Phyllanthus*.

Again, as with *Hymenocardia* and *Uapaca*, Mueller (1866) had *Bischofia* as constituting a subtribe of his Tribe Phyllantheae, and Pax & Hoffmann (1922, 1931) also kept *Bischofia* as the sole representative of their subtribe Bischoffiinae. Hutchinson (1969) followed Hurusawa (1954) in upgrading this to tribal status, but included with *Bischofia* genera now considered as having nothing to do with it, namely those belonging to the group generally known as the Paivaeusinae, of which more later. Webster (1975) also maintains this at the tribal level, as the last tribe of his subfamily Phyllanthoideae. Bentham & Hooker (1880) again accorded no special status to *Bischofia*, but placed it in an unnamed subdivision of their Tribe Phyllantheae between *Oldfieldia* and *Piranhea*, both members of the Paivaeusinae. Airy Shaw (1965) considered his new family Bischofiaceae as probably related to the Staphyleaceae (Sapindales), and especially to *Tapiscia*, but at the same time acknowledged that it differed from this in many respects, notably in its apetaly, absence of disc, far fewer ovules per loculus and long-reflexed stigmas. It also has a different basic no. ($x = 7$ instead of 13) (Hans, 1973).

In the last century Blume (1826) and George Don (1832) had felt that *Bischofia* should be placed near the Rutaceae or the Anacardiaceae (both Sapindales). Decaisne (1842) was probably the first to transfer *Bischofia* to the Euphorbiaceae, without giving any reasons for doing so, and until Airy Shaw (1965) all later workers followed him.

Peraceae

The fifth and last of the controversial families segregated from the Euphorbiaceae and briefly recognized at Kew is the Peraceae. This is again a monogeneric family, containing only the genus *Pera*. *Pera* comprises about forty species of trees or shrubs ranging more or less throughout the New World Tropics, from Mexico and the West Indies to Tropical South America. The indumentum is stellate or lepidote, as in many Euphorbiaceae, but such an indumentum is found in many other families as well, and a great many undisputed Euphorbiaceae have a simple indumentum or even none at all. The often highly-reduced, sometimes achlamydeous, flowers may be either monoecious or dioecious and they are arranged in small axillary unisexual or bisexual involucrate clusters. The involucre consists of one or two small free outer and two larger variously connate spathaceous inner bracts which have the appearance of young calyx-lobes so that the young inflorescence resembles a flower-bud. No disc is present in either sex. The styles are fused to form a sessile peltate disc. The fruit is a tardily-dehiscent trilocular drupiform capsule with a woody endocarp, fleshy or spongy mesocarp and a smooth epicarp which becomes much-wrinkled when dried. When the valves split, they do not spring back elastically as in most schizocarpic Euphorbiaceae, and usually remain attached at their base to the pedicel, whereas in most Euphorbiaceae with incomplete dehiscence the valves remain attached to the apex of the columella. Furthermore, in *Pera* the slender columella splits longitudinally into three parts, whereas in all the dehiscent Euphorbiaceae, the columella remains as a solid unit. Thus in many macromorphological features *Pera* departs from the euphorbiaceous norm. Palynologically, however, *Pera* shares a number of features with *Ricinus* and other genera, such as the very narrow colpi, but it also has a number of distinguishing characteristics, and Punt (1962) regarded the pollen of *Pera* as constituting a distinct type. Anatomically, on the other hand, the wood of *Pera* would not appear to be aberrant for the Euphorbiaceae.

Mueller (1866) had *Pera* as the sole representative of a subtribe of his Tribe Acalypheae, and Pax & Hoffmann (1931) had it as the sole representative of their Tribe Pereae, which they placed between the tribes Dalechampieae and Clutieae in their subfamily Crotonoideae, which included all the uniovulate

Euphorbiaceae except for the stenolobe genus *Ricinocarpos* and its allies. Hutchinson (1969) and Webster (1975) also have *Pera* in a group of its own at the tribal level, the latter as the last tribe of his subfamily Acalyphoideae. Bentham & Hooker (1880) have *Pera* in the Tribe Crotoneae, which is more or less equivalent to Pax's Crotonoideae; however, they have it right at the end of their system, as the last genus of their subtribe Hippomaneae (which nowadays we should call Hippomaninae if accorded subtribal status). Klotzsch (1859) appears to have been the first to establish the family Peraceae, but the only subsequent major contributor to euphorbialean classification to have kept *Pera* in a family of its own has been Airy Shaw (1966, 1973).

Whether or not it is right to follow Airy Shaw in the recognition of the Stilaginaceae, Hymenocardiaceae, Uapacaceae, Bischofiaceae and Peraceae as families distinct from the Euphorbiaceae, only time and results from other disciplines will tell.

Androstachydaceae

Airy Shaw (1965) also regarded the Afro-Malagasy genus *Androstachys* as constituting a distinct family, but Kew has never accepted it. The genus has some anomalous features. The branching is decussate, and the leaves are opposite. Opposite leaves *are* known in several euphorbiaceous genera, for example *Erismanthus*, *Excoecaria* and *Mallotus*, but alternate leaves are the general rule in the family. The leaves of *Androstachys johnsonii* Prain are slightly peltate, entire and densely white-tomentose beneath, and this combination of features alone renders it extremely distinctive. One of the most curious vegetative features of *Androstachys* lies in the nature of the stipules, which are large, intrapetiolar (as, for example, those of *Pilea* in the Urticaceae) and connate into an oblong sheath enclosing the terminal bud which is flattened at right-angles to the petiolar axis.

For many years the genus was only known from one species in southeast tropical Africa and Madagascar, but in 1970 Airy Shaw described four new species from Madagascar: *A. merana* Airy Shaw, *A. imberbis* Airy Shaw, *A. viticifolia* Airy Shaw and *A. rufibarbis* Airy Shaw. They all differ from *A. johnsonii* in having digitately 3–7-foliolate leaves as opposed to simple ones, and in them the stipules are adnate to the petiole, rather than being free from it. In *A. merana*, the leaflets actually appear to spring out of the stipular sheath, whereas in the other three there is a stretch of free petiole between the sheath and the leaflets. Leroy (1975 *in scheda*) regards these as constituting a separate genus, *Stachyandra*.

The male flowers of *Androstachys* have been variously interpreted. They are more or less cernuous, and are disposed in shortly pedunculate axillary triads. The calyx has two or three sepals in the lateral flowers of each triad, and five in the terminal one, and the sepals are spirally arranged instead of being arranged in whorls. No disc is present in either the male or female flower. The stamens are numerous and are disposed on an elongate axis with the filaments very short and the athers linear with each theca apiculate. On account of the spiral arrangement of the sepals, these have been considered as being bracteoles, the staminal column a partial inflorescence axis and the anthers as vestiges of reduced flowers on the axis.

The pollen grains are inaperturate, but they seem to approximate quite closely to those of the genus *Stachystemon* in the Tribe Caletieae.

The style is elongate, and there are three simple spreading stigmas. The fruit is more or less typically euphorbiaceous, showing the usual dehiscence pattern for the family, and it is perhaps for this reason that Androstachydaceae is not recognized in the Kew Herbarium.

Androstachys was not known to the 19th century euphorbiologists. It was first described by Prain (1908), and Pax & Hoffmann (1931) placed it with *Toxicodendron* (i.e. *Hyaenanche*) on account of the non-alternate leaves and the absence of a female disc, as the only other representative of their subtribe Toxicodendrinae of the Tribe Phyllantheae. Hutchinson (1969) also placed it with *Hyaenanche* in his Tribe Hyaenancheae, along with *Longetia*, *Dissiliaria*, *Mischodon* and *Paradrypetes*, again primarily on the grounds of vegetative macromorphology. However, Webster (1975) placed it with *Petalostigma* in his Tribe Petalostigmateae of his subfamily Oldfieldioideae.

Airy Shaw (1965) regarded the male structure of *Androstachys* as "unique, not only in the family Euphorbiaceae, but in the Angiosperms as a whole".

The anatomy resembles that of *Microdesmis* in the Pandaceae in many respects. Metcalfe & Chalk (1950) were not able to fit it into any of their main euphorbiaceous groups.

Concerning Pandaceae, Picrodendraceae, Androstachydaceae, Bischofiaceae, Stilaginaceae, Hymenocardiaceae and Uapacaceae, Gibbs (1974), says, "We know so little of the genera named by Airy Shaw that we can ignore them here; but it would be nice to study their chemistry".

Paivaeusinae

The other main problem with regard to aberrant and anomalous euphorbioid groups concerns that of genera collectively known by Pax's subtribal name of Paivaeusinae. This includes the genera *Oldfieldia*, *Paivaeusa*, *Aristogeitonia* and *Piranhea*; *Paivaeusa* is generally considered to be congeneric with *Oldfieldia*. Nowadays, *Celaenodendron* and an as yet undescribed genus related to it are also included in this group, and *Mischodon* and *Hyaenanche* are often associated with it as well.

Oldfieldia was described by Bentham & Hooker (1850) as a monotypic genus from the West African primary forests. It occurs in Sierra Leone, Liberia, Côte d'Ivoire and Cameroun and was named after Mr Oldfield, one of the early collectors of the Sierra Leone interior. *Oldfieldia africana* Benth. & Hook.f. is a large forest tree often over 100 ft in height with a clear bole of some 50 ft (a nightmare for collectors!) and was in great demand for shipbuilding. Its wood was known in the timber trade as 'African Oak' or 'African Teak'.

Mueller (1866) excluded *Oldfieldia* from his treatment of the Euphorbiaceae for De Candolle's *Prodromus* and said of it: "doubtfully ascribed to the Euphorbiaceae, differing from this family in the loculicidal dehiscence of the fruit; it pertains rather to the Sapindaceae". As far as dehiscence patterns go, however, there may be several types even in the same genus: thus, different species of *Croton* may be septicidally or loculicidally dehiscent or even subindehiscent, and the same applies to *Jatropha* as well. The most striking feature of *Oldfieldia* is unquestionably the foliage: the leaves are opposite and

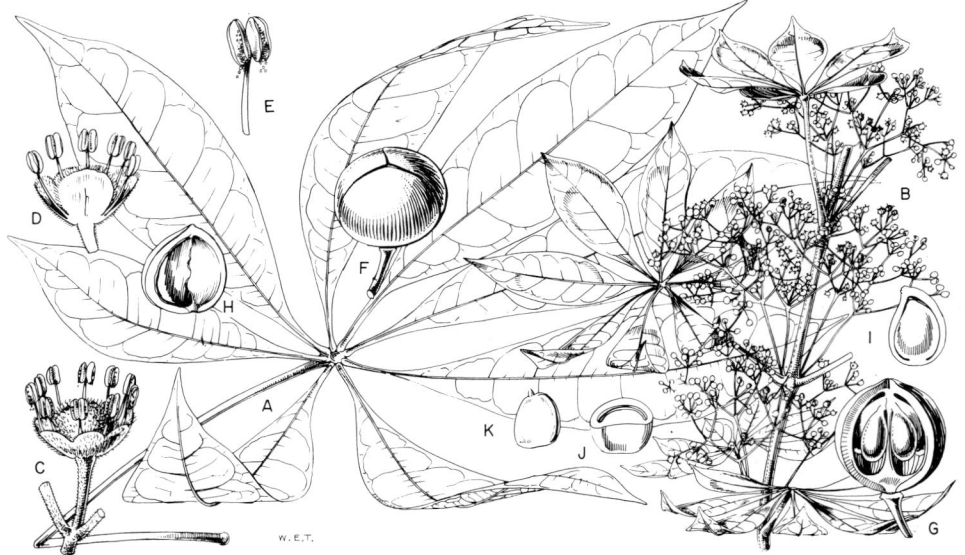

Figure 8. *Oldfieldia africana* Benth. & Hook. fil. Drawing made in January 1926, by William Edward Trevithick, published in the *Flora of West Tropical Africa*, by J. Hutchinson & J. M. Dalziel. In Vol. 1 (2) of the 2nd edition (1958), revised by R. W. J. Keay, it appeared as fig. 130 on p. 369. A. Leaf. B. Male inflorescence. C. Male flower with rudimentary ovary. D. Vertical section of male flower. E. Stamen. F. Fruit. G. Fruit with valve removed. H. Fruit-valve. I. Vertical section of seed. J. Transverse section of seed. K. Embryo.

palmate, and superficially resemble those of *Aesculus* in the Hippocastanaceae, which is a segregate family from the Sapindaceae! As with *Uapaca* and *Androstachys*, the wood-anatomy of *Oldfieldia* does not appear to conform to any known euphorbiaceous pattern.

Bentham & Hooker (1880) placed *Oldfieldia* along with *Piranhea* in their Tribe Phyllantheae, but with no infratribal designation. Hutchinson (1969) placed Pax's Paivaeusinae with *Bischofia* in Hurusawa's Tribe Bischoffieae on account of the compound leaves of all the genera concerned. Airy Shaw (1966, 1973) tentatively suggested a possible link between Paivaeusinae and the Picrodendraceae, and thought that they should perhaps constitute a separate family intermediate between the Euphorbiaceae and the Picrodendraceae. Webster (1967) and Köhler (1965) established the subfamily Oldfieldioideae as distinct from the subfamily Phyllanthoideae, thus recognizing two subfamilies of biovulate Euphorbiaceae as Pax had done, but on a different basis: of Pax's four subfamilies there was one biovulate and one uniovulate in each of his two infrafamilial but supersubfamilial groups, the Platylobeae and the Stenolobeae, respectively. Webster (1975) merged the Picrodendraceae with the Euphorbiaceae, establishing it as a tribe of his subfamily Oldfieldiodeae, so that Airy Shaw's view of the relationships of these groups was given the formal treatment which he was somewhat reluctant to give it, albeit at a different level than that which he first mooted. Subsequent work on *Picrodendron* by Hayden and others (1977, 1984) and by Hakki (1985) has tended to confirm the euphorbiaceous assimilation of this genus.

The genus *Paivaeusa* was named by Welwitsch in honour of Baron do Castello de Paiva, a botany professor at Oporto, and published by Bentham & Hooker (1867) for a species in Angola which was subsequently found to extend into southern Zaïre, Zambia and Tanzania. This species, *Paivaeusa dactylophylla* Welw. ex Oliv., was described by Oliver (1868) the following year. *Paivaeusa* was originally assigned to the Burseraceae (Sapindales), and only later deemed to be euphorbiaceous. Léonard (1956) considered *Paivaeusa* not to be generically separable from *Oldfieldia*. *Oldfieldia dactylophylla* (Welw. ex Oliv.) J. Léon. is a much smaller tree than *O. africana*, and prefers a drier habitat in open woodland. Another genus *Cecchia*, described by Chiovenda (1932) for a Somali plant, had also been merged with *Oldfieldia* by Milne-Redhead (1949). With the description of a new Zaïrean species by Léonard (1956), *Oldfieldia* is now held to comprise four tropical African species.

Aristogeitonia (which means 'good neighbour') is a genus described by Prain (1908) for an Angolan tree, collected by Gossweiler, with trifoliolate or sometimes unifoliolate leaves, which has generally been associated with *Oldfieldia*. *Aristogeitonia limoniifolia* Prain has not, to my knowledge, so far been found outside Angola, but in the early 1960s Peter Greenway found a second species (in Tanzania) which Airy Shaw (1972) named *Aristogeitonia monophylla* Airy Shaw, as it had unifoliolate leaves. A third species has been recently found to occur in Madagascar (Radcliffe-Smith, in press). The stipules are inserted on the petiole, and as this arrangement is found in the Euphorbiaceae only in the genus *Mischodon*, it suggested to Airy Shaw that *Mischodon* should be included in the Paivaeusinae. If this were done with the group retained at the subtribal level, then Mueller's earlier subtribal name Mischodontinae would have to be used.

However, *Mischodon*, a monotypic genus from South India and Sri Lanka, is different in other respects. There is no indication that it is unifoliolate, the leaves are produced in whorls of up to four, whereas those of *Aristogeitonia* are alternate, the indumentum is of a different character, and the male inflorescence is diffuse and elongate, whereas in *Aristogeitonia* the flowers are fasciculate. Figure 9 of *Aristogeitonia monophylla* photographed in Kenya's Coastal Province shows the fruits. For comparison, Fig. 10 shows *Mischodon zeylanicus* Thw. Pax & Hoffmann (1922, 1931) placed *Michodon* in their subtribe Dissiliariinae along with *Longetia* and *Dissiliaria*, but these also have fasciculate flowers. Hutchinson (1969) also associated these three genera, as mentioned previously, in his Tribe Hyaenancheae without designating subtribes, whilst Webster (1975) has his Tribe Hyaenancheae with four separate subtribes, the Mischodontinae, Hyaenanchinae, Paivaeusinae and Dissiliariinae as a kind of Alexandrian solution to this particular Gordian Knot!

The flowers of *Aristogeitonia* are polymorphic. In addition to the usual male and female flowers, there are also pseudohermaphrodite flowers produced, in which the stamens are longer than are those of the male flowers but the anthers appear not to contain fertile pollen. Furthermore, the ovaries of the pseudohermaphrodite flowers are perfectly glabrous, whereas those of the normal female flowers are sparsely adpressedly puberulous, and the styles are longer than those of the normal female flowers at a comparable stage of development.

Hyaenanche (*Toxicodendrum* Thunb., not *Toxicodendron* Mill., which is *Rhus*) is a monotypic genus from South Africa with verticillate leaves like *Mischodon*

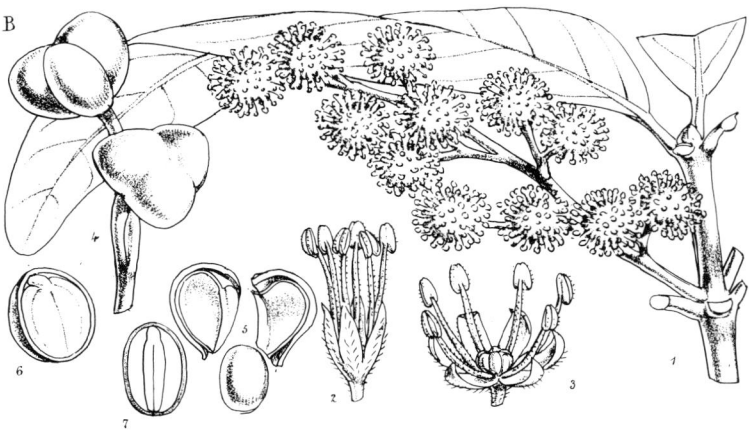

Figures 9 & 10. Fig. 9. *Aristogeitonia monophylla* Airy Shaw. A fruiting branch, photographed by the author at Mwarakaya in Kenya's Coast Province on 24 November 1971. Fig. 10. *Mischodon zeylanicus* Thw. Drawing by George Henry Kendrick Thwaites and Harmanis de Alwis published in *Hooker's Journal of Botany and Kew Garden Miscellany*, 6, t. 10B (1854). 1. Male flowering-branch. 2, 3. Male flowers. 4. Young fruits. 5. Two fruit-valves and a seed. 6, 7. Vertical (dorsiventral and bilateral) sections of seeds. Based on material obtained by the describing author near Ooma Oya, on the lower Badulla road some 40 km. from Kandy, Sri Lanka.

(Fig. 11) but unlike that genus it lacks a female disc. Mueller (1866) placed *Hyaenanche* in a subtribe of its own on account of the absence of a pistillode or central disc in the male flowers. Its subsequent taxonomic history has been touched upon already in the discussion of *Androstachys* and *Mischodon*.

The Paivaeusinae has three Neotropical representatives in addition to the Palaeotropical ones just described. *Piranhea* is a genus of one or two species from Venezuela, Brazil and Guyana which is closely related to *Aristogeitonia*, differing

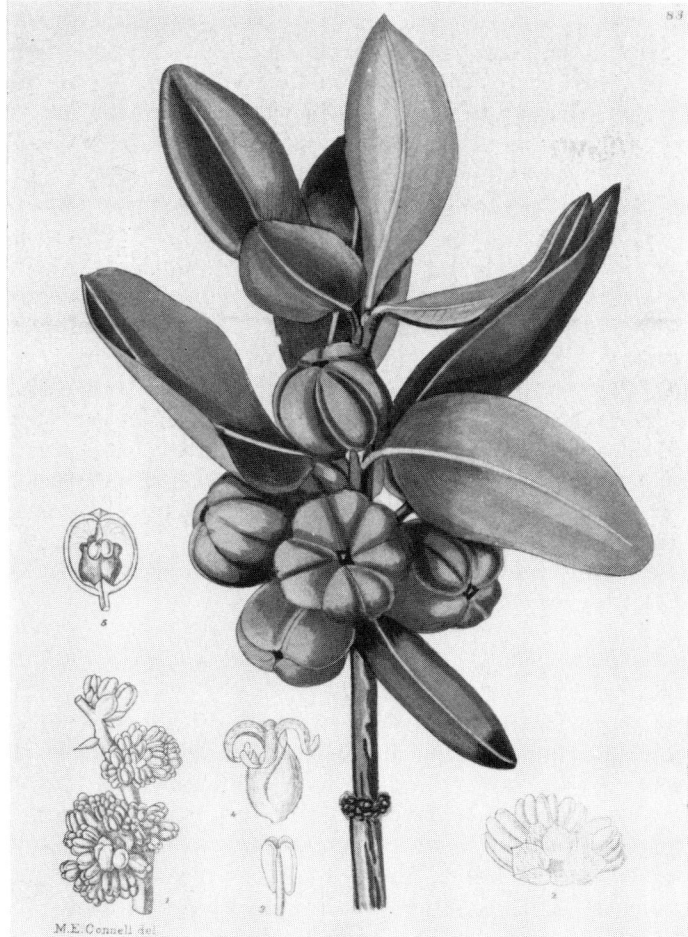

Figure 11. *Hyaenanche globosa* (Gaertn.) Lamb. et Vahl. A partly-coloured illustration by M. E. Connell, published in *The Flowering Plants of South Africa*, 21, t. 837 (1941). Fruiting branch with enlargements and dissections. 1. Male inflorescence. 2. Group of male flowers. 3. Stamen. 4. Female flower. 5. Vertical section of young fruit showing two developing ovules. Based on material obtained by Mr John van Niekerk of Puts, van Rhynsdorp.

from it in that the stipules are not epipetiolar, the leaves are consistently trifoliolate, the inflorescences are shortly interruptedly racemose and the anthers are introrse. It was described by Baillon (1866) from material obtained by Spruce at Manaos on the Amazon in 1851, and its name is derived from the vernacular names Piranha-uba and Piranheira, so called because its seeds are used as a fish-bait. As with *Oldfieldia*, the wood anatomy does not appear to conform to any of the usual euphorbiaceous patterns, and furthermore it is different from that of the other members of the Paivaeusinae.

Celaenodendron is a Mexican genus of this affinity which was described by Standley (1927) on the basis of material collected by Ortega in the State of Sinaloa. Its name is a rendering into Greek of the Mexican name, Palo Prieto. It too is trifoliolate, and has very long pedicellate fruits. It is represented in the Kew Herbarium only by an isotope of the sole species, *C. mexicanum* Standl. In its wood anatomy it shares some features in common with *Pogonophora*, an otherwise somewhat anomalous and rather isolated acalyphoid genus.

In 1972, in the State of Minas Gerais, Brazil, a second species of *Celaenodendron*, or perhaps a new genus allied to it, was discovered by Ratter, da Fonseca and de Castro in deciduous forest on limestone. Its vernacular name, Piraieira Preta is something of a hybrid between those of *Piranhea* and *Celaenodendron*! It is apparently a very hard-wooded tree some 20–25 m tall which damaged the edge of a machete blade used on it. It too has compound leaves, which may be either trifoliolate or 5-foliolate, and long-stalked loculicidally dehiscent fruits. Its bark has been compared to that of *Eucalyptus* and Guava. Flowering material of this tree is still awaited before it can be formally described.

The pollen morphology of *Oldfieldia*, *Paivaeusa*, *Aristogeitonia*, *Piranhea*, *Hyaenanche* and *Mischodon* suggest a close relationship between them, and Punt (1962) placed them all in a group which he referred to as the "*Aristogeitonia*-type", in which the grains have very small apertures situated in the equatorial plane, unlike those of *Stachystemon*. However, the pollen of other genera can be referred to this type also, for example *Tetracoccus* (considered by Pax and Hutchinson as being related to *Drypetes* but now placed by Webster with *Mischodon*), *Paragelonium*, *Petalostigma* and *Longetia*.

CONCLUSION

This review of the controversial genera of the Euphorbiaceae shows that we are still far from agreement as to the most satisfactory mode of classification to adopt for them. The problem lies, as with all taxonomic problems, in there being more levels of similarity and distinction in nature than can be adequately reflected in the categories of our hierarchical structure, so that whatever system we eventually settle for is bound to be a compromise, and the more so as data from the various newer disciplines superimposes further, and sometimes rather different, networks of relationships upon those already considered to exist.

DEDICATION

This paper is dedicated to the memory of the late Herbert Kenneth Airy Shaw, 1902–1985.

ACKNOWLEDGEMENTS

I wish to put on record my thanks to the following: the Director of the Botanical Research Institute, Pretoria, Republic of South Africa, for permission to reproduce M. E. Connell's illustration of *Hyaenanche globosa* which was published as Plate 837 in *The Flowering Plants of South Africa*, 21 (1941); to the Bentham-Moxon Trustees, who hold the copyright to the plates published in Hooker's *Icones Plantarum* for permission to reproduce Plates 758 (*Microdesmis puberula*) & 844 (*Bischofia javanica*) from that work, as well as Plate 10B (*Mischodon zeylanicus*) from Hooker's *Kew Journal*, 6, and also to the Crown Agents, publishers of the *Flora of West Tropical Africa*, for permission to reproduce Figures 130 & 132 from Vol. 1(2) of the second edition of that work, which were illustrations prepared by W. E. Trevithick of *Oldfieldia africana* and *Hymenocardia acida* respectively.

REFERENCES

AGARDH, C. A., 1825. Stilaginaceae. In *Aphorismi Botanici, 14:* 199–200.
AIRY SHAW, H. K., 1965. Diagnoses of new families, new names, etc., for the seventh edition of Willis's 'Dictionary'. *Kew Bulletin, 18 (2):* 249–273.
AIRY SHAW, H. K., 1966. *J. C. Willis: A Dictionary of the Flowering Plants and Ferns,* 7th edition. Cambridge: Cambridge University Press.
AIRY SHAW, H. K., 1970. The genus *Androstachys* Prain in Madagascar. *Adansonia, series 2, 10 (4):* 519–524.
AIRY SHAW, H. K., 1972. A second species of the genus *Aristogeitonia* Prain (Euphorbiaceae) from East Africa. *Kew Bulletin, 26 (3):* 495–498.
AIRY SHAW, H. K., 1973. *J. C. Willis: A Dictionary of the Flowering Plants and Ferns,* 8th edition. Cambridge: Cambridge University Press.
BAILLON, H., 1858. *Étude Générale du Groupe des Euphorbiacées.* Paris: Victor Masson.
BAILLON, H., 1866. Sur Deux Euphorbiacées Brésiliennes. *Adansonia, 6:* 231–238.
BENTHAM, G., 1878. Notes on Euphorbiaceae. *Botanical Journal of Linnean Society, 17:* 185–267.
BENTHAM, G. & HOOKER, J. D., 1850. African oak (or teak). *Hooker's Journal of Botany & Kew Gardens Miscellany, 2:* 183–186.
BENTHAM, G. & HOOKER, J. D., 1867. *Genera Plantarum, 1 (3):* 993. Londini: L. Reeve & Co.; Williams & Norgate.
BENTHAM, G. & HOOKER, J. D., 1880. *Genera Plantarum, 3 (1):* 239–340. Londini: L. Reeve & Co.; Williams & Norgate.
BLUME, C. L., 1826. *Bischofia. Bijdragen tot de flora van Nederlandsch Indie, Stukke 17:* 1168. Batavia: Ter Lands Drukkerij.
BURMANN, J., 1737. *Thesaurus Zeylanicus:* 22, t. 10.
CHIOVENDA, E., 1932. *Flora Somala, 2:* 397–399. Modena: R. Orto Botanico.
CRONQUIST, A., 1981. *An Integrated System of Classification of Flowering Plants:* 729–740. New York: Columbia University Press.
DECAISNE, J., 1842. *Bischofia* in D'Orbigny, A.C.V.D., *Dictionnaire Universel d'Histoire Naturelle, 2:* 639. Paris: Abel Pilon et Cie.
DON, G., 1832. *Bischofia. A General History of the Dichlamydeous Plants,* or *A General System of Gardening and Botany, 2:* 69. London: J. G. & F. Rivington; J. & W. T. Clarke; Longman & Co.; T. Cadell; J. Richardson, etc.
ENDLICHER, S. L., 1837. Antidesmaceae. In *Genera Plantarum, 1 (4):* 287. Wien: Fr. Beck Universitatis Bibliopolam.
FORMAN, L. L., 1966. The reinstatement of *Galearia* Zoll. & Mor. and *Microdesmis* Hook. f. in the Pandaceae. *Kew Bulletin, 20 (2):* 309–318.
GIBBS, R. D., 1974. *Chemotaxonomy of flowering plants, 3:* 1351. Montreal & London: McGill-Queen's University Press.
HAKKI, M. I., 1985. Floral morphology, anatomy and relationship of *Picrodendron baccatum* (L.) Krug & Urban (Euphorbiaceae). *Botanische Jahrbücher für Systematik, Pflanzengeschichte und Pflanzengeographie, 107 (1–4):* 379–394.
HANS, A. S., 1970. Polyploidy in *Antidesma. Caryologia, 23:* 321–327.
HANS, A. S., 1973. Chromosomal conspectus of the Euphorbiaceae. *Taxon, 22 (5/6):* 591–636.
HAYDEN, W. J., 1977. Comparative anatomy and systematics of *Picrodendron,* Genus Incertae Sedis. *Journal of the Arnold Arboretum, 58:* 257–279.
HAYDEN, W. J., GILLIS, W. T., STONE, D. E., BROOME, C. R. & WEBSTER, G. L., 1984. Systematics and palynology of *Picrodendron:* further evidence for relationship with the Oldfieldioideae (Euphorbiaceae). *Journal of the Arnold Arboretum, 65:* 105–127.
HURUSAWA, I., 1954. Eine nochmalige Durchsicht des herkömmlichen Systems der Euphorbiaceen im weiteren Sinne. *Journal of Faculty of Science, University of Tokyo, Sect. III (Botany), 6 (6):* 209–342.
HUTCHINSON, J., 1959. *Families of flowering plants, 1, 2nd edition:* 179, 184. Oxford: Oxford University Press.
HUTCHINSON, J., 1969. Tribalism in the family Euphorbiaceae. *American Journal of Botany, 56 (7):* 738–758.
KLOTZSCH, F., 1859. Linne's natürliche Pflanzenklasse Tricoccae. *Monatsbericht der Königlichen Akademie der Wissenschaften zu Berlin:* 236–254. Berlin: Gedruckt in der Druckerei der Königlichen Akademie der Wissenschaften.
KÖHLER, E., 1965. Die Pollenmorphologie der biovulaten Euphorbiaceae und ihre Bedeutung für die Taxonomie. *Grana Palynologica, 6:* 26–120.
LÉONARD, J., 1956. Observations sur les genres *Oldfieldia, Paivaeusa* et *Cecchia. Bulletin du Jardin Botanique de L'Etat, Bruxelles, 26:* 335–343.
LÉONARD, J. & MOSANGO, M., 1985. Hymenocardiaceae. In P. Bamps (Ed.), *Flore d'Afrique Centrale:* 1–16. Bruxelles: Goemaere.

LEROY, J.-F., 1975. *Stachyandra* [Unpublished genus established *in scheda* only, based on the 4 multifoliate *Androstachys* species of Madagascar].
METCALFE, C. R. & CHALK, L., 1950. *Anatomy of the Dicotyledons, 2:* 1207–1235. Oxford: Clarendon Press.
MILNE-REDHEAD, E. W. B. H., 1949. Tropical African plants XX. *Kew Bulletin, 3 (3):* 456–457.
MUELLER OF AARGAU, J., 1866. Euphorbiaceae. In A. De Candolle (Ed.), *Prodromus Systematis Universalis Regni Vegetabilis, 15 (2):* 89–1286. Paris: Sumptibus Victoris Masson et Filii.
MUELLER OF AARGAU, J., 1869. Daphniphyllaceae. In A. De Candolle (Ed.), *Prodromus Systematis Universalis Regni Vegetabilis, 16 (1):* 1–6.
OLIVER, D. 1868. Burseraceae. In *Flora of Tropical Africa, 1:* 328. London: Reeve & Co.
PAX, F. & HOFFMANN, K., 1922. Euphorbiaceae—Phyllanthoideae—Phyllantheae. In A. Engler (Ed.), *Das Pflanzenreich, IV. 147. xv (Heft 81):* 281–316. Leipzig: Verlag von Wilhelm Engelmann.
PAX, F. & HOFFMANN, K., 1931. Euphorbiaceae. In A. Engler & H. Harms (Eds), *Die Natürlichen Pflanzenfamilien, 19c,* 2nd edition: 11–233. Leipzig: Verlag von Wilhelm Engelmann.
PIERRE, L., 1896. Plantes du Gabon. *Bulletin Mensuel de la Société Linnéenne de Paris, 158:* 1255.
PIERRE, L., 1987. Sur le genre *Plagiostyles* (footnote). *Bulletin Mensuel de la Société Linnéenne de Paris, 167:* 1327.
PRAIN, D., 1908. Diagnoses Africanae XXVI. *Bulletin of Miscellaneous Information, Kew, 1908:* 438.
PUNT, W., 1962. Pollen morphology of the Euphorbiaceae with special reference to taxonomy. *Wentia, 7:* 1–116.
STANDLEY, P. C., 1927. In R. S. Ferris, Preliminary report on the flora of the Tres Marias Islands. *Contributions from the Dudley Herbarium of Stanford University, 1 (2):* 76–77.
VAN TIEGHEM, Ph., 1897. Sur les Buxacées: Simmondsia. *Annales des Sciences Naturelles, Botanique, 8ᵉ Séries, 5:* 290–301.
WEBSTER, G. L., 1967. The genera of Euphorbiaceae in the Southeastern United States. *Journal of the Arnold Arboretum, 48:* 303–430.
WEBSTER, G. L., 1975. Conspectus of a new classification of the Euphorbiaceae. *Taxon, 24 (5/6):* 593–601.

Problems of distinction among succulent *Euphorbia* species from eastern tropical Africa

SUSAN CARTER

The Herbarium, Royal Botanic Gardens, Kew, Richmond, Surrey TW9 3AB

Received April 1986, accepted for publication September 1986

CARTER, S., 1987. **Problems of distinction among succulent *Euphorbia* species from eastern tropical Africa.** Identification of herbarium specimens of succulent *Euphorbia* species is especially difficult due to distortion of important characters upon drying. Problems encountered in collecting and preserving material result in poor representation in the herbarium, and true variation and distribution of closely related taxa are thus often established only from field studies. These problems are exposed and some solutions given in four groups of species from east and northeast Africa: the non-spiny *E. schimperi–nubica* group, the single-spined *E. triaculeata* group, the shrubby *E. heterochroma–stapfii* group and the tree *E. nyikae–kibwezensis* group.

ADDITIONAL KEY WORDS:—Non-spiny species – single-spined species – shrubs – trees – variation.

CONTENTS

Introduction	67
Investigation of closely-related taxa	68
Non-spiny *E. schimperi–E. nubica* group	68
Single-spined *E. triaculeata* group	69
Shrubby *E. heterochroma–E. stapfii* group	71
Tree *E. nyikae–E. kibwezensis* group	74
General conclusions	77
Appendix: species referred to with type data	77
References	78

INTRODUCTION

A major problem to be faced in the preservation of herbarium material of succulent plants is that of desiccation, and the subsequent shrinkage that takes place during the drying process, so that the possibility of positive identification of specimens is greatly reduced.

The identification of many stem-succulents, such as the Cacti and the succulent *Euphorbia* species, relies largely upon stem characters, but because of stem distortion, it is often quite impossible to name herbarium specimens beyond a basic grouping. As a result, herbarium specimens give, with few exceptions, little indication of the true nature of the plant they represent. Small portions of material preserved in spirit provide more accurate information about stem characters, and photographs, especially of habit and stem detail, are invaluable in the study of material which greatly changes in character on drying, but photographs are rarely taken as an accompaniment to a specimen, except by a specialist.

Because of the extreme problems involved in making herbarium specimens of these plants, non-specialist collectors will generally avoid making them altogether: this leads inevitably to the misrepresentation in the herbarium of their distribution. Variation within the distribution of a species or group of species is often determinable only from research into living plants in the field. Collectors' notes do not always record features important for determination, and it is thus perhaps more important with succulent plants than with most other major types to obtain first-hand knowledge of their characteristics from living material. Even in cultivation these characteristics can change, being influenced all too easily by the unnatural growing conditions.

N. E. Brown's treatment of the species, (Brown, 1911–1913) provided about 60 names for succulent *Euphorbia* species from east and northeast tropical Africa, but over half of the descriptions were based on single specimens. Although some areas are still undercollected, an increasing number of herbarium specimens has since become available for study. Brown's 60 names, of which 15 or more have since been put into synonymy, have proved to be relatively few compared with what now appear to be distinct species. Many are obviously distinct, but problems of variation and distribution still necessitate field research. Inevitably, some names provide only a blanket-cover for groups of closely related species which I have attempted to distinguish morphologically, sometimes with great difficulty even after intensive field study.

INVESTIGATION OF CLOSELY RELATED TAXA
Non-spiny E. schimperi–E. nubica *group*

A non-spiny aggregate, with leafless cylindrical pencil-like branches, has long been known as either *E. nubica* N. E. Brown or *E. schimperi* Presl. *Euphorbia schimperi* was the first to be described by Presl (1844), from a specimen collected in the southern Arabian peninsula, and then *E. nubica* by N. E. Brown (1911) on the basis of two specimens from Eritrea. Brown considered his species to be distinct from *E. schimperi* by its more erect branching and longer umbel-rays. Since then, in other material collected from more-or-less the same areas, the length of the umbel-rays has been variable, responding to differences in season, habitat, and stage of development, but the mode of branching does seem to hold some significance. Without thorough investigation at the type localities, it is difficult to establish which of several variants occurring in Arabia, Ethiopia and Kenya correspond to the type of the species. Differences between them appear to lie in mode of branching, colour and thickness of stems and branches, and especially in habit, all of which are characters not easily apparent from herbarium specimens. In northern Somalia there are two, possibly three, distinct forms; but which, if any, of these is *E. nubica*, and which *E. schimperi* is open to question. One is an erect shrub with thick bright green branches: this may be the true *E. nubica*, but should it be bright green when most members of the complex are somewhat glaucous? A more common scrambling plant, with rather thinner glaucous spreading branches, seems distinct on habit alone: certainly it is taller than plants described from the Arabian peninsula, but this could be because there it has not been collected from a similar habitat. The identities of *E. schimperi* and *E. nubica* must be established before the status of these Somali plants can be determined.

A study of herbarium specimens indicated that there might be a distinct form widespread throughout northern Kenya and into Ethiopia. The branches of this variant are widely spreading and conspicuously thinner and more brittle than those taken from shrubs further north—characters that were confirmed to be constant in the wild. Another constant distinction which characterizes this variant is shown by the central sessile cyathium of the inflorescence umbel. In common with other species in this complex, this develops first; and, as it is quickly deciduous, it is rarely present on herbarium specimens. On flowering plants in the field, it is equal in size to the lateral cyathia and possesses five involucral glands. Other variants produce central cyathia which are larger than the laterals and have seven or eight involucral glands. The Kenyan plant is undoubtedly a distinct species, and has been named *E. calamiformis* Bally & S. Carter (Carter, 1985).

Unidentified shrubs in the same complex occur spasmodically throughout Somalia, but until now there have been no herbarium specimens at Kew. In Somalia, in 1985, I found several groups of small, densely branched plants, each with a different appearance. The question of whether the smallest ones (perhaps no more than 25 cm high) have either simply been grazed, resulting in stunted growth, or represent a distinct endemic species can be answered only after further field investigation (Figs 1 & 2).

Single-spined E. triaculeata *group*

In the spiny *Euphorbia* species paired spines are produced from a horny pad or spine-shield surrounding the leaf-scar. One little-known complex in which species are less difficult to differentiate is the group with 'single' spines related to *E. triaculeata* Forsskål (1775). Seedlings of this group initially produce paired spines, but partially fused spines soon appear, and spine fusion is complete on plants only 2–3 cm high.

Euphorbia triaculeata originates from the Arabian peninsula, near the Red Sea Coast. There is no problem of identification here, for it still occurs at the type locality; and what appears to be the same plant is found nearby in Djibouti, with a fairly short central stem and long spreading basal branches.

An apparently closely-related variant occurs throughout northern Kenya and into Ethiopia. Until recently, this was also treated as *E. triaculeata* but it is certainly distinct, with an extremely short, very stout central stem which branches from the apex near ground level. When making herbarium specimens, collectors invariably have removed the side branches, and as this particular feature of an abbreviated main stem and 'medusoid growth' was not mentioned in any collector's notes, the Kenyan plant has been consequently identified as a sturdy variant of the Arabian species. Field investigation showed this mode of branching to be an important diagnostic feature which, combined with the almost procumbent branches, stronger spinescence and larger cyathia, provided ample evidence for the erection of a new species, *E. kalisana* S. Carter (1982).

Figures 1 & 2. Fig. 1. *Euphorbia* sp., an erect shrub from Eritrea, possibly *E. nubica* N. E. Brown. Fig. 2. *Euphorbia* sp., a tall scrambling plant with spreading branches from northern Somalia, related to *E. schimperi* Presl.

Several variants in this group have been collected, between Arabia and Kenya, from eastern Ethiopia and northern Somalia, which bear only superficial resemblance to *E. triaculeata*. With some there is no problem, differences of habit, size, branch detail and spinescence all being sufficient to distinguish them.

The situation is different with some other variants: too little is known of the plants in the field, and there is almost no preserved material available for study. The question arises as to whether a small short-stemmed plant collected from central northern Somalia would develop into a larger sprawling plant, like those occurring much further east; and whether this sprawling habit differs sufficiently from that of *E. triaculeata* to merit distinction. There is, for example, a densely tufted plant with much narrower branches from the foot of the north facing escarpment in Somalia, which is obviously distinct on those two features alone.

Figures 3 & 4. Fig. 3. *Euphorbia triaculeata* Forsskål, with sub-erect spreading branches to about 50 cm high, from Djibouti. Fig. 4. *Euphorbia kalisana* S. Carter from Kenya, with semi-prostrate variegated branches and very large spines.

The dry eastern regions of Ethiopia and the whole of Somalia are rich in endemic species. Probably these variants all represent endemic species with limited distributions, but it would be unwise to describe them as such without further investigation and the collection of suitable material (Figs 3 & 4).

Shrubby E. heterochroma–E. stapfii *group*

A widespread complex of closely related spiny shrubs occurs in East Africa, principally in Kenya. *Euphorbia heterochroma* Pax (1895) was described from a specimen collected in Tanzania, and *E. stapfii* Berger (1907) from cuttings of a plant cultivated in the Entebbe Botanic Garden, Uganda. In 1911, N. E. Brown considered them to be synonymous, despite their wide geographical separation, and united them without explanation. He examined nine specimens altogether, but since then, 10 times that number have been accumulated in the Kew herbarium alone, all of which have been named *E. heterochroma*. These can be divided geographically into several variants. Field investigations confirmed the existence of some which were distinct, and others which were not.

Plants occurring at the type locality of *E. heterochroma* are up to 2 m high with a procumbent habit and four-angled, quite strongly variegated stems and branches. The spines are small and always absent from the margins of the upper branches, and the cyathia are yellow, producing buff-coloured capsules. This form has a limited distribution among the foothills of the Usambara and Pare Mountains in northeast Tanzania.

Study of herbarium specimens indicated that a possibly identical plant occurs in southeast Kenya. Only on one or two specimens did collectors' notes remark that stem variegation was less obvious, or that mature cyathia and capsules were often flushed with red, but we now know these features to be constant. Furthermore, the habitat of these plants seems to be restricted to the slopes of rocky inselbergs. This variant seems to merit subspecific status since, apart from distribution, the only consistent differences in colouring and habitat preference are slight and are usually undetectable on herbarium specimens without adequate data.

The task of identifying *E. stapfii* correctly has proved to be extremely difficult. The species was typified from a plant cultivated in the Entebbe Botanic Garden, and was said to occur commonly around Entebbe. The few collections made subsequently from this area all originate from cultivated plants, usually used for hedging. The swampy regions north of Lake Victoria seem unlikely to provide the dry rocky habitats required by all other plants in this shrubby complex growing in the wild, and, until a proper study has been made in Uganda, we shall not know whether it occurs wild around Entebbe or whether it was originally brought there from a wild population elsewhere. I have seen only one herbarium specimen that indicates a nearby wild locality. This was collected in 1927 from the southwest slopes of Mt Elgon, and is an unbranched specimen identical to collections from the Entebbe area. Until the status and nature of these shrubs in both areas are much better understood, the name *E. stapfii* can be applied only with the greatest caution.

Other variants of this shrubby complex occur over all the drier regions of northern Kenya, northeast Uganda and southern Ethiopia. Several of these, with restricted distributions, can be separated as distinct taxa at species level, including, for example, *E. vulcanorum* S. Carter (Carter, 1982) in Kenya: this variant produces a low, compactly branched shrub, with narrow five-angled branches and yellow cyathia.

One of the principal features of this complex is the way in which spine-length decreases as the shrub increases in height. This characteristic is often lacking in specimens collected from the lava plains surrounding Lake Turkana in northern Kenya. Field investigation proved this sturdy shrub, *E. tescorum* S. Carter (Carter, 1982) to be about 1.5 m high, with spreading five- to eight-angled branches which are heavily armed, and with yellow cyathia producing rosy-red capsules. Within its distribution there are slight differences. Taller, more branching shrubs occur east of the lake; and among the hills on the west side is a population with intensely variegated stems which are constricted at irregular intervals to a more marked extent than any elsewhere. There is a corresponding difference in spine-length along each segment and the spinescence is the heaviest known. However, such differences appear only on the younger, more vigorous plants, and no firm overall characters have been found which can be used consistently to separate these variants.

One group of herbarium specimens which appeared to represent a single taxon, originated from two areas within Kenya. Shrubs from both areas were described as erect, usually with five-angled branches, red cyathia and darker red capsules. Some specimens had a strong spinescence, whilst on others, usually flowering, the spines were much diminished. It was suspected that two taxa were involved, and this proved to be the case in the field, for the living plants from the two areas are distinct. Those from the north, on the eastern side of the Rift Valley, are more sturdy, especially as young plants, with thicker branches, particularly at the base of the plant, stouter spines on wider horny margins along the branch angles, and spine-pairs generally more closely spaced.

However, these sturdier vegetative features can sometimes be found on plants growing in exposed places among the southern populations deep in the Rift Valley. In both areas, spine-length decreases on the upper, usually flowering branches, but this is more noticeable in the southern plants, where branching is also more lax. All the plants have usually five-angled branches, but in the south they are often six-angled and in the north often four-angled.

Apart from these somewhat variable differences in habit and branching, the most obvious distinction between the two taxa is in the colour of their

Figures 5 & 6. Fig. 5. *Euphorbia* sp., an erect shrub with sturdy branching, from the southern Rift Valley of Kenya, related to *E. heterochroma* Pax. Fig. 6. Close view of a seedling (right) from the same population as in Fig. 5 with 5-6-angled stem and branches; and an upper branch (left) with spine-length much diminished.

inflorescences. The cyathia of the southern plants are bright red, with the glands at first yellow-based, turning bright red. Mature cyathia and glands of the northern plant are dark crimson. Although some overlap occurs, these character differences can be used to distinguish the two taxa at specific level.

One group of shrubs in this complex remains, from the lower slopes of the mountain ranges along the border of Uganda and Kenya, and extending eastwards into the Rift Valley Province of Kenya. Herbarium specimens showed that these plants had four-angled branches, but were otherwise extremely variable, sometimes erect to about 1.5 m high with spines to about 1 cm long, sometimes subscandent to 3.5 m with much shorter spines which were occasionally obsolete. Investigation of living populations proved these variants to comprise one variable species. Habit, thickness of the branches, spinescence, and colour of the cyathia, glands and capsules, all vary, not only in a single population, but in some cases on one individual. Spine-length diminishes as plants increase in height. Plants in the open develop as erect shrubs to 1.5 m high, but become subscandent to 3.5 m in denser vegetation, and in the latter the spines on the uppermost branches are usually obsolete.

Inflorescences of plants from the western mountain ranges are yellow, but occasionally the involucres are tinged with red. In plants further east and separated from the former by the Kerio Valley, a small but consistent difference occurs, with shrubs less scandent, usually less than 2 m in height, with red cyathia bearing glands which are at first dull yellow but rapidly turn entirely red, and producing capsules a darker red than the western form. Such lesser differences than between the other taxa in this complex, together with the close geographic proximity, suggest separation at subspecific level. (Figs 5 & 6).

Tree E. nyikae–E. kibwezensis *goup*

Undoubtedly, the most difficult problems are presented by groups of tree species occurring in Kenya and northeast Tanzania. Herbarium specimens present a wide range of vegetative characters, but all have long been named as either *E. nyikae* Pax (1895) or *E. kibwezensis* N. E. Brown (1912), with no apparent reasons for differentiation between the two. Before field work began, the possibility of distinguishing more than one species seemed highly improbable.

Euphorbia nyikae is the oldest name which can be applied, a species described by Pax from a single specimen from northeast Tanzania. N. E. Brown (1912) considered this to be identical with specimens involving other species, so his, fuller description of *E. nyikae* was a composite one. This has led to much confusion, especially as both he and Pax described several species, all from the same region, which nevertheless involved only eight specimens altogether! Since then more material has been collected, but even this is not entirely representative of what can be found in the field.

Plants of *E. nyikae* have all disappeared under cultivation at the type locality, but others from a few kilometres away match Pax's description and leave no doubt as to the identity of the species. It is a smallish tree, with mostly two-winged branches, slender, separated spine-shields, and usually solitary cymes. A drawing at Kew of the holotype shows a piece of branch resembling specimens originating from coastal populations of plants which must be identified as *E. neovolkensii* Pax (1904). This was described by Pax from a specimen collected on Zanzibar, which was judged by N. E. Brown to be identical with the type specimen of *E. nyikae*. Individuals from these coastal populations develop a different habit, as shrubs or shorter bushy trees, with the lower branches persistent instead of deciduous, spine-shields forming a normally continuous horny margin along the branch angles, and with up to seven cymes clustering at each flowering eye.

Intermediates between these two taxa can sometimes be found in specimens from either area, indicating a very close relationship, such that *E. nyikae sensu stricto* seems to be an inland, dense bushland form of a more widespread species. A reduction of the shrubby coastal species (*E. neovolkensii*) to varietal level seems to be merited.

Pax (1903) described a related species from the same area as *E. bussei* which was also reduced by N. E. Brown to synonymy under *E. nyikae*. Plants growing at the type locality match precisely Pax's description and also the drawing, in Kew, of the holotype. The three-angled branches, deeply constricted into somewhat rounded segments, spine-shields forming a fairly robust continuous horny margin, and five to eight cymes densely clustered at each flowering eye, distinguish the species from typical *E. nyikae* occurring in the same area.

However, there are no differences between the descriptions of *E. bussei* and *E. mbaluensis*, Pax (1904), both described from single specimens, or between trees undoubtedly representing the two species at their respective type localities, again only a few kilometres apart. The latter species should thus be reduced to synonymy under *E. bussei*.

The name *E. kibwezensis* N. E. Brown (1912) has become a blanket-cover for this tree complex in Kenya. There is no problem in identifying the true species, for it is still common around the type locality of Kibwezi in southeast Kenya. In their habitat, these trees are generally more sturdy than the Tanzanian *E. bussei*, with more robust spination and larger, subsessile cyathia. These differences persist in populations to th north-west, but in plants from areas between the type localities, there is often some overlap in the distinguishing features, which suggests the reduction of *E. kibwezensis* to a variety of *E. bussei*.

However, the complete picture is not nearly so straightforward. Vegetative characters shown by herbarium specimens are extraordinarily diverse, and seem to bear very little relation to trees in habitat. Within a population branch characteristics depend very much upon the age of the individual, and the extent to which it is sheltered by its neighbours and other vegetation. The older the tree and the more exposed its branches, the more pronounced and regular is branch segmentation.

It is not surprising that Pax and N. E. Brown described so many species, when so few specimens were at hand for accurate comparison, and when they had no personal experience of the variation exhibited by living plants.

A further complication is that of possible hybridization where apparent species occur together. The variant of *E. nyikae* (*E. neovolkensii*) normally occurs as a shrub or small tree with persistent branches, only the older exposed individuals sometimes shedding their lower branches. Principally two-winged, branches of mature trees can be three- or even four-winged with segmentation and spination varying considerably. Older robust specimens therefore resemble weaker plants of *E. bussei*, and it is possible that hybridization may occur between the two.

Figures 7 & 8. Fig. 7. A shrubby tree from a coastal population in Kenya, representing *Euphorbia neovolkensii* Pax. Fig. 8. Branches from individuals of a single population in Kenya of *Euphorbia bussei* Pax, exhibiting extreme variability.

In contrast, the spination of young plants of both variants of *E. bussei* is usually very robust, often bearing a similarity to weaker plants of another, quite different, shrubby species, *E. breviarticulata* Pax (1904). Confusion is most likely in areas where both species grow together, and, again, the possibility of hybridization cannot be ruled out. However, *E. bussei* can usually be distinguished from *E. breviarticulata* by the presence of a central stem and by the shorter length of its longest spines.

Herbarium specimens of this complex originating from the Rift Valley of Kenya and northern Tanzania are often indistinguishable from those of the Kibwezi plants, but a few fruiting specimens indicated that capsule size might be significantly different. In living material the capsules of these plants are, indeed, larger and also less deeply lobed, being almost triangular before hardening, prior to dehiscence. Although the tall trunk, topped by a relatively small crown of loosely placed branches, is a description which could apply equally to both trees, there are subtle differences. The most reliable distinctions are found in the four- to five-angled stoutly winged branches, with more-or-less regular segmentation of the Rift Valley plants, as opposed to the predominantly three- to four-angled branches, with noticeably thinner wings and irregular segmentation of the Kibwezi ones. These are variable characters, but together with the constant character of larger, differently shaped capsules, and also larger cyathia, they provide enough reasons for differentiating the Rift Valley variant as a distinct species. It occurs in dense stands, except at the northern end of its range, where trees are usually scattered, or form small groups. Mature trees here are more bulky, with a broader flattened crown on a shorter stouter trunk, and the cymes are more crowded at each flowering eye. Some intermediates occur in populations further south suggesting that varietal status is appropriate for the bulky trees.

All the species and varieties suggested for this difficult complex are obviously very closely related, reflecting a graduation from the shrubby trees of *E. nyikae* and the *E. neovolkensii* variant, occurring in humid low altitude coastal regions, to the sturdy Rift Valley trees, occurring in arid regions at higher altitudes. It is essential for collectors to note details of habit and branch characteristics, otherwise, since distributions are reasonably distinct, locality offers the only positive method for identifying herbarium specimens.

Even now, the situation is not entirely clear, for there are a few specimens of this complex from isolated localities, which do not fit into any distribution pattern. They may prove to represent distinct taxa or be used to unite those which are currently regarded as different (Figs 7 & 8).

GENERAL CONCLUSIONS

Pax and N. E. Brown had only a limited number of specimens available for study which sometimes led them to an inaccurate assessment in distinguishing species. Since then the accumulation of material from an area much wider than that covered by the species they described has often led to the loose application of some names. Our knowledge of inter- and intra-specific relationships has undoubtedly increased, gained from field observations of variation and distribution of wild populations. The results of future field studies and the further accumulation of herbarium material will always lead to a greater understanding of the true situation and help towards elucidating problems which still exist.

The suggested taxonomic rearrangements and descriptions of new taxa mentioned in this paper will be published in *Kew Bulletin* (Carter, in press; Carter & Gilbert, in press).

APPENDIX: SPECIES REFERRED TO, WITH TYPE DATA

E. breviarticulata Pax, 1904. TYPE: Tanzania, W. Usambaras, Kwai, *Engler 1184b*.
E. bussei Pax, 1903. TYPE: Tanzania, Magofu near Handeni, *Busse 317*.
E. calamiformis Bally & S. Carter, 1985. TYPE: Ethiopia, Moyale–Mega road, *Gillett 14174*.

E. heterochroma Pax, 1895. TYPE: Tanzania, SE Kilimanjaro, Himo River, *Volkens 1759*.

E. kalisana S. Carter, 1982. TYPE: Kenya, Lokitaung–Lodwar road, *Carter & Stannard 198*.

E. kibwezensis N. E. Brown, 1912. TYPE: Kenya, Kibwezi, *Scheffler 223*.

E. mbaluensis Pax, 1904. TYPE: Tanzania, Usambaras, Mbalu Mountain, *Engler 1472c*.

E. neovolkensii Pax, 1904. TYPE: Zanzibar, *Werth*.

E. nubica N. E. Brown, 1911. TYPES: Sudan, Nubia coast, Bent; & Eritrea, Accrur, *Schweinfurth & Riva 1083*.

E. nyikae Pax, 1985. TYPE: Tanzania, near Magila, *Volkens 51*.

E. schimperi Presl., 1844. TYPE: Yemen Arab Republic, Mount Cara, *Schimper 878*.

E. stapfii Berger, 1907. TYPES: Uganda, Entebbe Botanic Garden, *Dawe*; & *Brown 227*.

E. tescorum S. Carter, 1982. TYPE: Kenya, Lodwar–Likitaun road, *Carter & Stannard 219*.

E. triaculeata Forsskål, 1775. TYPE: Yemen Arab Republic, Musa, *Forsskål*.

E. vulcanorum S. Carter, 1982. TYPE: Kenya, Marasabit, *Carter & Stannard 663*.

REFERENCES

BERGER, A., 1907. *Sukkulente Euphorbien*. Stuttgart.
BROWN, N. E., 1911–13. Euphorbiaceae. Euphorbia. In Thiselton-Dyer (Ed.), *Flora of Tropical Africa, 6(1):* 470–603, 1035–1042. London: L. Reeve & Co.
CARTER, S., 1982. New succulent spiny Euphorbias from East Africa. *Hooker's Icones Plantarum, 39* (3).
CARTER, S., 1985. New species and taxonomic changes in *Euphorbia* from east and northeast tropical Africa and a new species from Oman. *Kew Bulletin, 40:* 809–825.
CARTER, S., in press. New species and taxonomic changes amongst the succulent trees of *Euphorbia* subg. *Euphorbia* from East Africa. *Kew Bulletin*, in press.
CARTER, S. & GILBERT, M. G., in press. *Euphorbia heterochroma, E. stapfii* and related taxa. *Kew Bulletin*, in press.
FORSSKÅL, P., 1775. *Flora Aegyptiaco-Arabica*. Copenhagen.
PAX, F., 1895. *Die Pflanzenwelt Ost-Afrikas und der Nachbargebiete, C*. Berlin.
PAX, F., 1903. Euphorbiaceae africanae, VI. *Engler, Botanische Jahrbucher, 33:* 276–291.
PAX, F., 1904. Monographische Ubersicht uber die afrikanischen Arten aus der Sektion Diacanthium der Gattung Euphorbia. *Engler, Botanische Jahrbucher, 34:* 61–85.
PRESL, K. B., 1844. *Botanische Bemerkungen*. Prague.

Members of Euphorbiaceae in primitive and advanced societies

R. E. SCHULTES

Botanical Museum, Harvard University, Cambridge, Massachusetts, U.S.A.

Received August 1986, accepted for publication October 1986

SCHULTES, R. E., 1987. **Members of Euphorbiaceae in primitive and advanced societies.** The family contains a number of species used by man in a wide variety of ways. Often use is restricted to a small group of people, but in a few instances the plant has been a major crop and the product used throughout the world, e.g. *Hevea*, which yields rubber, used extensively in every civilized country. Cassava from *Manihot* is almost as widespread, being one of the staple foods of tropical regions. Species with more occasional use have constituents with interesting properties of potential future economic application.

ADDITIONAL KEY WORDS:—Amazon Indians – *Caryodendron* – cassava – *Croton* – *Elaeophorbia* – *Euphorbia* – *Hevea* – *Jatropha* – *Manihot* – rubber.

CONTENTS

Introduction	79
Useful euphorbiaceous genera in the northwest Amazon	81
Elaeophorbia	81
Euphorbia	82
Caryodendron	82
Croton	83
Jatropha	85
Manihot	86
Hevea	89
References	94

INTRODUCTION

Considering the Euphorbiaceae's probable polypheletic origin, cosmopolitan distribution, 300 genera with more than 6000 species and extreme botanical and chemical diversity, it is logical to expect that the family has given mankind numerous major economic plants, and that in primitive societies the uses of species number in the hundreds, if not thousands, and span an extraordinarily wide range of utilitarian applications.

The uses of euphorbiaceous plants in our own society run a broad gamut of categories: foods, various kinds of poisons, medicines, sundry types of oils, waxes, rubbers, varnishes, compounds for paints and other industrial materials.

We may consider as major economic plants the following: *Aleurites fordii* Hemsl. (tung oil); *A. moluccana* willd. (candlenut or lumbang oil); *A. montana* E. H. Wilson (mu tree); *Cnidoscolus* sp. (chilte rubber); *Croton tiglium* L. (croton oil); *Euphorbia antisyphylitica* Zucc. (candellila wax); *E. intisy* Drake (intisy rubber); *Hevea brasiliensis* (Willd. ex Adr. Juss.) Muell. Arg. (Pará rubber); *Manihot esculenta* Crantz (cassava, tapioca); *M. glaziovii* Muell. Arg. (Ceará rubber); *Ricinus communis* L. (castor bean); *Sapium sebiferum* Roxb. (vegetable tallow). (Hill, 1952; Purseglove, 1968.)

This brief consideration of the economic botany and ethnobotany of the Euphorbiaceae is selective in so far as the thousands of ethnobotanical reports are concerned. As a consequence, several different categories of euphorbiaceous plants are discussed: 1, genera employed by Indians in the north-west Amazon; 2, species of *Elaeophorbia* and *Euphorbia*, with uses in primitive societies that suggest the practical need of chemical and pharmacological study for potential medicinal compounds; 3, *Caryodendron orinocense* Karst., a wild oil-rich tree worthy of domestication as a new tropical plantation crop; 4, *Croton* and *Jatropha* species, biodynamic plants with native uses that have recently yielded new chemical compounds of medicinal promise; and 5, the two most important economic plants of the family: *Manihot esculenta*, cassava or the source of tapioca, one of the 13 species of plants that feed the world population, and *Hevea brasiliensis*, the Pará rubber tree, that has changed the world in a century.

Scientists have long been interested in those toxic substances in the Euphorbiaceae, esters of the diterpene alcohols, phorbal, resiniferonol and ingenol, that have skin-irritant and tumour-inducing activities. They occur in a number of genera: *Aleurites, Baliospermum, Croton, Euphorbia, Exoecaria, Hippomane, Hura, Jatropha, Micrandra, Ostodes, Sapium, Stillingia* and *Synadenium*. More than 100 compounds structurally related to diterpene alcohols are known from these 13 euphorbiaceous genera. It is perhaps significant that these chemical irritants are known only from two sections of the Euphorbiaceae (in accord with the classification suggested by Webster, 1975).

It would seem logical to orientate chemical and pharmacological investigation not so much by random choice but more by taking advantage of the presumed knowledge of the properties of plants that primitive societies have discovered over the millennia by trial and error. In these primitive societies man is not usually going to utilize any plant that has no biodynamic activity: the fact that a plant is physiologically active indicates the presence of one or more active chemical constituents. Only a small percentage of the half-million species of plants have ever been even superficially examined as to their chemical constitution. If chemists were to attempt to procure and analyse the 80 000 species in the Amazon, for example, the task would never be finished. One effective short cut would be to concentrate on those species known to have native uses as medicines, toxins or narcotics. This ancient and intimate knowledge of plant properties is fast disappearing, even in one generation, with the ever encroaching acculturation and civilization of primitive societies the world over. Much of great value to our phytochemists and pharmacologists is rapidly disappearing, and the knowledge acquired over centuries will forever become extinct (Plotkin & Schultes, 1983).

USEFUL EUPHORBIACEOUS GENERA IN THE NORTH-WEST AMAZON

An example of the importance of biodynamic euphorbiaceous plants may be appreciated by the following noted genera which are employed by Indians in the small north-western part of the Amazon from which information has been gathered during 40 years of ethnopharmacological research in that region.

A species of *Alchornea* is believed to be effective in treating diarrhoea. The milky latex of *Chamaesyce* (often treated as a subgenus of *Euphorbia*) is applied to fungal infections between the toes ('athletes' foot'). The reddish juice from several species of *Croton* is applied to clean and help cure infected cuts and sores, ulcers and boils. One species of *Euphorbia* is considered to have purgative properties, and the latex of another is applied to fungal infections of the feet. *Hieronyma* is believed to be effective against a skin eruption called chandra. The sap of the petiole of a species of *Jatropha* is applied to sore gums of teething children, and another species is prepared as a bath used to reduce fevers and headache. The seed oil of *Mabea* is rubbed into the scalp to delay loss of hair. *Manihot* has, according to the natives, anti-diarrhoea properties. The latex of *Micrandra* is spread on sore gums and mucous membranes of the mouth and is valued as a styptic for the umbilical cord. *Nealchornia* is ichthyotoxic. *Phyllanthus* in cultivated as a fish poison, and several species are considered to be insect repellants, to be efficaceous in decoction in treating kidney troubles and as a wash for reducing headaches (Schultes, unpublished field notes, 1941–1985).

ELAEOPHORBIA

In Africa and Madagascar, a common custom involves the administration of justice—that is, the establishment of guilt or innocence—through the use of preparations based on highly toxic plants. At least 25 species in eight families are known to be so employed. The most important sources are leguminous and apocynaceous species, but two of the minor ordeal poisons belong to the Euphorbiaceae (Robb, 1957).

The introduced South American *Manihot esculenta* is used to a certain extent throughout tropical Africa, but especially in the Congo Basin. The active principle may be the cyanogenic glycoside which, when the root is prepared for food, must be removed.

However, perhaps the more interesting euphorbiaceous ordeal poison is *Elaeophorbia drupifera* (Thonn.) Stapf. This plant is abundant in parts of western Africa, but it seems to be used in criminal trials only in the Ivory Coast. The natives express the caustic white latex and mix it with water, or they prepare the plant as a decoction. The usual manner of use is ingestion, but occasionally it may be spread on the eyes of the accused and rubbed in with the fingers, in which case damage to the cornea constitutes evidence of guilt. The latex contains the toxic constituent, but the active principle has not yet been elucidated.

While it would seem rather futile to investigate such an apparently dangerous ordeal poison, the discovery of the medicinally extremely useful cholinergic alkaloid physostigmine from an even more pernicious African ordeal poison, *Physostigma venenosum* Balf., the Calabar bean of Nigeria, should indicate that we cannot prejudge the cause of any biodynamism in plants was of no value in modern medicine. Consequently, this toxic euphorbiaceous species deserves chemical and pharmacologic study.

EUPHORBIA

A promising ethnopharmacological lead that should be more thoroughly investigated is the use by Mayan Indians of Guatemala of *Euphorbia lancifolia* Schlecht. The vernacular name *ixbut* comes from early Mayan languages of the Pokom group, *ix* meant woman, but indicated an increase in the flow or volume of liquid, in other words a galactogogue. The leaves and branchlets are boiled in water which is allowed to cool, sugar is added, and the drink is taken thrice daily. *Ixbut* tea should not be used, the natives state, when it starts to ferment (Rosengarten, 1978).

In 1949, experiments were carried out in collaboration with Merck & Co. Of 86 postpartem women tested, 45 showed increased milk production, and in several cases mothers who had previously been unable to nurse were able to do so after several days of drinking *ixbut* tea. The plant was introduced to Cuba in 1945 as a galactogogue and it was reported that *ixbut* could increase lactation in nursing mothers up to 100%. A study carried out in Mexico between 1949 and 1951 gave inconclusive and somewhat negative results, but the plant material used was about a week old and so may have lost its active ingredient(s). The report on this Mexican study suggested that the lactogenic properties may have been exaggerated, but the striking effects claimed in Guatemala by lay people cannot be totally denied on the basis of the material tested in Mexico.

Ixbut has been employed in Guatemala as a supplement to cattle feed, and a commerically available cattle galactogogue called Galac-Latex, in which the principal ingredient was *ixbut*, enjoyed a brief period on the market. Studies carried out in El Salvador and in south-eastern Mexico indicated substantial increase in milk productions in cows; experiments in Guatemala on goats showed no signs of toxicity and a modest increase in milk production.

The active principle in *Euphorbia lancifolia* has never been isolated and until it is found any decisions concerning its value in human or veterinary medicine cannot be made. Yet *ixbut* remains, on the basis of native use for centuries and on the evidence of preliminary experiments, a promising ethnopharmacological subject worthy of further phytochemical and pharmacological investigation.

CARYODENDRON

The euphorbiaceous genus *Caryodendron* has three species, all from tropical South America, one of which, *C. orinocense*, represents a most promising neotropical plant for domestication as new oil-rich crop. Growing wild in the Orinoquia of Venezuela and Colombia, and the westernmost Amazonia of Colombia, Ecuador and Peru, this tree attains a height in forests, where it must compete with thick vegetation, of 70 ft and possesses a dense crown that may measure some 40 ft in diameter. It also grows in open plains or grassland areas or llanos. It is known by a number of vernacular names, principally *inchi* (from the Ketchwa term for the groundnut, *Arachis hypogaea* L.) in Colombia, Ecuador and Peru (Martínez, 1970); and *palo de nuez* (nut tree) and *tacay* in Venezuela (Schnee, 1973). One especially oil-rich variety with particularly large fruits and seeds is called *ninacuru inchi* (fire-fly inchi, because fire-flies congregate in its crown).

The fruit is a typical trilocular capsule. The epicarp is greyish and chartaceous. The epicarp and mesocarp, when ripe and dry, make up a single layer that is easily removed. The endocarp is hard and woody. The seeds are ivory-coloured when fresh, but turn dark cream when rancidity sets in. The endosperm tastes like groundnuts, but when rancid it becomes both bitter and pungent. Most of the endosperm comprises oil-bearing cells. The seeds (without husk) constitute up to 35.8% of the weight of the fruit.

The seeds are viable for only a short period. One experimental planting in Colombia gave 97% germination, whereas another had only 5%. The trees grow approximately $3\frac{1}{2}$ ft a year, and they begin to bear fruit in 6 or 8 years. In the humid tropics, the seeds rapidly become rancid but, if collected immediately upon falling from the trees and taken to a cool, dry climate, a year will pass before they become rancid (Pérez-Arbeláez, 1956; Patiño, 1967).

Caryodendron orinocense is deemed superior to the African oil palm (*Elaeis guineensis* Jacq.) in its ecological adaptability, its agronomic characteristics and in the quantity and quality of its oil.

Caryodendron has been neglected and is not included in the major books on oleaginous plants. Historically, the earliest report appears to be that of an explorer of the Orinoco River (Gumilla, 1745) who wrote that the fruit known as *cumana* and which the Indians called *aboy* yielded an oil like that of olives in colour and taste.

CROTON

With 750 tropical and subtropical species in both hemispheres, *Croton* is ethnobotanically one of the most fascinating genera of the Euphorbiaceae. The extraordinary spectrum of uses to which its species are put in primitive societies offers a cogent argument for intensive phytochemical research (Altschul, 1973).

The tumour-inducing properties of the fixed oil in the seed of *Croton tiglium* were discovered in the 1940s. It has been suggested that the unusual frequency of oesophageal cancer in China may possibly be due to the common use of this species in folk medicine for a variety of ills. Another species, *C. flavens* L., has been indicated in the same kind of cancer in Curacao.

One chemical compound, phorbol myristate acetate (PMA), a phorbol diester and a constituent of croton oil, has been found to be the most active skin-irritant, inducing skin tumours in mice. However, croton oil itself is not carcinogenic, but when applied together with a subeffective carcinogen, cancer can be induced. Therefore, phorbol esters represent one of the most valuable agents in studying the carcinogenic process at the cellular level (Berenblum, 1941). Fourteen different phorbol diesters are known from *C. tiglium*. These compounds appear to be found only in the Euphorbiaceae; and they are well represented in *Croton*, particularly in *Euphorbia* (Kinghorn, 1979).

This use of phorbol diesters as a tool in pharmacological studies represents an example of one valuable application of a toxic substance from a species employed for a very different purpose in primitive societies, one of the hallmarks of ethnopharmacological research (Kinghorn, 1985; Reis & Lipp, 1982).

Several species have become important economic plants in advanced societies: *Croton cascarilla* Benn. and *C. eluteria* Benn. of the West Indies, the bark of which yields a tonic; the Asiatic *C. tiglum*, source of the strongest known purgative, croton oil, from the seed; *C. lacciferus* L. of Sri Lanka and India, from the lac of which a varnish is made; and several Brazilian species that produce a red dragon's-blood resin (Le Cointe, 1934; Uphof, 1968).

But this list of commercially valuable uses of *Croton* cannot compare with the uses provided by a great number of species in primitive societies in both the Old and New Worlds:

Croton alamosanus N. E. Rose: the resin relieves toothache (Mexico) and the leaves are made into a kind of tea (south-east Asia). *C. cajucara* Benth.: the plant is considered to be febrifugal (Brasil). *C. caudatus* Geisel: a preparation of the roots is taken to relieve constipation (India, Malaysia). *C. ciliato-glandulosus* Ort.: valued as a febrifuge and purgative for man and as a galactogogue for goats (Mexico). *C. cortesianus* HBK.: the caustic sap is applied for treating skin diseases (Mexico). *C. crassifolius* Geisel: the root is used to heal wounds by mixing with wine (Hainan). *C. dioidus* Char.: the plant is believed to be anti-rheumatic and to calm hysteria (Mexico). *C. draco* Schlecht: the red sap is thought to heal ulcers and is applied as a remedy for a disease of horses' hoofs (Mexico). *C. echinocarpus* Muell. Arg.: this species is a strong drastic and anthelminthic and is also used in treating eye diseases (Brazil). *C. glabellus* L.: the Witoto Indians crush the leaves to poultice infected cuts and sores (Colombia). *C. gossypiifolius* Vahl: the red sap is spread on cuts (Venezuela). *C. guatemalensis* Lotsy: the bark is valued as a remedy for chills and to prepare a bath for treating rheumatism (Guatemala). *C. humilis:* employed in treating skin and urinary diseases (Brazil). *C. macrostachyus* Hochst. ex Del.: the juice of the leaves is anthelminthic (Ethiopia). *C. niveus* Jacq.: a tea is tonic for intermittent fevers (tropical America). *C. oblongifolius* Roxb.: the powdered seeds are utilized as an insecticide and for stupefying fish (India). *C. oligandrum* Pierre ex Hutchinson: the scented bark is a remedy for colic (Gabon): *C. palanostigma* Klotzsch: the sap is applied to ulcers and boils to reduce pain (Brazil). *C. populifolius* Mill.: an alcoholic maceration of the bark is considered to be anti-rheumatic (Venezuela). *C. poilanei* Gagnep.: the bark is reputedly used to treat eye diseases (Cambodia). *C. reflexifolius* HBK.: this plant is said to be febrifugal and tonic and to relieve symptoms of colds. *C. repens* Schlecht.: this species is accepted as a treatment for dysentery and stomach-ache (Central America). *C. setigerus* Hook.: the stems and leaves are macerated for ichthyotoxic use (south-western U.S.A.). *C. texensis* (Klotzsch) Muell. Arg.: the Pawnee Indians bathe sick babies in a decoction; the Zuni drink a tea of the whole plant for upset stomach; the Hopi value it as an emetic and other tribes ascribe insecticidal properties to it (south-western U.S.A.). *C. tiglium:* the seeds are employed as a fish poison (tropical Asia). *C. xalapensis* HBK.: the mucilage is applied for tooth-ache, sore gums, cleaning teeth and the sap is considered good for dressing wounds (Mexico, Guatemala).

By far the greater number of species are employed in the form of resin or sap to hasten the healing of wounds and ulcers. Other frequent uses in primitive medicine are as purgatives, anthelminthics and fish poisons.

JATROPHA

The genus *Jatropha*, native to tropical and sub-tropical regions of both hemispheres as well as to North America and South Africa, comprises some 175 species of herbs, shrubs and small trees.

Numerous species find a wide spectrum of folk uses in the many regions where this genus occurs. *Jatropha curcas* L., an American species early introduced to the Old World, is the source of curcas oil, but is also valued as a strong purgative and for use in making candles and soap and in illumination and lubrication. In El Salvador, the latex is applied for eruptions on the lips (herpes?) and in front of the ears to treat eye inflammation (conjunctivitis?) (Burkill, 1935; Uphof, 1968). The related, *J. gossypiifolia* L. has been employed in various parts of the world in treating leprosy, venereal diseases and ulcers, as a cataplasm for swollen breasts, an emetic, emmenagogue, stomachic and febrifuge. In Mexico and the south-western U.S.A. bark of *J. spathulata* (Ortega) Muell. Arg. is employed in tanning leather and is the source of a red dye; another species is used to clean and tighten teeth and to invigorate the hair; in Yucatan the roots of *J. gaumeri* Greenman have been employed in treating snakebite. The latex and sap of numerous members of the genus are applied to wounds and cuts to hasten healing. A Peruvian species has the wide reputation of being an aphrodisiac. The roots of a Brazilian species are valued in treating dropsy. The ancient Mayan Indians boiled the young leaves of a local species, *J. aconitifolia* Mill., and ate them as a vegetable. Many representatives of *Jatropha* are caustic and toxic, causing severe dermatitis (Altschul, 1973; von Reis & Lipp, 1982).

It is obviously of interest to note that extracts of *Jatropha gossypiifolia* and related species have enjoyed folk uses in sundry areas in treating 'cancer'. Recent chemical and biological investigations have indicated that a novel macrocyclic diterpenoid, which has been called jatrophone, displays inhibitory activity *in vitro* against cells from human carcinoma and against four animal tumour systems and that it was significantly effective against lymphocytic leukaemia. Jatrophone is currently being considered as a candidate for Phase I human testing in cancer patients (Kupchan *et al.*, 1970).

MANIHOT

The human race is fed mainly by 12 or 13 plants: rice, wheat, maize, soybean, common bean, groundnut, sugar cane, sugar beet, white potato, sweet potato, cassava, banana (including the plaintain) and coconut. The cassava plant not only enjoys outstanding popularity, but it also ranks as one of the most complex cultigens known; an allotetraploid, it has never been found in the wild. It is also the only toxic plant that has become a major food plant (Burkill, 1935; Schery, 1947; Cook, 1982).

All 171 species of *Manihot* are American, and there is no doubt that *M. esculenta* is a New World plant. But there is little agreement concerning its background or its place of origin, and it is not even possible to say whether this plant originated from one or several species (Sauer, 1952; Jennings, 1976).

During its long history, this cultigen has developed nearly one hundred races. Customarily, these races are divided into two large assemblages, bitter cassava and sweet cassava, depending upon the amount and location of the toxic constituent, a cyanogenic glycoside related to hydrocyanic (prussic) acid; free hydrocyanic acid may occasionally also be present.

In the races called sweet cassava, the toxic constituents are localized in the bark of the underground parts of the plant, and this bark is naturally peeled off in preparing the tuberous root for food. The bitter cassavas have the poisonous elements present, often in relatively high concentrations, throughout the plant, particularly in the starch component of the root. It is impossible to establish a demarcation between these two vague races, since the cyanogenic content may vary with individual cultivars. The plants of sweet cassava usually have light green leaves, whereas those of the bitter races tend to have dark green leaves, often somewhat reddish or purplish. The sweet cassavas tend to be more widespread than the bitter types and appear to be more closely associated with the older civilizations.

But the many races developed have other characteristics, differing in length of growth, mealiness, size and colour of the root, colour of the stems and even parts of the leaves, shape and hairiness of the leaves, and a host of other easily visible differences. Because of this great diversity, the botanical synonymy of *Manihot esculenta* has grown to be extensive; two of the most commonly used binomials are *M. aipi* Pohl for the strains with low concentrations of the glycoside, and *M. utilissima* Pohl for the strains with high concentrations of the poisonous constituents. The vernacular terminology for this most useful plant is extraordinarily extensive; for, as the plant spread to the Old World tropics, the New World names were rarely adopted. The most frequently employed common names in the non-aboriginal South and Central American languages are cassava, manioc and yuca in Spanish-speaking parts; *macaxera* and *mandioca* in Brazil (Rogers & Appan, 1973).

Cassava obviously has a very long history as a food crop in the Americas. There is evidence that cassava flour was traded in north-west South America in the second and third millennia BC. Ancestral types of our modern cassava may have been amongst the first food plants of tropical America (Mangelsdorf *et al.*, 1964). Archaeological evidence indicates that it was cultivated at least 2500 years ago in Mexico; starch grains from caves in Tamaulipas (Mexico) which are dated from 900–200 BC (MacNeish, 1958; Callen, 1967) appear to be from cassava. Specimens have been found in profusion in many archaeological sites in coastal Peru, suggesting that it was a major food crop in ancient times in that region. There are also representations of the tuber on clay pottery and textiles, and remains have often been found in Peruvian mummy bundles (Towle, 1961). Griddles and other indirect evidence has placed cassava at the mouth of the Amazon in AD 1300, but it must be borne in mind that in hot humid areas such as the Amazon the preservation of plant material in archaeological form is very rare (Schultes, 1984b).

One indication of the great age of cassava as a major food source is found in the appearance of this plant in the origin myths of many Indian tribes. For example, one myth of the natives of north-west Amazon, holds that the tribe originated in the Milky Way. One man and one woman came down in a dug-out canoe drawn by an anaconda. With them, they had three plants: coca (*Erythroxylon coca* Lam. var. *ipadu* Plowman, the source of cocaine) caapi (*Banisteriopsis caapi* (Spruce ex Griseb.) Morton, a sacred hallucinogen) and cassava.

All early European explorers and travellers to tropical America mentioned cassava, which, at least in the humid forested areas, was the mainstay of the carbohydrate intake in regions where cereals could not be cultivated. As early as the late 17th century, a long bibliography had grown up concerning the crop; and Sir Hans Sloane in his *Catalogus Plantarum in Insula Jamaica* (1696) and his

Natural History of Jamaica (1707–1725), extolled cassava as offering "the most general use of any provision all over the West Indies, especially the hotter parts, and used to victual ships". But it was not the favourite provision for ships stores: there is associated with the toxic glycoside an enzyme that, upon wilting of the tuber, can liberate free hydrocyanic acid; this action rarely occurs in growing or fresh material, but a stale tuber can become poisonous, deterring many sea-voyagers from storing them in preference to white potatoes, sweet potatoes and yams. This peculiarity may have been one reason for the relatively slow spread of *Manihot esculenta* to the Old World tropics. If ships' masters had stored cassava meal, the chemical reaction would not have been possible; but meal could not give rise in new locations to living plants.

How did natives circumvent the toxicity of this plant, the carbohydrate basis of many of their diets? Little needs to be done beyond peeling in the case of sweet cassavas. The Amazon natives grate the tuber, soak the mash overnight in a dug-out canoe; in the morning the wet mash is pressed into a tubular wicker squeezer called a tipi-tipi, which is squeezed under pressure. The resulting partly dried residue is crushed, broken up, spread on a flat stone or metal plate and subjected to a hot fire. The result is a coarse meal (the consistency of sawdust) known as farinha or fariña. It is eaten in this form or else as unleavened bread. The heat is apparently sufficient to break down the toxic materials that might have survived the overnight soaking. The toxic milky juice that is squeezed out of the overnight mash is concentrated by boiling to a thick syrup. This syrup is widely employed as a condiment in many sauces in the West Indies, and known as *tucupí* amongst the Amazonian Indians.

The starch content of the tubers averages about 26%, but in some strains it may reach 74%; they are 1.5–2.0% protein, but have practically no fat. Consequently, a diet made primarily of cassava tends to be unbalanced (unless well supplemented with other plant or animal elements), a situation that is widespread, especially in the humid tropics.

Cassava is used as a food in many ways, especially in the Old World tropics. Sweet cassavas may be boiled, mashed or fried—much like potatoes. In the Amazon and other humid, forested parts of the American tropics, the most frequent method of using bitter cassava, once it is detoxified, is as a coarse meal which may be eaten in dry form or mixed with water or fruit juices; or it can be baked into an unleavened bread on a flat 'oven', usually a large stone. Small pieces of the tuber may be salivated, collected in a dug-out canoe and left to ferment. In Africa and south-east Asia, there are many other ways, some of them rather sophisticated, of preparing the tuber as a food. The product known in western cultures as tapioca is one of these products. To make tapioca, the tubers are peeled, washed and grated; this preliminary procedure is followed by repeated washings and changes of the water with squeezing until no more starch can be removed; the starch contained in the collected waters is allowed to settle, and the water is drained off, leaving pure starch. All traces of cellulose are removed in the process. The starch is made into small flakes which are slightly heated. Tapioca is a Brazilian invention, but it seems that it is now more frequently prepared in home culinary in south-east Asia. Commercial tapioca factories follow exactly the same steps, but machinery is used.

If we go beyond the applications of *Manihot esculenta* as human food, we encounter many additional uses of this plant: as an animal food, as a source of glucose and alcohol, added to mal in brewing, in preparing a seasoning for rice.

An ethnobotanically most interesting food-use amongst the Brazilian Indians was reported by many European travellers: eating the *leaves* of some of the strains as food. It appears to have died out now in Brazil, or at least to be extremely rare, but it is a very common custom today in Malaysia, where the leaves are sold in many native markets.

Manihot esculenta did not spread rapidly in the Old World. It arrived at the end of the 16th Century on the west coast of Africa, and towards the end of the 18th Century on the east coast, an event due primarily to Portuguese shipmasters plying between Brazil and Africa. It was introduced into Sri Lanka in 1796 and into India about 1800. It had apparently reached Java somewhat earlier. There has been a suggestion that it had earlier reached the Philippine Islands from Mexico (Burkill, 1935). Until about 1850, the importance of cassava in south-east Asia was not significant, but from that time onwards its place as an accepted plant steadily grew, until now in Malaysia and Indonesia it has taken its place as a major crop. It was accepted in this region first as a famine reserve plant: it was drought resistant, did not suffer unduly from locust attack and, most important of all, tolerated poor husbandry.

We meet a much more confused picture concerning the origin and local spread of *Manihot esculenta* and its many strains in the New World tropics. All species of the genus can be intercrossed, but some may be productively isolated under natural conditions. All species apparently have 36 chromosomes, but there is evidence of polyploidy in *M. esculenta*.

Older writers on the origin of cultivated plants all indicated a South American origin. One modern student (Renvoize, 1972), on the basis of ecological and anthropological data, suggested that the savannahs of Venezuela should be considered. A Brazilian botanist (Ducke, 1946), thoroughly familiar with the Amazon, stated that nothing is known with certainty, while Renvoize (1972) suggests that Mexico and Central America must be considered the place of origin of the species. Still other investigators believe that several species have been involved in the origin of cassava: possibly *Manihot aesculifolia* Pohl, *M. pringlei* S. Wats. and *M. rubricaulis* I. M. Johnston, its closest relatives (Rogers, 1963, 1965). *Manihot ringlei* is unusual in wild species in having the toxic glycosides. Variability in the cultivated *M. esculenta* is very extensive and may have resulted from hybridization with several wild species. Many of these colonize disturbed areas near cultivated plots where the opportunities for gene exchanges would be available, leading to hybrids that could have produced new cultivated and wild forms.

A recent meticulous and critical evaluation of the numerous theories on the origin of *Manihot esculenta* by Mangelsdorf *et al.* (1964) has concluded that the sweet cassava strains were probably first domesticated in Central America, where there is no evidence of early cultivation of bitter cassava. On the other hand, bitter cassava appears to have first been cultivated in northern South America. There is much 'varietal' diversity in Brazil, where, with the centre of diversity of other *Manihot* species, there could have been favourable conditions for hybridization and the creation of new types of cassava. But this study suggests that there is little evidence that bitter cassava was first domesticated in that region (Renvoize, 1972).

Perhaps the most thorough modern study is by Rogers & Appan (1973) who propose two geographical centres of origin and diversification: they suggest that the two types of cassava have come from one or from several species. The proponents of this opinion seemed later to renounce it, but other writers on the origin of cultivated plants have seemed to accept the idea that the sweet and bitter strains of cassava originated separately and developed independently. It has further been suggested that the great food value of cassava was early recognized in both putative centres of origin: Central America and the Amazon. It has even been postulated that before the food value of the starchy tuber was recognized the bitter cassava might possibly have been employed as a fish poison, utilizing the ichthyotoxic properties of the water with which the natives extracted the cyanogenic glycoside.

If cassava originated in two centres, there would probably have been no need to postulate long routes of early migration from South America to Mexico or vice versa. Intercommunication and trade amongst pre-colonial American Indians was well established: both types of cassava could therefore have diffused over wide areas.

After careful evaluation of the evidence offered to date, it is still not possible to state definitely where *Manihot esculenta* or the two general types originated and were domesticated. It is hoped that future interdisciplinary studies of such an important plant may shed more light on this enigma.

HEVEA

By whatever measure is used, *Hevea brasiliensis*, one of the most recently domesticated economic plants, stands out significantly from at least two points of view. Firstly, there can be little doubt that no other species has so drastically changed human life around the world in the short space of a century (Schultes, 1984a,b). The domestication of this Amazonian tree literally altered the course of civilization. Secondly, this single species now furnishes 98% of all the world's needs of natural rubber, an achievement not duplicated by any other plant source of an economic product (Polhamus, 1962; Compagnon, 1986).

Rubber has certainly been known to man since he began to roam the fields and forests, for this polymerized isoprene is found in the latex of at least 7000 species in sundry families. However, it seems that rubber had few if any significant uses in any of the Old World civilizations. And even in the New World, its utilization by aboriginal peoples was minor. In Peru, Yucatan and Mexico, the Indians esteemed rubber from the moraceous genus *Castilla* in magico-religious ceremonies. Early European chroniclers, especially Columbus, spoke of the use of rain capes, balls and syringes. Amazon Indians valued *Hevea brasiliensis* primarily as the source of edible, carbohydrate-rich seeds, not for its latex (Schultes, 1956).

As early as 1755, King John of Portugal tried to encourage a rubber shoe industry at Belém at the mouth of the Amazon River but, because the process of vulcanization had not been discovered, the product was so useless that the industry failed.

In 1832, MacIntosh found that rubber could be dissolved in naphtha, a discovery that led to the establishment of small industries in Europe and the U.S.A. Most of these industries failed, because the rubber was sticky in heat and brittle in cold.

Then in the late 1830s, Goodyear in the U.S.A. and Hancock in England independently discovered vulcanization, the chemical insertion of sulphur molecules into the isoprene chain. Rubber immediately became a truly useful product, no longer sticky but plastic and elastic; and demand for it grew vertiginously. The only practical source was the wild *Hevea* in Amazonian forests, supplies were frequently scarce and the quality was poor. Yet from the 31 tons produced in 1827, the world market used 2600 tons in 1857. This dramatic increase in demand was fulfilled by the virtual enslavement of the Indians.

The work of tapping wild trees, separated one from the other at usually appreciable distances in the inhospitable jungle, required the rubber-workers to live for 4 or 5 months a year in shacks far from their homes and agricultural plots, suffering from exposure and malnutrition, falling prey to tropical diseases and, in very many areas, enduring mistreatment and even torture or death for not bringing in an assigned quota of latex (Hardenburg, 1912; Collier, 1968). They lived in perpetual debt to unscrupulous 'rubber barons' who resided in cities and towns, often in sumptuous luxury. This nefarious forest industry survived until the Asiatic plantations came into production in the early years of the 20th Century. When the supply of rubber from plantations, cheap and of high quality, became available, the forest industry was killed, and a multitute of Indians and other tappers were freed from the inhuman conditions under which they had had to live and die for nearly a century.

The botanical source of *Hevea* rubber was elucidated when, in 1775, the French botanist Fusée Aublet (1775) described the genus and a species *H. guianensis* Aubl. from French Guiana. He detailed its exploitation and added significant ethnobotanical observations, such as the native use of the seeds as food. Forty-nine years later, Carl Willdenow (in Jussieu, 1824: 40, t.12, fig. 38B, 1–6) described a second species, *H. brasiliensis* (Willd. ex Andr. Juss.) Muell. Arg., from a collection made at the mouth of the Amazon. The epoch explorations of the British plant-collector Richard Spruce in the Brazilian Amazon between 1849 and 1855 provided a number of species new to science: *H. benthamiana* Muell. Arg., *H. lutea* (Benth.) Muell. Arg. (now considered to be a variety of *H. guianensis*), *H. pauciflora* (Spruce ex Benth.) Muell. Arg., *H. rigidifolia* (Spruce ex Benth.) Muell. Arg., and *H. spruceana* (Bth.) Muell. Arg. Spruce's Amazonian collecting coincided with the period during which the 'rubber boom' began. Later studies by several botanists have added still other species to *Hevea*: *H. camargoana* Murça Pires, *H. camporum* Ducke, *H. microphylla* Ule and *H. nitida* Muell. Arg. Of the 10 species now recognized, only three, *H. guianensis* and its varieties *lutea* (Spruce ex Bentham) Ducke & Schultes and *marginata*, *H. benthamiana* and *H. brasiliensis*, yield usable rubber, the latex of all of the other species being too rich in resins or too low in caoutchouc. *Hevea brasiliensis* is the source of the highest grades of rubber. Some of the local variants or strains of this species, the only species domesticated and the basis of all plantations, yield rubber superior to others.

Domestication of *Hevea brasiliensis* came almost fortuitously at just the right time. The success in domesticating the quinine tree in India in the 1850s convinced the British government that rubber should be considered as a candidate for a new crop for the humid tropical colonies. The dynamo behind the pro-*Hevea* campaign was a government geographer. Sir Clement Markham, who in the 1870s enlisted the enthusiastic support of Sir Joseph Hooker, director of the Royal Botanic Gardens, Kew.

Several early attempts to introduce *Hevea* seed from the Amazon failed: in 1871 and twice in 1875. The long, slow trip from the Amazon to Kew proved too much for the viability of these delicate seeds rich in sugars that rapidly fermented, killing the embryos.

Henry Wickham, an Englishman who had travelled widely for a number of years in the Amazon and Orinoco, had sent rubber seeds to Kew without success. But he was well known to Sir Joseph Hooker, who encouraged him to continue his efforts. Wickham realized that the earlier failures were the result of slow transportation. In 1876, a steamer bringing industrial material to Manáos in the Brazilian Amazon found no return cargo. According to Wickham: "I determined to plunge for it. I had no cash on hand. The seed was even then beginning to ripen. I knew that Capt. Murry must be in a fix, so I wrote chartering the ship." He had Indians collect 70 000 seeds near Santarém at the mouth of the Rio Tapajóz in the eastern Amazon. The steamer picked up the seeds, raced downstream, called at the Brazilian customs official who, realizing the delicacy of plants for delivery to "Her Britannic Majesty's own Royal Botanic Gardens" immediately despatched the ship. Of the 70 000 seeds, 2800 germinated at Kew, a rate which would be considered astonishingly high even in the native home of the rubber tree.

There had been apathy or even opposition in certain governmental circles to the plan to domesticate *Hevea brasiliensis*, but the enthusiasm of Markham and Hooker and the devastation in Asia at that time of the *Coffea arabica* L. plantations by the fungal disease *Hemilea vastatrix* must receive much credit for the attempt to go ahead with the project (Wycherly, 1968).

Young trees were eventually sent from Kew to Ceylon, whence they later went to Singapore and other tropical parts of the Empire. The domestication of the tree that has so thoroughly revolutionized life styles around the world would not have been possible without a chain of botanical gardens throughout the British possessions. So far as is known, all cultivated trees of *Hevea brasiliensis* in Asia and Africa are descendants from the Wickham introduction (Burkill, 1935; Barlow, 1978).

Tales concerning the 'British seed steal' are rife. The story, endlessly repeated in the population literature and accepted as absolute truth in Brazil, is wholly without foundation, even though Wickham himself, who lived to a great age, mischievously enjoyed the circulation of the story in England in his later years. Sir Henry Wickham, knighted for his efforts, broke no Brazilian law, since at that time there was no law prohibiting the export of seeds, and several prior exportations had been effected. Furthermore, the ship's cargo was duly passed by Brazilian customs, and there had been no attempt in the Amazon to conceal the collection of the rubber seeds.

The whole history of economic botany has been one of exchange of useful plants, and most of the world's principal plantation crops are produced in regions far from their original homes. In fact, Brazil's major agricultural industries are based on plants introduced from other areas: soybean from China, rice and jute from India, sugar from south-eastern Asia, the African oil palm, cacao from Colombia and Ecuador and, above all, coffee from Ethiopia.

Coming from a single locality in an area occupied by *Hevea brasiliensis* half the size of the U.S.A., the Wickham seeds represented but one of many variants or ecotypes; and of this single ecotype, the seed came from a limited number of trees. Thus, the germ plasm on which plantation scientists had to work was extremely limited. Yet the improvements in yield that have been brought about in less than a century are almost unbelievable: the first plantations (from seed material) in Sri Lanka yielded 450 pounds of dry rubber per acre per year. Since clonal grafting was instituted, yields greater than 3000 pounds per acre are now not uncommon.

The story of the rubber tree's domestication includes the names of many famous personages: Aublet, Spruce, Macintosh, Goodyear, Hancock, J. D. Hooker, Markham and Wickham. But once the tree was planted in Asia, the names of two other scientists are outstanding in the creation of the great plantations: Ridley and Cramer. Fortunately, Wickham lived long enough to see the results of his endeavours establish a new and vast industry, and he also saw how it re-shaped the world.

Ridley (1928, 1955), director of the Botanic Gardens in Singapore, found only eight of the original nine trees and 1000 young plants from the original introduction still in existence when he assumed directorship of the Garden in 1888. He immediately raised 8000 additional plants from seed imported from Sri Lanka. He was an indefatigable experimenter, and his earliest innovations had to do with tapping methods. Trees in the Amazon were slashed in a great variety of makeshift techniques, usually to the detriment of the trees. It was Ridley who instituted the cutting off of thin layers of the bark every 2 or 3 days to make a slanting channel for the latex to travel down to a cup, avoiding injury to the cambium, since in *Hevea* the latex-bearing vessels are external to the cambium. With numerous modifications over the years, this is still the practice in all plantations. Amongst other discoveries, he demonstrated the advantages of tapping in the early morning, instead of in the afternoon. It was Ridley's enthusiasm and constancy that, aided by the fall in world prices of tea, poor results with cacao, the coffee disease and especially the advent of the motor car, convinced planters to go into rubber (Fairchild, 1928).

The Dutch, in the meanwhile, were busy with innovations in the Dutch East Indies. Cramer, whose total stock was likewise descended from the Wickham seeds, made several unsuccessful attempts to introduce material of other species from Brazil, looking ahead to the need for genetic experiments. With great opposition from planters, he managed to substitute bud-grafting of clonal material instead of planting seeds; the best of the first clones almost doubled the average yield of trees planted from seed, and the best of the first generation of bred clones almost doubled the yield again. He carried out analysis of variation in yield of latex, predicting that vegetative selection, cloning and genetic selection or breeding, would inevitably lead to increases in yield, all aspects basic to today's plantation management and scientific efforts toward improvement (Wilson, 1943; Dijkman, 1951).

Rubber research institutes in several countries are actively experimenting along avenues of research, especially in view of the fast-developing new applications of natural rubber which portend greatly increased demands in the not-too-distant future. Amongst some of the goals that are sought are a search for dwarfism, induced mutations, artificial polyploids and the use of tissue culture. Naturally, an interest in still higher yields and disease resistance remain major goals in the improvement programme.

There was a brief and sporadic resurgence of the silvatic Amazon rubber industry in the 1940s, when the Japanese overran most of south-east Asia, cutting off the greater part of the world's supply of natural rubber. The governments of the U.S.A. and of the South American countries where *Hevea* grows stimulated the extraction of rubber for the war effort. Natural rubber was necessary for many war-related operations such as the manufacture of tyres for heavy aircraft, for which the artificial elastomers available were inferior.

Several generations of Amazonian natives had had no experience in rubber tapping. Foresters and botanists were sent in to teach modern methods of cutting. Navigation and, in more remote areas, small airstrips were provided to supply the urgent need for natural rubber. Food, clothing, medicines and other necessities were made available for the tappers who were not mercilessly exploited, as in earlier years. In comparison with the pre-war production of rubber from south-east Asia, the amount procured in the few years of duration of the fighting was very small, but it represented a very important contribution to the war effort, even though the quality of the rubber was far inferior to that produced by the Old World plantations. Three species were exploited: *Hevea brasiliensis* (which gave the best rubber); *H. benthamiana* (the rubber was usually as good as that from *H. brasiliensis*) and *H. guianensis* (yielding an inferior, yet still usable rubber), the most widespread species (Schultes, 1970).

When peace was restored, south-east Asia again rapidly fulfilled the world's needs, and in most parts of South America, the production of rubber from wild trees gradually diminished or disappeared.

One of the fortunate by-products of this short-lived revival of the Amazon rubber tapping was that the technical people who were sent in to stimulate the industry were able to study *Hevea* in great detail, to collect germ plasm of many of the non-commercial species for future genetic work and to find in the forest wild, élite trees of *H. brasiliensis* that potentially seemed to be high yielders of latex and/or to exhibit possible disease resistance (Schultes, 1977). A search for alternative sources of natural rubber in the Euphorbiaceae and in other families was also stimulated at this time.

What will the future of natural rubber be now that germ plasm of the whole spectrum of the genus *Hevea*, material from the 10 species and all of the many ecotypes from the wild, is available for scientifically directed programmes of improvement? The domestication of *Hevea brasiliensis* marks a revolutionary milestone in human history. What applied science and human ingenuity have accomplished on such a limited germ plasm base is astounding, although perhaps not widely recognized by the world population that has been so drastically affected by the event.

One of the great Colombian novels describes the misery and suffering of the poor Amazonian rubber tappers of the early part of this century. It should be read by every modern scientist and planter concerned with today's plantation industry (Rivera, 1946). The title, *La Voragine* (*The Vortex*), refers to the mysterious swallowing up of man in the often nefarious vastness of the Amazon jungle. One magnificent passage describes the magnetic hold of the rubber tree in those days: "I have been a rubber tapper. I am a rubber tapper. I have lived in the muddy swamps in the solitude of the forests with my crew of malaria-ridden men cutting the bark of the trees that have white blood like that of the gods".

Considering what changes *Hevea brasiliensis* latex has wrought for the good of mankind, perhaps we can call it the "blood of the gods".

REFERENCES

ALTSCHUL, S. VON REIS, 1973. *Drugs and Foods from Little Known Plants. Notes in Harvard University Herbaria.* Cambridge, Mass.: Harvard University Press.
AUBLET, J. B. C. F., 1775. Histoire des plantes de la Guiane française. 4 vols. London & Paris.
BARLOW, C., 1978. *The Natural Rubber Industry: its Development, Technology and Economy in Malaysia.* Oxford: Oxford University Press.
BERENBLUM, I., 1941. The mechanism of carcinogenesis. A study of the significance of coarcinogenic action and related phenomena: *Cancer Research, 1:* 807–814.
BURKILL, I. H., 1935. *A Dictionary of the Economic Products of the Malay Peninsula.* London: Crown Agents for the Colonies.
CALLEN, E. O., 1967. Analysis of the Tehuacan coprolites. In D. S. Byers (Ed.), *The Prehistory of the Tehuacan Valley, 1:* 271–289.
COLLIER, R., 1968. *The River that God Forgot.* London: Collins.
COMPAGNON, P., 1986. *Le Caoutchouc Naturel-Biologie, Culture, Production.* Paris: Editions G.-P. Maisonneuve & Larose.
COOK, J. H., 1982. Cassava—a basic energy source in the tropics. *Science, 218:* 755–762.
DIJKMAN, M. J., 1951. *Hevea, Thirty Years of Research in the Far East.* Coral Gables, Florida: University of Florida Press.
DUCKE, A., 1946. Plantas de cultura precolombiana na Amazônia brasileira. Notas sobre as especies formas espontaneas que supostamente les terian dado origem. *Boletin del Instituto Agronômico, no. 8.*
FAIRCHILD, D., 1928. Dr Ridley of Singapore and the beginnings of the rubber industry. *Journal of Heredity, 195:* 193–263.
GUMILLA, J., 1745. *El Orinoco Ilustrado y Defendido, Historia Natural, Civil y Geográfica de este Gran Río.* Madrid: Supremo Consejo de la Inquisición.
HARDENBURG, W. E., 1912. *The Putumayo—The Devil's Paradise.* London: T. Fisher Unwin.
HILL, A. F., 1952. *Economic Botany,* 2nd edition. New York: McGraw-Hill.
JENNINGS, D. L., 1976. Cassava. *Manihot esculenta* (Euphorbiaceae). In N. W. Simmonds (Ed.), *Evolution of Crop Plants:* 81–84. London: Longmans.
JUSSIEU, A. H. L. DE, 1824. De Euphorbiacearum generibus medicique earundum viribus tentamen, tabalis aeneis. Paris.
KINGHORN, A. D., 1979. Carcinogenic irritant Euphorbiaceae. In A. D. Kinghorn (Ed.), *Toxic Plants:* 137–159. New York: Columbia University Press.
KINGHORN, A. D., 1985. In A. A. Seawright *et al.* (Eds.), *Plant Toxicology:* 357–365. Yeerongpilly, Australia: Queensland Poisonous Plant Committee, Queeensland Department of Primary Industry.
KUPCHAN. S. M., SIGEL, C. W., MATZ, M. J., SAENZ RENAULD, J. A., HALTIWANGER, R. C. & BRYAN, R. F., 1970. Jatrophone, a novel macrocyclic diterpenoid tumor inhibitor from *Jatropha folia. Journal American Chemical Society, 92:* 4476–4477.
LE COINTE, P., 1934. *Arvores e Plantas Uteis. A Amazônia Brasileira.* Belém do Pará, Brazil: Livraria Classica.
MACNEISH, R. S., 1958. Preliminary archaeological investigations in the Sierra de Tamaulipas, Mexico. *Transactions of the American Philosophical Society, 48:* 1–210.
MANGELSDORF, P. C., MACNEISH, R. & WILLEY, G. 1964. Origins of agriculture in Middle America. In Wachope (Ed.), *Handbook of Middle American Indians, 1:* 427–445. Austin, Texas: University of Texas Press.
MARTÍNEZ, S., J. B., 1970. *El Inchi (Caryodendron orinocense Karst.), Oleaginosa Nativa de America Tropical.* Pasto, Colombia: Universidad de Nariño.
PATIÑO, V. M., 1967. *Plantas Cultivadas y Animales Domésticos en America Equinoccial: Fibras, Medicinas, Misceláneas, 3:* 373. Cali, Colombia: Imprenta Departamental.
PÉREZ-ARBELÁEZ, E., 1956. *Plantas Utiles de Colombia,* 3rd edition. Bogotá: Librería Colombiana Camacho Roldán.
PLOTKIN, M. J. & SCHULTES, R. E., 1983. Tropical forests as sources of new biodynamic compounds. *Environmental Awareness, 6:* 9–10.
POLHAMUS, L. G., 1962. *Rubber: Botany, Production and Utilization.* New York: Interscience Publishers.
PURSEGLOVE, J. W., 1968. *Tropical Crops. Dicotyledons.* London: Longmans, Green and Co. Ltd.
REIS[Altschul], S. von & LIPP, F., 1982. *New Plant Sources for Drugs and Foods from the New York Botanical Garden Herbarium.* Cambridge, Mass.: Harvard University Press.
RENVOIZE, B., 1972. The area of origin of *Manihot esculenta* as a crop plant—a review of the evidence. *Economic Botany, 26:* 352–360.
RIDLEY, H. N., 1928. History and evolution of the cultivated rubber industry. *Bulletin of the Rubber Growers' Association, 10:* 45–49.

RIDLEY, H. N., 1955. Evolution of the rubber industry. *Proceedings of the Institution of the Rubber Industry, 24:* 114–122.
RIVERA, J. E., 1946. *La Voragine.* Bogotá: Biblioteca Popular de Cultura Colombiana.
ROBB, G. L.k, 1957. The ordeal poisons of Madagascar and Africa. *Botanical Museum Leaflets, Harvard University, 17(10):* 265–316.
ROGERS, D. J., 1963. Studies of *Manihot esculenta* and related species. *Bulletin Torrey Botanical Club, 90:* 43–54.
ROGERS, D. J., 1965. Some botanical and ethnological considerations of *Manihot esculenta. Economic Botany, 19:* 369–377.
ROGERS, D. J. & APPAN, S. G., 1970. Untrapped genetic resources for cassava imporvement. In *Proceedings 2nd International Symposium Tropical Root and Tuber Crops, Hawaii:* 72–75.
ROGERS, D. J. & APPAN, S. G., 1973. *Manihot, Manihotoides* (Euphorbiaceae) in *Flora Neotropica,* New York: New York Botanical Garden, Monograph 13.
ROSENGARTEN, F., Jr., 1978. A neglected Mayan galactogogue—*ixbut* (*Eupharbia lancifolia*). *Botanical Museum Leaflets, Harvard University, 26 (9–10):* 277–309.
SAUER, C. O., 1952. *Agricultural Origins and Disperals.* New York: American Geographical Society.
SCHERY, R. W., 1947. Manioc—a tropical staff of life. *Economic Botany, 1:* 20–25.
SCHNEE, L., 1973. *Plantas Comunes de Venezuela.* Caracas: Universidad Central de Venezuela.
SCHULTES, R. E., 1956. The Amazon Indian and evolution in *Hevea* and related genera. *Journal of the Arnold Arboretum, 37:* 123–148.
SCHULTES, R. E., 1970. The history of taxonomic studies in *Hevea. Regnum Vegetabile, 71:* 229–293.
SCHULTES, R. E., 1977. Wild *Hevea:* an untapped source of germ plasm. *Journal of the Rubber Research Institute Sri Lanka, 54:* 227–257.
SCHULTES, R. E., 1984a. The tree that changed the world in one century. *Arnoldia, 44:* 2–16.
SCHULTES, R. E., 1984b. Amazonian cultigens and their northward and westerward migration in pre-Columbian times. *Papers of the Peabody Museum of Archaeology and Ethnology, 76:* 21–37.
TOWLE, M. A., 1961. *The Ethnobotany of Pre-Columbian Peru.* New York; Wenner-Gren Foundation.
UPHOF, J. C T., 1968. *Dictionary of Economic Plants,* 2nd edition. Lehre, Germany: Verlag von J. Cramer.
WEBSTER, G., 1975. Conspectus of a new classification of the Euphorbiaceae. *Taxon, 24:* 593–601.
WILSON, C. M., 1943. *Trees and Test Tubes.* New York: Henry Holt and Co.
WOODROFFE, J. F. & SMITH, H. H., 1915. *The Rubber Industry of the Amazon and how its Supremacy can be Maintained.* London: John Bale, Sons and Danielson, Ltd.
WYCHERLY, P., 1968. Introduction of *Hevea* to the Orient. *The Planter, 44:* 504 ff.

Fuel oils from euphorbs and other plants

MELVIN CALVIN

Department of Chemistry and Lawrence Berkeley Laboratory, University of California, Berkeley, California 94720, U.S.A.

Received June 1986, accepted for publication September 1986

CALVIN, M., 1987. **Fuel oils from Euphorbs and other plants.** The increasing energy costs of finding petroleum, together with the sure knowledge that its supply is finite, has prompted us to seek other sources of liquid hydrocarbon for both fuel and material. We have turned to annually renewable plant sources such as seed oils, an obvious source, with palm oil as the most productive. Sugar cane used to produce ethanol is another fuel source already in use.

We have examined non-food plants which can be grown on marginal soil for their productivity, particularly the genus *Euphorbia*. All species of this genus produce a latex which can be converted into useful fuel and other material, including precursors for what might be a valuable anti-tumor agent. Euphorbias and other similar plants require repeated planting and harvesting of the entire plant, which constitutes a drain on the soil. Trees can be long-term sources for hydrocarbon-like materials with a single planting. Examples are: the genus *Copaifera* which can be tapped for sesquiterpenes, the genus *Pittosporum* which bears fruits rich in terpenes and can be harvested annually. Finally, there are algae whose oil productivity is already of interest.

It seems possible to modify genetically the terpene biosynthetic pathways in plants to improve both the quality and quantity of the oils produced from them.

ADDITIONAL KEY WORDS:—Algae – *Botryococcus braunii* – *Copaifera multijuga* – Cyanobacteria – Euphorbiaceae – *Euphorbia lathyris* – ingenol – Leguminosae – Pittosporaceae – *Pittosporum resiniferum* – seed oils – sesquiterpene – triglycerides – triterpenes.

CONTENTS

Introduction	97
Hydrocarbon-producing herbs	99
Biosynthetic Pathways	103
Hydrocarbon-producing trees	103
Oil production by algae	106
Genetic engineering	106
Conclusion	107
Acknowledgements	109
References	109

INTRODUCTION

With the oil embargo of the early 1970s we were forced to concern ourselves with matters of fuel resources (Calvin, 1976, 1977, 1979, 1980, 1983a, 1984; Calvin, Nemethy, Redenbaugh & Otvos, Hoffman, 1983; Lipinsky, 1981; McLaughlin, Kingsolver & Hoffman, 1983). The price of oil today (1986) is an aberration which also will not last very long and should not divert us from developing domestic sources along renewable avenues.

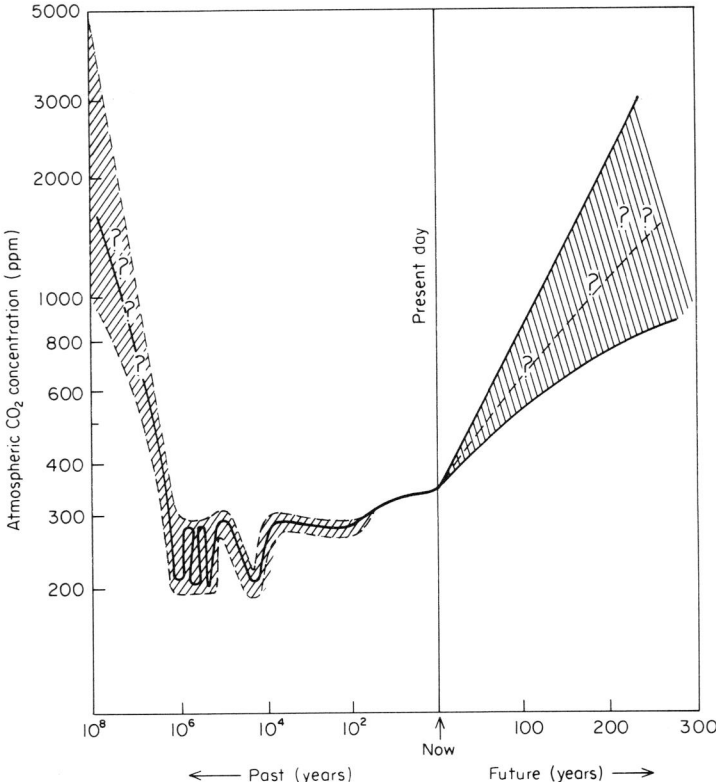

Figure 1. Concentration of CO_2 in the earth's atmosphere, past and future (redrawn, with permission, from Gammon, Sundquist & Fraser).

The first thing suggested by many economists is that we should turn our attention to using coal in a more efficient and environmentally satisfactory manner. There is an environmental consequence of the use of any fossil fuel, but especially coal, which cannot be eliminated: that is the fact that when fossil carbon is burned enormous amounts of carbon dioxide are produced. Even within the last 50 years when most of our fossil combustion has been not carbon (coal) but hydrocarbon (oil/gas) the CO_2 level in the atmosphere has risen. This is a consequence which would accelerate if coal was used as a major fuel source, either indirectly in the form of liquid fuel, or directly. Burning coal produces approximately twice as much CO_2 per unit energy as hydrocarbon, where both carbon and hydrogen are burned.

In the last few years we have been dominated by hydrocarbons rather than coal, and even under those circumstances the rate of production of CO_2 in the atmosphere has been roughly twice as great as the rate at which it is being removed by both the oceans and the biosphere (Sundquist & Broecker, 1985; MacCracken & Luther, 1985). This global trend is illustrated in Fig. 1 which shows the variation of CO_2 concentration in the atmosphere during the past geological time scale as well as a modern human one. One hundred years ago the CO_2 concentration in the atmosphere was approximately 290 ppm and the

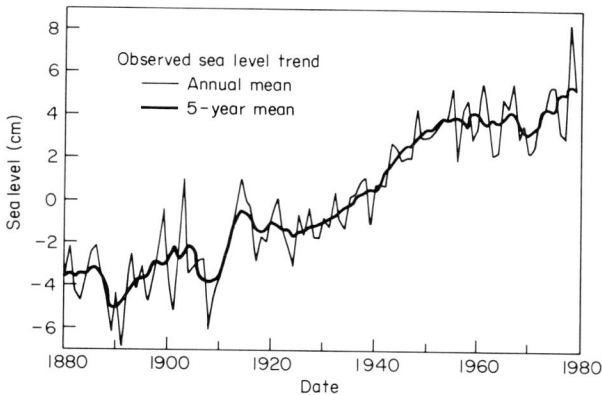

Figure 2. Global mean sea level trend based on tide gauge measurements.

concentration today is 315 ppm. The expectation is that the CO_2 concentration will continue to rise.

It is possible to detect the warming consequences of rising CO_2 concentration by using a device which integrates the temperature fluctuation over a long period of time. There are two ways this can be done, one by examining the size of the polar ice-caps and glaciers from satellite pictures and also from surface measurements: these indicate that the Antarctic ice-cap has decreased substantially in the last 100 years. If substantial quantities of ice have melted from the polar ice-caps and glaciers, there should be a rise in sea-level (Fig. 2) which would thus be related to increased combustion of fossil fuel (Gormitz, Lebedeff & Hansen, 1982). The evidence is available that the earth's temperature is increasing and the consequences of that increase are measurable. As a result of the increased temperature and loss of coastal areas due to higher ocean levels, there will be very severe world-wide consequences for agriculture: the agricultural pattern across the surface of the earth will change markedly. For example, the plains of the United States of America might no longer be capable of growing grain; northern Canada and the Soviet Union would become the chief sources of grain for the world. It will be very difficult for the human race to adjust such an enormous change in agricultural patterns in the 20–30 years which will elapse for this rise in sea-level to manifest itself catastrophically.

HYDROCARBON-PRODUCING HERBS

We must look for our liquid fuel (hydrocarbons) in the form of a renewable resource that can be grown each year. There is one plant whose cultivation for the purpose of renewable fuel and materials is already underway, i.e. the sugar cane in Brazil (Geller, 1985). In 1975, the Brazilians produced 700 million litres of ethanol from sugar cane and in 1985 the production was 7 billion litres. The Brazilians are producing more than 20% of their total liquid energy needs from alcohol and they are beginning to create a chemical industry, the sucrochemical industry, based on this energy source.

Table 1. Oil content and average oil yield for some oil crops

Oil Crop (Location)	Oil content (wt. %)	Average oil yield (kg ha^{-1})
Palm oil (Malaysia)	20	3475
Copra (Philippines)	65–68	800
Peanuts (U.S.A.)	45–50	790
Safflower (U.S.A.)	30–35	762
Sunflower (U.S.A.)	40–45	589
Rapeseed (Canada)	40–45	409
Soybean (U.S.A.)	18–19	319
Corn kernel (U.S.A.)	4–8	254
Flaxseed (U.S.A.)	35–42	230
Sesame (India)	45–50	220
Cottonseed (U.S.A.)	18–20	140

It seemed worthwhile to look for plants which would carry out the reduction of carbon dioxide all the way to hydrocarbon instead of half-way, to carbohydrate (Calvin, 1977, 1979, 1980, 1983a, 1984, 1985; Lipinsky, 1981; McLaughlin *et al.*, 1983; Nemethy, 1984). This process is exemplified in the seed oils which have substantial possibilities. The oil content and average oil yield for some oil crops is shown in Table 1, with palm oil the most productive. Peanuts, safflower and sunflower are the most important seed oils in the United States of America, and all of them produce triglycerides. The oil content of the seeds represents a reasonable opportunity for renewable fuels (American Society of Agricultural Engineers, 1982; Shultz & Morgan, 1984; Princen, 1983; Harrington, 1986). Sunflower cultivation has been expanded in the United States of America as a real possibility to provide diesel fuel for farm machinery. The oil itself, a triglyceride (glycerine with three fatty acid chains attached to it), as it comes from the seed is not a very satisfactory diesel fuel, but by treating it with methanol the fatty acids can be transesterified so the byproduct is free glycerine and the methyl esters of the fatty acids. This latter material can be used directly as diesel fuel.

There are also plants which take the initially produced carbohydrate and instead of converting it into fatty acids and glycerides (such as the seed oils) convert it into terpenes. The most important commercial plant of this type is *Hevea brasiliensis* (Willd. ex Adr. Juss.) Muell. Arg., a member of the *Euphorbiaceae* which makes polyisoprene rubber. Other members of the *Euphorbiaceae* produce hydrocarbons, especially the genus *Euphorbia* which has 2000 species of all sizes which grow throughout the world. Every *Euphorbia* species contains a latex which is an emulsion of about 30% terpenes in water. The latex hydrocarbon is largely a C_{30} triterpenoid which can be cracked like oil to make high octane gasoline (Weisz & Marshall, 1979). The zeolite catalyst cracking of the crude oil from *Euphorbia lathyris* L. resulted in the usual group of products, similar to those obtained from standard cracking of petroleum, such as olefins, paraffin, aromatics and nonaromatics. This information confirms the desirability of the products of *E. lathyris* as possible raw materials to substitute for crude oil.

We started experimental cultivation of Euphorbias in California about 10 years ago and *E. lathyris* (Fig. 3) was the first hydrocarbon producing plant

Figure 3. *Euphorbia lathyris*, Davis, California.

studied to test the hypothesis that plants could be grown for fuel and chemical content in marginally suitable land (Calvin, 1977, 1978, 1979; Nielsen, Nishimura, Otvos & Calvin, 1977). The entire *E. lathyris* plant is harvested, and from the harvest 8% of the dry weight is extracted as terpenes (oil), 20% of the dry weight as fermentable sugars, which leaves a residue of about 65% as lignocellulose. The terpenes can be cracked like crude oil and the sugar can be fermented like sucrose, while the lignocellulose can be used in a way similar to the bagasse of sugar cane. The products of the extraction of *E. lathyris* represent a new possibility for a future energy and materials source. The conceptual processing sequence to recover terpenoids and sugars is shown in Fig. 4 (Nemethy, Otvos & Calvin, 1979, 1981a, b). The oil from *E. lathyris* is black and tarry and resembles crude oil and consists mostly of triterpenes which are steroids and steroids esters (C_{30} compounds).

Some of the latex sterols of the *E. lathyris* latex (Fig. 5) are important in the pharmaceutical industry and could conceivably be of more value than the actual crude oil obtained. A very small percentage of the latex hydrocarbons

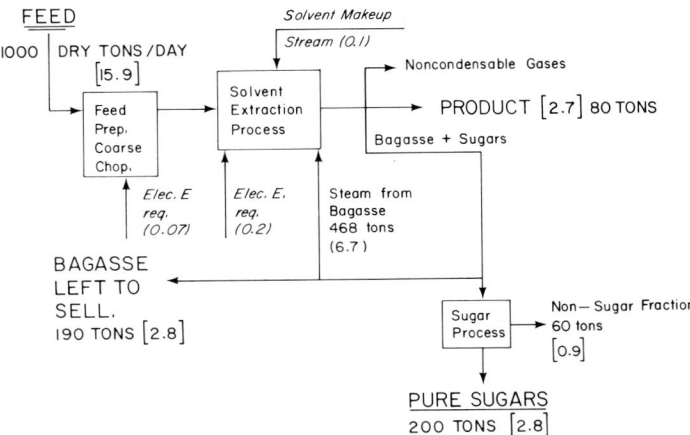

Figure 4. Conceptual processing sequence to recover terpenoids and sugars from *Euphorbia lathyris*.

consists of C_{20} components, some of which are related to ingenol (Fig. 6). Certain ingenol esters are potent stimulants for cell division and also have an irritant property which is harmful to the skin, mucuous membranes of the nose, eyes, etc. (Adolf & Hecker, 1975; Furstenberger & Hecker, 1985; Bissell,

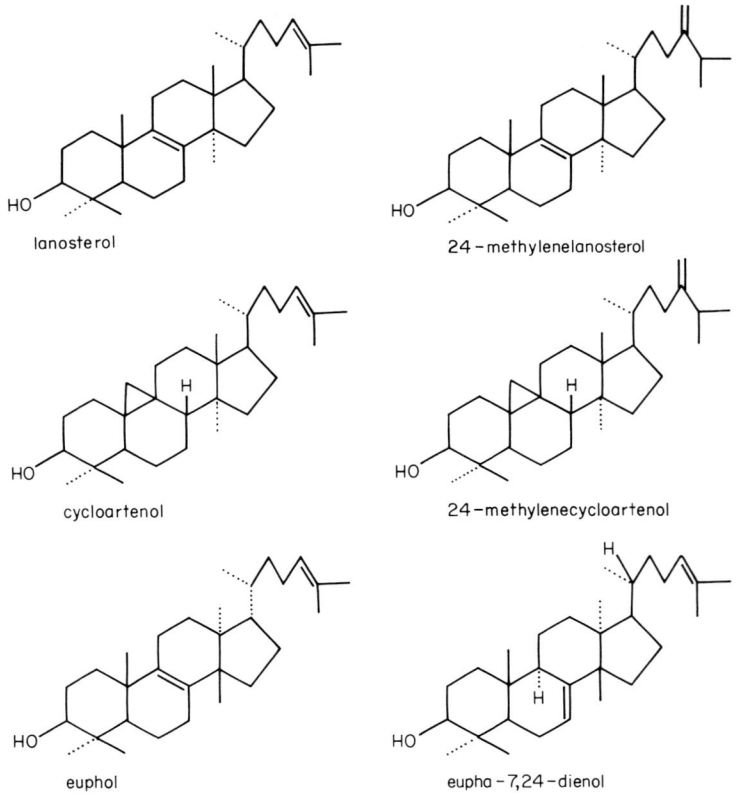

Figure 5. Latex sterols of *Euphorbia lathyris*.

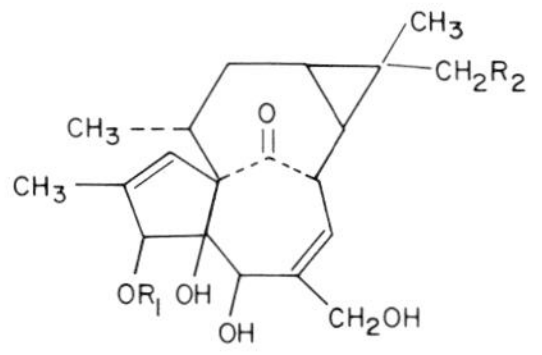

Figure 6. Ingenol esters found in *Euphorbia lathyris* (reproduced, with permission, from Adolf & Hecker, 1975).

Active Esters of Ingenol:

R_1 = hexadecanoate
R_2 = H

R_1 = 3-tetradeca−2,4,6,8,10−pentaenoate
R_2 = H

Inactive:

16-hydroxy ingenol: R_2 = OH

Nemethy, Riddle & Calvin, 1981). If the ingenol is converted into the dibenzoate ester it becomes an anti-tumor agent (Kupchan *et al.*, 1976). There is a definite relationship between the two biological effects. The stimulant molecule has the long chain on it which stimulates cell division. The binding sites are presumably the same for the dibenzoate dibenzoate, probably determined by the structure of the C_{20} skeleton. For the dibenzoates of the basic structure of ingenol there is no surface-active component and presumably the material binds to the binding site and reverses the stimulation of cell growth and thus has an effect on leukemic mice.

BIOSYNTHETIC PATHWAYS

It might be possible to modify the products of the Euphorbias by modifying their biosynthetic routes to triterpenes. The terpenes in hydrocarbon-producing plants and trees are probably made by a well-known biosynthetic pathway: sugar to pyruvic acid to mevalonic acid to isopentenylpyrophosphate (IPP). A portion of the unsaturated IPP is isomerized to dimethylallylpyrophosphate (DMAP), and the two isomers are combined. The allylic phosphate comes off DMAP and the resulting carbonium ion attacks the double bond of IPP, followed by proton loss, which results in exactly the same allylic structure as before. Eventually, if the process continues, rubber is the result.

A comparison of biosynthetic routes might be useful. In the case of *E. lathyris*, the sequence produces C_{15} compounds which dimerize to C_{30} followed by cyclization to produce triterpene steroids. However, in *Pittosporum* the route is to the C_{10} compounds at which cyclization occurs to create monoterpenes in the fruit. The biosynthetic method by which the 'diesel' oil from *Copaifera* is made is the same as that used by the *E. lathyris* up to the C_{15} step. *Copaifera* cyclizes the C_{15} farnesyl pyrophosphate, producing cyclic C_{15} compounds.

HYDROCARBON-PRODUCING TREES

One problem with using annual herbaceous plants as sources of hydrocarbons is soil erosion. It seemed better agronomic practice to grow trees (Seibert,

Figure 7. *Pittosporum resiniferum*, the Philippines.

Williams, Folger & Milner, 1986), and either harvest the fruit, or tap the tree for oil.

A tree that seems to be a likely candidate is *Pittosorpum resiniferum* Hemsl. (Fig. 7) a member of *Pittosporaceae* which grows in the Philippines. The fruit of this particular tree is rich in light oil, containing about 30% terpenes. The fruits

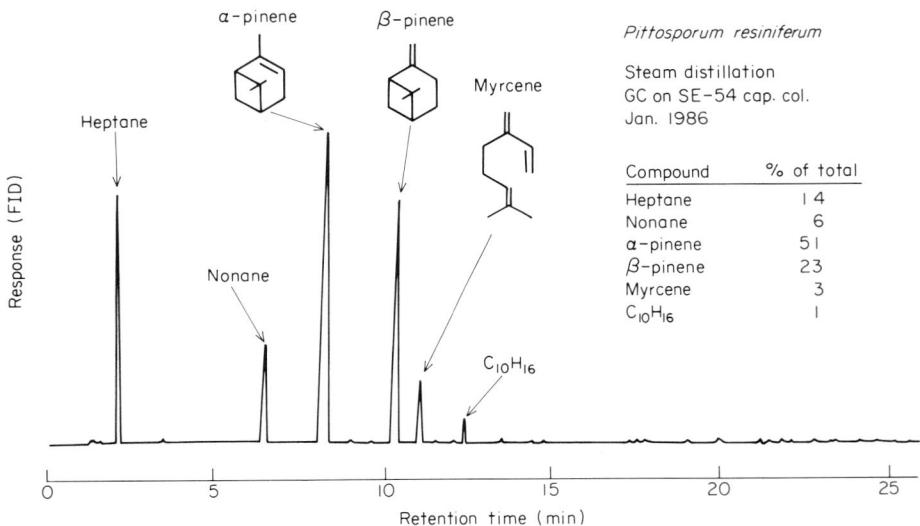

Figure 8. Gas chromatogram of components of *Pittosporum resiniferum*.

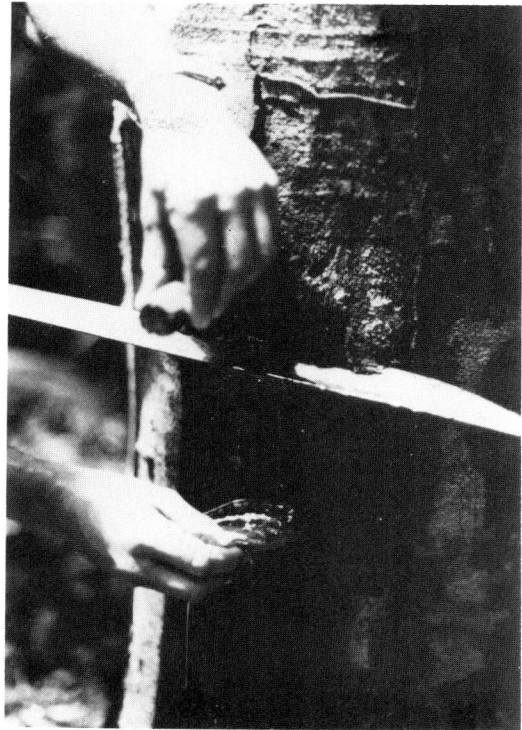

Figure 9. *Copaifera multijuga* showing oil flow, Manaus, Brazil.

can be picked and distilled and thus the terpene-like fuel extracted. The composition of the steam-distilled oil from *P. resiniferum* fruit contains four main components (Fig. 8), the most predominant being α-pinene and β-pinene, both C_{10} compounds (Nemethy & Calvin, 1982). The results indicate that the fuel properties of steam distilled oil from *P. resiniferum* fruits after hydrogenation are quite comparable with those of gasoline.

The *Copaifera multijuga* Hayne (a member of *Leguminosae*) grows in the Amazon region of Brazil and produces C_{15} hydrocarbons. The oil obtained by tapping is light yellow, very similar in appearance to olive oil (Alencar, 1982). A hole is bored horizontally in the trunk of a *C. multijuga* tree, into the heartwood (Fig. 9) and a bung is placed in the hole. The bung is removed at certain times of the year and the oil flows directly into a container. A single hole in a large tree may yield about 25 litres of oil in 24 h. The bung is reinserted and 6 months to a year later another 25 litres of oil is drained out.

The main components of the Copaiba oil, as it is commonly called, are caryophyllene, bergamotene and copaene, all cyclic C_{15} compounds (Wenninger, Yates & Dolinsky, 1967). This oil has also been subject to the Mobil zeolite catalyst process, as was the oil from *E. lathyris*, and the results indicate that it can be cracked into a useful suite of compounds. The characteristic that makes this species so attractive is the fact that the material from the tree can be used *directly* in a diesel engine without any further processing.

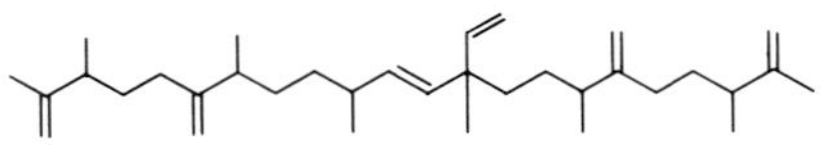

botryococcene

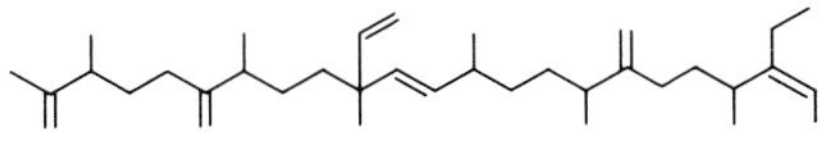

darwinene

Figure 10. Structures of botryococcene and darwinene.

OIL PRODUCTION BY ALGAE

There is another candidate for oil production which grows in many parts of the world where it occurs primarily in fresh waters. This is the unicellular green alga *Botryococcus braunii* Kuetz. which produces terpenoid oils (Wolf, 1983; Wolf, Nemethy, Blanding & Bassham, 1985). Colonies are often observed floating on the surface of undisturbed waters and this buoyancy is due to the large amounts of accumulated oil in the alga. The hexane-extracted *Botryococcus* oil is orange as a result of the presence of carotenoids. After the removal of the pigments a clear oil is obtained that contains a homologous series of unusual isoprenoids (Hillen et al., 1984). The structure of the C_{34} component (Botryococcene) (Cox et al., 1973) has been determined, and more recently the structure of the C_{36} compound (Darwinene) (Galbraigh, Hillen & Wake, 1983) has also been elucidated (Fig. 10). This alga is a very real candidate. We have learned how to grow it in the laboratory but we do not yet know how to regulate the algal metabolism to produce large amounts of oil.

GENETIC ENGINEERING

It would be very useful to be able to obtain from the Euphorbias a product similar to the Copaiba oil (Calvin et al., 1982). To achieve this it seems that a single gene would need to be transferred from the plant that cyclizes the C_{15} from farnesyl pyrophosphate into a plant species such as a *Euphorbia*, which has all the other enzymes already, but which goes on to create C_{30} materials. It appears that only one class of enzyme, the farnesyl pyrophosphate cyclase, is needed for this purpose. In other words, a single gene transplant from a donor cell of *Copaifera* to the acceptor cell of *Euphorbia* would be required. To perform this operation it would first be necessary to extract the required gene from a plant which has it and insert it into a plant that does not have it. One relevant possibility would be to take the protoplasts from the two plants (*Copaifera* and *Euphorbia*) and have a fusion/selection procedure for plant regeneration (Fig. 11) (Redenbaugh, personal communication; Calvin, 1983b). We have been able to reproduce a Euphorbia plant from a protoplast. A single protoplast in culture aggregates very quickly into a callus; the callus is adjusted with various hormones to form shoots and roots and thus regenerates a cloned plantlet.

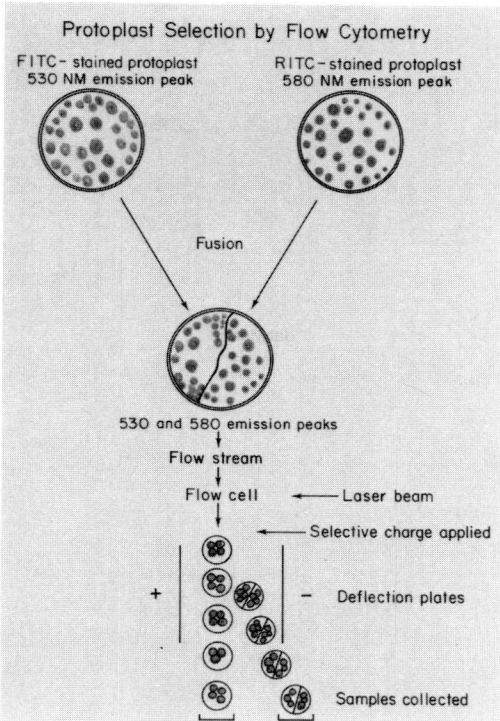

Figure 11. Protoplast selection by flow cytometry.

CONCLUSION

A summary of energy yields from different hydrocarbon producing plants is shown in Table 2 in terms of the yield of liquid fuel/acre/year/inch of water. This indicates the work that is underway with different plants of different types to develop the idea of using biomass as a significant component of total energy use, particularly in the United States of America. It has been predicted that, by the year 2000, biomass will represent approximately 6% of the total energy used in the United States of America. The idea of using plants to create hydrocarbon-like materials as a substitute for our current fuel and materials sources will become more important, especially in some of the less developed areas of the world which have a great deal of land not suitable for food production (Calvin, 1985). Various efforts are being made toward this end in Okinawa, Thailand, Australia and Spain, and attempts are underway to improve agronomic yields, develop small scale extraction plants, learn more about the composition of the plant oils and study possible ways of modifying the biosynthetic routes (particularly with cyanobacteria) to produce more desirable end products.

The idea of growing hydrocarbons is not new. It was first promulgated by the Italians in 1936 in Ethiopia (Frick, 1938): they were running out of oil and thought they would grow plants such as *Euphorbia abyssinica* J. F. Guel., extracting the oil from these plants and using it for their vehicles. This

Table 2. Comparison of energy yields for different crops

Process	Dry biomass yield, Tons (Tons acre^{-1} Year^{-1})	Liquid fuel yield (Btu acre^{-1} year^{-1})	Water Requirement (Inches Year^{-1})	Energy in Liquid Fuel (10^8 Btu) Acre^{-1} Year^{-1} per Inch of Water	Cellulosic residue (Btu acre^{-1} Year^{-1})*	Energy in Cellulose (10^6 Btu acre^{-1} inch^{-1})
Corn to ethanol	5	16 × 10^6 (0.64)	25	0.65	44.2 × 10^6 (3.4)	1.77
Sugar cane to ethanol	30	60 × 10^6 (2.4)	78	0.78	312 × 10^6 (24)	4
Energy cane to ethanol	35–50	65 × 10^6 (2.56)	48	0.35	400 × 10^6 (31)	8.2
Euphorbia lathyris to hydrocarbons and ethanol	8.5	20 × 10^6 (0.58) 17.3 × 10^6 (0.68)	25	0.82	79.6 × 10^6 (6.12)	3.2
Pittosporum resiniferum (fruits only) to hydrocarbons	7.8	50 × 10^6 (1.5)	~25	2.0	101 × 10^6 (7.8)	4.0
Jatropha curcas (seed only) to hydrocarbons	5.0	92 × 10^6 (2.2)	~25	3.6	36 × 10^6 (2.8)	1.45
Palm (fruit)	8.1	73 × 10^6	(~25)	(2.9)	(c. 1 ton)	

*Figures in parentheses are tons acre^{-1} year^{-1}.

development was not pursued because of the war. Similarly, the French in Morocco in 1940 developed plantations for growing *Euphorbia resinifera* Betg over a period of several years; however, this also was abandoned because of World War II (de Steinheil, 1940).

What is needed now is an effort on the part of the agricultural and energy community to commit itself to an energy agriculture which would have long term benefits for the entire world.

ACKNOWLEDGEMENT

The work described in this paper was supported in part by the Assistant Secretary for Conservation and Renewable Energy, Office of Renewable Energy, Biofuels and Municipal Waste Technology Division of the U.S. Department of Energy under Contract No. DE-AC03-76SF-00098.

REFERENCES

ADOLF, F. & HECKER, E., 1975. On the active principles of the Spurge family. III. Skin irritant and cocarcinogenic factors. *Zeitschrift für Krebsforschung, 84:* 325–341.
ALENCAR, J., 1982. Estudos Silviculturais de uma População Natural de *Copaifera multijuga* Hayne—Leguminosae na Amazonia Central. 2. Produção de Oleo-resina. *Acta Amazonia, 12:* 75–89.
AMERICAN SOCIETY OF AGRICULTURAL ENGINEERS, 1982. *Vegetable Oil Fuels*. Proceedings of the International Conference on Plant and Vegetable Oils as Fuels. Fargo, North Dakota: American Society of Agricultural Engineers.
BISSELL, M. J., NEMETHY, E. K., RIDDLE, L. & CALVIN, M., 1981. Testing for tumor promoters in *Euphorbia lathyris*. Analysis of possible health hazards. *Bulletin of Environmental Contamination & Toxicology, 27:* 894–902.
CALVIN, M., 1976. Photosynthesis as a resource for energy and materials. *Photochemistry and Photobiology, 23:* 425–444.
CALVIN, M., 1977. Hydrocarbons via photosynthesis. *Energy Research, 1:* 299–327.
CALVIN, M., 1978. Chemistry, population, resources. *Pure and Applied Chemistry, 50:* 407–426.
CALVIN, M., 1979. Petroleum plantations for fuel and materials. *BioScience, 29:* 533–538.
CALVIN, M., 1980. Hydrocarbons from plants: analytical methods and observations. *Die Naturwissenschaften, 67:* 525–533.
CALVIN, M., 1983a. New sources for fuel and materials. *Science, 291:* 24–26.
CALVIN, M., 1983b. The path of carbon: from stratosphere to cell. In F. Ahmad, K. Downey, J. Schultz & R. W. Voellmy (Eds), *Advances in Gene Technology: Molecular Genetics of Plants and Animals:* 1–35. New York: Academic Press.
CALVIN, M., 1984. Renewable fuels for the future. *Journal of Applied Biochemistry, 6:* 3–18.
CALVIN, M., 1985. Fuel oils from higher plants. *Annual Proceedings of the Phytochemical Society of Europe, 26:* 147–160.
CALVIN, M., NEMETHY, E. K., REDENBAUGH, K. & OTVOS, J. W., 1982. Plants as a direct source of fuel. *Experientia, 38:* 18–22.
COX, R. E., BURLINGAME, A. L., WILSON, D. M., EGLINTON, E. & MAXWELL, J. R., 1973. Botryococcene—a tetramethylated acyclic triterpenoid of algal origin. *Journal of the Chemical Society, Chemical Communications, 1973:* 284–285.
DE STEINHEIL, P., 1941. L'Euphorbe resinifère, plante à caoutchouc et resine vernis. *Revue Générale Caoutchouc, 18(2):* 54–56.
FRICK, G. A., 1983. A new source of gasoline. *Cactus and Succulent Journal, 10(9):* 60.
FURSTENBERGER, G. & HECKER, E., 1985. On the active principles of the Spurge family (*Euphorbiaeceae*). XI (1). The skin irritant and tumor promoting diterpene esters of *Euphorbia tirucaclli* L. originating from South Africa. *Zeitschrift für Naturforschung, 40c:* 631–646.
GALBRAITH, M. N., HILLEN, L. W. & WAKE, L. V., 1983. Darwinene: a branched hydrocarbon from a green form of *Botryococcus braunii*. *Phytochemistry, 22:* 1441–1443.
GELLER, H. W., 1985. Ethanol fuel from sugar cane in brazil. *Annual Review of Energy, 10:* 135–164.
GORNITZ, V., LEBEDEFF, S. & HANSEN, J., 1982. Global sea level trend in the past century. *Science, 215:* 1611–1613.
HARRINGTON, K. J., 1986. Chemical and physical properties of vegetable oil esters and their effect on diesel fuel performance. *Biomass, 9:* 1–17.

HILLEN, L. W., POLLARD, G., WAKE, L. V. & WHITE, N., 1984. Hydrocracking of the oils of *Botryococcus braunii* to transport fuels. *Biotechnology and Bioengineering, 24:* 193–205.
HOFFMAN, J. J., 1983. Arid lands plants for feedstocks for fuel and chemicals. *CRC Critical Review of Plant Science, 1:* 95–116.
KUPCHAN, S. M., UCHIDA, I., BRANFMAN, A. R., DAILEY, R. G. & FEI, Y. M., 1976. Antileukemic principles isolated from *Euphorbiaceae. Science, 191:* 571–572.
LIPINSKY, E. S., 1981. Chemicals from biomass: petrochemical substitution options. *Science, 212:* 1465–1471.
McCRACKEN, M. C. & LUTHER, F. M. (Eds), 1985. *The Potential Climatic Effects of Increasing Carbon Dioxide.* Washington, D.C.: Department of Energy.
McLAUGHLIN, S. P., KINGSOLVER, B. E. & HOFFMAN, J. J., 1983. Biocrude production in arid lands. *Economic Botany, 37:* 150–158.
NEMETHY, E. K., 1984. Biochemicals as an energy resource. *CRC Critical Review of Plant Science, 2:* 117–129.
NEMETHY, E. K. & CALVIN, M., 1982. Terpenes from *Pittosporaceae. Phytochemistry, 21:* 2981–2982.
NEMETHY, E. K., OTVOS, J. W. & CALVIN, M., 1979. Analysis of extractables from one Euphorbia. *Journal of the American Oil Chemists Society, 55:* 957–960.
NEMETHY, E. K., OTVOS, J. W. & CALVIN, M., 1981a. Hydrocarbons from *Euphorbia lathyris. Pure and Applied Chemistry, 53:* 1101–1108.
NEMETHY, E. K., OTVOS, J. W. & CALVIN, M., 1981b. Natural production of higher energy liquid fuels from plants, pp 405–419. In D. L. Klass & G. H. Emeret (Eds), *Fuels from Biomass and Wastes.* Ann Arbor, Michigan: Ann Arbor Science Press.
NIELSEN, P. E., NISHIMURA, H., OTVOS, J. W. & CALVIN, M., 1977. Plant crops as a source of fuel and hydrocarbon-like materials. *Science, 198:* 942–944.
PRINCEN, L. H., 1983. New oilseed crops on the horizon. *Economic Botany, 38:* 478–492.
SEIBERT, M., WILLIAMS, G., FOLGER, G. & MILNER, T., 1986. Fuel and chemical co-production from tree crops. *Biomass, 9:* 49–66.
SCHULTZ, E. B., Jr. & MORGAN, R. P., 1985. *Fuels and Chemicals from Oilseeds: Technology and Policy Options.* Washington, D.C.: American Association for the Advancement of Science.
SUNDQUIST, E. T. & BROECKER, W. D. (Eds), 1985. *The Carbon Cycle and Atmospheric CO_2: Natural Variations Archaean to Present.* Washington, D.C.: American Geophysical Union.
WENNINGER, J. A., YATES, R. L. & DOLINSKY, M., 1967. Sesquiterpene hydrocarbons of commercial copaiba and American cedarwood oils. *Journal of the American Oil Chemists Society, 50:* 1304–1312.
WEISZ, P. B. & MARSHALL, J. F., 1979. Catalytic production of high-grade fuel (gasoline) from biomass. Compounds by shape-selective catalysis. *Science, 206:* 57–58.
WOLF, F. R., 1983. *Botryococcus braunii*—an unusual hydrocarbon producing alga. *Applied Biochemistry and Biotechnology, 8:* 249–260.
WOLF, F. R., NEMETHY, E. K., BLANDING, J. H. & BASSHAM, J. A., 1985. Biosynthesis of unusual acyclic isoprenoids in the alga *Botryococcus braunii. Phytochemistry, 24:* 733–738.

Wood anatomy of the Euphorbiaceae, in particular of the subfamily Phyllanthoideae

ALBERTA M. W. MENNEGA

Institute of Systematic Botany, Rijksuniversiteit, Utrecht, The Netherlands

Received June 1986, accepted for Publication September 1986

MENNEGA, ALBERTA, M. W., 1987. **Wood anatomy of the Euphorbiaceae, in particular of the subfamily Phyllanthoideae.** The great variety in wood structure of the large family Euphorbiaceae makes it impossible to describe briefly a general wood pattern. Nevertheless, a more or less clear division into four anatomical groups can be made.

A short overview is given of the wood structure of the uni-ovulate subfamilies Acalyphoideae, Crotonoideae, and Euphorbioideae, following the classification by Webster. These subfamilies cannot be distinguished by their anatomy. The paper is mainly devoted to the bi-ovulate subfamily Phyllanthoideae. Within this subfamily, two groups can be recognized on the basis of their wood anatomy: the *Aporusa* type with a great number of primitive characters, and the *Glochidion* type, in which primitive features such as scalariform vessel perforation plates are absent. A short description of the 13 tribes is given as well as suggestions for rearrangement of the succession of the tribes. In several tribes some taxa are anomalous and, on anatomical evidence, exclusion of some genera, sometimes with assignment to another tribe, is suggested.

ADDITIONAL KEY WORDS:—Fibre types – parenchyma – Stilaginaceae – Uapacaceae.

CONTENTS

Introduction	112
Material	112
Characters used for establishing an evolutionary grouping of the tribes in the subfamily Phyllanthoideae	112
Short descriptions of the tribes of Phyllanthoideae	116
Tribe 8. Aporuseae (Lindl. ex Miq.) Airy Shaw	116
Tribe 7. Antidesmeae (Endl.) Hurusawa	116
Tribe 9. Drypeteae (Griseb.) Hurusawa	117
Tribe 1. Wielandieae Baill. ex Hurusawa	118
Tribe 2. Amanoeae (Pax & Hoffm.) Webster	119
Tribe 6. Spondianthea Webster	119
Tribe 3. Brideliaea Muell. Arg.	121
Tribe 11. Hymenocardieae (Muell. Arg.) Hutch.	121
Tribe 13. Bischofieae (Muell. Arg.) Hurusawa	122
Tribe 4. Dicoelieae Hurusawa	122
Table 5. Porantheeae (Muell. Arg.) Grüning	122
Tribe 10. Phyllantheae Dumort.	122
Tribe 12. Uapaceae (Muell. Arg.) Hutch.	124
Conclusions	124
Acknowledgements	125
References	125

INTRODUCTION

The large family Euphorbiaceae includes a great number of woody taxa, shrubs, trees, a few lianas, and some plants with an unusual habit, like the candelabra Euphorbias or the ericoid *Clutia*. The wood structure also shows a great variation. It is impossible to present a general diagnosis of an euphorbiaceous wood. However, at least three, perhaps four types of wood can be distinguished. Metcalfe & Chalk (1950), in the *Anatomy of the Dicotyledons*, recognized three types in the subfamily Phyllanthoideae: the *Aporusa* type, the *Glochidion* type, and a group of taxa designated "other genera"; in the subfamily Crotonoideae the structure is far more uniform, but different from the structure in Phyllanthoideae. In recent years new taxonomic treatments have been published, including those of Hutchinson (1969) and Webster (1975). The latter system is followed in the present paper, because the proposed classification of genera agrees better with the wood anatomical data than that of Hutchinson. Webster recognizes five subfamilies: Phyllanthoideae and Oldfieldioideae (comprising all the bi-ovulate genera), while the uni-ovulate genera are placed in Acalyphoideae, Crotonoideae and Euphorbioideae. With some exceptions (among others, *Acalypha*, *Clutia*, *Pogonophora*) the basic pattern of the wood is alike in the genera belonging to the three last-mentioned subfamilies. It is characterized by absence of scalariform vessel perforation plates; presence of medium to very large intervascular pitting, and similar vessel/ray pitting; presence of apotracheal diffuse or banded parenchyma; numerous narrow, heterocellular rays, often vertically fused; and non-septate, often wide and thin-walled fibres.

The subfamily Oldfieldioideae, comprising most of the "other genera" mentioned above, is not treated in the present paper because this has been studied extensively by Hayden (1980).

In this paper short descriptions are given of the tribes of the Phyllanthoideae composed as in Webster's Conspectus, but arranged in a different order, entirely based on the results of wood anatomy. The results are summarized in Table 1. In most tribes one or more genera did not seem well placed; sometimes a suggestion has been made for a better positioning, but sometimes this was not possible.

MATERIAL

The part of this paper dealing with the anatomy of the subfamily Phyllanthoideae is based on examination of slides and macerations of 141 specimens of 116 species and 35 genera. The information on the subfamilies Acalyphoideae, Crotonoideae and Euphorbioideae is founded on data from, respectively, 45, 39 and 17 genera.

The slides are housed in the Wood Anatomy Section of the Institute of Systematic Botany, University of Utrecht. Full details may be obtained on application to the author.

CHARACTERS USED FOR ESTABLISHING AN EVOLUTIONARY GROUPING OF THE TRIBES IN THE SUBFAMILY PHYLLANTHOIDEAE

The sequence of the tribes in the descriptive section, and as shown in Table 1, is entirely based on wood characters that may be considered as having a definite

Table 1. Subfamily Phyllanthoideae Aschers. A suggested rearrangement of Webster's tribes based on wood anatomical data

	Aporusa type		*Glochidion* type
8.	Aporuseae	6.	Spondiantheae
	Didymocistus excepted	3.	Bridelieae
	Martretia included	11.	Hymenocardieae
7.	Antidesmeae	13.	Bischofieae
	Antidesma excepted	4.	Dicoelieae
9.	Drypeteae	5.	Poranthereae
	Paradrypetes included	10.	Phylantheae
1.	Wielandieae	A.	Securineginae
	Astrocasia excepted	B.	Flueggeinae
	Discocarpus excepted		
	Actephila included		
2.	Amanoeae		
	Actephila excepted		

Place of tribe Poranthereae dubious.
Position of the genera *Antidesma*, *Astrocasia*, *Didymocistus*, *Discocarpus*, *Jablonskia*, and *Uapaca* (tribe 12. Uapaceae) undecided.

relation to phylogeny. Particularly in the case of the Phyllanthoideae, these features are evident in the vessels, the fibres and the parenchyma. In Table 2 the important characters of the vessels of all genera investigated are summarized. The most primitive type of wood shows vessels with scalariform perforation plates, long vessel elements, medium to large intervascular pitting and similar vessel/ray pitting; long, thick-walled, non-septate fibres with bordered pits on the tangential as well as the radial walls, and parenchyma in diffuse strands, or in short irregular bands one cell wide. (For the theoretical underlying principles the reader is referred to Chalk, 1983.) Slightly more advanced genera are considered to be those in which some of the vessel perforation plates are scalariform and others simple, in combination with very fine intervascular pitting. The next group comprises taxa with simple perforation plates and medium to large intervascular pits, with and without parenchyma; the fibres are septate or not. The combination of characters considered most advanced includes vessels with simple perforation plates, medium-sized vessel elements, small intervascular and vessel/ray pitting, absence of parenchyma or some paratracheal parenchyma, and septate fibres.

It is worth noting that often in tribes where most of the genera show diffuse or short bands of parenchyma and non-septate fibres one genus seems out of place by the absence of parenchyma and presence of septate fibres.

Several of the genera with the *Aporusa* type of wood structure have a special type of fibre wall thickening, first described by Bamber (1974). He noted two types in the subfamily Phyllanthoideae *sensu* Pax: type I with normal thickening of the walls, type II with very thick walls completely occluding the lumen, and showing abnormal birefringence. In some genera all fibres are of type II, in this paper referred to as Bamber type II; in others, both types occur, as in *Actephila* (see Fig. 4). Apart from the Phyllanthoideae, Bamber type II also occurs in some genera of the Oldfieldioideae, but not in the other three subfamilies, with the possible exception of *Pogonophora* (Giraud, 1983).

Table 2. Selected vessel characters of subfamily Phyllanthoideae

Tribes and sequence according to Webster's Conspectus (Webster, 1975)	Perforation plates			Intervascular pitting and vessel/ray pitting			Average element length		
	Scalariform		Exclusively simple	Similar c. 4 μm	Intervascular medium vessel/ray medium to large		< 350 μm	350–800 μm	>800 μm
	Exclusively	Partly							
1. Wielandieae									
Astrocasia			+	+				450	
Discocarpus			+	+				800	
Heywoodia	+			+				750	
Lachnostylis			+	+				700	
Pentabrachium	+			+					1200
Savia			+	+					
Wielandia		+		+				700	1200
Blotia	+			+					1350
2. Amanoeae									
Amanoa			+	+					1300
Actephila	+			+					1470
3. Brideliaeae									
Bridelia			+		+			550	
Cleistanthus			+	+					1000
4. Dicoelieae									
Dicoelia			+	+				700	
5. Poranthereae									
Andrachne			+	+			350		
6. Spondiantheae									
Spondianthus			+		+				1090
7. Antidesmeae									
Antidesma			+		+				920
Celianella		+			+			660	
Hyeronima		+			+				1250
Thecacoris	+				+				875

8. Aporuseae						
Aporusa	+					1550
Baccaurea		+				1500
Didymocistus		+	++			1450
Maesobotrya		++				1200
Protomegabaria	+		+++			1860
Richeria		++	+	++++		1640
Ashtonia				+++		1035
9. Drypeteae						
Drypetes		+	++			900
Putranjiva		+				1700
Neowawrea*		+		+	560	
10. Phyllantheae						
A Securineginae						
Chascotheca		+	+++		630	
Pseudolachnostylis			++		600	
Securinega			++		630	
B Flueggeinae						
Breynia			++++	++++	525	865
Glochidion			++++		770	865
Flueggea			++++		740	
Margaritaria			+		750	
Phyllanthus				+++		
Richeriella†			++			
Sauropus			+		800	
11. Hymenocardieae						
Hymenocardia			+	+	550	
12. Uapaceae						
Uapaca			+	+	750	
13. Bischofieae						
Bischofia			+	+		1040
i.s. *Jablonskia*	+			++		940
i.s. *Martretia*	+			++		850
i.s. *Paradrypetes*†					1200	

i.s. = *incertae sedis*, genera of an uncertain position in the subfamily.
*Neowawrea fits well in subtribe 10B Flueggeinae (see Hayden & Brandt, 1984).
†Small intervascular pitting and large vessel/ray pitting.

Another feature of the fibres in Euphorbiaceae in general is the frequent occurrence of gelatinous walls, but this character has no relation to phylogeny.

Characters of the rays have not been used in establishing a phylogenetic order because they are all rather similar, and of a primitive type.

SHORT DESCRIPTIONS OF THE TRIBES OF PHYLLANTHOIDEAE

The first five tribes to be described all belong to group A, the *Aporusa* type of Metcalfe & Chalk (1950). This type is mainly characterized by scalariform perforation plates in all or at least part of the vessels, long vessel elements, presence of parenchyma in diffuse strands or in short, narrow, irregular bands, non-septate, long and thick-walled fibres, and wide, high, heterogeneous rays (Figs 1–3).

Tribe 8. Aporuseae (Lindl. ex Miq.) Airy Shaw

The tribe comprises 7 genera, all represented by wood samples (Figs 1–3).

VESSELS with partly scalariform, partly simple perforation plates; vessel elements very long, 1–2 mm; intervascular pitting medium-sized, the vessel/ray pitting similar or larger. PARENCHYMA in diffuse strands or in short, scattered bands one cell wide; strands of 8–14 cells. RAYS uniseriate and multiseriate, (2–)6(–9) cells wide, the multiseriate part composed of many rows of procumbent cells, the extensions usually very high; total height 2–8 mm; sheath cells often present. FIBRES very thick-walled, Bamber type II, except in *Protomegabaria* and *Didymocistus*, non-septate, large bordered pits on radial and tangential walls; mean length 2900 µm.

The wood structure of the genera is remarkably similar, with the exception of *Didymocistus*. In this monotypic genus parenchyma is absent, fibres are thin-walled, wide, and partly septate, scalariform vessel perforation plates are rare.

Tribe 7. Antidesmeae (Endl.) Hurusawa

The tribe consists of 5 genera; wood samples were available for *Antidesma* and *Hyeronima*, but only pieces of twigs of *Celianella* and *Thecacoris*.

VESSEL perforation plates, partly scalariform, partly simple, but in *Antidesma* exclusively simple, element length on average 900 µm; intervascular pitting 5–7 µm; the vessel/ray pitting moderately large to large, oval, often elongate, in scalariform arrangement. PARENCHYMA as isolated strands, or in short, diffuse bands one cell wide; In *Antidesma* exclusively scarce paratracheal, and absent in *Celianella*. RAYS uniseriate and multiseriate, 2–4 cells wide, the middle part very high, often composed of mixed procumbent and square cells, the extensions of tall erect cells moderately to very high, vertically fused rays frequent; height up to 5 mm. FIBRES non-septate in *Hyeronima* and *Thecacoris*, partly septate in *Antidesma* and *Celianella*; thick-walled, with bordered pits in *Hyeronima*; thin-walled and often gelatinous in *Antidesma* and *Celianella*; simple pits on radial walls; mean length 1760 µm.

This tribe is not homogeneous in its structure. *Celianella* shows the same type of structure as *Didymocistus* regarding the absence of parenchyma and the occasional septate thin-walled fibres of moderate length. *Antidesma* shares the

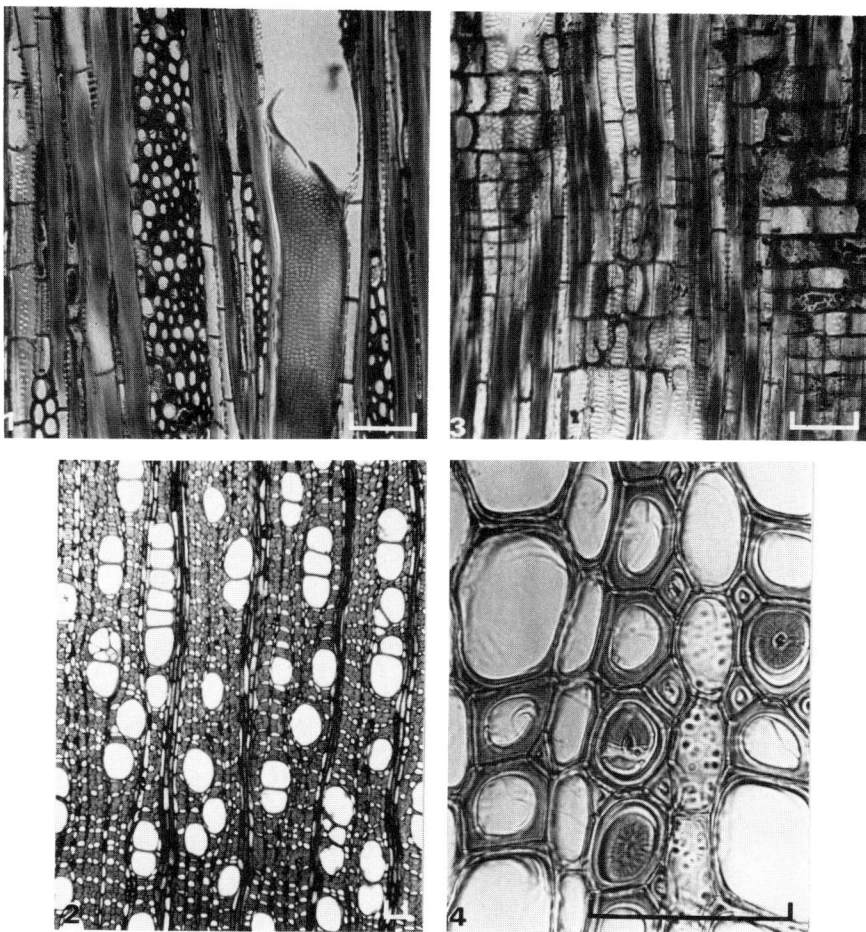

Figures 1–4. Fig. 1. *Aporusa microcalyx* tangential section showing a wide ray, a vessel with medium-sized intervascular pitting, part of a parenchyma strand with large pits. Fig. 2. Cross-section of *A. microcalyx* showing vessel grouping and fine irregular bands of parenchyma. Fig. 3. Radial section of *A. microcalyx* showing the large vessel/ray pitting. Fig. 4. *Actephila excelsa* showing normal thick-walled fibres, and three fibres with occluded lumen, Bamber's type II. Scale bars = 100 µm.

thin-walled, septate fibres with the foregoing genera, but it also lacks scalariform perforation plates. In Metcalfe & Chalk (1950), *Antidesma* was placed in group B with *Glochidion*-type of wood, associated with *Hymenocardia* and *Aporosella*. The wood shows affinities with *Spondianthus* (tribe 6, Spondiantheae) but also with *Bischofia* (tribe 13, Bischofieae). *Antidesma* is one of the genera of doubtful taxonomic position: Airy Shaw (1965) transferred it to the family Stilaginaceae, suggesting affinities with Icacinaceae. Such an affinity is not expressed in the wood anatomy. To me the wood structure fits well with Phyllanthoideae, but perhaps it should be placed in a tribe by itself.

Tribe 9. Drypeteae (Griseb.) Hurusawa

In Webster's classification the tribe comprises *Drypetes, Neowawrea*, and *Putranjiva*. No wood of *Neowawrea* was available to me, but from the extensive treatment by Hayden & Brandt (1984) of the genus, I am convinced that it was wrongly assigned to this tribe. I agree with their conclusion that it should be placed in subtribe 10B, Flueggeinae.

VESSEL perforation plates partly simple, partly scalariform. Vessel element length (770–)900(–2000) µm; intervascular pitting minute, 3–4 µm, the ray/vessel pitting similar. PARENCHYMA apotracheal in numerous short, often interrupted bands one cell wide; strands of 6–14 cells. RAYS uniseriate and multiseriate, 2–3 cells wide, the wide part often very high, and the uniseriate extensions also high; vertically fused rays frequent; height from 1.5 to 4 mm. FIBRES non-septate, very thick-walled, Bamber type II; small pits on radial walls, length 1575–3000 µm.

Drypetes and *Putranjiva* are anatomically similar. Some species of *Drypetes* show a marked resemblance to species of *Wielandia* and *Heywoodia*. The wood structure of *Paradrypetes* Kuhlm. is, according to descriptions by Milanez (1935) and by Matos Araujo & Mattos Filho (1984), almost identical to that of *Drypetes*, except for the large oval, scalariformly arranged vessel/ray pits in *Paradrypetes*. This combination of very small intervascular pits and large scalariform vessel/ray pitting is unusual in Euphorbiaceae, but it is also present in *Richeriella* (tribe 10B, Phyllantheae).

Tribe 1. Wielandieae Baill. ex Hurusawa

This tribe comprises 11 genera, wood specimens were available of the genera mentioned below. VESSEL perforation plates partly or exclusively scalariform in *Blotia*, *Heywoodia*, *Pentabrachium* and *Wielandia*, exclusively simple in *Astrocasia*, *Discocarpus*, *Lachnostylis* and *Savia*. Vessel elements over 1000 µm long in the genera with scalariform perforation plates (except *Heywoodia*), 800 µm or less in the other genera.

Vessels numerous, narrow, often arranged in long radial rows, intervascular pitting and vessel/ray pitting small in all genera. PARENCHYMA diffuse or in short irregularly distributed bands one cell wide, but very scarce and restricted to some paratracheal strands in *Discocarpus* and completely absent in *Astrocasia*. RAYS more or less alike in all genera, except for the very wide and high rays in *Blotia* (up to 120 µm = 11 cells wide and up to 8 mm high), the middle part of the rays 2–3 cells wide, usually very high, the extensions extremely high to moderately high; vertically fused rays frequent; height from 600 to 3000 µm. FIBRES non-septate, very thick-walled with a minute lumen (Bamber type II) in *Blotia*, *Heywoodia*, *Lachnostylis*, *Savia* and *Wielandia*. In *Astrocasia*, *Discocarpus* and *Pentabrachium*, fibres sometimes septate with thin walls; walls sometimes gelatinous; length 1300–2400 µm, shorter in *Astrocasia* (740 µm).

The taxa of this tribe show variation in structure. *Blotia* has the most primitive wood structure of all Euphorbiaceae as seen by the numerous bars of the scalariform perforation plates in combination with long vessel elements, long fibres with thick walls, wide and high rays, and diffuse parenchyma. Others, such as *Savia* and *Lachnostylis*, also show primitive features, but combined with more advanced characters, whereas *Discocarpus*, and in particular *Astrocasia*, seem out of place in this tribe.

According to Punt (1962) the pollen of *Discocarpus* represents a type of its own intermediate between that of *Amanoa* and that of *Antidesma*. Anatomically *Discocarpus* shows more affinities with *Antidesma* than with *Amonoa*. Punt commented furthermore that the pollen grains of *Astrocasia* are undoubtedly of the *Securinega*-type (a type that should now be called the *Flueggea* type, see Webster, 1984a). The wood structure shows affinities with *Breynia* of the subtribe Phyllanthoideae, but also with *Andrachne* of tribe Porantheroideae.

The wood structure of *Pentabrachium* closely resembles that of *Actephila* of the tribe Amanoeae.

Tribe 2. Amanoeae (Pax & Hoffm.) Webster

Amanoa and *Actephila* are the only two genera of Amanoeae.

VESSELS with scalariform perforation plates in *Actephila*, simple in *Amanoa*; element length (1030–)1400(–1570) μm; intervascular pitting very small to minute, the vessel/ray pitting similar. PARENCHYMA in diffuse, short bands one cell wide in *Amanoa*, absent or restricted to a few paratracheal cells in *Actephila*. RAYS uniseriate and multiseriate, mostly uniseriate in *Amanoa*, the multiseriates 2–4 cells wide; vertically fused rays present; the type of ray cells different in the two genera, erect cells scarce in *Amanoa*, abundant in *Actephila*, in the latter genus sheath cells also present, as well as perforated ray cells with scalariform perforations. FIBRES not septate and very thick-walled, Bamber type II, in *Amanoa*, septate and sometimes with very thick walls, Bamber type II, sometimes with moderately thick walls and a distinct lumen in *Actephila* (Fig. 4), the length in both genera about 2100 μm.

The structure of the wood of the two genera is so markedly different from the anatomical point of view that there is no evidence to support their retention in the same tribe. *Amanoa* has several features in common with *Savia*, whereas *Actephila* seems closely related to *Pentabrachium* as mentioned before. However, it should be noted that of the 35 species of *Actephila*, wood samples were available only of *A. excelsa* var. *javanica*.

The remaining 8 tribes belong to group B, the *Glochidion*-type of Metcalfe & Chalk (1950). This type is chiefly characterized by simple vessel perforation plates, absence of parenchyma (sometimes a few paratracheal strands), thin-walled septate fibres (Figs 5–8).

Tribe 6. Spondiantheae Webster

Spondianthus is the only genus of this tribe.

VESSEL perforation plates simple, though one vessel with scalariform perforation noted; element length on average 1090 μm, intravascular pitting large, c. 12 μm; vessel/ray pitting large and oval, or ±circular. PARENCHYMA mostly paratracheal, scarce, as incomplete or slightly aliform rings, occasionally some diffuse strands. RAYS uniseriate and multiseriate, 5–8 cells wide and very high, up to 3–4 mm, the middle part high and composed of procumbent and square cells, the extensions not very high. FIBRES sometimes septate, walls moderately thick and lumen wide, average length 2660 μm.

In Metcalfe & Chalk (1950), *Spondianthus* is considered as a rather problematic genus doubtfully belonging to the Euphorbiaceae. I quote: "... if it is correctly placed in the Euphorbiaceae, [it] falls into this group [*Glochidion*-type] on account of its septate fibres. It has the large rays typical of group A, the intervascular pitting is similar to that of *Antidesma*, and the occurrence of

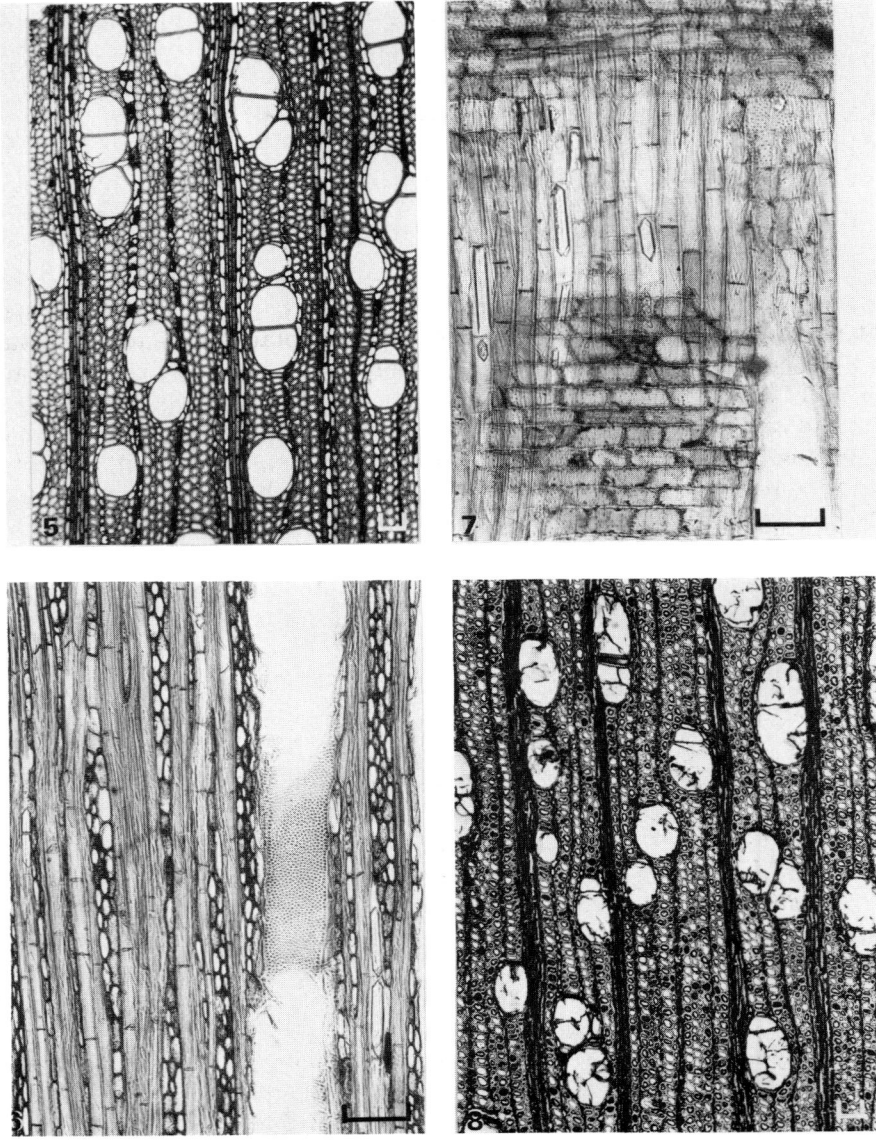

Figures 5–8. Fig. 5. *Glochidion borneense* cross section, parenchyma absent. Fig. 6. Tangential section of *G. obscurum* showing septate thin-walled fibres, multiseriate rays 2–3 cells wide. Fig. 7. Radial section of the same species with large elongate crystals in some of the septate fibres. Fig. 8. Cross-section of *Bischofia javanica*, showing resemblance to *Glochidion*; vessels with thin tyloses. Scale bars = 100 µm.

vasicentric parenchyma suggest *Bridelia*; only the pitting between vessels and rays cannot be matched, being large and usually circular". Having had the opportunity to see more material of groups A and B, I do not consider the vessel/ray pitting of *Spondianthus* so different from that of *Antidesma* or *Phyllanthus*. On account of the large rays, the long vessel elements and long fibres, this tribe seems to occupy a transitional position between groups A and B.

Tribe 3. Bridelieae Muell. Arg.

The tribe comprises the genera *Bridelia* and *Cleistanthus*.

VESSEL perforation plates exclusively simple, mean element length 550 μm in *Bridelia*, 1000 μm in *Cleistanthus*; intervascular pits 5–10 μm wide, vestured in *Bridelia*, 2–3 μm wide in *Cleistanthus*, the vessel/ray pitting in *Bridelia* large, rounded to oval, in *Cleistanthus* similar to the small intervascular pits, but occasionally unilaterally compound and scalariform. PARENCHYMA scarce, paratracheal, strands of 2–8 cells. RAYS uniseriate and multiseriate, 2–6 cells wide, the middle part heterocellular with scarce truly procumbent cells, the extensions of square and erect cells, high; vertically fused rays often present; height up to 3.5 mm, usually c. 1.5 mm. FIBRES septate, thin- to thick-walled, often with gelatinous walls, pits small with a narrow border, restricted to the radial walls; mean length 1240 μm for *Bridelia*, 1800 μm for *Cleistanthus*.

Since in all taxonomic classifications *Bridelia* and *Cleistanthus* are placed together, the anatomical differences in pitting and length of vessel elements and fibres are apparently more of diagnostic importance than an indication that the genera are unrelated.

Tribe 11. Hymenocardieae (Muell. Arg.) Hutch.

Hymenocardia is the only genus in this tribe.

VESSEL perforation plates simple; element length on average 550 μm; intervascular pitting 5–7 μm wide, the vessel/ray pits partly ± circular, partly large, elongate, often in scalariform arrangement. PARENCHYMA absent. RAYS uniseriate and multiseriate, 2–3 cells wide, with a very high middle portion and equally high extensions; vertically fused rays often present, height up to 3 mm. FIBRES septate, walls moderately thick, length on average 1300 μm.

Hymenocardia is a taxon of controversial position within the Euphorbiaceae. Airy Shaw (1965) considered the genus as not belonging to the family, because of its winged fruit and pollen structure. He placed the genus in a separate family Hymenocardiaceae, and suggested a relationship with Urticaceae, also drawing attention to a resemblance with *Holoptelea* of Ulmaceae. Other authors in their treatments of African floras followed his idea of a new family, for example Radcliffe-Smith (1973) and recently Leonard & Mosango (1985); (see also papers in this volume by Radcliffe-Smith and Webster). From the wood anatomical point of view a position in the Euphorbiaceae is not entirely unreasonable, a certain resemblance to *Spondianthus* exists. Metcalfe & Chalk (1950) placed *Hymenocardia* somewhere near to *Glochidion* and *Aporosella* (now included in *Phyllanthus*), but at the same time an alliance with three tribes of the Flacourtiaceae was mentioned. From my own data and from those of Dechamps, Mosango & Robbrecht (1985), which are in close agreement, a similarity to some genera of the tribe Casearieae (according to the treatment of Miller (1975)) is present, but discrepancies also exist. This similarity is not only true for *Hymenocardia*, but for several more genera with the *Glochidion* type of wood structure. A relationship with Urticaceae is anatomically not acceptable, and the wood of *Holoptelea* with its storied structure is entirely different.

Tribe 13. Bischofieae (Muell. Arg.) Hurusawa

A monotypic tribe, comprising *Bischofia* (Fig. 8).

VESSEL perforation plates simple; elements on average 1040 µm long; intervascular pitting large, 12 µm; ray/vessel pitting ± circular and large elliptic or irregular. PARENCHYMA absent. RAYS uniseriate and multiseriate, cells wide, the middle part high, extension low to high, sheath cells often present, up to 2 mm high. FIBRES septate, very wide (40 µm) and thin-walled, few pits on radial walls; length 2175 µm.

This is another genus of disputed placement in the Euphorbiaceae. Airy Shaw (1965) established the family Bischofiaceae to accommodate the genus, and suggested affinities with Staphyleaceae, a suggestion not endorsed by wood anatomy. Neither does the wood structure agree with that of genera of the subtribe Paiveusinae (subfamily Oldfieldioideae) as was pointed out by Hayden (1980). In my opinion *Bischofia* fits well in the *Glochidion* type of the Phyllanthoideae, either in a tribe of its own, or even in the subtribe Flueggeinae of the Phyllantheae.

Tribe 4. Dicoelieae Hurusawa

A monotypic tribe, comprising *Dicoelia*.

VESSELS with simple perforation plates, though in thin twigs some scalariform perforations present. Vessel element length on average 700 µm. Intervascular pitting very small, 3–4 µm, vessel/ray pits similar. PARENCHYMA almost absent, a few strands occasionally adjacent to vessels. RAYS uniseriate and 2–5-seriate, the latter with a high middle part and very high extensions, height up to 2 mm; rays often vertically fused. FIBRES septate and non-septate; moderately thick-walled, minute bordered pits on radial walls; length *c.* 1210 µm.

According to Airy Shaw, this is a very isolated genus with obscure affinities. The wood structure shows resemblance to *Cleistanthus*.

Tribe 5. Poranthereae (Muell. Arg.) Grüning

The three genera in this tribe are subshrubs or herbs. The only material available was the woody base of a specimen of *Andrachne aspera*.

VESSEL perforation plates simple; element length short, 350 µm; intervascular pits very fine, the vessel/ray pitting similar or slightly wider, unilaterally compound. PARENCHYMA scarce, diffuse, difficult to see. RAYS uniseriate and biseriate, most cells square and erect. FIBRES non-septate, with numerous small bordered pits on radial and tangential walls; length 580 µm on average.

The short vessel elements and fibres are probably related to the habit of the specimen. This habit makes interpretation of the data difficult. The absence of septate fibres makes a place amongst the tribes with *Aporusa* type of wood seem appropriate. The place I have given it seems doubtful.

Tribe 10. Phyllantheae Dumort.

This tribe is split into two subtribes: Securineginae and Flueggeinae.

Subtribe 10A. Securineginae Muell. Arg.

This subtribe comprises nine genera; wood samples of *Securinega, Chascotheca* and *Pseudolachnostylis* were available.

VESSEL perforation plates simple in *Securinega* and *Pseudolachnostylis*, exclusively scalariform with 8–15 bars in *Chascotheca*; mean element length *c.* 600 μm intervascular pitting very small, 2.5–4.0–6.0 μm, the vessel/ray pitting similar. PARENCHYMA abundant in bands 2–6 cells wide in *Pseudolachnostylis*, diffuse and in irregular bands one cell wide in *Securinega*, and absent in *Chascotheca*; strands of 8–12 cells. RAYS uniseriate and 2–3 cells wide, up to 5 cells wide in *Pseudolachnostylis*; extension low; in *Chascotheca* the extensions very high in *Securinega* moderately high; vertically fused rays frequent except in *Pseudolachnostylis*. FIBRES non-septate, thick-walled with small bordered pits on the radial and tangential walls in *Securinega* and *Pseudolachnostylis*; thin-walled, often with gelatinous walls, and some fibres septate, pits small, simple on the radial walls only in *Chascotheca*. Length on average 900 μm.

Chascotheca shows, like *Didymocistus* in Wielandieae, the same complex of characters differing from those of other members of the subtribe: scalariform perforation plates, absence of parenchyma, and partly septate fibres. *Securinega* and *Pseudolachnostylis* do not comply with the general aspect of the *Glochidion* type of wood structure by their variable amounts of parenchyma and the absence of septate fibres, features characteristic for the *Aporusa* type. In fact, Metcalfe & Chalk (1950) cited the wood of *Lingelsheimia*, another genus of this subtribe (not investigated here), as resembling *Drypetes* in group A, which also applies for the two genera *Securinega* and *Pseudolachnostylis*.

Subtribe 10 B. Flueggeinae Muell. Arg.

A large taxon comprising *Breynia, Flueggea, Glochidion* (Figs 5–7), *Margaritaria, Phyllanthus, Richeriella, Sauropus* and two more genera of which no wood was available. *Neowawrea*, incorporated in Drypeteae, should be included in this subtribe, as already mentioned.

VESSEL perforation plates exclusively simple; mean vessel element length 720 μm; intervascular pits medium-sized, 5–8 μm, except in *Breynia* and *Sauropus* where pits very small; the vessel/ray pitting large, ±circular or, often, narrow, oval, arranged in horizontal rows, again with the exception of *Breynia* and *Sauropus* with pitting similar to the intervascular pits. PARENCHYMA absent, or very scarce to scanty, paratracheal as a few strands adjacent to a vessel, strands of 2–8–10 cells. RAYS uniseriate and multiseriate, 2–3 cells wide (much wider in *Phyllanthus emblica*), the middle part not very high, composed of square and relatively short procumbent cells, extensions very high to high; vertically fused rays often present; total height from 1.5 to 4.5 mm, higher in *Phyllanthus emblica*. FIBRES septate, walls generally thin, but often gelatinous, small pits restricted to the radial walls; length on average 1340 μm.

The wood structure of the subtribe is remarkably uniform, except for the very small pits in *Breynia* and *Sauropus*. These two genera show a great similarity, which is in accordance with the remark of Airy Shaw (1975) in his treatment of the Euphorbiaceae of Borneo, that *Breynia*, *Sauropus* and *Synostemon* are scarcely distinct genera. Metcalfe & Chalk (1950) treat *Breynia* in the group of "other genera", but their description is entirely different from the wood structure of the three specimens of various species seen by me; the discrepancy is most probably due to incorrectly identified material.

Until recently much confusion existed between wood samples assigned either to *Securinega* or *Flueggea*. From the foregoing descriptions it will be evident that these taxa are far more different in their wood structure than any of the genera of the subtribe Fluegginae are among themselves. Webster (1984a), in his revision of *Flueggea*, has straightened out this confusing state. *Securinega* is now restricted to four species, three from Madagascar and one from Mauritius, all showing the same type of wood structure.

Tribe 12. Uapaceae (Muell. Arg.) Hutch.

A monotypic tribe of doubtful position, considered by Airy Shaw (1965) as belonging in a separate family having affinities with Anacardiaceae and Picrodendraceae.

VESSELS with simple perforation plates, though Normand (1955) mentioned an occasional scalariform perforation; element length *c*. 750 µm; intervascular pits 10 µm, vessel/ray pitting large elliptic and smaller roundish. PARENCHYMA paratracheal as incomplete rings, also in short bands and diffuse; strands of 4–6 cells. RAYS uniseriate and multiseriate, 3–6 cells wide, the middle part very high and heterocellular, the extensions also very high; rays often vertically fused; height 3.5–5 mm. FIBRES thick-walled, non-septate, simple pits restricted to the radial walls; mean length 1900 µm.

The wood structure is in several respects unusual for the Euphorbiaceae. Metcalfe & Chalk (1950) place *Uapaca* with the group "other genera", which comprises mainly the Oldfieldioideae. From a comparison of the wood structure of *Uapaca* with the data published by Hayden (1980) for the genera in the subfamily Oldfieldioideae no close affinities between the taxa are apparent. Giraud (1983) lists in her thesis a number of genera according to an evolutionary scala. In this list *Uapaca* is placed next to *Drypetes* and near to *Hieronyma*. The three genera have indeed some features in common, but I would hesitate to assign *Uapaca* to either Drypeteae or Antidesmeae.

CONCLUSIONS

Conclusions concerning the composition of the tribes have already been given, and the results are summarized in Table 1. It is apparent that the sequence of the tribes in Webster's classification, as indicated by the numbers of the tribes, is quite different from the sequence obtained when anatomical characters are used exclusively for classification. The author is well aware that the data of wood anatomy have to be interpreted together with data from other disciplines, but she considers it helpful to present the anatomical information as a tool for a definitive classification. Results in the past, such as the settling of marked anatomical differences noticed among species of *Flueggea* and *Securinega* by Webster's revision of *Flueggea* (Webster, 1984a), which brought uniformity to specimens previously described as belonging to either one genus or the other, strengthen the wood anatomist in the believe that wood structure can offer a positive contribution to classification.

The sequence of the tribes with woods of the *Aporusa* type has a sounder base than the sequence in the other group. This is mainly due to the fact that in the *Aporusa* type, anatomical characters, well-established as indicating a primitive or more advanced state, are present, whereas such characters are less obvious in the taxa with the *Glochidion* type of wood. Furthermore, the tribes or taxa are much more alike, and distinctions are not clear.

The position of Poranthereae is dubious, partly because material was not adequate, but also because the wood characters are intermediate between the *Aporusa* and *Glochidion* groups. This is also the case for the genera *Antidesma*, *Astrocasia*, *Didymocistus*, *Discocarpus* and *Jablonskia*, genera in which some of the fibres are septate and in which parenchyma is absent. The two last-mentioned genera also possess some vessels with scalariform perforation plates.

Tentatively *Antidesma* might be placed in a tribe by itself amongst the tribes with *Glochidion* wood, or in a family of its own as advocated by Airy Shaw (1965) who re-established the family Stilaginaceae. *Astrocasia* might be related to *Andrachne*, but lack of reliable material of the latter genus prevents a definite judgment. *Jablonskia*, a genus established by Webster (1984b), was provisionally placed by him somewhere near to Phyllantheae, subtribe Securineginae. In wood characters it shows more affinities with genera of the *Aporusa* group, but it does not fit in any of the tribes (Mennega, 1984).

The wood of *Uapaca* is in several respects different from that of the Euphorbiaceae. Perhaps this taxon might better be assigned to a family Uapacaceae in conformity with Airy Shaw's suggestion (Airy Shaw, 1965). Comments on the taxonomic position of *Uapaca*, as well as *Hymenocardia* and *Bischofia*, will be found in the papers by Radcliffe-Smith and by Webster in this volume.

Genera not listed by Webster (1975) are *Martretia* and *Paradrypetes*. *Martretia* fits remarkably well in the tribe Aporuseae, a position corresponding to the place next to *Aporusa* in Pax & Hoffman (1931).

Paradrypetes resembles *Drypetes* very closely in structure, as is apparent from the description of Matos Araujo & Mattos Filho (1984), except for a different type of vessel/ray pitting. Accordingly, I have included the genus in the tribe Drypeteae.

ACKNOWLEDGEMENTS

I am greatly indebted to Miss Mary Gregory and to Dr D. F. Cutler for reading and carefully correcting the preliminary manuscript, and for their helpful comments.

REFERENCES

AIRY SHAW, H. K., 1965. Diagnoses of new families, new names etc., for the seventh edition of Willis's 'Dictionary'. *Kew Bulletin, 18:* 249–273.
AIRY SHAW, H. K., 1975. *The Euphorbiaceae of Borneo*. London: H.M.S.O.
BAMBER, R. K., 1974. Fibre types in wood of Euphorbiaceae. *Australian Journal of Botany, 22:* 629–634.
CHALK, L., 1983. Wood anatomy, phylogeny, and taxonomy. In C. R. Metcalfe & L. Chalk (Eds), *Anatomy of the Dicotyledons*, 2nd edition, *2:* 108–125. Oxford: Clarendon Press.
DECHAMPS, R., MOSANGO, M. & ROBBRECHT, E., 1985. Etudes systématiques sur les Hymenocardiaceae d'Afrique: la morphologie du pollen et l'anatomie du bois. *Bulletin du Jardin Botanique National de Belgique, 55:* 473–485.

GIRAUD, B., 1983. 'Xylologie et Phylogénie des Euphorbiacées arborescentes.' Thèse de Doctorat d'Etat à l'Université Pierre et Marie Curie, Paris.
HAYDEN, W. J., 1980. 'Systematic Anatomy of Oldfieldioideae.' Ph.D. Thesis, University of Maryland.
HAYDEN, W. J. & BRANDT, D. 1984. Wood anatomy and relationships of *Neowawrea* (Euphorbiaceae). *Systematic Botany, 9:* 458–466.
HUTCHINSON, J., 1969. Tribalism in the family Euphorbiaceae. *American Journal of Botany, 56:* 738–758.
LEONARD, J. & MOSANGO, M., 1985. Hymenocardiaceae. In Bamps, P. (Ed.) *Flore d'Afrique Centrale:* 1–16. Bruxelles: Jardin Botanique National de Belgique.
MATOS ARAUJO, P. A. & MATTOS FILHO, A. de., 1984. Estrutura das madeiras brasileiras de Dicotiledôneas. XXVI. Euphorbiaceae. *Rodriguesia, 36 (59):* 25–40.
MENNEGA, A. M. W., 1984. Wood structure of *Jablonskia congesta* (Euphorbiaceae). *Systematic Botany, 9:* 236–239.
METCALFE, C. R. & CHALK, L., 1950. *Anatomy of the Dicotyledons*. Oxford: Clarendon Press.
MILANEZ, F. R., 1935. Anatomia de *Paradrypetes ilicifolia*. *Arquivos do Instituto de Biologia Vegetal. Rio de Janeiro, 2:* 133–156.
MILLER, R. 1975. Systematic anatomy of the xylem and comments on the relationships of Flacourtiaceae. *Journal of the Arnold Arboretum, 56:* 20–102.
NORMAND, D., 1955. *Atlas des bois de la Côte d'Ivoire, 2*. Nogent-sur-Marne, France: Centre Technique Forestier Tropical.
PAX, F. & HOFFMAN, K., 1931. Euphorbiaceae. In A. Engler, *Die natürlichen Pflanzenfamilien, Band 19c:* 11–233. Leipzig: W. Engelmann.
PUNT, W., 1962. Pollen morphology of the Euphorbiaceae with special reference to taxonomy. *Wentia, 7:* 1–116.
RADCLIFFE-SMITH, A., 1973. A new variety of *Hymenocardia acida* (Hymenocardiaceae) In F. N. Hepper (Ed.), Tropical African plants: XXXIII. Kew Bulletin, 28: 319–326.
WEBSTER, G. L., 1975. Conspectus of a new classification of the Euphorbiaceae. *Taxon, 24:* 593–601.
WEBSTER, G. L., 1984a. A revision of *Flueggea* (Euphorbiaceae). *Allertonia, 3:* 259–312.
WEBSTER, G. L., 1984b. *Jablonskia*, A new genus of Euphorbiaceae from South America. *Systematic Botany, 9:* 229–235.

A survey of pollen morphology in Euphorbiaceae with special reference to *Phyllanthus*

W. PUNT

Laboratory of Palaeobotany and Palynology, State University of Utrecht, Utrecht, The Netherlands

Received May 1986, accepted for publication September 1986

PUNT W., 1984. **A survey of pollen morphology in Euphorbiaceae with special reference to *Phyllanthus*.** The Euphorbiaceae is a very large family, but the number of publications containing descriptions of pollen grains relatively low. Especially the category of monographic papers is poorly represented. The Webster system is discussed in relation to pollen morphology. It appears that four subfamilies are characterized by a basic pollen type; the subfamily Acalyphoideae is an exception to this generalization. Webster's contribution to the classification of the subgenus *Phyllanthus* is illustrated with pollen morphological results.

ADDITIONAL KEY WORD:—taxonomy.

CONTENTS

Survey of pollen morphology in Euphorbiaceae	127
Floras	128
Technical papers	128
Monographs	128
Pollen morphology in relation to the system of Webster	130
Some notes on the pollen morphology of *Phyllanthus*	134
Conclusion	140
Acknowledgements	140
References	140

SURVEY OF POLLEN MORPHOLOGY IN EUPHORBIACEAE

The Euphorbiaceae form a cosmopolitan family comprising some 7000 species in about 300 genera. It is the fourth largest family after the Orchidaceae, Compositae and Gramineae. Not only is it one of the largest families, but also one of the most interesting in its pollen morphology, with an enormous variety of pollen types. The number of individuals studying pollen of the Euphorbiaceae is remarkably low. Relatively few have ever published descriptions or illustrations of pollen grains of the family, and during the preparation of this paper it became quite clear that not only is the number of papers very low but also the quality of many of them is poor. These publications are scattered throughout the literature and most of them give information on only a few taxa. The publications can be divided into three categories: Floras, technical papers and monographs.

Floras

Characteristically Floras, usually local Floras, show one or a few representative plates of the Euphorbiaceae, for example: Blasco et al. (1980), Bonnefille (1971a, b), Bonnefille & Riollet (1980), Caratini & Guinet (1974), Diaz Zavaleta & Palacios Chavez (1980), Heusser (1971), Huang (1972), Markgraf & d'Antoni (1978), Marticorena (1962) and Salgado-Labouriau (1971).

It is remarkable that pollen Floras with Euphorbiaceae pollen in them were published only between 1970 and 1980. Since 1980 no pollen Flora of a tropical region has appeared.

Technical papers

A few technical papers show excellent illustrations of Euphorbiaceae. However, in these the treatment of pollen grains was quite incidental and served purposes other than describing pollen morphology; e.g. in Lynch & Webster (1975). This paper contains beautiful micrographs of *Manihot*, *Jatropha*, *Croton* and *Euphorbia*.

Other papers dealt with terminology: e.g. Thanikaimoni, Caratini, Nilsson & Grafström (1984).

Monographs

Monographs on pollen morphology are usually connected with taxonomy. Some of the publications emphasize the taxomomy and these pollen descriptions are given merely to support the taxonomic results. The remainder of the publications, the minority, are fundamentally taxonomic in interest, but lay special emphasis on pollen morphology. These are often used to develop, or support, phylogenetic theories.

a. *Pollen morphology as an incidental product of taxonomic research.*

Examples in this field of study are: Dehgan & Webster (1979), on *Jatropha*; Hamilton (1976), identifying Urticales pollen; Köhler (1967), on *Phyllanthus*; Long Huo & Yu Chong-Hong (1984), Martin (1974), on fossil pollen; Miller & Webster (1967), on *Tragia*; Sykes (1969), on *Homalanthus*; Webster & Webster (1972), on *Dalechampia scandens*; and Weber-El Ghobary (1985) and Ybert (1975), on *Manihot*.

In all these publications it is clear that the taxonomic part is more important than pollen morphology which is only used as an additional, albeit very welcome, character, but nevertheless of minor importance. This situation is regrettable because many more data from pollen could be used for studies of fundamental pollen morphology. Of course, the number of trained pollen morphologists is far too small to improve this situation at the moment.

Attention should be drawn to the special connection between taxonomy and pollen morphology in *Phyllanthus* as recorded by Brunel & Roux. Brunel (1975a) produced a thesis in which pollen of *Phyllanthus* was studied in connection with the taxonomy of some African species. After this, Brunel & Roux (1975b, 1976, 1977, 1980, 1981a, b, 1982, 1984) studied taxa not only from Africa but also from Thailand and Madagascar.

b. Surveys.

In the Euphorbiaceae four examples of surveys are available. Erdtman (1952) started the investigation of the Euphorbiaceae with a large survey but his descriptions are minimal and his figures are scanty. His survey gives only a small impression of the large variation of pollen types in the Euphorbiaceae.

The survey by Khan (1968) is far from complete (37 out of 300 genera were studied) and is mainly based on Indian material. The pictures are drawings (palynograms) and his survey is of low quality.

A well presented survey of the Phyllanthoideae by Köhler (1965) was made under the supervision of Professor Erdtman. The quality of the plates is excellent and even at this early stage a theory was presented on the phylogeny of the pollen types.

The survey by Punt (1962) was primarily designed as an additional study to the taxonomy of the Euphorbiaceae, and much emphasis was therefore laid on the use of pollen in delimiting taxa. It was the first survey of its kind to be carried out and shows the characteristic teething problems of such early publications. For example, only drawings were given and too few taxa were studied.

These surveys have today been superseded because they are too superficial and give too little information on basic features. They give some information but this often turns out to have been biased too much in one direction. A good example is found in the information that both Köhler and myself gave about the pollen of *Zimmermannia*. Both surveys suggested that pollen of *Zimmermannia* taxa has a peculiar, verrucate ornamentation. However, in the excellent monograph on *Zimmermannia* pollen by Poole (1981) it appeared that only half of the species show verrucate pollen; the other half are simply reticulate or vermiculate. It is clear that in large taxa one has to be especially careful in extrapolating from results based on a small sample.

c. Pollen morphology as a basis for taxonomy.

These monographs contribute more to taxonomy as they are based on a comprehensive sampling of the taxa. It is thus possible not only to give full information about the taxon, but also to provide the taxonomist with a more reliable basis for understanding its interrelationships and phylogeny.

Examples of this kind of monograph are extremely scarce in Euphorbiaceae. A good example is the paper by Poole (1981) who suggested that *Zimmermannia* might be related to the genus *Meineckia*. A new monograph in the Euphorbiaceae might result from a study of the pollen grains of all the species of *Meineckia*. This would surely reveal most interesting information, and give taxonomists a new impulse to re-study the relationships of *Meineckia* in the subfamily Phyllanthoideae.

Another example in this genre is the recent monograph by Mosango & Robbrecht (1985) of the genus *Hymenocardia*, where pollen of the five species has been studied. It appears that they resemble each other very closely and the genus is remarkably homogeneous (eurypalynous). On the other hand, it is clear that no particular feature, or trend, points to a relationship with other pollen types in the subfamily Phyllanthoideae or, for that matter, any other pollen type in the Euphorbiaceae. In fact, the pollen of *Hymenocardia* resembles much more that found in Moraceae, Urticaceae and Ulmaceae.

Apart from my monograph on *Phyllanthus* (Punt, 1967, 1972, 1980), these two papers (Poole, 1981; Mosango & Robbrecht, 1985) describe only 12 species out of 7000, a pitifully small proportion for such an important family!

POLLEN MORPHOLOGY OF THE EUPHORBIACEAE IN
RELATION TO THE SYSTEM OF WEBSTER (1975)

When connecting pollen morphology with taxonomy the choice of taxonomic system is most important. Following the earlier work of Mueller (Mueller & Boisser, 1866) and Baillon (1858), Bentham & Hooker (1880) suggested a useful system. Subsequently, Pax & Hoffmann (1931) did a great deal of work in the Euphorbiaceae but the classification they proposed was far from natural. After Pax, Airy Shaw (1965, 1980) and Hutchinson (1969) suggested new systems. Webster (1975) proposed a new classification which, in my opinion, represents the best concept we have today: he recognizes five subfamilies, three more than the traditional two (Phyllanthoideae and Crotonoideae). I should like to add some information to this system of classification which is the result of studying pollen morphology.

I. Subfamily Phyllanthoideae

In this subfamily two distinct lines of related pollen types may be distinguished. The first line, representing a rather primitive series of pollen types, has 3-colporate, more or less reticulate, prolate pollen without much ornamentation. It is found in representatives of *Aporusa* (Fig.1), *Hyeronima* (Fig. 5), *Antidesma*, *Securinega*, *Actephila* etc. The second line, starting with *Wielandia* (Fig. 8) and continuing with genera like *Pseudolachnostylus*, *Amanoa* (Fig. 9) and *Bridelia* (Fig. 4) is distinctly, often coarsely reticulate and more or less oblate.

II. Subfamily Oldfieldioideae

This subfamily is remarkably homogeneous in its pollen types. In fact all the types represented are polyaperturate; either zonoaperturate with six or more apertures, or pantoaperturate with five or more. They are often ornamented with conspicuous spines. Examples are the genera *Mischodon*, *Tetracoccus* (Fig. 2), *Hyaenanche* (Fig. 7) and *Austrobuxus* (Figs 3 & 6).

III. Subfamily Acalyphoideae

Unlike the other subfamilies, the Acalyphoideae shows a large variation in its pollen types, which suggests it should be interpreted as heterogeneous. The pollen types might be linked in a phylogenetic scheme. Examples of pollen types found are: *Doryxylon*, *Alchornea* (Fig. 10), *Argythamnia* (Fig. 11), *Mallotus* (Fig. 12), *Acalypha*, *Dalechampia* and *Plukenetia*.

IV. Subfamily Crotonoideae

The pollen morphology of this subfamily is characterized by one particular feature, the 'croton-pattern' (Fig. 13). This characteristic pattern is found in different pollen types, which although they vary in several features, such as apertures, shape and size, nevertheless always show a basic similarity. Examples of pollen types found are: *Manihot*, *Croton*, *Domohinea* and *Klaineanthus* (Fig. 13). There are a few genera which have pollen which is not distinctly crotonoid, but

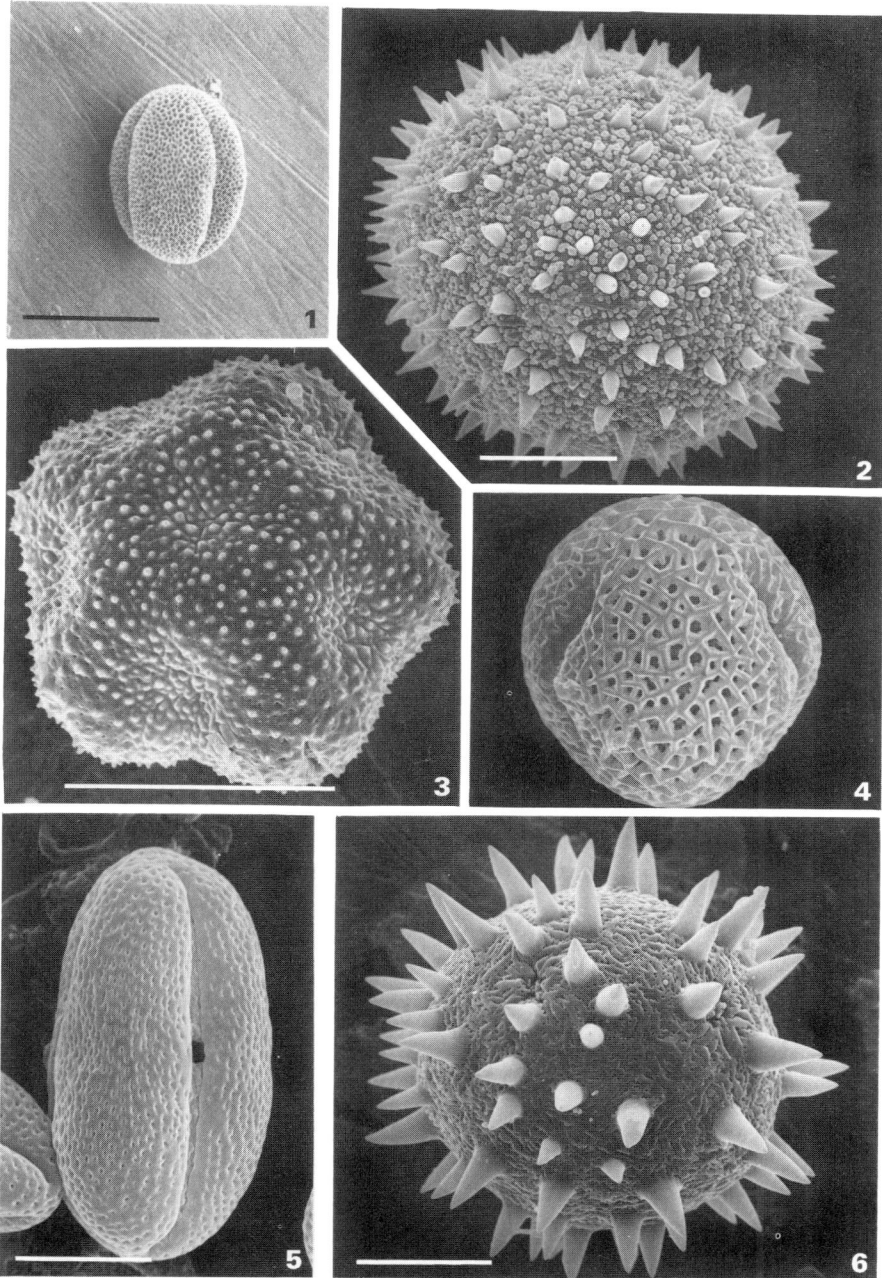

Figures 1–6. Fig. 1. *Aporusa falcifera* Hook. fil., *Achmad 1253* (U). Fig 2. *Tetracoccus ilicifolius*, Coville & Gilmay, *Gilmay 3054* (L). Fig. 3. *Austrobuxus buxoides* (Baill.) v. Steenis, *McKee 4895* (L). Fig. 4. *Bridelia lancifolia* Jabl., *Ramos 10* (U). Fig. 5. *Hyeronima laxiflora* (Tul.) Muell. Arg., *B.B.S. 264* (U). Fig. 6. *Austrobuxus nitidus* Miquel, *Berkhout 981* (L). Scale bars = 10 µm.

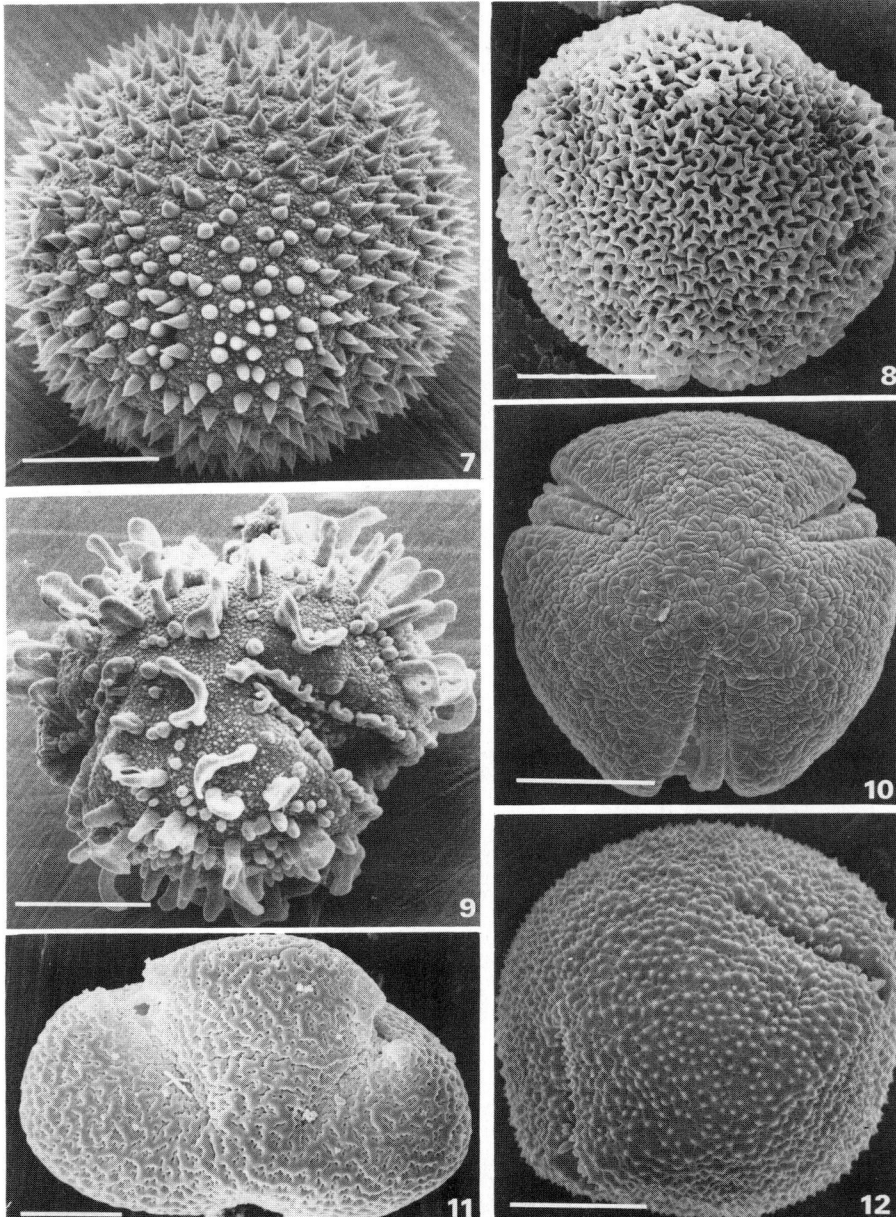

Figures 7–12. Fig. 7. *Hyaenanche capensis* (Thunb.) Lamb. & Vahl, *D'Alleizette 6422* (L). Fig. 8. *Wielandia elegans* Baill., *Fosberg 52202* (L). Fig. 9. *Amanoa guianensis* Aubl., *Silva 2421* (U). Fig. 10. *Alchornea schomburgkii* Klotzsch, *Ramos et al. 6337* (U). Fig. 11. *Argythamnia argothamnoides* (Bert. ex Spreng.) Ingram, *Stoffers 396* (U). Fig. 12. *Mallotus barbatus* Muell. Arg. *J. & M. S. Clemens 3618* (U). Scale bars = 10 μm.

Figures 13–18. Fig. 13. *Klaineanthus gaboniae* Pierre, *Zenker 583* (U). Fig. 14. *Euphorbia cooperi* N. E. Brown & Berger var. *ussanguensis* (N. E. Brown) Leach (K). Fig. 15. *Hevea brasiliensis* (Willd. ex Juss.) Muell. Arg., *Ducke 21958* (U). Fig. 16. *Phyllanthus maderaspatensis* L., *De Wilde 5903* (WAG). Fig. 17. *Phyllanthus caroliniensis* Walt., *Shinners 24808* (U). Fig. 18. *P. emblica* L., *Docters van Leeuwen 101* (U). Scale bars = 10 μm.

even in these cases, the pattern is certainly related to it, for example *Hevea* (Fig. 15) and *Micrandra*.

V. Subfamily Euphorbioideae (Fig. 14)

From what little has been published about the pollen of this subfamily, it can be seen to be very consistent in shape, ornamentation and size.

The published material is relatively sparse but it is nevertheless possible on the evidence of pollen morphology to support Webster's idea that the Euphorbiaceae should be divided into five subfamilies. Each subfamily is more or less characterized by a basic pollen type; the subfamily Acalyphoideae is an exception to this generalization.

SOME NOTES ON THE POLLEN MORPHOLOGY OF *PHYLLANTHUS*

Following Webster's contributions (Webster, 1970; Webster & Airy Shaw, 1971), Radcliffe-Smith has stimulated the study of *Phyllanthus* species occurring in Africa. Webster's most radical innovation in the classification of the *Phyllanthus* species was his grouping of the sections in subgenera. In 1965 he recognized eight subgenera and a ninth in 1967. The subgeneric classification is based on the stem branching pattern, pollen morphology and floral characters.

1. *Phyllanthus* subgenus *Isocladus*

This subgenus contains four sections.

Section *Isocladus*. This section probably contains the most primitive pollen types of *Phyllanthus*: 3-colporate, simply reticulate and slightly prolate (e.g. *Phyllanthus maderaspatensis*, see Fig. 16 and *P. polygonoides*).

Section *Loxopodium*. Pollen in this section is more advanced in being 4-colporate, finely reticulate to more or less punctate, distinctly prolate, but still with a simple reticulum (*P. caroliniensis*, Fig. 17). *Phyllanthus urinaria* of the subgenus *Phyllanthus* also has 4-colporate pollen, but the endoaperture is different and its ornamentation is bi-reticulate.

Section *Macraea*. Pollen of section *Macraea* is much more advanced in being pantocolporate, or syncolpate forming an areolate pattern (*P. leonensis*, see Fig. 19, and *P. virgatus*). The tectum has a simple reticulum and the pollen shape is ellipsoid. It resembles the areolate pollen grains of the subgenera *Botryanthus* and *Xylophylla*, but the position of the endoapertures is different. In subgenus *Macraea* the endoapertures are situated in the middle of the colpus, whereas in subgenera *Botryanthus* and *Xylophylla* the endoapertures are under the point of coalescence of the colpi.

Section *Ceramanthus* (synonym *Anisolobium*). Pollen grains in this section are also advanced, but unlike those of section *Macraea*, they show colpi of reduced length or sometimes pori; the reticulum closely resembles that of pollen in section *Macraea* (e.g. *P. welwitschianus*, see Fig. 22, and *P. albodiscus*).

2. *Phyllanthus* subgenus *Kirganelia*

During a study of the *Phyllanthus* species of New Guinea (Punt, 1980) and Africa (Meewis & Punt, 1983) a large survey of the pollen types of this subgenus was obtained. The African species were especially interesting for the information they gave about mutual relationships and evolutionary trends. Webster recognized three sections in this subgenus.

Section *Floribundi*. This section comprises the species: *P. nummulariifolius* (Figs 20, 26), *P. muellerianus* and *P. tenellus*. The pollen of *P. muellerianus* does not resemble that of the other two. The pollen of *P. tenellus* and *P. nummulariifolius* shows features of that of a group of African species classified under the subgenus *Phyllanthus*, including *P. hutchinsonianus*, *P. friesii*, *P. paxii*.

Section *Kirganelia* (syn. *Anisonema*). This includes *P. reticulatus*, *P. casticum* and many other representatives in Africa and Asia. Those species that occur on the continent of Africa show an interesting phylogenetic trend (*P. pinnatus*, Fig. 21, *P. engleri*, *P. tessmannii*, Fig. 23, *P. klainei* and *P. dinklagei*). See also Meewis & Punt (1983).

Section *Pentandrus*. *P. pentandrus* (Fig. 24) has pollen grains that are completely different from those of all other species in the subgenus. The pollen resembles that of the common pollen type in the subgenus *Phyllanthus*.

Brunel (1975a) gave a more extensive classification of the subgenus, but his taxonomic results were arrived at without comparing species with those from other regions. The evidence from pollen morphology sometimes conflicts with his taxonomic suggestions.

3. *Phyllanthus* subgenus *Cicca*

This small subgenus is characterized by its usually indehiscent fruits. According to Webster (1956–1958), there are several sections but only a few species. The American sections have pollen which is relatively primitive: 3-colporate, finely reticulate more or less spheroidal in shape and with circular endopori. The best known species is *P. acidus* from the section *Cicca*. Other sections are section *Ciccopsis*, with one species endemic to Cuba and section *Aporosella* with two species (*P. elsiae*, Fig. 25, and *P. chocoensis*).

Section *Emblica*, with the species *P. emblica* (Fig. 18), is from Asia and shows more advanced pollen grains: its pollen is 4–5-colporate, slight oblate but also finely reticulate and with a circular endoporus. The species is certainly related to the American taxa.

4. *Phyllanthus* subgenus *Phyllanthus*

This is the largest subgenus and contains the majority of the species. Its sections are often subdivided into subsections (Webster, 1956–1958; Brunel, 1975a).

The pollen grains of most of the species are well characterized and show the following features: 3-colporate, more or less prolate, distinctly bi-reticulate (Bor, 1979) and with a distinct endocolpus. More advanced pollen types do, of course, occur but they are clearly derived from the type described above. For example, in section *Urinaria* (Fig. 30), the pollen is 4(–5)-colporate with a circular endoporus. Examples of section *Phyllanthus* are *P. phillyreifolius*, *P. debilis*, *P. amarus*, etc. In Africa all the species of subsection *Swartziani* (e.g. *P. trichotepalus*, Fig. 31) have the basic pollen type, but subsection *Niruri* (Brunel, 1975b) with the majority of the American species, has 4- or 5-colporate pollen.

Some African species of subsection *Phyllanthus*, which is heterogeneous, show quite spectacular pollen grains (e.g. *P. paxii* (Figs 27 & 28), *P. holostylus*,

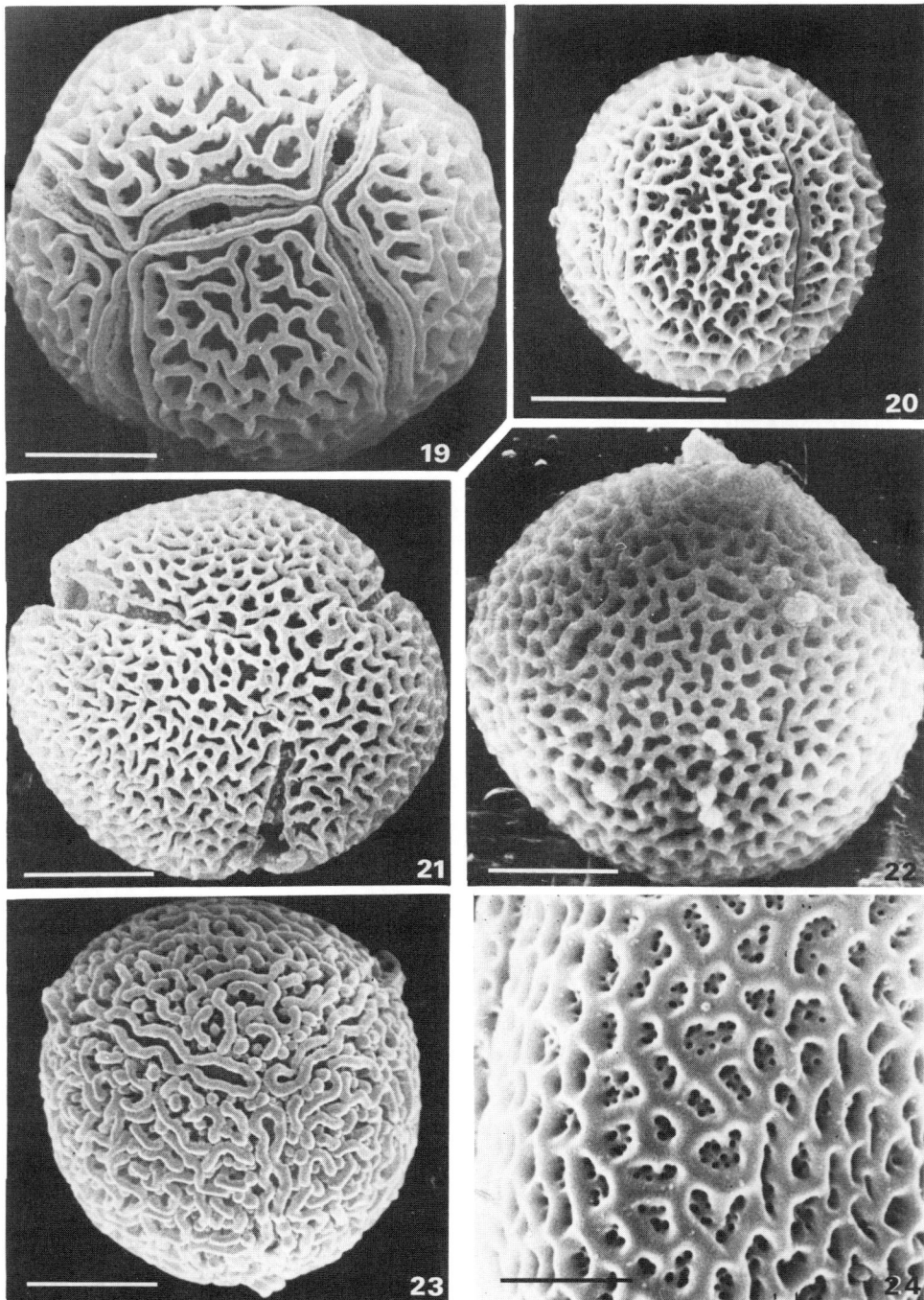

Figures 19–24. Fig. 19. *P. leonensis* Hutch., *De Wilde 888* (WAG). Fig. 20. *P. nummulariifolius* Poir., *De Witte 10623* (WAG). Fig. 21. *P. pinnatus* (Wight) Webster, *Polhill & Paulo 685* (K). Fig. 22. *P. welwitschianus* Muell. Arg., *Milne-Redhead 8848* (BR). Fig. 23. *P. tessmannii* Hertel, *Tessmann 710* (K). Fig. 24. *P. pentandrus* Schum. et Thonn., *Kotschy 303* (BR). Figs 19–23, scale bars = 10 μm; Fig. 24, scale bar = 2 μm.

P. hutchinsonianus and *P. friesii*). I am certain that they belong to a subsection of their own, but it is clear that in this section much taxonomic work has still to be done.

Some other sections of this subgenus, such as *Phyllanthodendron* and *Paraphyllanthus* are restricted to Asia and their pollen grains have not yet been properly studied.

In South America, section *Microglochidion* is well represented but its species show a great variety of pollen types. Many of these species are described by Jablonski (1967) from the Guyana Highlands and they are divided into 4 groups of which only the *Adianthoides* group was given a name. Examples of these species are: *P. vacciniifolius* (Fig. 29), *P. major*, *P. francavillanus*, *P. adenophyllus*, and the very interesting species *P. fluitans* which is one of the only two aquatic *Phyllanthus* species in the world, and the only free-floating one. It is a pity that this South American group of species has not yet been completely revised, but since the work of Jablonski (1967) nobody seems to have taken any further interest.

5. *Phyllanthus* subgenus *Eriococcus*

This is a well-circumscribed subgenus of the Palaeotropics. Its representatives show distinct, very advanced pollen of a single type with the following features: pantoporate, coarsely reticulate, pores form part of one lumen of the reticulum (*P. buxifolius*, *P. lamprophyllus*, Fig. 32).

6. *Phyllanthus* subgenus *Conami*

This subgenus is restricted to the Neotropics. Webster (1956–1958) divided it into two sections, *Nothoclema* and *Apolepis*.

Section *Apolepis* consists of a single species, *P. orbiculatus* (Fig. 33). Its pollen has no ectoapertures and its ornamentation consists of simple elements (pilate). There are endoapertures which are arranged in a zone.

Section *Nothoclema* has a considerable number of species in South America and their pollen is also inaperturate and pilate, but there are more endoapertures and they are usually arranged at random (e.g. *P. acuminatus*, Fig. 35, and *P. brasiliensis*).

7 and 8. *Phyllanthus* subgenus *Botryanthus* and subgenus *Xylophylla*

These two subgenera are undoubtedly related to one another (Webster, 1956–1958), but differ principally in the branching of the stem. Species of subgenus *Botryanthus* are not specialized in their branching, whereas those in subgenus *Xylophylla* have a phyllanthoid branching pattern (Webster, 1956–1958). However, both subgenera show the highly specialized 'areolate' pollen type; e.g. *P. latifolius* (Fig. 34), *P. salviifolius* and *P. mimisoides*. Their areolate pollen has syncolpate ectoapertures and the endoapertures are situated under the points of coalescence of the colpi. This kind of pollen resembles that in subgenus *Macraea* although pollen of this subgenus has endoapertures in the middle of the ectocolpi.

The problem of the presence of identical pollen in subgenera which have a major difference in their stem branching can only be resolved when other lines of evidence are examined.

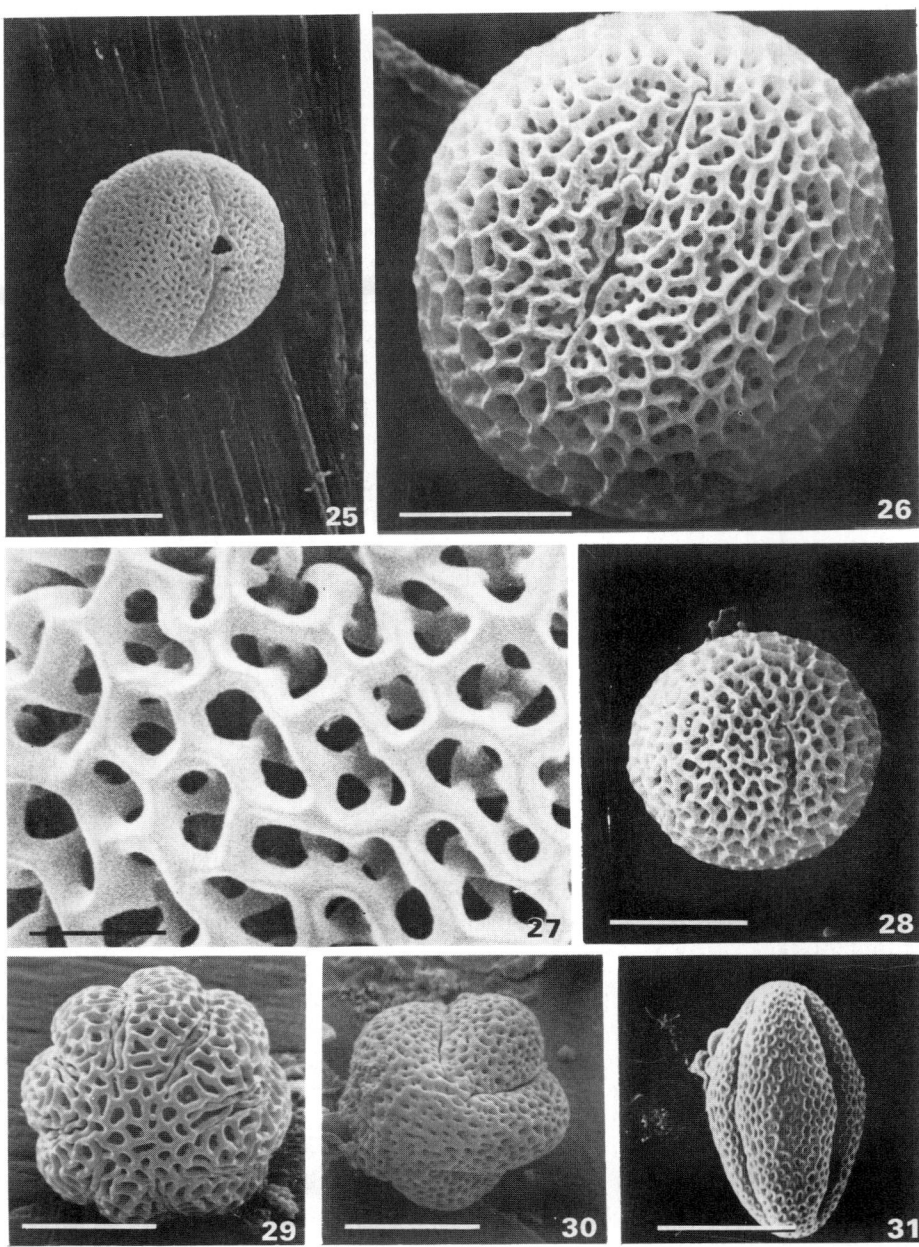

Figures 25–31. Fig. 25. *P. elsiae* Urban, *Geyskes 154* (U), scale bar = 10 μm. Fig. 26. *P. nummulariifolius* Poir., *Morton s. n.* (WAG), scale bar = 5 μm. Fig. 27. *P. paxii* Hutch., *Lewalle 6582* (K), scale bar = 2 μm. Fig. 28. *P. paxii* Hutch., *Lewalle 6582* (K), scale bar = 10 μm. Fig. 29. *P. vacciniifolius* Muell. Arg., *Sandwith 1315* (U), scale bar = 10 μm. Fig. 30. *P. urinaria* L., *Hoogerwerf 252* (L), scale bar = 10 μm. Fig. 31. *P. trichotepalus* Brenan, *Lewalle 2460* (K), scale bar = 10 μm.

POLLEN MORPHOLOGY IN EUPHORBIACEAE 139

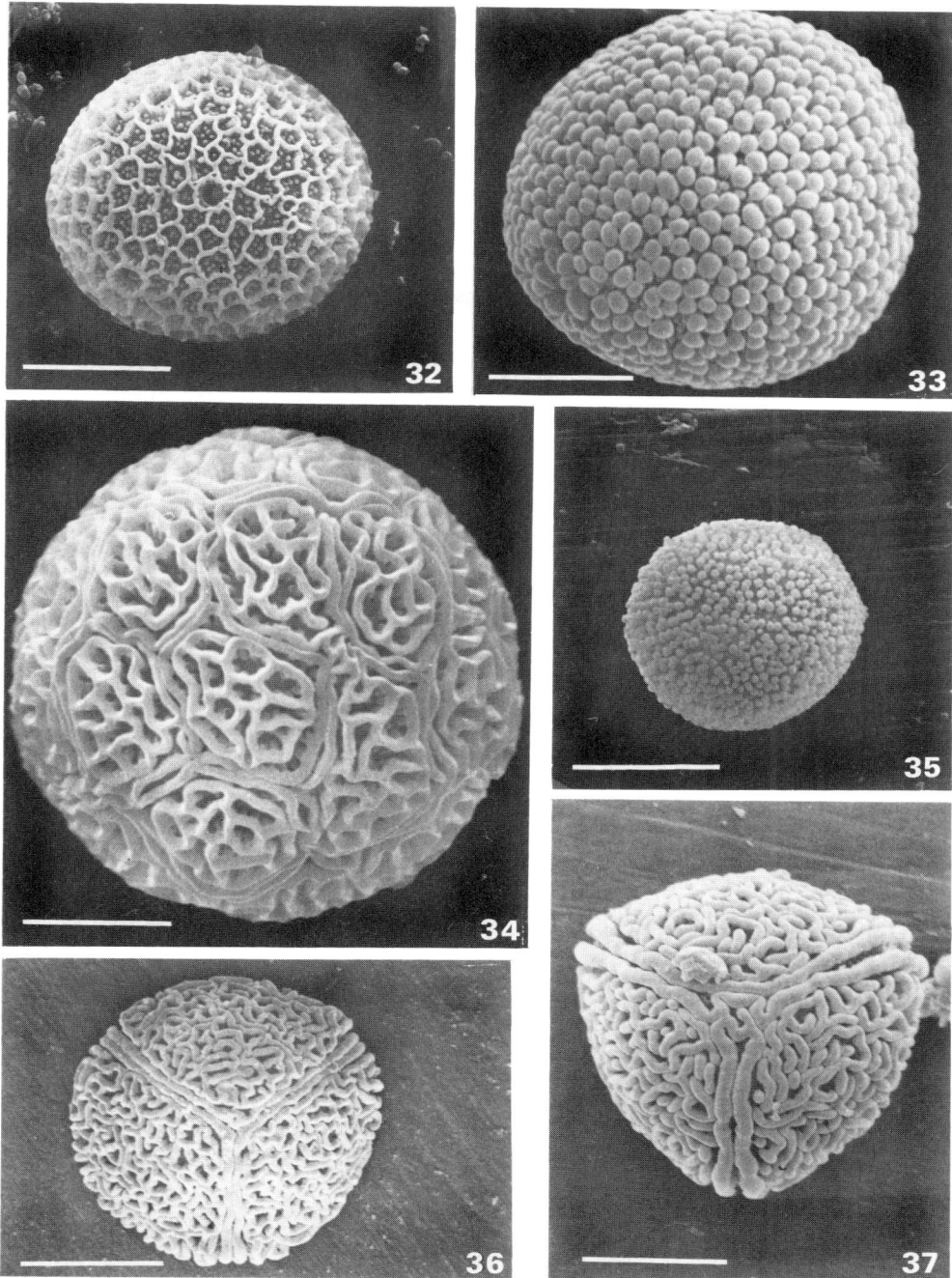

Figures 32–37. Fig. 32. *P. lamprophyllus* Muell. Arg., *Pleyte 52* (L). Fig. 33. *P. orbiculatus* L. C. Richard, *Prance et al. 15852* (U). Fig. 34. *P. latifolius* Sw., *Webster et al. 4875* (U). Fig. 35. *P. acuminatus* Vahl, *Archer 2355* (U). Fig. 36. *P. tenuirhachis* J. J. Smith, *Cult. Hort. Bog. VII. E. 14* (L). Fig. 37. *P. merinthopodus* Diels, *Ridsdale, n.g.f. 3511* (K). Scale bars = 10 μm.

9. *Phyllanthus* subgenus *Gomphidium*

This polymorphic subgenus, described by Webster (1956–1958), is confined to the Old World. The majority of the species are endemic to New Caledonia with smaller numbers in Australia, New Guinea, Fiji and Madagascar. The sectional boundaries are still poorly defined but the pollen is well characterized by a number of features such as the oblate shape, marginated colpi and often triangular polar view. The ornamentation is usually distinctly reticulate (e.g. *P. tenuirhachis*, Fig. 36) and sometimes vermiculate (e.g. *P. merinthopodus*, Fig. 37). Two pollen types have been recognized (Punt, 1980) which differ mainly in the length of the colpi. In the *P. acinacifolius* type the colpi do not anastomose at the poles and the reticulum is slightly irregular (e.g. *P. flaviflorus*, *P. bourgeoisii*). In the *P. aeneus* type the colpi are more or less anastomosed at the poles and the reticulum is very irregular.

CONCLUSION

The genus *Phyllanthus* shows much variation in pollen types. We know a reasonable amount about these pollen types and are beginning to find out their phylogenetical significance. However, there is a great deal of work still to be done, especially in South American and Australian species.

ACKNOWLEDGEMENTS

The author is grateful to Dr R. J. W. M. Van Der Ham and Dr W. Berendsen for their technical assistance with the SEM; the directors of the Rijksherbarium, Leiden, State University, Leiden, and the Electron-microscopical Structure Analysis Section of the Department of Botany, State University, Utrecht for their hospitality in using the SEM. He wishes to express his sincere thanks to Dr G. C. S. Clarke for his critical reading of the manuscript.

REFERENCES

AIRY SHAW, H. K., 1965. Diagnoses of new families, new names etc., for the seventh edition of Willis's "Dictionary". *Kew Bulletin, 18:* 249–273.
AIRY SHAW, H. K., 1980. The Euphorbiaceae of New Guinea. *Kew Bulletin Additional Series, 8:* 1–253.
BAILLON, H., 1858. *Étude générale du groupe des Euphorbiacées*. Paris: Victor Masson.
BENTHAM, G. & HOOKER, J. D., 1880. *Genera Plantarum, 3:* 239–340.
BLASCO, F., CARATINI, C., FREDOUX, A., GIRESSE, P., MOGUEDET, G., TISSOT, C. & WEISS, H., 1980. *Mangroves d'afrique et d'Asie. Travaux Documents Géographie Tropicale no 39*. Bordeaux: C.N.R.S.
BONNEFILLE, R., 1971a. Atlas des pollens d'Éthiope. Principales espéces des forêts de montagne. *Pollen et Spores, 13:* 15–72.
BONNEFILLE, R., 1971b. Atlas des pollens d'Éthiope. Pollens actuels de la basse vallée de l'Omo, récoltes botaniques 1968. *Adansonia sér. 2, 11:* 463–518.
BONNEFILLE, R. & RIOLLET, G., 1980. *Pollens des Savannes d'Afrique Orientale*. Paris: C.N.R.S.
BOR, J., 1979. Pollen morphology and the bi-reticulate exine of the *Phyllanthus* species (Euphorbiaceae) from Mauritius and Reunion. *Review of Palaeobotany and Palynology, 27:* 149–172.
BRUNEL, J. F., 1975a. *Contribution à l'étude de quelques* Phyllanthus *africaines et à la taxonomie du genre* Phyllanthus *L. (Euphorbiaceae)*. Strasbourg. Thèse 3e cycle, U.L.P.
BRUNEL, J. F., 1975b. Notes sur les Phyllanthoideae (Euphorbiaceae) ouest-africaines. I. *Phyllanthus niruri* L., une espèce à exclure des flores de l'Afrique occidentale. *Bulletin, Société botanique de France, 122:* 153–162.
BRUNEL, J. F. & ROUX, J., 1976. Notes sur les Phyllanthoideae (Euphorbiaceae) ouest-africaines. II. A propos du polymorphisme du *Phyllanthus sublanatus* Schum. et Thonn. *Bulletin, Société botanique de France, 123:* 365–375.

BRUNEL, J. F. & ROUX, J., 1977. Notes sur les Phyllanthoideae (Euphorbiaceae) ouest-africaines. III. A propos de la position systématique du *Phyllanthus dinklagei* Pax. *Bulletin, Société botanique de France, 124:* 217–225.

BRUNEL, J. F. & ROUX, J., 1980. Phyllantheae de Madagascar. I: à propos de deux *Phyllanthus* L. de la sous-section *Swartziani* Webster. *Adansonia, sér. 2, 20:* 393–403.

BRUNEL, J. F. & ROUX, J., 1981a. Phyllantheae de Madagascar. II: à propos du "complex" *Phyllanthus nummulariaefolius* Poir.—*Phyllanthus tenellus* Roxb. *Bulletin du Muséum National d'Histoire Naturelle, 4e sér., 3:* 185–199.

BRUNEL, J. F. & ROUX, J., 1981b. *Phyllanthus* subsect. *Odontadenii* (Euphorbiaceae) au bord du fleuve Congo (Afrique de l'Ouest). *Willdenowia, 11:* 69–90.

BRUNEL, J. F. & ROUX, J., 1982. Phyllantheae de Madagascar. III: Une nouvelle espèce du genre *Meineckia* Baillon. *Bulletin du Muséum National d'Histoire Naturelle, 4e sér., 4:* 79–84.

BRUNEL, J. F. & ROUX, J., 1984. South-east Asian Phyllantheae II. Some *Phyllanthus* of subsect. *Swartziani*. *Nordic Journal of Botany, 4:* 469–473.

CARATINI, C. & GUINET, P., 1974. *Pollen et Spores d'Afrique Tropicale. Travaux Documents Géographie tropicale no. 16.* Bordeaux: C.N.R.S.

DEHGAN, B. & WEBSTER, G., 1979. Morphology and infrageneric relationships of the genus *Jatropha* (Euphorbiaceae). *University of California Publications in Botany, 74:* 1–73.

DIAZ ZAVALETA, G. & PALACIOS CHAVEZ, R., 1980. Contribucion al conocimiento de la morfologia de los granos de polen de los generos mas comunes de la familia Euphorbiaceae de Mexico. *Boletin de la Sociedad Botanica de Mexico, 39:* 25–62.

ERDTMAN, G., 1952. *Pollen morphology and plant taxonomy, Angiosperms.* Stockholm: Almqvist and Wiksell.

HAMILTON, A. C., 1976. Identification of East African Urticales pollen. *Pollen et Spores, 18:* 27–66.

HEUSSER, C. J., 1971. *Pollen and Spores of Chile.* Tucson: University of Arizona Press.

HUANG, T. C., 1972. *Pollen Flora of Taiwan.* Taipei, Taiwan: Ching-Hwa Press.

HUTCHINSON, J., 1969. Tribalism in the family Euphorbiaceae. *American Journal of Botany, 56:* 738–758.

JABLONSKI, E., 1967. Botany of the Guyana Highland – Part VII. Euphorbiaceae. *Memoirs of the New York Botanical Garden, 17:* 80–190.

KHAN, H. A., 1968. Contributions to the pollen morphology of the Euphorbiaceae. *Journal of Palynology, 4:* 21–35.

KÖHLER, E., 1965. Die Pollenmorphologie der biovulaten Euphorbiaceae und ihre Bedeutung für die Taxonomie. *Grana Palynologica, 6:* 26–120.

KÖHLER, E., 1967. Über Beziehungen zwischen Pollenmorphologie und polyploidiestufen im Verwandtschaftsbereich der Gattung *Phyllanthus* (Euphorbiaceae). *Feddes Repertorium Specierum Novarum Regni Vegetabilis, 74:* 159–165.

LONG HUO & YU CHENG-HONG, 1984. Intrageneric variation of pollen types in the genus *Phyllanthus* L. *Acta Botanica Sinica, 26:* 247–251.

LYNCH, S. & WEBSTER, G., 1975. A new technique of preparing pollen for scanning electron microscopy. *Grana Palynologica, 15:* 127–136.

MARKGRAF, V. & D'ANTONI, H. L., 1978. *Pollen Flora of Argentina.* Tucson: University of Arizona Press.

MARTICORENA, C., 1962. Morfologia de los granos de polen Euphorbiaceae, Melphigiaceae Chilenas. *Gayana, 5:* 3–12.

MARTIN, H., 1974. The identification of some tertiary pollen belonging to the family Euphorbiaceae. *Australian Journal of Botany, 22:* 271–291.

MEEWIS, B. & PUNT, W., 1983. Pollen morphology and taxonomy of the subgenus *Kirganelia* (Jussieu) Webster (genus *Phyllanthus*, Euphorbiaceae) from Africa. *Review of Palaeobotany and Palynology, 39:* 131–160.

MILLER, K. I. & WEBSTER, G. L., 1967. A preliminary revision of *Tragia* (Euphorbiaceae) in the United States. *Rhodora, 69:* 241–305.

MOSANGO, M. & ROBBRECHT, E., 1985. In R. Dechamps, M. Mosango & E. Robbrecht. Études systématiques sur les Hymenocardiaceae d'Afrique: la morphologie du pollen et l'anatomie du bois. *Bulletin du Jardin Botanique National de Belgique, 55:* 473–485.

MUELLER, J. & BOISSIER, E. 1866. Euphorbiaceae. In A. P. De Candolle (Ed.), *Prodromus Systematis Naturalis Regni Vegetabilis, 15:* 189–1261.

PAX, F. & HOFFMANN, K., 1931. Euphorbiaceae. In A. Engler und K. Prantl (Eds), *Die Natürlichen Pflanzenfamilien. 19c:* 11–233.

POOLE, M. M., 1981. Pollen diversity in *Zimmermannia* (Euphorbiaceae). *Kew Bulletin, 36:* 129–138.

PUNT, W., 1962. Pollen morphology of the Euphorbiaceae with special reference to taxonomy. *Wentia, 7:* 1–116.

PUNT, W., 1967. Pollen morphology of the genus *Phyllanthus* (Euphorbiaceae). *Review of Palaeobotany and Palynology, 3:* 141–150.

PUNT, W., 1972. Pollen morphology and taxonomy of section *Ceramanthus* Baillon s.l. of the genus *Phyllanthus*. *Review of Palaeobotany and Palynology, 13:* 213–228.

PUNT, W., 1980. Pollen morphology of the *Phyllanthus* species (Euphorbiaceae) occurring in New Guinea. *Review of Palaeobotany and Palynology, 31:* 155–177.

ROUX, J., 1984. *Phyllanthus brunelii* sp. nov. (Euphorbiaceae) from Thailand. *Nordic Journal of Botany, 4:* 47–51.

SALGADO-LABOURIAU, M. L., 1971. *Contribuição a palinologia dos cerrados.* São Paulo, Academia Brasileira de Ciencias.

SYKES, W. R., 1969. *Homalanthus* in New Zealand. *New Zealand Journal of Botany, 7:* 302–307.

THANIKAIMONI, G., CARATINI, C., NILSSON, S. & GRAFSTRÖM, E., 1984. Omniaperturate Euphorbiaceae pollen with striate spines. *Bulletin du Jardin Botanique National de Belgique, 54:* 105–125.

WEBER-EL GHOBARY, M. O., 1985. Pollen morphology of four succulent species of *Euphorbia* (Euphorbiaceae). *Anales de la Asociación de Palinólogos de Lengua Española, 2:* 75–86.

WEBSTER, G. L., 1956–1958. A monographic study of the West Indian species of *Phyllanthus*. *Journal of the Arnold Arboretum, 37, 38 & 39.*

WEBSTER, G. L., 1967. The genera of Euphorbiaceae in the Southeastern United States. *Journal of the Arnold Arboretum, 48:* 303–430.

WEBSTER, G. L., 1970. A revision of *Phyllanthus* (Euphorbiaceae) in the continental United States. *Brittonia, 22:* 44–76.

WEBSTER, G. L., 1975. Conspectus of a new classification of the Euphorbiaceae. *Taxon, 24:* 593–601.

WEBSTER, G. L. & AIRY SHAW, H. K., 1971. A provisional synopsis of the New Guinea taxa of *Phyllanthus* (Euphorbiaceae). *Kew Bulletin, 26:* 85–109.

WEBSTER, G. L. & WEBSTER, B. D., 1972. The morphology and relationships of *Dalechampia scandens* (Euphorbiaceae). *American Journal of Botany, 59:* 573–586.

YBERT, J.-P., 1975. Observation du pollen de *Manihot* sp. (Euphorbiaceae) au microscope électronique à balayage. *Bulletin, Société Botanique de France, 122, Colloque Palynologie:* 131–133.

Laticifers in Euphorbiaceae—a conspectus

PAULA J. RUDALL

Jodrell Laboratory, Royal Botanic Gardens, Kew, Richmond, Surrey TW9 3AE

Received June 1986, accepted for publication September 1986

RUDALL, PAULA, 1987. **Laticifers in Euphorbiaceae—a conspectus.** Despite the multitude of references on laticifers in Euphorbiaceae, much of the work concentrated on economically important, and often highly specialized genera, particularly *Euphorbia* and *Hevea*, the evolution of the laticifer in the family is poorly understood. In this paper, laticifers, in a representative sample of genera of Euphorbiaceae, are investigated with respect to morphology and distribution. This information is supplemented with a review of existing literature. It seems probable that the articulated and non-articulated types are not as divergent as commonly supposed, since both may undergo intrusive growth, although more ontogenetic studies are needed to clarify the relationship between them.

ADDITIONAL KEY WORD:—Taxonomy.

CONTENTS

Introduction	143
Material and methods	145
Observations and discussion	145
Non-articulated laticifers	149
Articulated laticifers	156
Function of laticifers	158
Taxonomic and evolutionary considerations	159
Acknowledgements	161
References	161

INTRODUCTION

The term laticifer is defined as "a specialized cell or a row of cells containing latex" (Fahn, 1979), although latex may also be present in unspecialized parenchyma cells. Laticifers are known to occur in about 20 often unrelated plant families, including both Monocotyledons and Dicotyledons (Metcalfe, 1967), indicating that they are polyphyletic in origin. These cell types are usually considered to be highly advanced. Mahlberg, Field & Frye (1984) found fossil laticifers in Eocene brown coal deposits.

In the large cosmopolitan family Euphorbiaceae, laticifers occur in many genera (see Table 2). However, an examination of the vast literature on laticifers in this family reveals that whereas extensive investigations have been made on the economically important genera *Euphorbia*, *Jatropha*, *Hevea* and, to a lesser extent, *Manihot*, relatively little is known about laticifers in other genera of

Euphorbiaceae, many of which are underexploited. Furthermore, investigations have suffered from a confusion of terminology and interpretation of laticifer types (Table 1), with the result that some of the very early work is often overlooked.

The classification of laticifers that is now most widely used (De Bary, 1884) is based on structure and development. Articulated laticifers consist of chains of cells whose adjoining walls may sometimes break down, forming tubes or vessels. Non-articulated laticifers arise from single initial cells at an early stage in seedling development. The cells grow intrusively into intercellular air spaces between neighbouring cells, by wall elongation and successive mitoses unaccompanied by cytokinesis. Many authors (e.g. Mahlberg, 1959) have likened the non-articulated laticifers to a parasitic cell or fungal hypha ramifying throughout the plant tissue. Both articulated and non-articulated types may be anastomosing (branched) or non-anastomosing (unbranched), but in Euphorbiaceae non-articulated laticifers are generally branched.

The most extensive review of laticifers in Euphorbiaceae was that of Pax (1884), who used both anatomical and morphological information for his subsequent classification of the family (Pax & Hoffman, 1931; see Table 2). However, other authors (Scott, 1886; Solereder, 1908) have since disputed his interpretation of articulated laticifers. Metcalfe (1967, 1985), in reviewing the literature on laticifers in general, suggested that the occurrence of two different types (articulated and non-articulated) in the Euphorbiaceae, is not surprising in view of the fact that the family is very large and possibly polyphyletic, implying that the two types are of different origin. However Kakkar & Paliwal (1972) and Dehgan & Craig (1978) noted both articulated and non-articulated types of laticifer in *Jatropha*, whereas most previous observers had recorded only non-articulated laticifers. The discovery is significant as it is the only undisputed record in Euphorbiaceae of the two types occurring in the same genus, and indeed the same plant. Dehgan & Craig (1978) suggested that a re-examination of other genera, particularly in the Crotonoideae, would be useful in establishing evolutionary relationships within the family.

Table 1. Summary of terms used for laticifer types

Author	Articulated laticifers	Non-articulated laticifers
de Bary, 1884	Articulated laticiferous tubes	Non-articulated laticiferous tubes
Pax, 1884	Articulated laticiferous vessels ("gegliederte Milchröhren") or articulated sacs ("gegliederte schläuche"), with or without partial resorption of walls	Non-articulated or inarticulated laticiferous tubes ("Ungegliederte Milchröhren")
Solereder, 1908	Laticiferous vessels	Laticiferous cells
Metcalfe & Chalk, 1950	Laticiferous vessels, Unsegmented vessels	Laticiferous cells, Laticiferous canals, Latex tubes
Metcalfe, 1967	Articulated laticifers	Non-articulated laticifers, Unsegmented laticifers
Fahn, 1979 & others	Articulated laticifers: anastomosing or non-anastomosing (or laticiferous vessels)	Non-articulated laticifers: branched or unbranched (or laticiferous cells)

The aim of this investigation is to outline the distribution of laticifers in the Euphorbiaceae, both by a review of existing literature, and by new observations on material available. Recent interest in non-articulated laticifers has focussed on the large genus *Euphorbia* and its close relatives in the tribe Euphorbieae (Table 2), particularly on the taxonomic and evolutionary aspects of laticifer starch grain morphology (e.g. Mahlberg, 1975, 1982; Biesboer & Mahlberg, 1981) and laticifer differentiation (e.g. Mahlberg & Sabharwal, 1968; Bruni, Vannini & Fasulo, 1978; Fineran, 1983). Apart from the records of Pax (1884), Dehgan & Craig (1978) and the present paper, observations on non-articulated laticifers in other genera in the family are relatively sparse, and often consist of purely descriptive accounts for pharmacognostic purposes (e.g. López Guillén & Kiyán de Cornelio, 1974). Few authors since Solereder (1908) have attempted to relate laticifer distribution to classification within the entire family. It is hoped that this information will contribute to taxonomic and evolutionary applications of laticifer morphology and will be useful to economic botanists and biochemists in selecting new material for research.

MATERIAL AND METHODS

Most of the material examined was obtained from the living collections, Royal Botanic Gardens, Kew. Nearly all plants with laticifers exude a milky liquid at the broken end of the stem or leaf, although some (e.g. *Aleurites*) exude a thick, clear, viscous latex. Living material was fixed in FAA and stored in 70% alcohol. In cases where no living material was available to represent a particular tribe (using the system of Pax & Hoffman, 1931), dried material from the Kew Herbarium was used, although the preservation of dried material is usually inadequate for the critical study of laticifers. Stems and leaves were sectioned using a Reichert sliding microtome, stained with safranin and Alcian blue, and mounted in Euparal after dehydration through an alcohol series. Other sections were stained in chlor–zinc–iodine to test for the presence of starch granules, and mounted in glycerol. For observation of apical meristems, material of a few species was embedded in Paraplast, sectioned using a Reichert rotary microtome, stained in safranin and Alcian blue, and mounted in Canada Balsam after dehydration through an alcohol series.

Microscope slides already present in the collection at the Jodrell Laboratory at the onset of this investigation, many of them prepared for examination for Metcalfe & Chalk (1950) and Metcalfe (1967) using standard techniques, were also examined for this work. However, in some slides the stains used (safranin and haemotoxylin) had faded or deteriorated so that it was impossible to determine whether laticifers were present.

Photographs were taken using a Leitz Dialux 20 photomicroscope and M 400 photomacroscope.

OBSERVATIONS AND DISCUSSION

Table 2 presents a summary of present and previous observations on laticifers in Euphorbiaceae. The genera are arranged according to Pax & Hoffman (1931), since this is based to some extent on anatomical characters, but other systems are also discussed. The literature includes most of the works relevant to

Table 2. Occurrence of laticifers in genera of Euphorbiaceae, arranged according to Pax & Hoffmann, 1931. (Asterisk indicates material examined in this investigation.)

Subfamily/tribe	Genera	Observations
(Series Platylobeae) A. PHYLLANTHOIDEAE		
1. Phyllantheae	*Andrachne* L. *Aporusa* Bl. *Breynia* J. R. & G. Forst. *Drypetes* Vahl *Hieronima* Allem. *Hyaenanche* Lamb. & Vahl *Phyllanthus* L. *Securinega* Comm. ex Juss.	Laticifers absent; except one record of laticifers in wood rays in *Hieronoma* (Heimsch, 1942), not confirmed here or by Mennega (pers. comm.)
2. Bridelieae	*Bridelia* Spreng.	
B. CROTONOIDEAE		
3. Crotoneae	*Croton* L. *Crotonopsis* Michx. *Julocroton* Mart.	Non-articulated laticifers present (Léandri, 1939; Pax, 1884; Solereder, 1908)
4. Chrozophoreae	*Aleurites* J. R. & G. Forst.	Laticifers present (Belin-Depoux & Clair-Maczulajtys, 1974, 1975; Solereder, 1908). "Articulated sacs" recorded by Pax (1884)
	Garcia Rohr	"Articulated sacs" only, recorded by Pax (1884)
5. Joannesieae	*Hevea* Aubl.	Articulated laticifers present (Ashplant, 1928; Boblioff, 1923; Bryce & Campbell, 1917; Calvert, 1887; Fahn, 1979; Frey-Wyssling, 1931; Gomez & Thai, 1967; Gunnery, 1935; Lleras & Medri, 1978; Medri & Lleras, 1980; Metcalfe, 1967; Petch, 1911; Quisumbing, 1927; Scott, 1884b, 1886; Solereder, 1908; Spencer, 1939; Tranchard, 1979)
	Joannesia Vell.	Non-articulated laticifers recorded by Solereder (1908), Pax (1884)
6. Acalypheae	*Acalypha* L.	Non-articulated laticifers recorded by Metcalfe & Chalk (1950), but not seen in three species examined here
	Macaranga Thou.	Non-articulated laticifers recorded by Solereder (1908)
	Mallotus Lour. *Mareya* Baill.	Laticifers absent ("articulated sacs" recorded by Pax 1884)
	Mercurialis L.	Non-articulated laticifers recorded by Solereder (1908) and "unusual laticiferous cells" recorded by Bergfeld-Gaertner (1964); neither confirmed here
	Plukenetia L. *Ricinus* L.	Non-articulated laticifers recorded by Solereder (1908) but not confirmed here ("articulated sacs" recorded by Pax, 1884)
7. Pachystromatae	*Pachystroma* (Klotzsch) Muell. Arg.	Non-articulated laticifers present

Subfamily/tribe	Genera	Observations
8. Dalechampieae	*Dalechampia L.	Laticifers absent, although recorded in wood rays (Heimsch, 1942; Mennega, pers. comm.)
9. Pereae	*Pera Mutis	Laticifers absent, although recorded in wood rays (Mennega, pers. comm.)
10. Cluytieae	*Baloghia Endl. Clutia Ait.	Non-articulated laticifers present Non-articulated laticifers present (Solereder, 1908)
	*Cnidoscolus Pohl.	Articulated laticifers present
	*Codiaeum A. Juss.	Non-articulated laticifers present (Pax, 1884; Gaucher, 1902; Rao & Tewari, 1960)
	*Jatropha L.	Articulated, non-articulated and idioblastic laticifers present (Cass, 1968; de Castells, Ormond & Braconi, 1984; Dehgan & Craig, 1978; Kakkar & Paliwal, 1972; Pax, 1884; Rao & Malaviya, 1964; Solereder, 1908)
	Ostodes Bl.	Non-articulated laticifers present (Solereder, 1908)
	Trigonostemon Bl.	Non-articulated laticifers present (Pax, 1884)
11. Manihoteae	*Manihot Mill.	Articulated laticifers present (Calvert & Boodle, 1887; de Mendonça, 1982, 1983a; Scott, 1884a; Solereder, 1908; Tobler, 1920); also non-articulated laticifers reported by Pax (1884)
12. Gelonieae	Adenocline Turcz.	Non-articulated laticifers present (Solereder, 1908)
	*Chaetocarpus Thw.	Laticifers absent
	Cunuria Boull.	Laticifers present in wood rays (Mennega, pers. comm.; Record & Hess, 1943)
	Micrandra Benth.	"Rows of laticiferous sacs" recorded by Solereder (1908)
	Nealchornea Huber	Laticifers present in wood rays (Record & Hess, 1943)
13. Hippomaneae	Actinostemon Mart. ex Klotzsch.	Non-articulated laticifers present (de Mendonça, 1983b; Solereder, 1908)
	Adenopeltis Bert. ex A. Juss.	Non-articulated laticifers present (Pax, 1884); Solereder, 1908)
	Colliguaja Molina	Non-articulated laticifers present (Solereder, 1908; Summers, 1910)
	*Excoecaria L.	Non-articulated laticifers present (Pax, 1884; Santos, 1932; Solereder, 1908)
	Hippomane L.	Non-articulated laticifers present (Solereder, 1908); also in wood rays (Heimsch, 1942; Record & Hess, 1943)

Table 2. Continued

Subfamily/tribe	Genera	Observations
	*Homalanthus A. Juss.	Non-articulated laticifers present
	Hura L.	Non-articulated laticifers present (Pax, 1884; Solereder, 1908)
	*Mabea Aubl.	Non-articulated laticifers present (Lopéz Guillen & Kiyán de Cornelio, 1974; Pax, 1884; Solereder, 1908). Also in wood rays (Mennega, pers. comm.)
	Maprounea Aubl.	Non-articulated laticifers present (Pax, 1884; Solereder, 1908)
	Sapium P. Br.	Non-articulated laticifers present (Groom, 1889); also in wood rays (Mennega, pers. comm.)
	Sebastiania Spreng.	Non-articulated laticifers present (Pax, 1884; Solereder, 1908). Also in wood rays (Mennega, pers. comm.)
	Senefeldera Mart.	Non-articulated laticifers present (Solereder, 1908)
	Stillingia Garden ex L.	Non-articulated laticifers present (Holm, 1911; Pax, 1884; Solereder, 1908)
14. Euphorbieae	Anthostemma Juss.	Non-articulated laticifers present (Solereder, 1908)
	*Elaeophorbia Stapf	Non-articulated laticifers present; also in wood rays (Mennega, pers. comm.)
	*Endadenium Leach	Non-articulated laticifers present
	*Euphorbia L.	Non-articulated laticifers present (Biesboer & Mahlberg, 1981; Carlquist, 1970; Datta, Datta & Roy, 1984; de Bary, 1884; Fineran, 1982, 1983; Heimsch, 1942; Jimenez & Caballero, 1978; Lommasson, 1962; Mahlberg, 1975, 1982; Mahlberg & Sabharwal, 1968; Milanez, 1952a, b, 1954; Milanez & Machado, 1956; Milanez & Neto, 1956; Paliwal & Kakkar, 1969; Rao, Menon & Malaviya, 1968; Rosowski, 1968; Uhlarz & Kunschert, 1975)
	*Monadenium Pax	Non-articulated laticifers present (Biesboer & Mahlberg, 1981)
	*Pedilanthus Neck. ex Poit.	Non-articulated laticifers present (Biesboer & Mahlberg, 1981; Dressler, 1957)
	*Synadenium Boiss.	Non-articulated laticifers present (Biesboer & Mahlberg, 1981; Solereder, 1908)
(Series Stenolobeae) C. PORANTHOIDEAE		
15. Poranthereae	Poranthera Rudge	Laticifers absent (Pax, 1884)
16. Caletieae	Caletia Baill. Pseudanthus Sieb. ex Spreng. Stachystemon Planch.	

Table 2. Continued

Subfamily/tribe	Genera	Observations
D. RICINOCARPOIDEAE		
17. Riconocarpeae	*Ricinocarpus* Desf.	
	Bertya Planch.	
	Beyeria Miq.	Laticifers absent (Pax, 1884)
18. Ampereae	*Amperea* A. Juss.	
	Monotaxis Brongn.	

the present investigation, but it would be impractical to list here all the numerous references on the subject. Much of the earlier literature has been summarized by Metcalfe (1967) and Solereder (1908), and some of the less accessible works are therefore not given here. I have included a representative sample of the genera observed by Pax (1884) in his major review of laticifers in Euphorbiaceae, but have not listed all the many genera in the Crotonoideae (in Webster's Acalyphoideae) which he recorded as having "articulated sacs".

Selected genera in the subfamily Phyllanthoideae, where laticifers are absent, were examined for the purposes of this survey. However, in general I have left out references to absence of laticifers, except for a few in the subfamily Crotonoideae (Table 2).

Non-articulated laticifers

Morphology and distribution: Non-articulated laticifers are typically branching tubes with thin smooth walls and no dividing transverse walls (Figs 1, 4, 9, 10, 13). There is some variation in laticifer morphology; for example the degree of branching varies even within different species of *Euphorbia*, the most profuse being in the more succulent species (Fig. 7). In a few species (e.g. *E. abdelkuri* Balf. f., Figs 2–4) laticifers may be thick-walled, and sometimes distinct striations may be observed on the inner surface. According to Solereder (1908) the thick walls which occur in laticifers of some species are secondary formations. Some authors (e.g. Dressler, 1957) have recorded lignification of laticifer walls, and Rao & Tewari (1960) reported that the long branched sclereids in the mature leaves of *Codiaeum variegatum* were formed by lignification of the walls of non-articulated laticifers. In *Baloghia lucida* laticifers have a knobbled appearance due to the presence of short projections between many of the surrounding parenchyma cells (Figs 9, 10), and laticifer wall thickness is very uneven since walls are often thinner at the tips of side branches or projections. Laticifers are often similar in diameter to the surrounding cells, particularly in the inner cortex and phloem region of stems, but becoming increasingly narrower towards the epidermis and throughout the pith and leaf mesophyll. Very narrow laticifers may be difficult to recognize since they can occupy intercellular spaces with very little displacement of surrounding cells, and can consequently easily be overlooked in poorly stained or poorly preserved material, such as herbarium material.

Unusually-shaped starch grains have long been known to occur in laticifers of some genera of Euphorbiaceae (Solereder, 1908). In general these are only found within the tribe Euphorbieae, but Groom (1889) recorded rod-shaped

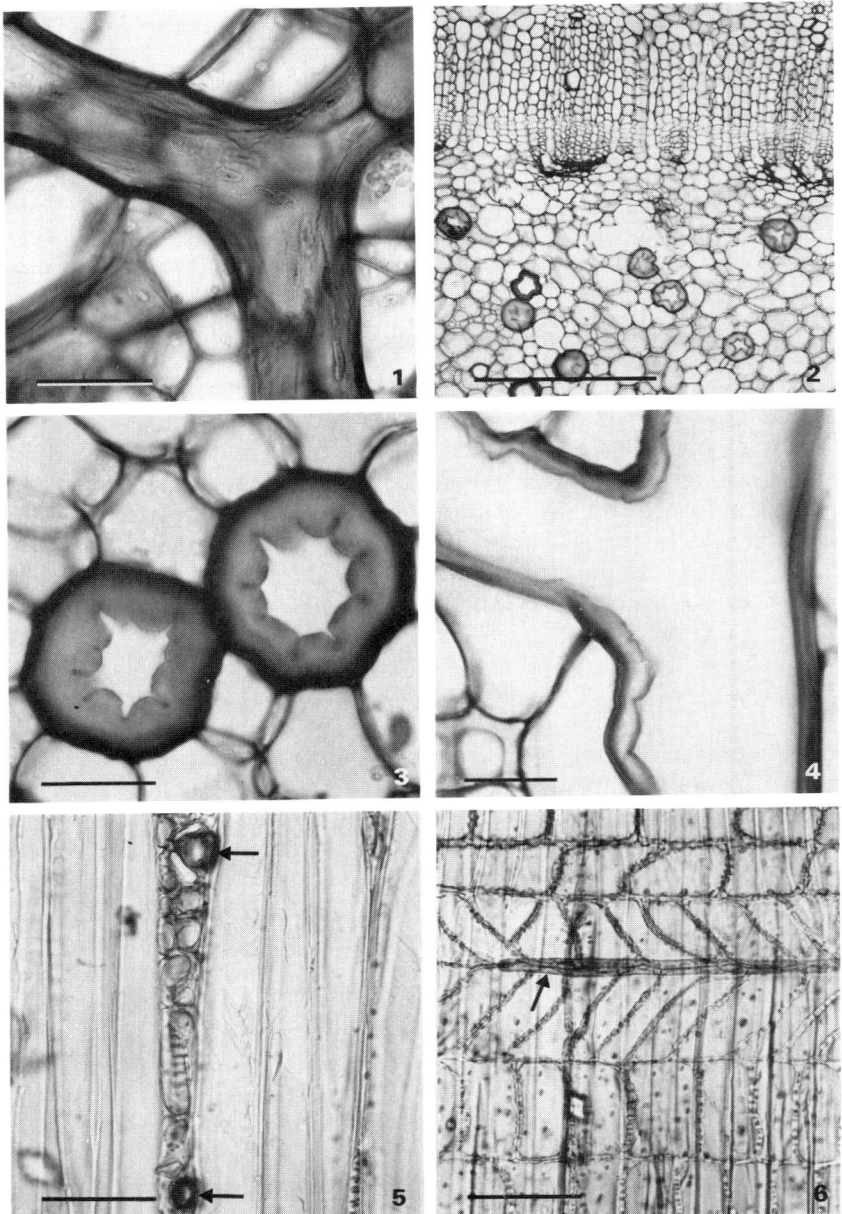

Figures 1–6. Fig. 1. *Euphoriba mitis* laticifer LS; scale bar = 50 μm. Figs 2–4. *Euphorbia abdelkuri*, stem: Fig. 2. TS; scale bar = 0.5 mm. Fig. 3. TS laticifer; scale bar = 50 μm. Fig. 4. LS laticifer; scale bar = 0.1 mm. Figs 5 & 6. *Croton panamensis*, wood; scale bars = 50 μm. Fig. 5. TLS. Fig. 6. RLS. Laticifers arrowed.

starch grains in *Sapium* sp., Solereder (1908) cited earlier records of rod-shaped starch grains in *Hippomane*, and we have found them in *Excoecaria*, all three genera being included in the tribe Hippomaneae, which suggests that this tribe should be further examined for this character. The laticifer starch grains differ

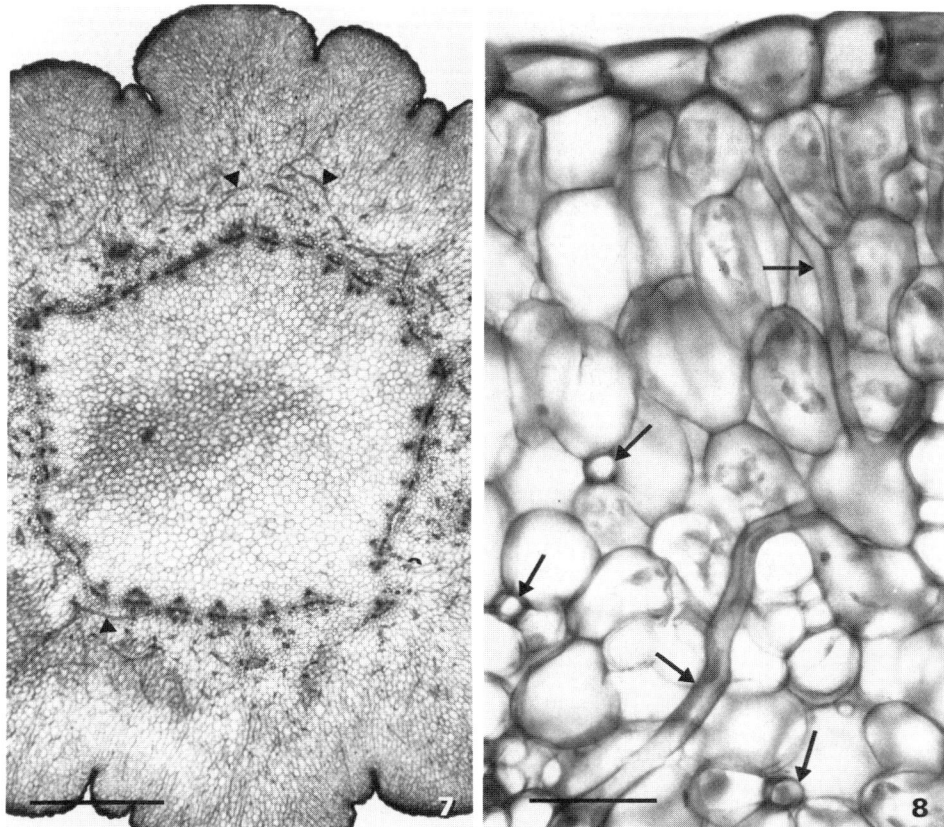

Figures 7 & 8. Fig. 7. *Monadenium ellenbeckii*, stem TS; scale bar = 1 mm. Fig. 8. *Euphorbia pugniformis*, leaf TS; scale bar = 50 μm. Laticifers arrowed.

in morphology from the more conventionally-shaped starch grains in surrounding parenchyma cells, and may be rod-shaped (Fig. 12) or variously lobed or osteoid (Fig. 11). Mahlberg (1975) and Biesboer & Mahlberg (1981) have convincingly demonstrated an evolutionary series based on differences in laticifer starch grain morphology. The rod-shaped grains are considered to be the most primitive, with the increasingly complex osteoid and lobed types being more highly specialized. Biesboer & Mahlberg (1981) related differences in starch grain morphology to gross morphological characters, and proposed phylogenetic relationships within *Euphorbia* from these data.

Non-articulated laticifers often ramify throughout the primary vegetative ground tissues (stem cortex and pith, and leaf mesophyll) of the plants in which they occur, although they are most common in the inner cortex and phloem, and never occur solely in the pith. The extend almost to the shoot apex, and in the first leaf primordia (Figs 21, 22). In the present study laticifers were never observed passing through the cambium, confirming the observations of Fineran (1982) and others, and indicating that medullary laticifers are connected to cortical ones only at the leaf gaps at the node, as previously reported (e.g. Schaffstein, 1932). In both leaves and stems laticifers tend to follow the course of

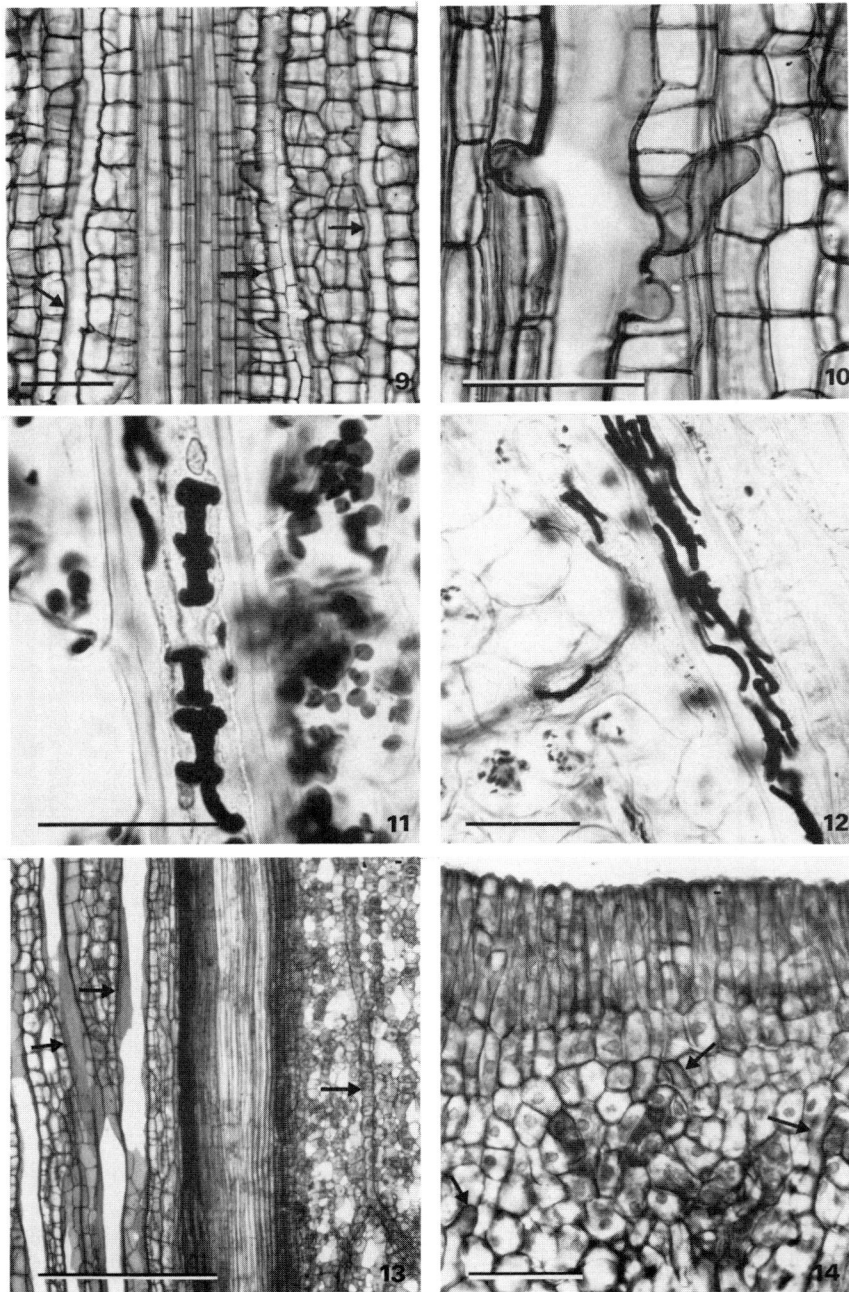

Figures 9–14. Figs 9 & 10. *Baloghia lucida*, stem LS. Fig 9. Scale bar = 0.1 mm. Fig. 10 scale bar = 50 μm. Figs 11 & 12. Laticifer starch grains; scale bars = 50 μm. Fig. 11. *Monadenium ellenbeckii*, osteoid laticifer starch grains (round starch grains in adjacent cells). Fig. 12. *Pedilanthus tithymeloides*, rod shaped laticifer starch grains. Fig. 13. *Euphorbia balsamifera*, stem LS; cortex (left), vascular tissue (centre), pith (right); scale bar = 0.5 mm. Fig. 14. *Hevea brasiliensis*, LS extra floral nectary with laticifers associated with vascular supply close to secretory epidermal cells; scale bar = 50 μm. Laticifers arrowed.

the veins, and also ramify outwards towards the epidermis (Fig. 16). Laticifers are frequently observed adjacent to the inner periclinal walls of the epidermal cells, particularly on the adaxial side of leaves, above the palisade tissue (Fig. 15). Groom (1889) reported that in species of *Euphorbia* and *Sapium* the subepidermal tubes branch copiously before ending blindly. According to Fahn (1979), laticifers never penetrate and end between epidermal cells, but this phenomenon was observed here in *Baloghia lucida* (Figs 17, 18) and by Gaucher (1902) in *Codiaeum irregulare*.

Laticifers are recorded mainly in primary vegetative parts of the plant (leaves, stems and sometimes roots). However, Rao & Malaviya (1964) also recorded laticifers in the flower pedicels, perianth lobes and fruits of *Jatropha*.

Laticifers in secondary xylem: Although it is generally accepted that laticifers may occur in wood of Euphorbiaceae, there are relatively few actual records. Heimsch (1942), Record & Hess (1943), Uhlarz & Kunschert (1975) and Carlquist (1970) all recorded laticifers in wood rays in various genera, although some of these records were not confirmed here, indicating that their presence may be variable. Mennega (personal communication) reports that laticifers are fairly common in woods of genera in the tribes Hippomaneae and Euphorbieae, particularly the cactoid species of *Euphorbia*. In this investigation, narrow, intrusive laticifers were observed in the wood rays of *Croton panamensis* (Figs 5, 6). The ray cells on either side of the laticifer (Fig. 6) are often displaced obliquely as though under some stress when the laticifer grew between them, a phenomenon which was not observed elsewhere, but is perhaps due to the tightly packed nature of the secondary xylem cells. From the illustrations in Carlquist (1970), and Uhlarz & Kunschert (1975), the laticifers have clearly greatly influenced the form of the surrounding ray cells, to the point where they resemble radial secretory canals. This suggests that in these cases the laticifers may have entered the secondary xylem across the vascular cambium rather than from the pith, although there no observations to substantiate this. Fahn (1979) suggested that laticifer growth occurs mainly in maturing tissues, or among cells which have not lost their ability to divide.

Development: The development of non-articulated laticifers in *Euphorbia* is well documented, although it has been the subject of much controversy. Both Metcalfe (1967) and Fahn (1979) comprehensively summarized the dispute as to whether these laticifers are of multicellular origin, and therefore articulated (Milanez, 1952a, b, 1954; Milanez & Machado, 1956; Milanez & Neto, 1956) or of unicellular origin, corresponding to the classical non-articulated type (Mahlberg, 1959; Mahlberg & Sabharwal, 1968). Most workers have accepted the latter theory, since they could find no evidence of cell fusion (e.g. Bruni, Vannini & Fasulo, 1978). According to Mahlberg & Sabharwal (1968) and Cameron (1936) a ring of 12 laticifer initials originates in the cotyledonary node immediately below the shoot apex. The initials, which are at first associated with the procambial strands, form branches which penetrate between surrounding cells, and eventually ramify throughout the plant tissue in the same way as the non-articulated laticifers of other Dicotyledons, such as *Nerium oleander* (Mahlberg, 1961, 1963). The laticifers are coenocytes, since successive mitotic divisions are unaccompanied by cytokinesis (Mahlberg & Sabharwal, 1968). Bruni *et al.* (1978) in essence agreed with the findings of previous workers

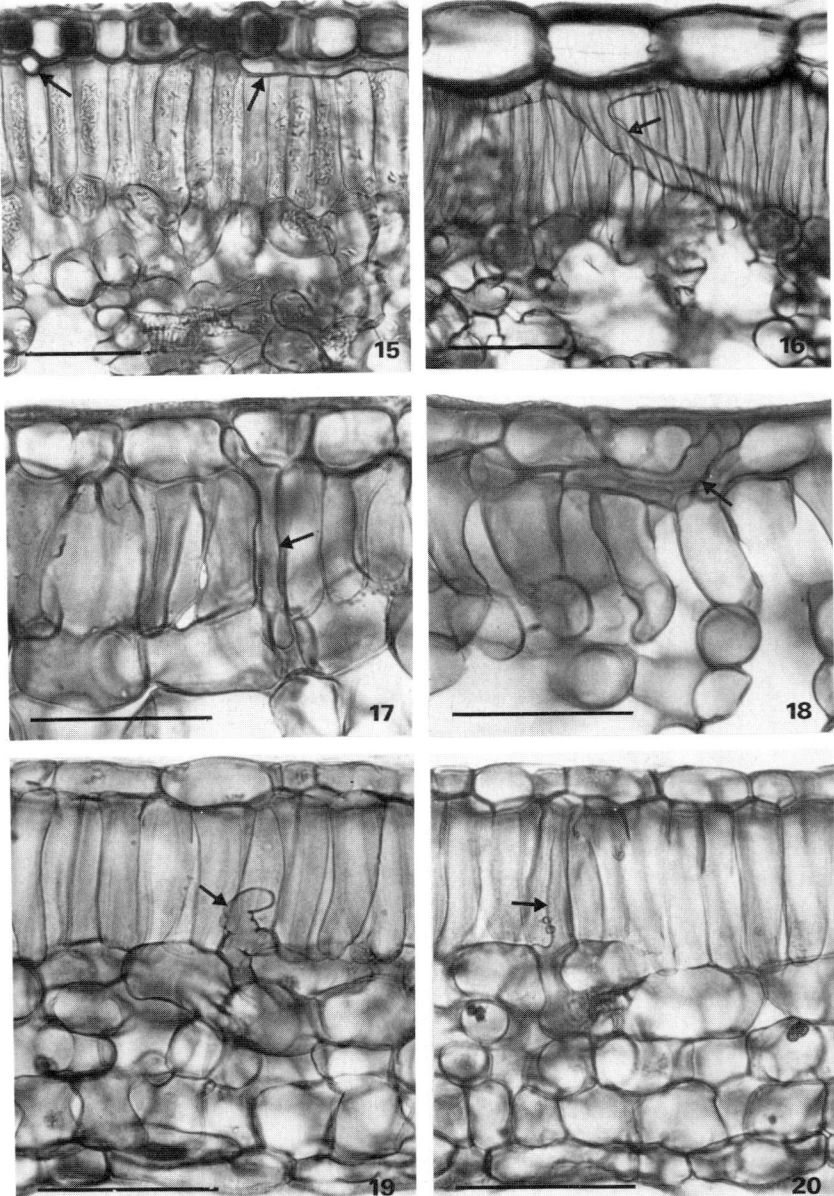

Figures 15–20. Leaf transverse sections with laticifers arrowed. Adaxial side with palisade tissue uppermost. Fig. 15. *Euphorbia characias*. Fig. 16. *Jatropha integerrima*. Figs 17, 18. *Baloghia lucida*, laticifers protruding between adaxial epidermal cells. Figs 19, 20. *Hevea brasiliensis*. Scale bars = 50 μm.

on the development of laticifers in embryos of *Euphorbia marginiata*, although they found 24 laticifer initials rather than 12. They considered that hormonal influences govern the intrusive growth of the laticifers, with the development of both laticifers and vascular elements probably being under common control.

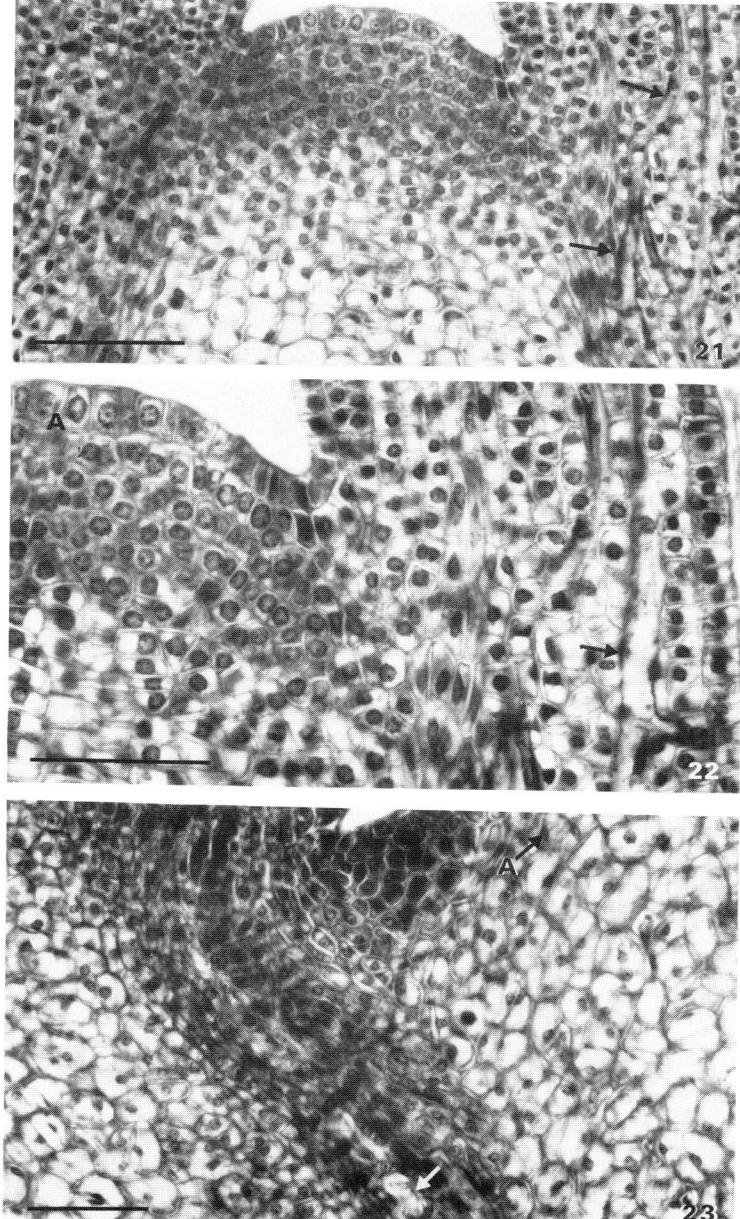

Figures 21–23. Longitudinal sections of shoot apical meristems with laticifers arrowed. (A = apex). Figs 21, 22. *Baloghia lucida* (lacticifers non-articulated). Fig. 21 scale bar = 50 μ. Fig. 22 scale bar = 50 μm. Fig. 23. *Cnidoscolus chayamansa* (laticifers articulated); scale bar = 50 μm.

Development of non-articulated laticifers in other genera of Euphorbiaceae has been little studied. However, it has been assumed that the mode of differentiation is essentially similar since the morphology of mature non-articulated laticifers shows relatively little variation. Rao & Malaviya

(1964) studied the development of the non-articulated laticifers of *Jatropha*, and showed that it differed from that of *Euphorbia* only in that new laticifer initials sometimes arise independently in the leaves.

Articulated laticifers

Morphology and distribution: Apart from the records by Pax (1884) of "articulated sacs" ("gegliederte Schläuche") or "articulated laticiferous vessels" ("gegliederte Milchröhren") in many genera, articulated laticifers have previously been recorded in only three genera of Euphorbiaceae: *Hevea*, *Manihot* and *Jatropha* (Table 2). As Scott (1886) and Solereder (1908) indicated, there is a distinct difference between the anastomosing laticiferous vessels recorded in *Hevea* and *Manihot*, and the series of cells containing latex to which Pax (1884) referred. Indeed, Solereder strongly disputed that Pax's "articulated sacs" could be interpreted as laticifers at all, and pointed out that the laticiferous nature of the mostly brown contents of these cells had not been demonstrated either by Pax or other observers. Pax (1884) generally referred to these cells as "articulated sacs" rather than laticiferous vessels, and did not examine material of *Heva*. Since these "articulated sacs" consist of unbranched axial chains of cells which are virtually identical in cell wall structure to the cells of the surrounding tissue, they are extremely difficult to identify. The granular contents are often either lost in microscopic preparations or indistinguishable from the tanniferous contents of other cells. Consequently, in this investigation I have followed Solereder's example and have recognized only the typical anastomosing articulated vessels, of the type found in *Hevea*, although the articulated laticifers of *Jatropha* (Dehgan & Craig, 1978) may represent a different type.

Dehgan & Craig (1978), in an extensive review of laticifers in *Jatropha*, found both non-articulated and articulated types in leaves of most species (and also idioblastic laticiferous cells which intergrade with laticifers). Previously most authors (e.g. Rao & Malaviya, 1964) had reported only non-articulated laticifers in this genus, although Kakkar & Paliwal (1972) also observed both types. In *Jatropha* the two types are apparently fairly similar, being distinguished by the intensity of staining and the presence of dividing transverse walls in the articulated laticifers. The only material available in this investigation was of *J. integerrima*, in which I have observed only non-articulated laticifers (Fig. 16), although Dehgan & Craig (1978) found both types, indicating that the relative distribution is variable. De Castells, Ormond & Braconi (1984) confirmed the presence of non-articulated, articulated and "idioblastic" laticifers in *J. gossypifolia*, although from their illustrations the articulated laticifers resemble the "articulated sacs" described by Pax (1884).

Anastomosing articulated laticifers were observed here in *Hevea* and *Manihot* (as also previously recorded, e.g. by Scott, 1884a, b, 1886) and in *Cnidoscolus chayamansa* (Figs 24–27) and are basically very similar in the three genera. They occur in the phloem of both stems and leaves. The articulated laticifers in the phloem have a characteristic appearance caused by the breakdown of the intervening walls of the component cells (Figs 25, 26). This anastomosis results in a complex network of laticiferous vessels.

The articulated laticifers of *Hevea*, *Manihot* and *Cnidoscolus* also follow the course of the veins in the leaves. However, it is a significant and little-reported

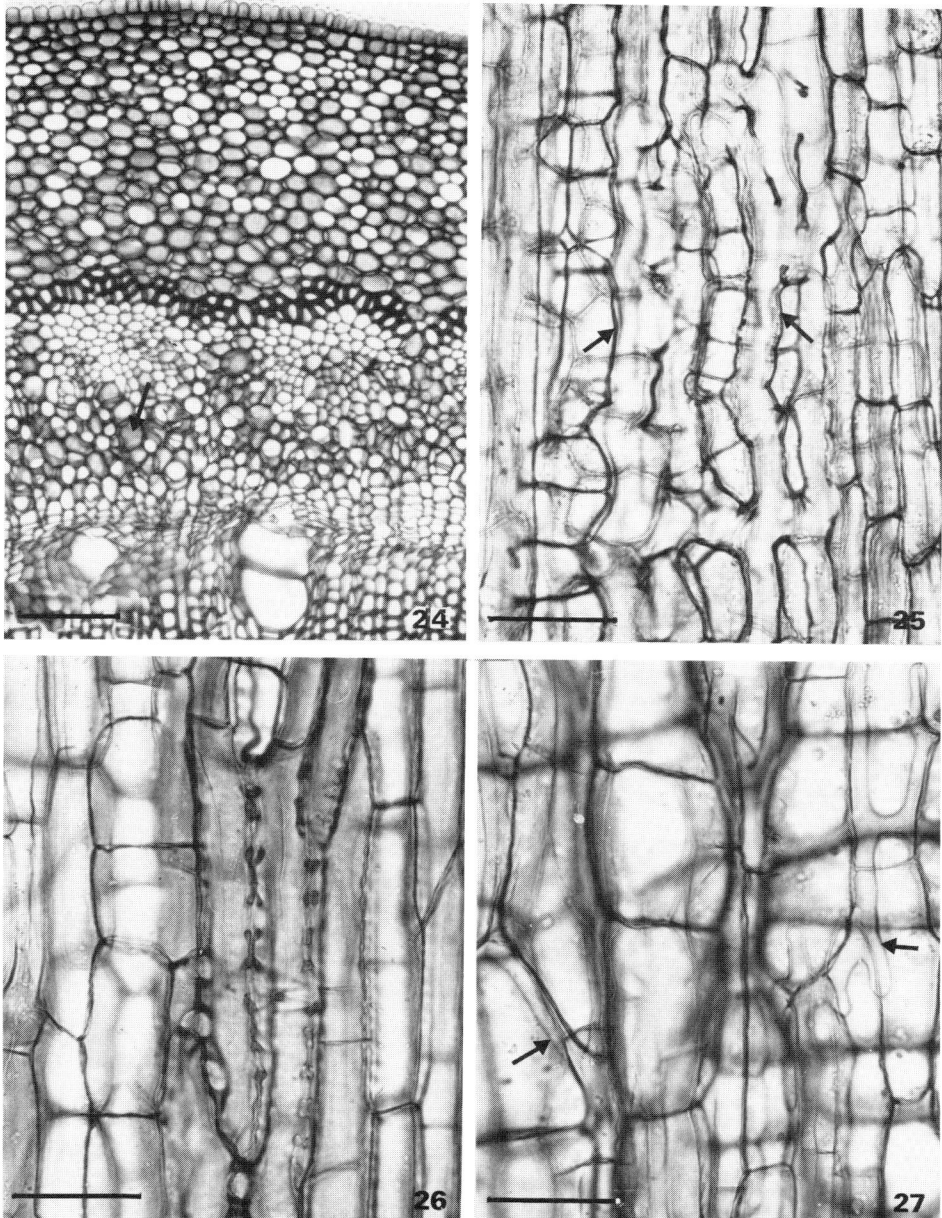

Figures 24–27. Articulated laticifers in stems. Figs 24 & 25. *Hevea brasiliensis*. Fig. 24. TS cortex; scale bar = 0.1 mm. Fig. 25. LS phloem region; scale bar = 50 μm. Figs 26 & 27 scale bars = 50 μm. *Cnidoscolus chayamansa*. Fig. 26. LS phloem region. Fig. 27. LS pith region.

fact that in all three genera branches of these laticifers ramify throughout the mesophyll (Figs 19, 20). These narrow branches lack septa and are virtually indistinguishable from non-articulated laticifers in genera such as *Euphorbia*, although usually with somewhat less extensive ramification. Ramifying branches of this type are also present in the stem pith and cortex of *Manihot* and

Cnidoscolus (Fig. 27), and other investigators (e.g. Calvert, 1887; Quisumbing, 1927) have seen them in the outer cortex and pitch of *Hevea* stems, although none were observed here (or by Bryce & Campbell, 1917). Calvert (1887) noted that laticifers in the inner stem cortex of young seedings of *Hevea* send out branches to the pith and later to the outer cortex, the medullary laticifers being connected to the cortical ones at the nodes. Boblioff (1923) observed cells with an "amoeboid-like form" in the latex vessels of the leaves and the inner integument of the seed of *Hevea*. He regarded these cells as "aberrancies" in the formation of the laticifers as they are capable of apical intrusive growth, and pointed out that they represent an intermediate form between the non-articulated laticifers of other genera, which develop by intrusive growth from a single initial cell, and the typical articulated laticifers of *Hevea*. The fact that recent studies have overlooked this intrusive growth is perhaps because most later investigations on articulated laticifers have been carried out on bark of mature trees in connection with commercial latex production (e.g. Gunnery, 1935; Trancard, 1979), and the occurrence of laticifer branches in the outer cortex of *Hevea* stems is variable. Quisumbing (1927) found that laticifers are far more common in the inner cortex and phloem region, and pointed out that this is consistent with the observations of rubber tappers in Ceylon, that the greatest latex yield is from this area of the stem.

Development: There have been relatively few studies on development of articulated laticifers, although examination of laticifers in the phloem of mature stems of *Hevea*, *Manihot* and *Cnidoscolus* indicates that cell fusion has taken place, as the remains of the lateral and transverse walls are apparent, giving the characteristic appearance of the articulated laticifers (Fig. 25). Laticifers are formed in phloem near the apex (Fig. 23), and also in secondary phloem from the vascular cambium. Scott (1886) and Calvert (1887) studied development of laticifers in *Hevea* seedlings, and showed that laticifers are first apparent at an early stage, before the vascular tissue is well developed. Examinations of laticifer development in *Jatropha* (Rao & Malaviya, 1964; Cass, 1968) have not included articulated laticifers, and have found no evidence of cell fusion in this genus.

Function of laticifers

There have been a variety of functions ascribed to laticifers. They are commonly supposed to be a secretory tissue (Fahn, 1979), although the precise advantage to the plant of the latex produced and stored is unknown. Many early workers, such as Haberlandt (1914), considered that laticifers serve as conductors of assimilated food material. Haberlandt cited as evidence the fact that they are closely connected with both vascular tissue and palisade mesophyll (the primary photosynthetic tissue) and also the results of earlier workers which showed that changes do take place in the relative constituents of latex, such as starch. Scott (1886) considered that since the laticiferous tissue is well developed before the protoxylem in *Hevea* seedlings, the laticifers have a role in the transport of food from the endosperm to the developing embryo. However, as Groom (1889) and others (e.g. Spilatro & Mahlberg, 1985) have pointed out, the latex starch does not appear to act as a food reserve in mature leaves, and also the anatomy of the laticifers could equally well support the now unpopular theory of other workers that the laticifers have an excretory function. In this context it is interesting to note that laticifers were observed here closely associated with the extra floral nectaries in leaves of *Hevea* (Fig. 14), extending almost to the secretory cells. However, this could perhaps be explained by the relative profusion of vascular tissue, which usually has associated laticifers. Similarly, Belin-Depoux & Clair-Maczulajtys (1974, 1975) reported laticifers close to the extra floral nectaries in leaves of *Aleurites*; de Castells *et al.* (1984) found copious laticifers associated with the vascular tissue to large glands on the leaf margins in *Jatropha*, and Genç & Rauh (1984) noted non-articulated laticifers reaching the subepidermis of floral nectaries of *Euphorbia*.

Haberlandt also pointed out two important "subsidiary" functions of the latex, in wound healing by coagulation, and also as a defence or deterrent against predation by herbivores. Both these roles have been demonstrated, and current opinion (Fahn, 1979) generally considers them to be the probable main functions of the laticifers. However, since in some genera laticifers are abundant in young tissues but later become lignified, whereas in others (e.g. *Hevea*) they are continually produced throughout the life of the plant, it seems possible that the prime function of the latex may vary between taxa.

Taxonomic and evolutionary considerations

In the system of Pax & Hoffmann (1931) (Table 2), laticifers are limited to the subfamily Crotonoideae, apart from an unconfirmed and somewhat doubtful record in the wood rays of *Hieronima*. However, of the 12 tribes of Pax's Crotonoideae, laticifers are present in only nine, and of variable occurrence in two of these (Acalypheae and Gelonieae). Of the alternative taxonomic treatments for Euphorbiaceae, that of Webster (1975) provides the most coherent grouping for this character (Table 3). In Webster's system, genera of all the tribes of Pax's Crotonoideae which lack laticifers (Chrozophoreae, Dalechampieae, Pereae, Acalypheae and *Chaetocarpus* of Pax's Gelonieae) are transferred to the subfamily Acalyphoideae, sometimes under new tribal grouping. Webster's subfamily Acalyphoideae in general lacks laticifers, in common with his two most primitive subfamilies, Phyllanthoideae and Oldfieldioideae. The only records of laticifers in the Acalyphoideae are in *Clutia*, *Acalypha* (unconfirmed in three species examined here) and in the wood rays of

Table 3. Distribution of laticifers in Euphorbiaceae according to Webster (1975)

Subfamily	
1. Phyllanthoideae	13 tribes; laticifers absent
2. Oldfieldioideae	4 tribes; laticifers absent
3. Acalyphoideae	19 tribes; representative genera of only 7 tribes recorded, but laticifers generally absent, except for a few scanty records (*Clutia*, *Acalypha*, and in wood rays in *Dalechampia* and *Pera*)
4. Crotonoideae	11 tribes; representative genera of 9 tribes recorded; laticifers present in all tribes except Ricinocarpeae, and variable in Aleuritidae. Articulated laticifers in 3 tribes
5. Euphorbioideae	5 tribes; laticifers present in all

Dalechampia and *Pera*, and these records need further investigation. Laticifers are present in all genera of Webster's Euphorbioideae, and all except members of two tribes of his Crotonoideae: the Ricinocarpeae (according to Pax's observations of 1884), and the Aleuritidae (*Aleurites* and *Garcia*), which also warrant further investigation.

Articulated laticifers are restricted to Webster's Crotonoideae. The record here of typical articulated laticifers in *Cnidoscolus chayamansa* is interesting, but predictable as it confirms current views on the systematic position of this genus. Although Pax & Hoffman (1931) included *Cnidoscolus* in the same tribe as *Jatropha* (Cluytieae: Table 2), Miller & Webster (1962), using evidence from anatomy, cytology and pollen morphology, pointed out that *Cnidoscolus* shows more affinities with *Manihot* than *Jatropha*. Subsequently Webster (1975) assigned *Cnidoscolus* and *Manihot* to the same tribe, Manihoteae, separate from *Jatropha*. Since articulated laticifers of this type have been recorded only in *Hevea*, *Manihot* and *Cnidoscolus*, this suggests a close relationship between the tribes Micrandreae (which includes *Hevea*) and Manihoteae of Webster (1975). It would be worthwhile to examine other related genera for articulated laticifers or intermediate forms, particularly *Micrandra*, in which Solereder (1908) recorded rows of elongated laticiferous sacs, which he regarded as significantly different from Pax's "articulated sacs".

The obvious similarities, both morphological and chemical, between non-articulated and articulated laticifers, indicate a common ancestor for these cell types in Euphorbiaceae. Both types have the capacity for intrusive growth between surrounding cells, the main differences being in developmental origin. Scott (1886) suggested that in the Euphorbiaceae the two types were both derived from forms with "articulated sacs". In support of this he pointed out that a similar evolutionary series, from closed sacs to true vessels formed from fused cells, could have taken place in the Papaveraceae. If this theory is correct, the capacity for intrusive growth may have evolved twice in latex-containing cells in the Euphorbiaceae. However, intrusive growth is not unique to laticifers among plant cells; for example astrosclereids grow intrusively in maturing leaves. Indeed, since Rao & Tewari (1960) reported that in *Codiaeum variegatum* the astrosclereids are actually formed by lignification of laticifers in young leaves, and Rao & Malaviya (1964) suggested that sclereids in leaves of many Euphorbiaceae may be sclerized laticifers, the relationship between the two cell types seems worth investigating. In other Dicotyledon families (e.g. Menispermaceae: Wilkinson, 1986) laticifers and astrosclereids have been found to be often almost mutually exclusive in related genera.

Clarification of the developmental pathways of laticifers in genera less highly evolved than *Euphorbia* and *Hevea* are needed for a better understanding of the evolutionary relationships of these cell types. It has hitherto been assumed that development of non-articulated laticifers is similar in genera of Euphorbiaceae other than *Euphorbia*, but the report by Rao & Malaviya (1964) that in *Jatropha* new non-articulated laticifer initials sometimes arise independently in the leaves, is comparable with other reports of intrusive growth of articulated laticifers into the mesophyll in *Hevea*, *Manihot* and *Cnidoscolus*, although in *Jatropha* the new initials arise in the spongy mesophyll. Dehgan & Craig (1978) suggested that much evidence indicates *Jatropha* to be primitive and ancestral to other genera in the Crotonoideae. We can speculate that both articulated and non-articulated laticifers evolved at some point either in the Acalyphoideae, or in the Crotonoideae in a genus related to *Jatropha*.

ACKNOWLEDGEMENTS

I am grateful to Dr D. F. Cutler and Miss M. Gregory of the Jodrell Laboratory for critically reading the manuscript, and to Mr A. Radcliffe-Smith of the Herbarium, Royal Botanic Gardens, Kew, for advice on taxonomic aspects. Mr T. J. Lawrence of the Jodrell Laboratory cut many of the sections and took the photographs of starch grains.

REFERENCES

ASHPLANT, H., 1928. Investigations into *Hevea* anatomy. *Bulletin of the Rubber Growers Association, 10:* 484–490.
BELIN-DEPOUX, M. & CLAIR-MACZULAJTYS, D., 1974. Introduction à l'étude des glandes foliaires de l'*Aleurites moluccana* Willd. (Euphorbiacée). I. La glande et son ontogénèse. *Revue Générale de Botanique, 81:* 335–351.
BELIN-DEPOUX, M. & CLAIR-MACZULAJTYS, D., 1975. Introduction à l'étude des glandes foliaires de l'*Aleurites moluccana* Willd. (Euphorbiacée). II. Aspects histologiques et cytologiques de la glande pétiolaire fonctionnelle. *Revue Générale de Botanique, 82:* 119–155.
BERGFELD-GAERTNER, H., 1964. Entwicklungsgeschichtliche und cytologische Untersuchungen der 'Von Hanstein' schen Zellen' bei *Mercurialis annua* und *perennis* L. *Zeitschrift für Botanik, 52:* 291–302.
BIESBOER, D. D. & MAHLBERG, P. G., 1981. Laticifer starch grain morphology and laticifer evolution in *Euphorbia* (Euphorbiaceae). *Nordic Journal of Botany, 1:* 447–457.
BOBLIOFF, W., 1923. *Anatomy of* Hevea brasiliensis. Zurich.
BRUNI, A., VANNINI, G. L. & FASULO, M. P., 1978. Anatomy of the latex system in the dormant embryo of *Euphorbia marginata* Purch.: correlative analysis of the histological organisation and topographic distribution of laticifers. *Annals of Botany, 42:* 1099–1108.
BRYCE, G. & CAMPBELL, L. E., 1917. On the mode of occurrence of latex vessels in *Hevea brasiliensis*. *Ceylon Department of Agriculture Bulletin, 30:* 1–22.
CALVERT, A., 1887. The laticiferous tissue in the stem of *Hevea brasiliensis*. *Annals of Botany, 1:* 75–77.
CALVERT, A. & BOODLE, L. A., 1887. On laticiferous tubes in the pith of *Manihot glaziovii* and on the presence of nuclei in this tissue. *Annals of Botany, 1:* 55–62.
CAMERON, D., 1936. An investigation of the latex systems in *Euphorbia marginata*, with particular attention to the distribution of latex in the embryo. *Transactions and Proceedings of the Botanical Society of Edinburgh, 32:* 187–194.
CARLQUIST, S., 1970. Wood anatomy of Hawaiian, Macaronesian, and other species of *Euphorbia*. In N. K. B. Robson, D. F. Cutler & M. Gregory (Eds), *New Research in Plant Anatomy* (Supplement 1 to *Botanical Journal of the Linnean Society, 63*). London: Academic Press.
CASS, D. D., 1968. Observations on the ultrastructure of the non-articulated laticifer of *Jatropha podagrica* (Euphorbiaceae). *Experientia, 24:* 961–962.
DE CASTELLS, A. R. C., ORMOND, W. T. & BRACONI, A., 1984. Contribuição ao estudo da biologia de *Jatropha gossypifolia* L. (Euphorbiaceae). I. Laticíferos e Glândulas. *Revista Brasileira de Biologia, 2:* 149–158.
DATTA, S. K., DATTA, K. & ROY, S. K., 1984. Bark anatomy of *Euphorbia* with histo-pharmacognostic evaluation. *Bangladesh Journal of Botany, 13:* 36–44.
DE BARY, A., 1884. *Comparative Anatomy of the Vegetative Organs of the Phanerogams and Ferns*. [English translation.] Oxford: Clarendon Press.
DEHGAN, B. & CRAIG, M. E., 1978. Types of laticifers and crystals in *Jatropha* and their taxonomic implications. *American Journal of Botany, 65:* 345–352.
DE MENDONÇA, M. S., 1982. Contribuição ao estudo da anatomia foliar de *Manihot dichotoma*. *Resumos XXIII Congresso nacional de Botanica, Sociedade botanica de Brasil*, Maceio 1982, 60 [Abstract].
DE MENDONÇA, M. S., 1983a. Estudo de plantas laticiferas. I. Aspectos anatômicos o distribução de vasos laticiferos de *Manihot caerulescens* Pohl. *Acta Amazonica, 13:* 501–517.
DE MENDONÇA, M. S., 1983b. *Actinostemon* Klotsch. (Euphorbiaceae)—morfologia e taxonomia. I. *A. communis* (Mull. Arg.) Pax var. *spathulatus* Mull. Arg. *ATAS da Sociedade Botânica do Brasil, Rio de Janeiro, 1:* 87–96.
DRESSLER, R., 1957. The genus *Pedilanthus* (Euphorbiaceae). *Contributions of the Gray Herbarium, 182:* 1–188.
FAHN, A., 1979. *Secretory Tissues in Plants*. London: Academic Press.
FINERAN, B. A., 1982. Distribution and organisation of non-articulated laticifers in mature tissues of Poinsettia (*Euphorbia pulcherrima* Willd.) *Annals of Botany, 50:* 207–220.

FINERAN, B. A., 1983. Differentiation of non-articulated laticifers in Poinsettia (*Euphorbia pulcherrima* Willd.). *Annals of Botany, 52:* 279–293.

FREY-WYSSLING, A., 1931. Étude sur la relation existant entre le diamètre des tubes à latex et la production du caoutchouc dans l'*Hevea brasiliensis*. *Bulletin économique de l'Indochine, 34:* 341–374.

GAUCHER, L., 1902. Recherches anatomiques sur les Euphorbiacées. *Annales Science Naturelle, série 8, 15:* 161–309.

GENÇ, Z. & RAUH, H., 1984. Vergleichend-anatomische Untersuchungen an den Honigdrüsen einiger *Euphorbia*-arten. I. Lichtmikroskopische Untersuchungen. *Tropische und subtropische Pflanzenwelt, 45,* 9–53.

GOMEZ, J. B. & THAI, C. K., 1967. Alignment of anatomical elements in the stem of *Hevea brasiliensis*. *Journal of the Rubber Research Institute, Malaya, 20:* 91–99.

GROOM, P., 1889. On the function of laticiferous tubes. *Annals of Botany, 3:* 157–169.

GUNNERY, H., 1935. Yield production in *Hevea*—a study of sieve-tube structure in relation to latex yield. *Journal of the Rubber Research Institute, Malaya, 6:* 8–20.

HABERLANDT, G., 1914. *Physiological Plant Anatomy.* [English edition.] London: MacMillan & Co.

HEIMSCH, C., 1942. Comparative anatomy of the secondary xylem of the 'Gruinales' and 'Terebinthales' of Wettstein with reference to the taxonomic groupings. *Lilloa, 8:* 83–198.

HOLM, T., 1911. Medicinal plants of North America. 48. *Stillingia sylvatica* L. *Merck's Reportonium, 20:* 36–38.

JIMENEZ, M. S. & CABALLERO, A., 1978. Laticiferos y morfologia de los granos de almidón presentes en el látex de euforbias canarias. *Vieraea, 8:* 113–124.

KAKKAR, L. & PALIWAL, G. S., 1972. Studies on the leaf anatomy of *Euphorbia*. IV. Terminal idioblasts. *Journal of the Indian Botanical Society. 51:* 118–126.

LÉANDRI, J., 1939. Les 'Croton' de Madagascar et des îles voisines. *Annales du Musée Colonial de Marseille, 47:* 1–88.

LLERAS, E. & MEDRI, M. E., 1978. Comparação anatômica entre folhas diplóides e poliplóides do hibrido *Hevea brasiliensis* x *benthamiana*. *Acta Amazonica, 8:* 565–575.

LOMMASSON, R. C., 1962. The latex system of some native spurges. *Proceedings of the Iowa Academy of Sciences, 69:* 147–151.

LOPÉZ GUILLÉN, J. E. & KIYÁN DE CORNELIO, I. 1974. Plantas medicinales del Perú. V. *Biota, 10:* 76–104.

MAHLBERG, P. G., 1959. Develoment of the non-articulated laticifer in proliferated embryos of *Euphorbia marginata* Pursh. *Phytomorphology, 9:* 156–162.

MAHLBERG, P. G., 1961. Embryogeny and histogenesis in *Nerium oleander*. II. Origin and development of the non-articulated laticifer. *American Journal of Botany, 48:* 90–99.

MAHLBERG, P. G., 1963. Development of non-articulated laticifers in the seedling axis of *Nerium oleander*. *Botanical Gazettee, 124:* 224–231.

MAHLBERG, P. G., 1975. Evolution of the laticifer in *Euphorbia* as interpreted from starch grain morphology. *American Journal of Botany, 62:* 577–583.

MAHLBERG, P. G., 1982. Comparative morphology of starch grains in latex from varieties of Poinsettia, *Euphorbia pulcherrima* Willd. (Euphorbiaceae). *Botanical Gazette, 143:* 206–209.

MAHLBERG, P. G. & SABHARWAL, P. S., 1968. Origin and early development of non-articulated laticifers in embryos of *Euphorbia marginata*. *American Journal of Botany, 55:* 375–381.

MAHLBERG, P. G., FIELD, D. W. & FRYE, J. S., 1984. Fossil laticifers from Eocene Brown Coal Deposits of the Geiseltal. *American Journal of Botany, 71:* 1192–1200.

MEDRI, M. E. & LLERAS, E., 1980. Aspectos da anatomia ecológica de folhas de *Hevea brasiliensis* Müell. Arg. *Acta Amazonica, 10:* 463–493.

METCALFE, C. R., 1967. Distribution of latex in the Plant Kingdom. *Economic Botany, 21:* 115–127.

METCALFE, C. R., 1987. *Anatomy of the dicotyledons. Vol. III. Magnoliales, Illiciales, and Laurales.* 2nd edition. Oxford: Clarendon Press.

METCALFE, C. R. & CHALK, L., 1950. *Anatomy of the Dicotyledons.* Oxford; Clarendon Press.

MILANEZ, F. R., 1952a. Sôbre os nucleos dos laticiferos de *Euphorbia phosphorea* Mart. *Rodriguesia, 15:* 163–179.

MILANEZ, F. R., 1952b. Ontogênese dos latićíferos do caule de *Euphorbia phosphorea* Mart. *Arquivos do Jardin Botanico, Rio de Janeiro, 12:* 17–35.

MILANEZ, F. R., 1954. Origem das ramificações dos latićíferos do caule de *Euphorbia phosphorea* Mart. *Arquivos do Jardin Botanico, Rio de Janeiro, 13:* 95–113.

MILANEZ, F. R. & MACHADO, R. D., 1956. Aplicação da microscopia electrônica do estudo dos latićíferos embrionários de *Euphorbia pulcherrima* Willd. *Rodigúesia, 18–19:* 425–440.

MILANEZ, F. R. & NETO, H. M., 1956. Origem dos latićíferos do embrião de *Euphorbia pulcherrima* Willd. *Rodrigúesia, 18–19:* 351–395.

MILLER, K. I. & WEBSTER, G. L., 1962. Systematic position of *Cnidoscolus* and *Jatropha*. *Brittonia, 14:* 174–180.

PALIWAL, G. S. & KAKKAR, L., 1969. Laticifers in the stems and leaves of certain *Euphorbia* species. In B. M. Johri, R. N. Kapil & A. Rashid (Eds), *Seminar on Morphology, Anatomy and Embryology of Land Plants.* Delhi: Univeristy of Delhi.

PAX, F., 1884. Die anatomie der Euphorbiaceae in ihrer Beziehung zum System derselben. *Botanische Jahrbücher, 5:* 384–421.

PAX, F. & HOFFMANN, K., 1931. Euphorbiaceae. In A. Engler (Ed.), *Die Natürlichen Pflanzenfamilien, ed. 2, 19C:* 11–233.
PETCH, T., 1911. *The physiology and diseases of* Hevea brasiliensis. London.
QUISUMBING, E., 1927. The occurrence of laticiferous vessels in the mature bark of *Hevea brasiliensis. University of California Publications in Botany, 13:* 319–332.
RAO, A. R. & MALAVIYA, M., 1964. On the latex-cells and latex of *Jatropha. Proceedings of the Indian Academy of Sciences, 60B:* 95–106.
RAO, A. R. & TEWARI, J. P., 1960. On the morphology and ontogeny of the foliar sclereids of *Codiaeum variegatum* Blume. *Proceedings of the National Institute of Science of India, 26B:* 1–6.
RAO, A. R., MENON, V. K. & MALAVIYA, M., 1968. The laticifers and latex of *Euphorbia tirucalli* L. *Proceedings of the Indian Academy of Sciences, B, 67:* 61–7.
RECORD, S. J. & HESS, R. W., 1943. *Timbers of the New World.* New Haven: Yale University Press.
ROSOWSKI, J. R., 1968. Laticifer morphology in the mature stem and leaf of *Euphorbia supina. Botanical Gazette, 129:* 113–120.
SANTOS, J. K., 1932. The laticiferous vessels and other anatomical structures of *Excoecaria agallocha. Philippine Journal of Science, 47:* 295–304.
SCHAFFSTEIN, G., 1932. Untersuchungen an ungegliederten Milchröhren. *Botanisches Centrablatt, 49:* 197–220.
SCOTT, D. H., 1884a. On the laticiferous tissue of *Manihot glaziovii* (the Ceara rubber). *Quarterly Journal of Microscopical Science, 24:* 193–203.
SCOTT, D. H., 1884b. Note on the laticiferous tissue of *Hevea spruceana. Quarterly Journal of Microscopical Science, 24:* 204–206.
SCOTT, D. H., 1886. On the occurrence of articulated laticiferous vessels in *Hevea. Journal of the Linnean Society* (Botany), *21:* 566–573.
SOLEREDER, H., 1908. *Systematic anatomy of the Dicotyledons.* [English edition.] Oxford: Clarendon Press.
SPENCER, H. J., 1939. On the nature of the blocking of the laticiferous system at the leaf base of *Hevea brasiliensis. Annals of Botany, 3:* 231–235.
SPILATRO, S. R. & MAHLBERG, P. G., 1985. Composition and structure of nonutilizable laticifer starch grains of *Euphorbia pulcherrima* Willd. *Botanical Gazette, 146:* 26–31.
SUMMERS, F., 1910. The leaf of *Colliquaya odorifera*, Molin. *New Phytologist, 9:* 320–325.
TOBLER, F., 1920. Zur Zenntnis des Milchsaftes von *Manihot glaziovii. Bericht der Deutschen Botanischen Gesellschaft, 38:* 159–165.
TRANCARD, J., 1979. Cellules à tannins et tubes lactifères dans le liber d'*Hevea brasiliensis* Müll. Arg. *Revue de Cytologie et de biologie végétales. Le Botaniste, 2:* 1–6.
UHLARZ, H. & KUNSCHERT, A., 1975. Das Holz von *Euphoribia grandicornis* Goeb. *Beiträge zur Biologie der Pflanzen, 51:* 391–406.
WEBSTER, G. L., 1975. Conspectus of a new classification of the Euphorbiaceae. *Taxon, 24:* 593–601.
WILKINSON, H. P., 1986. Leaf anatomy of *Tinomiscium* and *Fibraurea* (Menispermaceae tribe Fibraureae) with special reference to laticifers and astrosclereids. *Kew Bulletin, 41:* 153–169.

Laticifers and the classification of *Euphorbia*: the chemotaxonomy of *Euphorbia esula* L.

P. G. MAHLBERG

Department of Biology, Indiana University, Bloomington, IN 47405, U.S.A.

D. G. DAVIS

Metabolism and Radiation Research Laboratory, USDA, SEA/AR, Fargo, ND 58105, U.S.A.

D. S. GALITZ

Botany Department, North Dakota State University, Fargo, ND 58105, U.S.A.

AND

G. D. MANNERS

Western Regional Research Laboratory, USDA, SEA/AR, Berkeley, CA 94710, U.S.A.

Received April 1986, accepted for publication August 1986

MAHLBERG, P. G., DAVIS, D. G., GALITZ, D. S. & MANNERS, G. D., 1987. **Laticifers and the classification of *Euphorbia*: the chemotaxonomy of *Euphorbia esula* L.** Articulated and non-articulated laticifer cells represent distinctive cell types of relatively recent origin and occur in only a few families. Both types are of separate phylogenetic origin, reflecting independent evolutionary trends in the Euphorbiaceae. Supra-generic groupings of this family can be segregated into three taxonomic units using the laticifer character; with either articulated laticifers, non-articulated laticifers, or no laticifers. Such units may reflect more natural assemblages than now represented in the classification of this family. Laticifers possess chemical and morphological features of potential application as taxonomic characters to aid in delimiting species and interpreting evolutionary trends. The triterpenoid profile from latex of *Euphorbia* species has been shown to be diagnostic for a taxon. The qualitative and quantitative composition show a high level of stability under diverse environmental and physiological conditions indicating a genetic basis for triterpenoid synthesis. Triterpenoid profiles of known accessions of European *E. esula* L. and related presumptive taxa from North America readily separated them into distinctive chemotaxa that include one for *E. esula* L., whereas morphological features were found inadequate for separating accessions to presumptive taxa. Identification of adventive spurges in North America requires diagnostic analyses of Eurasian leafy spurges for comparison. Laticifer characters used in conjunction with relevant morphological features will provide a broadened insight into phylogenetic relationships with the Euphorbiaceae.

ADDITIONAL KEY WORDS:—Taxonomy – triterpenoids.

CONTENTS

Introduction	166
Materials and methods	166
Plant materials	166
Histology	166
Gas–liquid chromatography (GLC)	167
Results and discussion	167
Laticifers	167
Latex triterpenoid profiles in the chemotaxonomy of *Euphorbia*	171
Acknowledgements	179
References	179

INTRODUCTION

Recent reports have provided new insights into the anatomy, taxonomy, physiology, and economic importance of members of the family Euphorbiaceae including the biomedical implications of compounds, such as the diterpenes, derived from them. Laticifer cells are the most unusual of cell types in flowering plants, representing prominent anatomical components of genera in the family. These cell types and their contents, the latex, can provide us with new characters and data to supplement existing taxonomic treatments for interpreting inter- and intra-familial relationships. Further, the occurrence of diverse secondary products synthesized within these cells may serve as a source for substances of present and future biomedical and economic importance.

MATERIALS AND METHODS

Plant materials

Populations of North American and European leafy euphorbias were obtained from Dr M. McCarty of the USDA Laboratory at Lincoln, Nebraska, and grown in a nursery of the Agronomy Department of North Dakota State University (NDSU) and in greenhouses at USDA Metabolism and Radiation Research Laboratory (MRRL) Fargo, North Dakota. Additional accessions were collected in Minnesota and North Dakota. All plants were grown in Fargo clay soil (outdoors or greenhouse) in proximity to one another to minimize effects of any environmental differences.

Leaf measurements were made on 15 mature mid-cauline leaves on at least three shoots collected from each accession. Vegetative shoots were collected in March, September and December from actively growing greenhouse-maintained plants at NDSU and MRRL, and prepared as vouchers prior to measurements. Measurements included leaf length, leaf width at the midpoint of the blade, and the greatest width above the midpoint and toward the base of the blade. Data for each accession were averaged for tabulation. Our data were compared with the published data of Ebke & McCarty (1983) from field-grown plants.

Histology

Light and transmission electron microscopic procedures have been described elsewhere (Mahlberg, 1961, 1975; Mahlberg & Sabharwal, 1967, 1968; Nessler & Mahlberg, 1977; Vertrees & Mahlberg, 1978). Rubber particles from latex of *Euphorbia tirucalli* L. were isolated by fractionation, evaporated by freeze-drying and sputter-coated with gold–palladium in preparation for examination by scanning electron microscopy (Groeneveld, Furr & Mahlberg, personal communication).

Gas-liquid chromatography (GLC)

Latex (5–15 drops) exuded from several shoots was collected in spectroscopic grade acetone in acetone-washed vials. The supernatant containing the triterpenoids was decanted after 6–15 h into a fresh vial to remove excess coagulants, and the acetone was evaporated to facilitate shipment of samples to Bloomington, Indiana. To remove remaining particulate matter, each sample was reconstituted with 2 ml acetone and either centrifuged in a clinical centrifuge for 5 min or filtered through acetone-washed Whatman No. 1 paper and decanted into a fresh vial before being evaporated to dryness. Centrifugation was the preferred procedure. The residue in each vial was resuspended in 1 ml acetone as a stock solution. An aliquot (50–300 μl) of the acetone solution was transferred to a fresh vial and evaporated to dryness over nitrogen. The residue was resuspended in 100 μl acetone containing 0.5 mg ml^{-1} ml 4-androsten-3,17-dione (androstenedione) as an internal standard (IS). One microlitre of this sample typically was injected onto the chromatographic column. Duplicate samples of each accession were analyzed, and two or more injected runs of each sample were performed for comparison of chromatograms.

GLC analyses were performed on a Hewlett-Packard 5710A gas-liquid chromatograph equipped with a flame ionization detector and operated by programming the oven temperature from 240 to 310°C at 4°/min. Nitrogen was used as the carrier gas (20 ml min^{-1} flow rate). The injection port temperature was 250°C; the detector temperature was 350°C. Glass columns (2 mm ID × 2.43 m) were treated with 5% dimethyldichlorosilane in toluene and packed with 3% OV-1 on 100/120 mesh Supelcoport. Individual compounds were quantified on a Hewlett-Packard 3380A integrator with data on detected peaks expressed as percentage area.

Some components were identified at Berkeley, California, using published methods (Manners & Davis, 1984). Triterpenoids, as used here, include free and acetate forms of sterols and triterpenes.

Voucher specimens of each accession were deposited in herbaria at both the MRRL and Indiana University. Importantly, plants were not destroyed during the collection of latex. Living populations of each accession were maintained at the NDSU and MRRL nursery and greenhouses.

RESULTS AND DISCUSSION

Laticifers

Laticifers represent a salient anatomical character for many genera of the Euphorbiaceae and, in conjunction with morphological and other anatomical features, can contribute to the recognition of the natural phylogeny for genera of this family. Laticifers represent specialized cells genetically linked in evolution with distinctive plant form. Two types of laticifers are recognized, the articulated form consisting typically of numerous cells superimposed into a vessel arrangement that may be interconnected into an anastomosing network

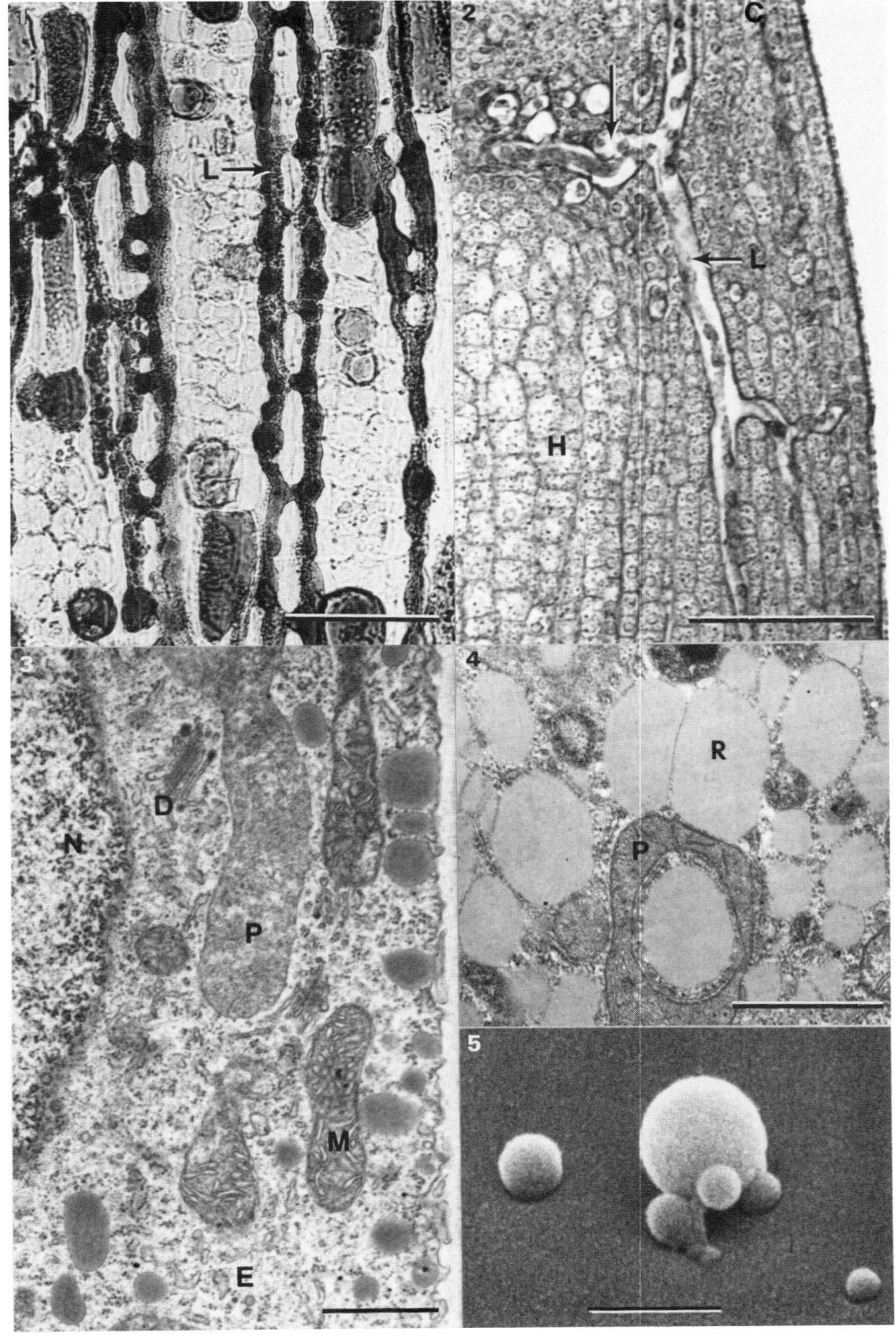

as in *Hevea* (Fig. 1), and the non-articulated form in which an individual cell undergoes intrusive growth, often branching during its growth, to form a very long cell as in *Euphorbia* (Fig. 2). The latter type, in fact, is the longest of biological cells, and in *Euphorbia* extends the entire length of the plant from the root tip to the shoot tip. The nucleus of the initial divides mitotically, as do the daughter nuclei, during the development of the non-articulated laticifer to form a multinucleated cell (Mahlberg & Sabharwal, 1967). Studies to the present time indicate that the laticifer when present in one member of a genus will occur in all the other taxa.

The distinctive development of both types of laticifers is indicative that they have evolved independently of each other and may reflect polyphyletic origins within vascular plants (Mahlberg, 1975). As evident from the few studies of these cell types in the Euphorbiaceae, the articulated laticifer is associated typically with the phloem, whereas the non-articulated form is derived from ground meristem tissue and penetrates intrusively throughout most tissues of the plant body during growth (Gomez, 1982; Mahlberg & Sabharwal, 1968). Similar developmental patterns for these cells are evident in other plant families (Mahlberg, 1961; Vertrees & Mahlberg, 1978; Nessler & Mahlberg, 1977). Laticifers are interpreted to be recently evolved cells types, compared to vascular elements, because they are so specialized and occur in so few families. Yet the fossil record indicates that the non-articulated form already was prominently developed in an arborescent plant of the Eocene (Mahlberg, 1975; Mahlberg, Field & Frye, 1984). It is pertinent to note, therefore, that both types of laticifers are stable, salient anatomical features, and after their appearance in a plant group they are conjoined in the evolution of taxonomic units, be it within the Euphorbiaceae or in the few other families that possess laticifers.

Pax (1884) recognized the importance of the laticifer in the taxonomy of the Euphorbiaceae. In Mueller's scheme (1873, 1874) he distinguished between those tribes lacking laticifers in the subfamily Phyllanthoideae from tribes possessing the articulated laticifer as contrasted to tribes possessing the non-articulated form distributed in the subfamily Crotonoideae (Table 1). More recently Webster (1975) included presence or absence of laticifers and latex colour, among other characters, to recognize five subfamilies of the Euphorbiaceae, although he did not utilize the laticifer character to distinguish between different groups.

Various authors recognize the family as being a wholly unnatural grouping (Cronquist, 1968; Pax, 1884; Takhtajan, 1969; Webster, 1975). The presence or absence of laticifers can contribute to an interpretation of the

---

Figures 1–5. Fig. 1. Articulated laticifer (L) in secondary phloem of *Hevea brasiliensis* Muell. showing anastomosed files of cells with rubber particles. Scale bar = 100 µm. Fig. 2. Non-articulated laticifer (L) in mature embryo of *Euphorbia marginata* Pursh. showing bifurcatively branched growth pattern. Branches from laticifer initial (arrow) at cotyledonary node developed into cotyledon (C) and into hypocotyl region (H). One branch has penetrated to proximity of epidermis. Scale bar = 100 µm. Fig. 3. Non-articulated laticifer in developing embryo of *E. terracina* L. showing components within the dense ribosomal content of the cell. A large central vacuole is typically absent at this stage. D, dictyosome; E, endoplasmic reticulum; M, mitochondrion; N, nucleus; P, plastid. Scale bar = 1 µm. Fig. 4. Early stage in rubber particle (R) formation in nonarticulated laticifer of maturing embryo of *E. terracina* L. P, plastid. Scale bar = 1 µm. Fig. 5. Isolated triterpenoid-containing rubber particles from *E. tirucalli* L. Particles differ in size. Scale bar = 0.5 µm.

Table 1. Classification of the Euphorbiaceae after Pax (1884)

I. Phyllanthoideae (laticifers absent; biovulatae)
  1. Caletieae Muell. Arg.
  2. Phylantheae Muell. Arg.
  3. Bridelieae Muell. Arg.
II. Crotonoideae (laticifers present; uniovulatae)
  A. Acalyphineae (articulated laticifers)
    1. Ricinocarpeae Muell. Arg.
    2. Acalypheae Muell. Arg.
    3. Dalechampieae Muell. Arg.
    4. Johannesieae, Garcieae, Heveeae Muell. Arg.
  B. Hippomanoineae (non-articulated laticifers)
    5. Hippomaneae Muell. Arg.
    6. Euphorbieae Muell. Arg.
    7. Crotoneae Muell. Arg.

phylogenetic relationships among genera comprising the Euphorbiaceae in conjunction with other accrued characters by segregating tribes on the basis of the laticifer character. Importantly, the laticifers represent conservative features that have evolved in synchrony with axial organization and, unlike some superficial morphological features, are less subject to variations induced by environmental conditions. Comparative anatomical and physiological studies of laticifers, in a manner analagous to tracheary elements (Cheadle, 1956), can contribute to the reinterpretation of evolutionary trends within this diverse group of genera. The treatment of the laticifers in this family by Pax is (1884) is reflective of an awareness of this interpretation and he showed remarkable insight into the nature of laticifers even though there was considerable controversy about the nature of laticifers in his time (Mahlberg, 1958).

Perhaps the continued recognition of laticifers within subfamilial units only perpetuates previous interpretations that the family represents an unnatural assemblage. My proposal corresponds in part to that of Pax (1884), but differs in that the subfamily Crotonoideae should be reorganized to group together those tribes that possess articulated laticifers as contrasted to those with non-articulated laticifers. Further, I propose that a more natural assemblage of this family would be represented by viewing the present Euphorbiaceae as three collateral families, one lacking laticifers including the Phyllanthoideae, a second possessing the articulated laticifer as in the Acalyphoideae and a third family composed of those present members possessing the non-articulated laticifer as represented in the subfamily Hippomanoideae of Pax. Additional investigations of genera within the present family are necessary to ascertain laticifer organization and distribution of laticifer types.

Latex, the cellular content of the laticifer, possesses a high osmotic pressure, is clear or coloured in appearance and typically is distinguished from other exudates by its coagulation upon exposure to air. At the structural level, as in *Euphorbia*, the non-articulated laticifer contains various organelles and membranes (Fig. 3). Latex, as from *Euphorbia*, is physiologically active in the synthesis and accumulation of diverse secondary products including triterpenes (Biesboer & Mahlberg, 1979; Groeneveld, 1976; Nemethy, Skrukrud, Piazza & Calvin, 1983; Ponsinet & Ourisson, 1968). Similar activities leading to the formation of other secondary products such as alkaloids and glycosides are present in laticifers in other genera and species. These substances are interpreted to serve a protective function against herbivory because they often are toxic to animals, including man.

One interpretation is that the occurrence of rubber in the laticifer in many plants is a consequence of coevolution with the appearance of toxic secondary products. The rubber serves as a sink for these often lipoidal substances which compartmentalizes them away from the metabolic stream, so enabling the laticifer to continue synthesizing them to unusually high concentrations without these compounds having adverse effects upon the cell or the plant. In the poinsettia (*E. pulcherrima* Willd.), for example, the triterpenoids and their esters compose 40% of the dry weight of latex (Spilatro & Mahlberg, 1986). These compounds can be extracted from the spheroidal rubber particles present in the latex (Fig. 4) from which they can be isolated as quantities of particles (Fig. 5) (Groeneveld, Furr & Mahlberg, in press).

Latex triterpenoid profiles in the chemotaxonomy of Euphorbia

The potential taxonomic value of the triterpenoids was recognized in previous studies (Anton, 1974; Ponsinet & Ourisson, 1965, 1968a,b). Extensive GLC analyses of latex from numerous leafy and succulent taxa of *Euphorbia* (in the sense of Pax) have shown that the triterpenoid profiles appear to be diagnostic for a taxon (Mahlberg & Pleszczynska, 1983; Mahlberg, Pleszczynska, Rauh & Schnepf, 1983). It also has been shown that these profiles are stable for a taxon when collected from different specimens of a species grown under different environmental conditions at different geographical locations worldwide as well as in specimens of normal and abnormal (cristata condition) morphogenetic physiology (Mahlberg, Pleszczynska & Furr, in press).

The stability of the profile within a taxon is evident from analyses of latex derived from different individuals of a single population and of different populations of the taxon. The triterpenoid profile for *E. maculata* L. was identical in latex collected from the parent plant and 17 seed-derived progeny growing under diverse climatic conditions ranging from very wet to extremely dry, and sampled at different times of the day during the 5 months growing season. In a study of *E. lathyris* L. derived from different sources in Europe, North America and South America, grown greenhouse conditions, the latex possessed a distinctive triterpenoid profile common to all accessions regardless of origin. The triterpenoid profile for a given species, therefore, possesses a stable profile among accessions of diverse origins, and is interpreted as typical for a species.

We have extended these latex triterpenoid analyses to *E. esula* L., leafy spurge and possibility-related taxa (Croizat, 1945), to aid in resolving taxonomic relationships among members of this complex. Eurasian plants were introduced into North America in imported grain during the early 1800s (Dunn, 1985) and have become troublesome weeds in extensive areas of western North America. The plant material, i.e. latex, is toxic to animals resulting in significant economic losses through reduction in animal production and acreage of utilizable grazing land (Messersmith & Lym, 1983). Proposed biocontrol programmes to control these weeds is predicated on knowledge of the ability to identify the species, to enable introduction of predatory organisms, such as

Table 2. Profile groups and sources of leafy spurge accessions

Group	Accession number	Origin of plant	Ebke & McCarty determinations
I	1	Krems, Austria	*E. esula* L.
	9	Cook County, Wyoming	*E. uralensis* Fisch.
	35	Kalkaska, Michigan	*E.* × *pseudovirgata* (Schur) Soó
	105	Becker County Minnesota	—*
	119	Becker County, Minnesota	—
II	7	Weiser, Idaho	*E.* × *pseudovirgata* (Schur) Soó
	15	Franklin, Manitoba	*E.* × *pseudovirgata* (Schur) Soó
	101	Sand Hills, North Dakota	—
	117	Becker County, Minnesota	—
III	67	Baker County, Oregon	*E.* × *pseudovirgata* (Schur) Soó

* These accessions collected separately from Ebke & McCarty materials.

insects, obligate on a given species. Several taxonomic evaluations of leafy spurges have resulted in diverse interpretations of the relationships of examined specimens, as reviewed by Ebke & McCarty (1983) and Radcliffe-Smith (1985). Both Dunn & Radcliffe-Smith (1980) and Ebke & McCarty (1983) emphasized the difficulty of identifying species differences between North American populations of the complex but both described morphologically separable species.

Ebke & McCarty (1983) examined 19 vegetative and floral morphological characters in 39 populations of leafy spurge including a population of *E. esula* obtained from Krems, Austria. They emphasized the absence or low level of correlations associated with nearly all characters for distinguishing between species among these populations. They reported that the ratios of the leaf width at the tip or at mid-leaf compared to the base of the leaf were the most useful characters for distinguishing between five recognizable taxa. On this basis they

Table 3. Triterpenoid components in chromatographic profiles of latex from ten accessions expressed as a percent of total triterpenoids*

Peak Number	Rention time†	Group I					Group II				Group III
		1	9	35‡	105	119	7	15	101	117	67
1	10.7	—	—	—	—	—	—	—	—	—	5
2	14.1	3	3	5	5	4	—	—	1	—	8
3	14.5	1	1	1	1	1	4	1	3	3	6
4	15.0	—	—	—	—	—	—	—	—	—	16
5	15.2	10	7	9	6	11	7	9	8	15	—
6	15.7	25	21	18	23	29	24	24	21	23	30
7	16.5	56	56	59	59	53	61	60	61	54	33
8	17.1	5	9	6	6	2	3	6	6	5	—
9	18.0	t	1	1	t	t	t	t	t	t	1

* Field-grown plants, except accession 35 which was grown in a greenhouse.
† Relative retention time corrected to IS of 6.50 min for all chromatograms.
‡ Average values from Table 4; data similar to field-grown plants.
§ Figures rounded to the nearest decimal.
—, Not detected; t, trace.

Table 4. Triterpenoid components in chromatographic profiles of latex from accessions 1 and 105 grown under field and greenhouse conditions

Accession	Condition	Retention time (min)*						
		14.1	14.5	15.2	15.7	16.5	17.1	18.0
1	A Field	3	1	10	25	56	5	t
	B Greenhouse	3	1	10	18	61	6	1
105	C Field	5	1	6	23	59	6	t
	D Greenhouse	7	1	5	17	62	6	1

* Relative RT corrected to IS of 6.50 min for all chromatograms; t, trace evident on the chromatogram.

separated *E. esula* L. from *E. uralensis* Fisch., and separated these from the remainder of the samples which they considered *E.* × *pseudovirgata* (Schur) Soó. Two taxa of these five, *E. agraria* M. Bieb. and *E. cyparissias* L., were interpreted as hybridizing with other taxa, thereby adding to the difficulty of identifying taxa in the complex. Radcliffe-Smith (1985) lists a key to species and hybrids based on leaf characteristics in an attempt to identify the various taxa, his interpretation of the number of taxa differing from that of Ebke & McCarty (1983).

Our GLC study of the latex triterpenoids included analyses of 10 spurge accessions, six accessions from the original Ebke and McCarty materials and four collections from local populations in Minnesota and North Dakota (Table 2). Triterpenoid profiles contained five to seven peaks, depending on the accession, with the percentage of each component varying from 0.1% to 61% of the total triterpenoids detected (Table 3). Compounds comprising less than 0.5% of the triterpenoids were not considered in the interpretation of the data.

Essentially identical triterpenoid profiles were observed in plants of the same accession regardless of whether they were grown in the field or greenhouse (Table 4; Fig. 6).

Accession 1, obtained originally from Krems, Austria, represents *E. esula* and all comparisons of GLC peaks are made relative to this accession (Table 3). Accession 1 had six peaks that represented 0.5% or more of the total triterpenoids. Peak 7 was the major component for all accession, and was identified as 24-methylene cycloartenol. Peak 6 was the second most abundant component and was identified as cycloartenol (Fig. 7). Together the two compounds were 81% of the total triterpenoids from accession 1. Peak 2 was identified as euphol, and comprised 0.5 to 8% of the total triterpenoids among the different accessions (Fig. 7).

The 10 different accessions can be separated into three different groups on the basis of qualitative and quantitative analyses of the triterpenoids (Fig. 8). Groups I and II have profiles that contained essentially all of the same major peaks, but differed in the relative quantities of the components. Group II is separated from Group I on the basis of the relative amounts of euphol (compound 2) and the as yet unidentified triterpenoid (compound 3) appearing at retention time (RT) 14.1 and 14.5, respectively. The relative amounts of these two constituents are the inverse of the same two triterpenoids in Group I. The ratios of euphol to compound 3 for Group I ranged from 3 to 5, but only 0

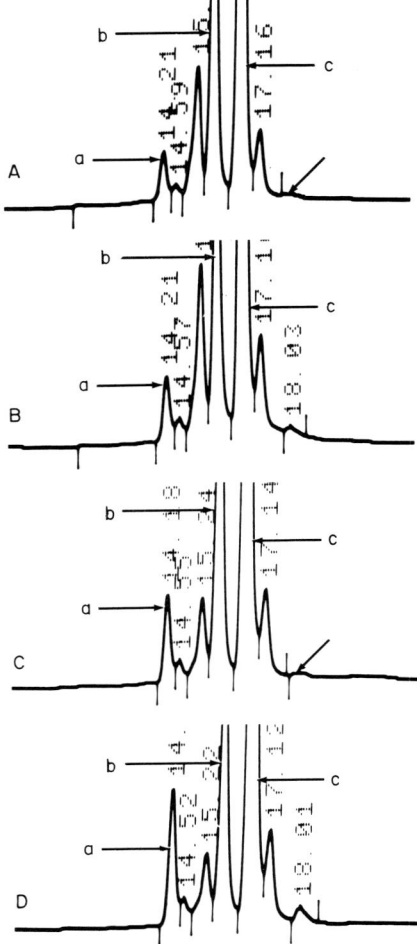

Figure 6. GLC profiles of latex samples from two accessions, 1 (*E. esula*) and 105, of Group I. Stability of the triterpenoid profile is illustrated between field (A) and greenhouse-grown (B) materials of accession 1, and field (C) and greenhouse-grown (D) materials of accession 105. a, Euphol; b, cycloartenol; c, 24-methylene cycloartenol. Angled arrow at right identifies a trace of the compound in field-grown specimens (A and C) which was detectable in measurable quantities at retention time 18.0 min in the greenhouse-grown plants (B and D).

to 0.17 for Group II. The relative amounts of the other triterpenoids (5, 6, 7 and 8) occur in similar yields in all of the accessions in Group I and Group II.

Accessions 9, 35, 105 and 119 are placed together with accession 1 to comprise Group I because they also contain six triterpenoids representing 0.5% or more of the total triterpenoids, and the quantities of the individual constituents agreed closely to each other (Table 3; Fig. 8).

Accession 67 has a profile distinctly different from those of the other nine accessions, and therefore is listed separately as Group III (Table 3; Fig. 8). Several characteristics distinguish this profile from the other two. First, there are seven triterpenoids compared with five or six for the other accessions; two of the seven triterpenoids are not observed in any of the other profiles. Second,

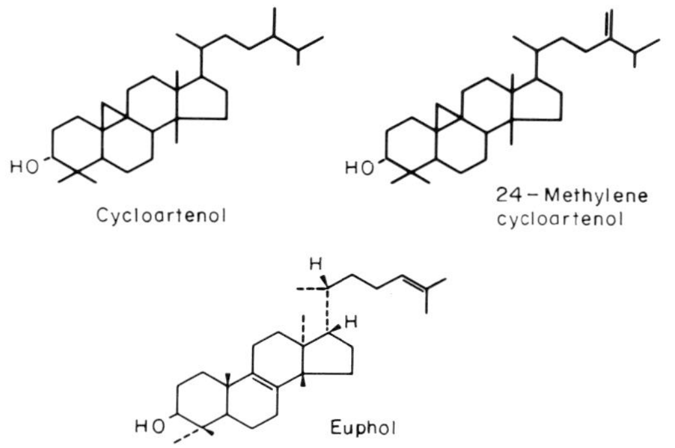

Figure 7. Chemical structures of the three identified triterpenoids from leafy spurge latex.

compound 1 represents 5% of the total triterpenoids in this accession; it is not observed in any of the other profiles. Third, the occurrence of high percentages of euphol and compound 3 (RT 14.1 and 14.5), 8% and 6%, respectively, of the total triterpenoids, compared to 0 to 5% (euphol) and 1% to 4% (peak 3) for the other accessions. Fourth, the ratio of 24-methylene cycloartenol to cycloartenol (compounds 7 and 6, respectively) is only 1.1, in contrast to ratios of 1.8 to 2.9 for the other nine accessions.

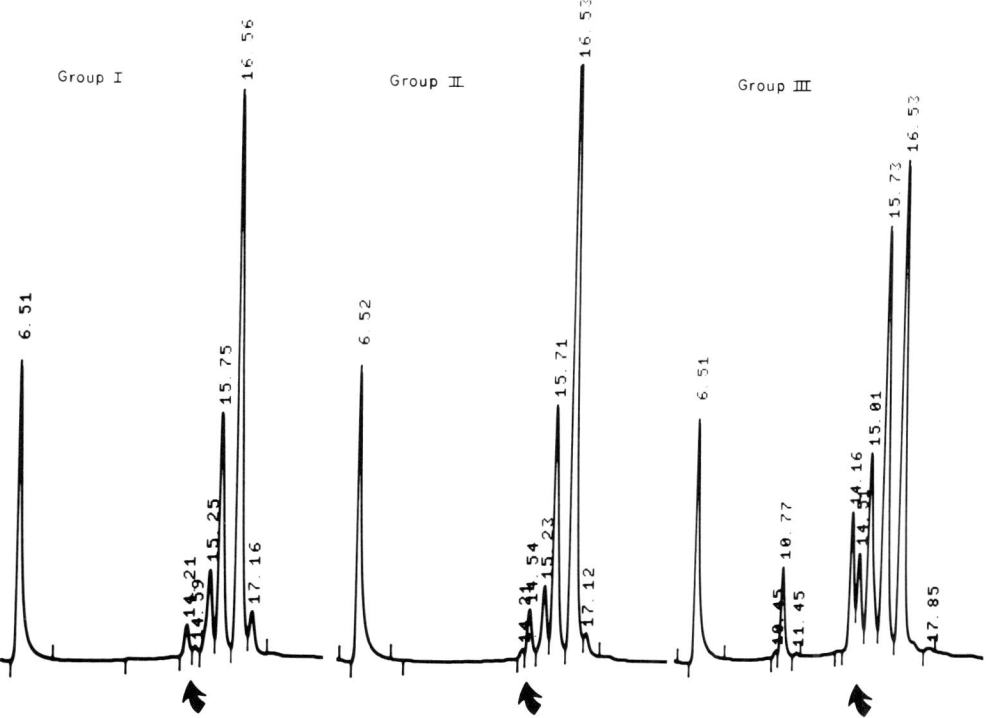

Figure 8. Characteristic GLC profiles of the latex triterpenoids for the three groups of accessions. Group I includes accessions 1, 9, 35, 105 and 119, Group II includes accessions 7, 15, 101 and 117. Group III includes accession 67. The peaks at retention time 6.50 to 6.52 are the internal standards. Retention time values represent time in minutes.

Table 5. Percent triterpenoid composition of latex from 13 different plants of accession 35 grown under greenhouse conditions*

Sample number	Retention time (min)†						
	14.1	14.5	15.2	15.7	16.5	17.1	18.0
1	6*	2	9	15	64	3	t
2	4	2	8	16	65	4	t
3	4	2	9	18	60	5	1
4	4	2	9	17	62	6	t
5	8	t	9	16	52	9	3
6	5	1	9	18	59	7	t
7	5	1	10	18	54	9	3
8	4	1	9	18	58	8	2
9	3	1	10	21	60	5	t
10	5	2	9	17	63	4	t
11	5	1	11	20	54	7	t
12	2	1	9	25	52	8	1
13	5	1	9	18	59	6	t
Average	4.6	1.3	9.2	18.2	58.6	6.2	0.7
S.D.	2.1	0.4	0.8	2.5	4.4	1.9	1.1

* All figures rounded to nearest decimal.
† Relative RT corrected to IS of 6.50 min for all chromatograms; t, trace evident in the chromatograms.

An analysis of variation for components within a profile of accession 35, in which we examined 13 different clones of a population, showed a high level of stability for each compound (Table 5). 24-Methylene cycloartenol (RT 16.5) was $58.6 \pm 4.4\%$ of the triterpenoids. Cycloartenol (RT 15.7) was $18.2 \pm 2.5\%$ of the triterpenoids. Euphol (RT 14.1) and the unidentified compounds at RT's of 14.5, 15.2, 17.1 and 18.0, also had consistently stable levels that reflected a possible homeostatic mechanism controlling their synthesis. All plants of accession 35 were cloned from an original collection of Ebke & McCarty (1983) and subsequently reproduced from roots or rooted cuttings. The relative stability of a given compound was reflected in the low deviation for each of the common peaks and was indicative of the high level of stability for triterpenoid biosynthesis within the laticifer cell in leafy spurge.

The above data for the 10 accessions show the taxa *E. esula*, *E. uralensis*, and *E.* × *pseudovirgata* from Ebke & McCarty (1983), as well as two additional accessions, to possess a profile in common under Group 1. Other accessions of reputed *E.* × *pseudovirgata* possessed the profile described in Group II, yet another accession of *E.* × *pseudovirgata* possessed a different profile taxon described as Group III. If the triterpenoid profile is distinctive for a taxon we should not expect to find *E. uralensis* included in Group I with *E. esula*. Further, we should not expect to find *E.* × *pseudovirgata* accessions distributed among the three profile groups.

Since leaf measurements, particularly the leaf width ratio for the tip with base and mid region with base, were employed by other authors (Ebke & McCarty, 1983; Dunn & Radcliffe-Smith, 1980; Radcliffe-Smith, 1985) to distinguish between taxa, we examined these characters in our accessions obtained from Ebke & McCarty and compared them with the corresponding data published by these authors (Table 6).

Table 6. Comparison of leaf measurements of leafy spurge accessions from Groups I, II, and III

Group	Accession	Ebke & McCarty determination†	Data source*	Leaf length (mm)	Tip	Mid	Base	Tip base	Mid base
I	1	*E. esula* L.	ND†	58	4.2	4.7	3.2	1.3	1.5
I	9	*E. uralensis* Fisch. ex fink	E*	60	5.3	5.0	3.1	1.7	1.6
			ND	56	5.3	5.4	3.6	1.5	1.5
			E	75	3.9	5.8	6.0	0.6	1.0
	35	*E.* × *pseudovirgata* (Schur) Soó	ND	49	4.4	4.7	3.4	1.3	1.4
			E	55	5.4	6.3	5.1	1.0	1.2
	105	—	ND	49	4.2	4.7	3.2	1.3	1.5
	119	—	ND	59	6.1	7.0	4.5	1.3	1.5
II	7	*E.* × *pseudovirgata* (Schur) Soó	ND	51	5.9	6.5	3.9	1.5	1.7
			E	59	5.8	7.4	5.9	1.0	1.2
	15	*E.* × *pseudovirgata* (Schur) Soó	ND	64	7.5	7.9	4.3	1.7	1.8
			E	54	3.8	5.0	3.9	1.0	1.3
	101	—	ND	59	4.0	4.2	2.9	1.4	1.4
	117	—	ND	48	7.0	7.3	4.8	1.4	1.5
III	67	*E.* × *pseudovirgata* (Schur) Soó [approaches *E. waldsteinii* (Sojak) Radcliffe-Smith]	ND	58	6.6	7.2	3.2	2.1	2.2
			E	55	6.4	9.6	7.0	0.9	1.4

—, Accession not from Ebke & McCarty collection.
* Data of Ebke & McCarty (1983) from field grown plants.
† North Dakota collection from Ebke & McCarty grown in greenhouses in North Dakota.

For *E. uralensis*, our data differ significantly from those of Ebke & McCarty in that the leaf on average was wider at the tip (1.5) and mid region (1.5) than at the base, whereas in the previous study the base (0.6) was wider than the tip region and equal (1.0) at the mid regions. For *E.* × *pseudovirgata* in Group III, the previous measurements of Ebke & McCarty (1983) indicated the leaf base (0.9) region to be wider than the tip portion, whereas our measurements show the tip (2.1) to be wider than the leaf base. For *E.* × *pseudovirgata* in Group I and II, Ebke & McCarty report leaf width to be equal along its length or wider at the tip and mid regions than at the leaf base, whereas our ratio data show the tip and mid regions to be wider than the base. These data indicate that leaf width ratios can vary on leaves derived from the same plant, and bring into question the utility of leaf data as diagnostic criteria for distinguishing between taxa of the spurges. We have found leaf data not to be reliable characters.

When leaf width analyses are applied to the spurges listed in Table 6, then all the specimens grown in North Dakota (ND), except accession 67 of Group III, would be classified as *E. esula*. The results of our chemotaxonomic study do not support such a designation, but indicate that purported *E. uralensis* and *E.* × *pseudovirgata* (accession 35) should be recognized as *E. esula*, whereas *E.* × *pseudovirgata* (accession 7, 15) along with accessions 101 and 117 form a distinctive Group II.

We interpret Group I as the chemotaxon for *E. esula* L. The triterpenoid profiles of accessions 9, 35, 105 and 119 also place them in this chemotaxon, and each is recognized as *E. esula*. The stability of such a profile, as for *E. esula* grown under diverse conditions and collected at different times of the year, is indicative of the utility of the triterpenoid profile as a reliable character to identify taxonomic relationships between spurge populations. Group II with accessions 7, 15, 101, and 117 and Group III with accession 67 also are recognized as distinctive chemotaxa. In the absence of both chemical and morphological data to support the identification of accessions 7, 15, 35 and 67 as *E.* × *pseudovirgata*, it is not a valid taxon in our study, and no species designations are associated with the Group II and III chemotaxa at the present time. It is anticipated that additional profile patterns, or chemotaxa, will emerge from subsequent triterpenoid studies of the spurges.

A more clear understanding of the taxonomy of adventive leafy spurges in North America can be gained only after studies of these plants in their Eurasian habitats. Analyses of GLC profiles in conjunction with utilization of relevant morphological data from Eurasian populations of *E. esula* and other species will establish the chemotaxa of Eurasian spurges which then can be utilized to document the adventive species in North America.

The factors that control the qualitative and quantitative differences of triterpenoids among species of *Euphorbia* have not been identified. Experimental studies in hybridization of the spurges are necessary to determine the bases of triterpenoid inheritance as individual and combinations of compounds that compose a profile. These studies also will help to clarify whether reputed hybrids exist and, if so, how hybridization affects the profile. An understanding of the mechanism regulating triterpenoid composition will provide insight into the evolutionary trends for components of the squalene pathway and more specifically into the evolutionary trends among the spurges.

ACKNOWLEDGEMENTS

The authors wish to thank Dr F. R. Turner and Mr K. Dusbabek for assistance in this study.

Mention of trademark or proprietary product does not constitute a guarantee or warranty of a product by USDA nor imply approval to exclusion of other suitable products.

REFERENCES

ANTON, R., 1974. Étude chimiotaxonomique sur le genre *Euphorbia* (Euphorbiacées). Thèse de Docteur, Université Louis Pasteur, Strasbourg.

BIESBOER, D. & MAHLBERG, P., 1979. The effect of medium modification and selected precursors on sterol production by short-term callus cultures of *Euphorbia tirucalli*. *Journal of Natural Products, 42:* 648–657.

CHEADLE, V., 1956. Research on xylem and phloem—progress in fifty years. *American Journal of Botany, 43:* 719–731.

CRONQUIST, A., 1981. *An integrated system of classification of flowering plants.* New York: Columbia University Press.

CROIZAT, L., 1945. *Euphorbia esula* in North America. *American Midland Naturalist, 33:* 231–243.

DUNN, P., 1985. Origins of leafy spurge in North America: In A. K. Watson (Ed.), *Leafy Spurge:* 7–13. Champaign: Weed Science Society.

DUNN, P. & RADCLIFFE-SMITH, A., 1980. The variability of leafy spurge (*Euphorbia* spp.) in the United States. *North Cen. Weed Control Conference Research Report, 37:* 48–53.

EBKE, D. & McCARTY, M., 1983. A nursery study of leafy spurge (*Euphorbia* spp.) complex from North America. *Weed Science, 31:* 866–873.

GOMEZ, J., 1982. Anatomy of *Hevea* and its influence on latex production. *Malaysian Rubber Research and Development Board Monograph no. 7.* Kuala Lumpur.

GROENEVELD, H., 1976. Biosynthesis of phytosterols and latex triterpenes in *Hoya australis* and *Hoya carnosa*. *Acta Botanica Neerlandica, 25:* 227–250.

GROENEVELD, H., FURR, M. & MAHLBERG, P., in press. Sites of *in vitro* triterpene synthesis in *Euphorbia* latex. *Acta Botanica Neerlandica.*

MAHLBERG, P., 1958. Ontogenetic and experimental studies on the laticifers of *Nerium oleander* L. and *Euphorbia marginata* Pursh. Doctoral dissertation, University of California, Berkeley.

MAHLBERG, P., 1961. Embryogeny and histogenesis in *Nerium oleander* L. II. Origin and development of the nonarticulated laticifer. *American Journal of Botany, 48:* 90–99.

MAHLBERG, P., 1975. Evolution of the laticifer in *Euphorbia* as interpreted from starch grain morphology. *American Journal of Botany, 62:* 577–583.

MAHLBERG, P., FIELD, D. & FRYE, J., 1984. Fossil laticifers from brown coal deposits of the Geiseltal. *American Journal of Botany 71:* 1192–1200.

MAHLBERG, P., PLESZCZYNSKA & FURR, M., in press. Comparison of latex triterpenoid profiles between normal and cristata taxa of *Euphorbia* (Euphorbiaceae). *Annals of the Missouri Botanical Garden.*

MAHLBERG, P. & PLESZCZYNSKA, J., 1983. Phylogeny of *Euphorbia* interpreted from sterol composition of latex, In J. Felsenstein (Ed.), *Numerical Taxonomy:* 500–504. Berlin: Springer.

MAHLBERG, P., PLESZCZYNSKA, J., RAUH, W. & SCHNEPF, E., 1983. Evolution of succulent *Euphorbia* as interpreted from latex composition. *Bothalia, 14:* 533–539.

MAHLBERG, P. & SABHARWAL, P., 1967. Mitosis in the nonarticulated laticifer of *Euphorbia marginata*. *American Journal of Botany, 54:* 465–472.

MAHLBERG P. & SABHARWAL, S., 1968. Origin and development of nonarticulated laticifers in the embryo of *Euphorbia marginata*. *American Journal of Botany, 55:* 375–381.

MESSERSMITH, C. & LYM, R., 1983. Distribution and economic impacts of leafy spurge in North Dakota. *North Dakota Farm Research, 40:* 8–13.

MANNERS, G. & DAVIS, D., 1984. Epicuticular wax constituents of North America and European *Euphorbia esula* biotypes. *Phytochemistry, 23:* 1059–1062.

MUELLER, J., 1873. Euphorbiaceae. *Flora Brasiliensis, 11:* 1–292.

MUELLER, J., 1874. Euphorbiaceae. *Flora Brasiliensis, 11:* 293–750.

NEMETHY, E., SKRUKRUD, C., PIAZZA, G. & CALVIN, M., 1983. Terpenoid biosynthesis in *Euphorbia* latex. *Biochimica et Biophysica Acta, 780:* 343–249.

NESSLER, C. & MAHLBERG, P., 1977. Ontogeny and cytochemistry of alkaloid vesicles in laticifers of *Papaver somniferum* L. (Papaveraceae.) *American Journal of Botany, 64:* 541–551.

PAX, F. 1884. Die Anatomie der Euphorbiaceen in ihrer Beziehung zum System derselben. *Engler Botaniche Jahrbucher, 5:* 384–421.

PONSINET, G. & OURISSON, G., 1965. Études chemio-taxonomique dans la famille des Euphorbiacées. I. Introduction générale et séparation et identification des triterpènes tétracycliques monohydroxylés naturels. *Phytochemistry, 4:* 799–809.

PONSINET, G. & OURISSON, G., 1968a. Les triterpènes des latex d'*Euphorbia*. Contribution à une étude chimo-systématique du genre *Euphorbia*. *Adansonia, 8:* 27–231.

PONSINET, G. & OURISSON, G., 1968b. Aspects particuliers de la biosynthèse des triterpènes dans le latex d'*Euphorbia*. *Phytochemistry, 7:* 757–765.

RADCLIFFE-SMITH, A., 1985. Taxonomy of North American leafy spurge. In A. K. Watson (Ed.), *Leafy Spurge:* 14–25. Champaign: Weed Science Society.

SPILATRO, S. & MAHLBERG, P., 1986. Latex and laticifer starch content of developing leaves of *Euphorbia pulcherrima* Willd. *American Journal of Botany 73:* 1312–1318.

TAKHTAJAN, A., 1969. *Flowering Plants, Origin and Dispersal*. [Translated from the Russian by C. Jeffrey.] Washington: Smithsonian Institution Press.

VERTREES, G. & MAHLBERG, P., 1978. Structure and ontogeny of laticifers in *Cichorium intybus* L. (Compositae). *American Journal of Botany, 65:* 746–771.

WEBSTER, G., 1967. The genera of Euphorbiaceae in the southeastern United States. *Journal of the Arnold Arboretum 48:* 303–361.

WEBSTER, G., 1975. Conspectus of a new classification of the Euphorbiaceae. *Taxon, 24:* 593–601.

New aspects of rubber biosynthesis

BERNARD L. ARCHER AND BRAIN G. AUDLEY

Tun Abdul Razak Laboratory, Brickendonbury, Hertford, Hertfordshire, SG13 8NL

Received May 1986, accepted for publication September 1986

ARCHER, B. L. & AUDLEY, B. G., 1986. **New aspects of rubber biosynthesis.** Following a review of the site of rubber biosynthesis in *Hevea brasiliensis* and *Parthenium argentatum*, evidence is given for the initiation of polyisoprene molecules from *trans*-terpenoid precursors including geranylgeranyl pyrophosphate. All *trans*-^{14}C-geranylgeraniol has been isolated from incubations of *H. brasiliensis* latex serum with ^{14}C-isopentenyl pyrophosphate. Gel-filtration chromatography of the serum yields very small rubber particles of high biosynthetic activity, and two proteinaceous fractions. One of these increases the biosynthesis of rubber and may contain the enzyme, isopentenyldiphosphate Δ-isomerase, whilst the other appears to inhibit rubber formation. The nature and molecular weight of the rubber formed *in vitro* is discussed and a mechanism for the *de novo* formation of rubber particles is suggested.

ADDITIONAL KEY WORDS:—dimethylallyltransferase – geranylgeranyl pyrophosphate – *Hevea brasiliensis* – isopentenyldiphosphate Δ-isomerase – *Parthenium argentatum* – polyisoprene – polyprenol – rubber transferase.

CONTENTS

Introduction	181
Materials and methods	182
Initiation of rubber biosynthesis	184
Biosynthesis of oligoisoprenoids in *H. brasiliensis* latex	187
Biosynthetic factors in the serum of *H. brasiliensis* latex	190
The molecular structure of rubber	191
The nature and molecular weight of rubber biosynthesized by washed rubber particles	192
Conclusions	194
Acknowledgements	195
References	195

INTRODUCTION

Rubber, or *cis*-1,4-polyisoprene, is formed in over 1800 species of plants distributed amongst 300 genera of seven families (Backhaus, 1985). Archer & Audley (1973) list some of the more important species, but today rubber is produced commercially only from *Hevea brasiliensis* Muell. Arg. (Euphorbiaceae), which yields 4 million tonnes of rubber from about 7.5 million hectares, per year.

In *H. brasiliensis*, latex is produced and stored in the inner cortex in laticifers, where it forms a major part of the cytoplasm. These specialized vessels are formed from chains of contiguous cells in the phloem, where they are arranged as sheaths concentric with the bark (Dickenson, 1969; Gomez & Moir, 1979). Between the vessels in each ring, there are anastomoses which allow withdrawal of latex from a large area of bark by means of a single tapping cut. Rudall (1987) has reviewed the anatomy of the laticifers in the Euphorbiaceae, whilst those found in other families have been discussed by Esau (1965).

At the turn of the century, *Parthenium argentatum* A. Gray, or guayule, was an important source of rubber but, although interest has recently been shown in growing this again in the arid regions of North Mexico, commercial production of rubber from this source is at present negligible. This is mainly because there are no laticifers in *P. argentatum* and in order to extract the rubber from the tissues, the whole bush has to be destroyed. In guayule, the rubber is produced in the cytoplasm of the parenchyma cells, but as the cells mature, the rubber particles appear to migrate into the vacuole, where they are stored (Backhaus & Walsh, 1983; Goss *et al.*, 1984). It is the ease with which rubber, in the form of latex, can be obtained on a routine basis from the trunk of *H. brasiliensis* which has made it pre-eminent as the commercial source of the natural product.

Rubber biosynthesis in *H. brasiliensis* and in *P. argentatum* has been reviewed most recently by Archer (1980), Benedict (1983) and Backhaus (1985). Early work in this field had shown that rubber particles are essential for the conversion of IPP into rubber, and it was proposed at the time that much of the observed uptake was due to elongation of existing rubber pyrophosphate chains (Archer *et al.*, 1963). McMullen & McSweeney (1966) had concluded that the enzyme necessary for the uptake of isopentenyl pyrophosphate (IPP)—rubber transferase, EC 2.5.1.20—was present on the surface of the rubber particles and in the serum of latex, and Archer & Cockbain (1969) described a method for its assay and isolation. Madhavan & Benedict (1984) have recently developed this technique for the separation of the enzyme from the leaves of guayule (*Parthenium argentatum*). The same authors also compared its activities in *Parthenium*, *Ficus* and *Euphorbia* but, as discussed below (see Conclusions), the assay method may not have been sufficiently well founded. Several authors have isolated prenyl pyrophosphate transferases, which catalyse the formation of polyprenols as far as C_{55}, from a number of sources. Examples of such work including that on compounds containing *cis*-double bonds, are given in papers by Kandutsch, Paulus, Levin & Bloch (1964), Oster & West (1968), Barnard & Popják (1981), Ishii, Sagami & Ogura (1986) and also Hemming (1983).

MATERIALS AND METHODS

Fresh H. brasiliensis *latex:* This was obtained from trees grown in a tropical greenhouse at Brickendonbury, Hertford. The white fraction was prepared by the procedure of Archer, Audley, Cockbain & McSweeney (1963).

Freeze-dried latex serum solids (CSS): These were prepared from *H. brasiliensis* latex as described by McMullen (1959) and supplied by the Rubber Research Institute of Malaysia, Kuala Lumpur, Malaysia. The rubber content (*c.* 5 mg g^{-1}) was measured by pyrolysis-GLC of a methylene chloride extract (McSweeney, pers. comm.).

Unlabelled pyrophosphates: Isopentenyl pyrophosphate (tri-lithium salt) and dimethylallyl pyrophosphate (*cyclo*-hexylammonium salt) were those used previously (Archer *et al.*, 1963). Neryl and geranyl pyrophosphates (ammonium salts) were prepared from the alcohols by the method described by Cornforth & Popják (1969). The compounds were generally used without purification and the allylic pyrophosphate content was determined after fractionation by TLC on silica gel H (Cornforth & Popják, 1969), followed by acid hydrolysis and phosphorus analysis. The *trans,trans*-farnesyl pyrophosphate was a gift from Dr G. T. Phillips, Shell Research, Sittingbourne, Kent, and geranylgeranyl pyrophosphates were donated by Prof. C. M. Allen, University of Florida, Gainesville, U.S.A.

1-^3H-Neryl and geranyl pyrophosphates. The alcohols were prepared at high specific activity by reducing citral (Koch Light, Haverhill, Suffolk) with NaB^3H_4 (Amersham International, Bucks,) separated by TLC (Cardemil, Vicuna, Jabalquinto & Cori, 1974) and phosphorylated as above. The pyrophosphates were purified by paper chromatography (Banthorpe, Doonan & Gutowski, 1977) and then immediately before use by TLC (Cornforth & Popják, 1969). Their specific activities were calculated from that of the NaB^3H_4, though it was later realized that the compounds lose tritium on storage (see Banthorpe, Christou, Pink & Watson, 1983).

1-^{14}C-Isopentenyl pyrophosphate and 2-^{14}C-mevalonolactone: These were obtained from Amersham International; the lactone was converted to the sodium salt before use.

Other material: Unlabelled mevalonolactone was obtained from Sigma, Poole, Dorset, and *trans,trans,trans*-geranylgeraniol was a gift from Prof. Y. Tanaka, Tokyo University of Agriculture and Technology, Tokyo, Japan.

Washed rubber particles (WRP) and other fractions: A solution of CSS (6 or 12%) was prepared by stirring the solids in 0.25 M Tris–HCl, pH 8.0, containing 1% Tween-20 and 5 mM 2-mercaptoethanol. Ten millilitres of this extract were then chromatographed on a 150 × 44 mm column of Ultrogel AcA-34 (LKB Instruments, S. Croydon, Surrey) with the same buffer mixture as eluant. The flow rate was 1 ml min^{-1} and detection was at 280 nm. A 150 × 16 mm column was used to obtain WRP from white fraction. All operations were at 4°C.

Concentration of washed rubber suspensions: A modification of the method of Anderson *et al.* (1979) with Bio-Gel P-6 (Bio Rad, Watford, Herts) was used. The dry gel was added to the suspension and after standing, the mixture squeezed in a syringe to recover the rubber; the procedure was repeated to yield a suspension of suitable concentration. A similar method was used to concentrate fractions B and C (see Fig. 4).

Purification of labelled rubber: Generally the method of Archer *et al.* (1963) was used, 'carrier' rubber in the form of latex being added when necessary. In some experiments, particularly those with tritiated compounds, the incubations were arrested by the addition of 0.2 M EDTA (pH 8) and the mixture first passed through 60 × 12 mm columns of Bio-Gel P-6 ('minicolumns') using a modification of the method of Fry, Courtland White & Goldman (1978). This enabled the removal of large amounts of radioactivity without coagulation of the rubber. In some cases (see below), the rubber eluted from the minicolumns was used directly. However, in other work the rubber, after the addition of carrier, was precipitated with ethanol and further purified as described above.

Effect of pH on solubility of rubber: ^{14}C-Rubber was prepared from WRP using a scaled-up version of the mixture given in Fig. 1A (incubation time, 18 h). After addition of EDTA, the rubber was purified on minicolumns equilibrated with ammonia, pH 10, and samples dried (by N_2). The films were incubated at various pH values for 1 h at 30°C and after adjusting to pH 12 (with KOH), again dried. Tetrahydrofuran (THF) containing 0.025% of the antioxidant, 4-methyl-2,6-di-*tert*.butyl phenol (BHT) was added and the mixtures allowed to stand overnight in the dark. After centrifugation, radioactivity in the samples was determined, and it showed that 90% or more of the ^{14}C-rubber had apparently dissolved. Aliquots of the 'solutions' were filtered through 30 × 10 mm columns of TLC silica (Merck Type 60) packed in THF, and radioactivity in the effluents determined. The solubility of the rubber was defined as the percentage of ^{14}C passing through the column.

Molecular weight of ^{14}C-rubber: ^{14}C-rubber was produced using conditions similar to those in Fig. 1A, incubation time 1 h. After addition of EDTA, the labelled material was purified on a minicolumn as above, then dried and treated with 1 M HCl for 1 h at 30°C. After evaporation of the acid, the film was taken up in THF containing BHT and a sample of the solution analysed by gel-permeation chromatography, fractions being collected at timed intervals: recovery of ^{14}C from the columns was 80%. Polystyrenes of known molecular weight were used to calibrate the chromatograph after suitable correction.

Determination of radioactivity: Scintillation counting was with 0.6% butyl-PBD in toluene. With aqueous rubber suspensions, etc., a mixture of 0.9% butyl-PBD and Metapol (2:1) was employed.

Gas–liquid chromatography (GLC): This was carried out on a column of 10% SP 1000 on Chromosorb W–AW DMCS, programmed from 170 to 250°C. The effluent was split, detection was by flame ionization, and the uncombusted fractions were collected.

High-pressure liquid chromatography (HPLC): Labelled geranylgeraniol was purified on silica using 2% isopropanol in hexane; detection was at 215 nm.

INITIATION OF RUBBER BIOSYNTHESIS

The first step in the biosynthesis of all the *trans*-isoprenoids from isopentenyl pyrophosphate (IPP) is the isomerization to dimethylallyl pyrophosphate (DMAPP) mediated by the enzyme isopentenyldiphosphate Δ-isomerase (EC 5.3.3.2), henceforth referred to as 'isomerase'. The reaction of this DMAPP with successive IPP molecules is then catalysed by transferase enzymes (e.g. dimethylallyltransferase, EC 2.5.1.1; geranyltransferase, EC 2.5.1.10) yielding C_{10}, C_{15}, etc. A similar mechanism for rubber biogenesis was proposed by Lynen (1969), who obtained evidence for the isomerase enzyme on the surface of the rubber particles present in *H. brasiliensis* latex. The requirement for isomerase in *Hevea* latex is clearly very small: owing to the high molecular weight of rubber (1×10^5 to 5×10^6), a simple calculation shows that only one

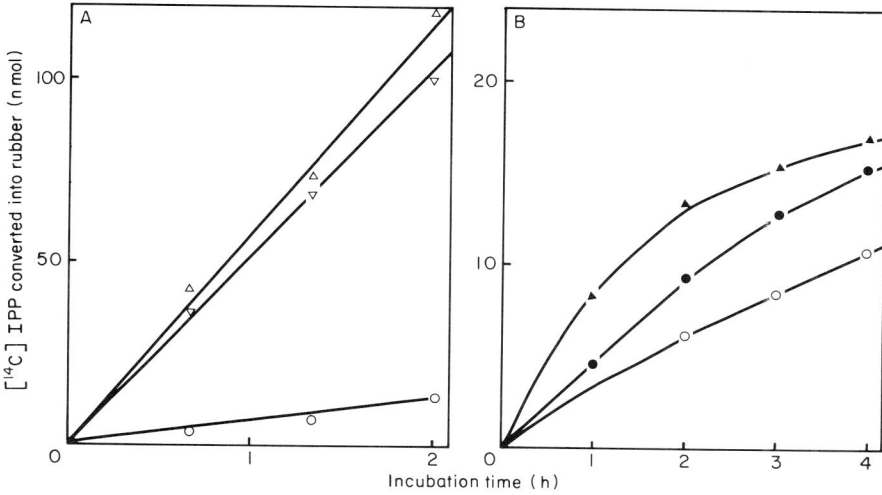

Figure 1. A. Effect of NPP and GPP on conversion of ^{14}C-IPP to rubber by WRP isolated from CSS. Incubations (25°C) contained WRP suspension, 100 μl (c. 300 μg rubber); 1-^{14}C-IPP, 200 nmol (46,000 dpm); NPP or GPP, 4 nmol; MgCl$_2$, 650 nmol: in 0.2 M Tris–HCl, pH 8, total volume, 130 μl. △, NPP; ▽, GPP; ○, control (no NPP or GPP). B. Effect of NPP on conversion of ^{14}C-IPP to rubber by WRP isolated from fresh latex white fraction. Incubations contained WRP suspension, 100 μl (c. 11 mg rubber); 1-^{14}C-IPP, 200 nmol (46,000 dpm); MgCl$_2$, 650 nmol; and 650 nmol (▲) or 20 nmol (●) NPP; in 0.2 M Tris–HCl, pH 8, total volume, 130 μl: ○ = control (no NPP).

starter molecule is needed for every 3×10^4 to 1×10^5 molecules of IPP polymerized.

In early experiments in this field, it was found that DMAPP, instead of accelerating the uptake of ^{14}C-IPP into rubber, caused a diminution in the rate (see Archer & Audley, 1967). However, as natural rubber is generally accepted to have all its double bonds in the *cis*-configuration, the *cis*-di-isoprenoid, neryl pyrophosphate (NPP) was suggested as a likely initiator for its biosynthesis. Therefore to circumvent potential problems with DMAPP, investigations on initiation were carried out with the C_{10}-compound.

Initial experiments in which 1-^{3}H-NPP and unlabelled IPP were incubated with either the white fraction of latex obtained from trees grown in the greenhouse or with reconstituted freeze-dried serum (CSS, see Methods), showed no significant incorporation of label into the resulting rubber. Other experiments in which these materials were incubated with ^{14}C-IPP and non-radioactive NPP, showed that the putative starter molecule had no effect on the rate of formation of labelled rubber. However, when washed rubber particles (WRP—see Methods) from CSS were substituted, the addition of NPP had a marked stimulatory effect on the uptake of ^{14}C-IPP, as shown in Fig. 1A, where the rate has been increased by a factor of 8.9. In the experiment shown in Fig. 1B, where the WRP from white fraction were used, the effect of adding the NPP was much less marked (note difference in scale and weight of rubber used). Possibly this may be due to the larger mean particle size in the fresh latex, resulting in more of the ^{14}C-IPP being used for chain extension in this case.

In the next experiment, WRP prepared from CSS were incubated with 1-^{3}H-NPP and cold IPP. The results shown in Fig. 2 demonstrate that a highly

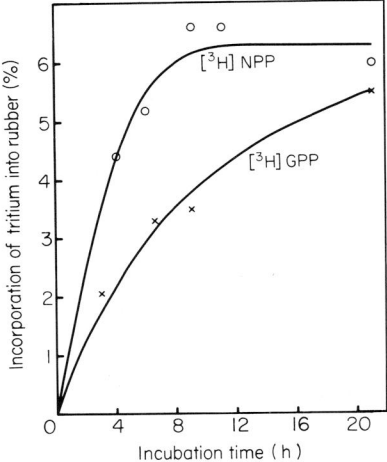

Figure 2. Utilization of ³H-NPP (O) and ³H-GPP (×) for rubber biosynthesis by WRP isolated from CSS. Incubations (25°C) contained. WRP suspension, 100 μl (c. 320 μg rubber); 1³-H-NPP or 1³-H-GPP, c. 0.65 nmol (670,000 dpm); IPP, 420 nmol; $MgCl_2$, 650 nmol; in 0.1 M Tris–HCl, pH 8, total volume, 130 μl.

significant uptake of tritium was obtained. All the control experiments were satisfactory (see Table 1), partly perhaps because the minicolumn technique was used (see Methods), which enabled the removal of large amounts of unreacted radioactive precursor before coagulation of the rubber. Gel-filtration chromatography of a solution of the rubber in toluene on a column of Bio-Beads showed that the label was in high molecular weight material (Fig. 3A), whilst ozonolytic degradation (McSweeney, 1968) to laevulinic acid (isolated as the 2:4-dinitrophenylhydrazone, which was crystallized to constant specific activity with a radioactive yield of 74%), confirmed that the labelled material was polyisoprene. Comparison of the specific activity of the water in the incubation mixture with that of the rubber itself, ruled out the remote possibility that ³H-exchange had occurred which might have lead to the production of labelled IPP.

These observations strongly indicated, but did not prove, that NPP is utilized *intact* for the formation of rubber, presumably by initiating the synthesis of new chains of polyisoprene.

Table 1. Control experiments: incorporation of labelled initiators (%) during various incubation times

	1-³H-NPP			1-³H-GPP
	4 h	6 h	9 h	21 h
IPP omitted	0.01	0.02	0.01	0.02
WRP omitted	0.004	0.005	—	0.003
WRP preheated*	0.005	0.001	—	0.008

*100°C for 3 min.

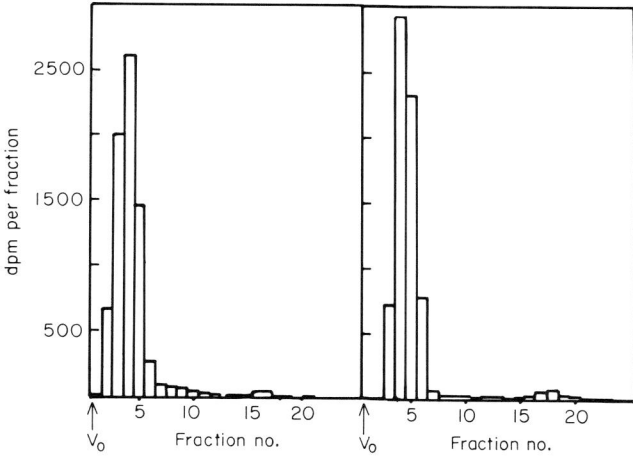

Figure 3. Gel-filtration chromatography of ^{3}H-labelled rubber in toluene on a column (420 × 22 mm) of Bio-Beads S-XI (exclusion limit, 14,000); load, c. 7000 dpm; recoveries, 103%. Rubber biosynthesized from: A, 1^3-H-NPP; B, 1^3-H-GPP. V_0, Breakthrough volume for authentic ^{14}C-rubber.

Trans-*allylic pyrophosphates as initiators of rubber synthesis:* According to earlier thinking on the structure of rubber, the *trans*-geranyl pyrophosphate (GPP) should not be utilized for rubber biosynthesis. However, when experiments were carried out with unlabelled GPP and with 1-^{3}H-GPP in incubations with WRP, results (see Figs 1A & 2) very similar to those with the *cis*-NPP were found. Again, gel-filtration chromatography (Fig. 3B) and a radiochemical yield of ^{3}H-laevulinic acid of 64% confirmed that high molecular weight rubber had been formed.

The data so far strongly suggest that the transferase enzyme system forming rubber has a comparable affinity for both *cis*- and *trans*-initiator molecules. However, the results do not rule out the possibilities that the GPP contained some NPP, or that GPP was isomerized to NPP before being converted to rubber. That the first hypothesis is more likely to be correct was shown by subsequent experiments with WRP in which it was found that *trans,trans*-farnesyl pyrophosphate (*t,t*-FPP), *trans,trans,trans*-geranylgeranyl pyrophosphate (*t,t,t*-GGPP) and *c,t,t*-GGPP were also powerful stimulators of rubber synthesis from ^{14}C-IPP.

Attempts were made to measure the Michaelis constants for a range of putative initiators, but these were not successful. One of the several possible reasons for this was that the compounds, as tested, were not pure. However, by measuring the ratio of the uptakes of label in the presence and absence of the initiator, a ranking was obtained as shown in Table 2. The precise order of enhancement obtained with a single batch of WRP depended on the concentration of the initiator, but generally increased as the chain length of the initiator was raised from 5 to 20 carbon atoms. A similar result was obtained by Baba & Allen (1978, 1980) by measuring K_m and V_{max} for similar compounds with the prenyltransferases isolated from *Lactobacillus plantarum* and *Micrococcus luteus*. One possible interpretation of the latest data is that the C_{20}-initiator has to be formed *via* DMAPP, before chain elongation to rubber can occur.

Table 2. Ranking of *in vitro* initiators of rubber synthesis: figures are ratios of ^{14}C-rubber produced* with and without initiator

	Initiator concentration (μm)†		
	0.1	0.5	2.0
C_{20} GGPP (c,t,t)‡	7.1	7.1	5.2
C_{20} GGPP (t,t,t)	6.1	7.0	4.5
C_{15} FPP (t,t)	5.3	6.5	4.4
C_{10} GPP (t)	0.9§	4.5	3.8
C_{10} NPP (c)	3.1	4.9	3.7
C_5 DMAPP	0.7	1.1	3.0

*Synthesized using WRP isolated from CSS.
†Estimated from purity data obtained by TLC and phosphorus analysis.
‡Each of the C_{20}-isomers was cross-contaminated.
§Probably erroneous.

BIOSYNTHESIS OF OLIGOISOPRENOIDS IN *H. BRASILIENSIS* LATEX

The biosynthesis of the polyisoprenoids containing up to 110 carbon atoms has been studied by several authors, who in a number of cases have isolated both the isomerase and transferase enzymes: for reviews on this subject see Hemming (1983) and Poulter & Rilling (1983). The formation in latex fractions of such compounds from IPP, without the addition of DMAPP, would be very strong evidence for the presence of the isomerase enzyme, and furthermore there is the possibility that the pyrophosphate esters of these compounds may act as precursors of rubber.

The general approach was to incubate ^{14}C-IPP with solutions of freeze-dried serum, decompose the pyrophosphates formed by the use of enzymes, and extract the labelled alcohols for chromatographic analysis. (see Methods). Although much ^{14}C-isopentenol was recovered, no significant amounts of ^{14}C-dimethylallyl alcohol could be detected. In one experiment, a trapping pool of FPP was added to the incubation; the subsequent GLC analysis showed that 4.9% of the label was associated with the *trans,trans*-farnesol and 0.7% with its other isomers. In a later experiment, these were identified as *cis,cis, cis,trans*, and *trans,cis*, in order of decreasing concentration. As no DMAPP has been added in these experiments, this is strong evidence for the presence of the isomerase in the serum. In the above experiment, when the column temperature was raised, a further sharp fraction containing 57% of the ^{14}C was eluted. Subsequent investigations, which included GLC of the ^{14}C-acetates, identified this material as the C_{20}-isoprenoid geranylgeraniol (GGOH), probably as a mixture of *t,t,t*-GGOH with small amounts of two of its other isomers. Using similar techniques, the formation of ^{14}C-nerol and ^{14}C-geraniol was demonstrated, but only in small quantities. A component eluting later than GGOH was also detected, this appeared not to be acetylatable and the possibility that it was a low molecular weight polyisoprenoid hydrocarbon cannot be excluded.

Final proof that a geranylgeranyl compound was formed in latex serum *via* DMAPP, was obtained by examining the distribution of label in the ozonolytic degradation products of the GGOH previously biosynthesized from 1-^{14}C and

Table 3. Distribution of ^{14}C label (%) amongst ozonolytic degradation products of ^{14}C-geranylgeraniol

*Labelled atom when 4-^{14}C-IPP is precursor
○Labelled atom when 1-^{14}C-IPP is precursor

Expt. No	Precursor	Acetone		Laevulinic acid	
		Predicted†	Observed	Predicted†	Observed
1	4-^{14}C-IPP	25	22	75	75
2	4-^{14}C-IPP	25	25	75	8
3	1-^{14}C-IPP	0	0.01	75	76
4	1-^{14}C-IPP	0	0.01	75	76

†Assuming no preformed allylic pyrophosphate pool.

4-^{14}C-IPP, following the method of Allen & Banthorpe (1981). As the latter labelled compound was not available, it was formed *in situ* from 2-^{14}C-mevalonate and adenosine triphosphate. In each experiment with reconstituted serum, the extracted alcohol was purified by HPLC on silica after the addition of carrier GGOH, and then ozonolysed (McSweeney, 1968). The decomposition products, acetone and laevulinic acid, were isolated as their tosylhydrazides and crystallized to constant specific activity; the third product, oxalic acid, was not examined. Table 3 reveals that the distribution of ^{14}C amongst the products, as calculated from the specific activities of the hydrazides, agrees closely with that predicted for GGOH and also clearly illustrates the labelling of the isopropylidene end-group, only when 4-^{14}C-IPP is the precursor. The presence of this radioactive group in the molecule is a strong indication that IPP isomerase does exist in *H. brasiliensis* latex, and also proves that ^{14}C-GGOH is not formed by reaction of ^{14}C-IPP with a preformed pool of allylic C_5-units, as can occur in certain examples of monoterpene biosynthesis (Allen *et al.*, 1976).

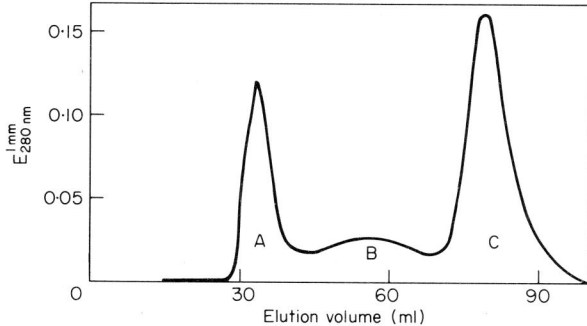

Figure 4. Gel-filtration chromatography of *Hevea brasiliensis* latex serum. Column (160 × 25 mm) packed with Ultrogel AcA-34; eluant, 0.25 M Tris–HCl, pH 8, containing 1% Tween-20 and 5 mM 2-mercaptoethanol; Load, 3 ml of a 12% solution of CSS in buffer; flow rate, 1 ml min^{-1}; detection, 280 nm. A. Rubber particles. B. Protein. C. Low molecular weight materials and protein.

The chromatographic technique of producing WRP (see Methods) enables other fractions of the CSS to be separated. Figure 4 shows a chromatogram from a 12% solution of the freeze-dried serum solids. The rubber particles are in fraction A with proteins in fraction B as expected; however fraction C, where the low molecular weight compounds are eluted, also contains some proteinaceous material.

Table 4 shows the data obtained when fraction C was added back to incubations of fraction A (i.e. WRP) and ^{14}C-IPP in the presence of Tween-20, 2-mercaptoethanol and Mg^{2+}. It is clear that fraction C accelerates the uptake of ^{14}C-IPP into rubber, but its effect is swamped by the presence of NPP. This suggests that there is an enzyme system in fraction C which synthesizes an initiator, although heating to 100°C does not entirely destroy the activity. Other experiments have indicated that one or more of these enzymes may be very labile, even when stored at 0°C in the presence of 2-mercaptoethanol. Clearly, this fraction may contain the IPP isomerase itself.

Table 5 shows the effect of adding fraction B to incubations of fraction A (WRP) and ^{14}C-IPP, in the presence of the standard cofactors. The slight inhibition shown by fraction B has been confirmed by other experiments. A much more marked effect of fraction B on the stimulation resulting from the addition of NPP is, however, clearly demonstrated and, furthermore, the addition of fraction C is now without influence. Other experiments have indicated that the inhibitor is a heat-denaturable protein with a molecular weight of about 5×10^4 and that it remains in solution when α-globulin, the major protein of *H. brasiliensis* latex serum (Archer & Cockbain, 1955), is precipitated at around pH 4. Iso-electric focusing indicates an iso-electric point of about 4.7. It has been shown that the inhibitor acts by blocking polymerization and not by degrading IPP or initiator; it does not act by chelating the essential cofactor of rubber biosynthesis, Mg^{2+}.

So far, the inhibitory activity has only been demonstrated with washed rubber particles of very small size, which have been prepared from serum in the

Table 4. Stimulation of incorporation of ^{14}C-IPP into rubber by fraction C

Fractions incubated*	Incorporation † (%)
A	0.6
A + C	2.7
A + NPP	6.2
A + C + NPP	6.0
A + C_H	1.6
A + C_H + NPP	5.6

*Fraction A is WRP peak as in Fig. 4, concentrated 2.6-fold. Fraction C is as in Fig. 4, concentrated 3.3-fold. Fraction C_H is C after heating to 100°C.

Incubations (30°C, 2.5 h) contained stated fraction, 50 µl; $MgCl_2$, 900 nmol; 1-^{14}C-IPP, 200 nmol (58,400 dpm); NPP, 3.6 nmol; in 0.2 M Tris–HCl, pH 8, total volume, 170 µl.

†Mean of two values which deviated from mean by ≯ 0.5 units.

Table 5. Inhibition of incorporation of ^{14}C-IPP into rubber by fraction B*

Fractions incubated	Incorporation (%)
A	0.32
A+B	0.24
A+NPP	4.60
A+B+NPP	0.29
A+B+C	0.21
A+B+C+NPP	0.31

*Fraction B is as in Fig. 4 concentrated eight-fold; for other conditions see Table 4.

presence of the surfactant Tween-20. Whether this protein has a physiological role has not been ascertained and its effect on the larger particles, where chain elongation rather than initiation could be the major process, is unknown. It is not impossible that it is involved in controlling the molecular weight of rubber, but this also has yet to be investigated.

THE MOLECULAR STRUCTURE OF RUBBER

The double bonds in natural rubber have generally been accepted as being all in the *cis-* configuration (see Archer, 1980), but the data on the initiators of rubber biosynthesis given above, suggest that this may not be completely true. The structures of several naturally occurring polyprenols of lower molecular weight than rubber, have been known for many years. For example, the betulaprenols from *Betula verrucosa* leaves have two *trans*-isoprene units, followed by 8–11 in the *cis* configuration, and the ficaprenols from *Ficus elastica* have three *trans* plus 5–9 *cis* residues in their molecules (Threlfall, 1980).

Tanaka, Sato & Kageyu (1982) have recently developed a ^{13}C-NMR method for establishing the structures of polyprenols and using it on two fractions of rubber ($\overline{Mn}$ 7.6×10^4 and 12×10^4) from *Solidago altissima*; the structure shown in Fig. 5 was deduced (Tanaka, Sato & Kageyu, 1983). This is the first direct evidence for the expected isopropylidene group at the ω-terminus in a rubber. The spectra indicated three or four *trans* double bonds, but as geranylgeraniol (C_{20}-*t,t,t*) occurs very much more frequently in nature than geranylfarnesol (C_{25}-*t,t,t,t*), the former number is more likely. In these two samples of rubber from *Solidago*, $n = 1000$ and 2200 respectively, which is in agreement with the observed molecular weights. A similar alignment of isoprene units was found for the polyisoprenes isolated from *Ficus elastica, Helianthus annuus*, (Tanaka, 1985) and *Parthenium argentatum* (Tanaka, pers. comm.).

In the case of a low molecular weight fraction of rubber from *H. brasiliensis*, the three *trans*-isoprene units were found together with the long *cis*-chain, but the ω-isopropylidene group was not detected (Tanaka, 1985). The presence of the three *trans*-units at the ω-end of the rubber molecule is in agreement with the biochemical data given above, which indicate that *trans, trans, trans*-geranylgeranyl pyrophosphate is the initiator of rubber biosynthesis (Archer,

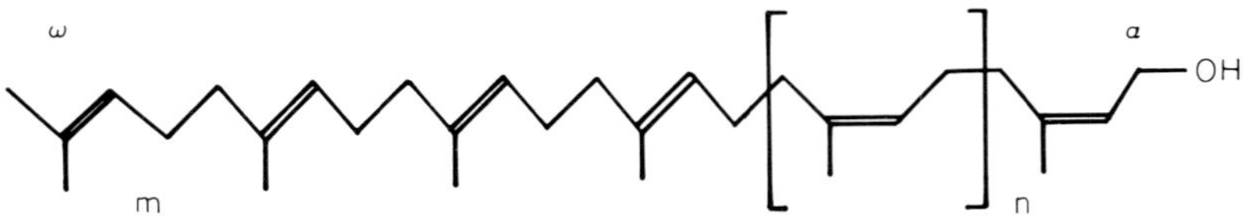

Figure 5. Structure of rubber (after Tanaka et al., 1983).

Audley & Bealing, 1982). The absence of the isopropylidene group in the ω-terminus of *H. brasiliensis* rubber may be due to the utilization of a derivative of geranylgeranyl pyrophosphate as the initiator, or less likely, to modification of the ω-terminus after polymerization.

Tanaka has also obtained information about the α-terminus of the polyisoprene molecule. In *Solidago*, it consisted entirely of primary hydroxyl groups, whereas in *Helianthus*, 60% of the hydroxyl groups were esterified with fatty acid. In *Hevea*, no primary hydroxyl groups could be found at all, although there was evidence for secondary or tertiary hydroxyls esterified with fatty acid or as lactones; such groups appeared at the rate of 1.1 per polymer chain. It is possible that this abnormal α-terminal unit may have resulted from the acid treatment used in the isolation of the rubber.

It would now appear that rubber obtained from *H. brasiliensis* is not quite such a unique compound, and may after all turn out to be simply a very high homologue of the polyprenol series.

THE NATURE AND MOLECULAR WEIGHT OF RUBBER BIOSYNTHESIZED BY WASHED RUBBER PARTICLES

Under suitable conditions, using the WRP-system from CSS described above, it was possible to obtain an acceleration in the rate of rubber formation of tenfold or more, on addition of an initiator. This implied that over 90% of the molecules being formed were new ones, rather than chain-extended ones, and thus it became feasible for the first time to attempt to measure the molecular weight of the material synthesized *in vitro*.

In these and subsequent experiments, the incubations were stopped with EDTA, the ^{14}C-rubber purified by the minicolumn technique (see Methods) and the resulting suspension dried. Initial attempts to measure the molecular weight of a sample of such rubber taken up in THF ended when the Micrel columns (polystyrene-divinyl-benzene copolymer particles, 10μm diam) on the gel-permeation chromatography (GPC) apparatus became blocked. Unlabelled rubber (detected at 215 nm) passed through as normal, but the ^{14}C was retained on the columns. If an open column packed with much larger cross-linked polystyrene particles (Bio-Beads S-X1, 50 μm diam) was used, both unlabelled and ^{14}C-rubber passed through quantitatively. The 'solutions' tested were perfectly clear to the naked eye and centrifugation (175,000 g 2 h) before GPC did not improve matters. Electron microscopy revealed the presence of swollen aggregates of rubber particles. Similar results were obtained using WRP from fresh serum which had not been freeze-dried.

Therefore, it appeared that the GPC columns were acting as filter beds: that is the biosynthesized rubber was not in true solution. Attempts were therefore made to solubilize it and for convenience the ability of the ^{14}C-rubber in THF to pass through a short column of fine silica after a given treatment, was adopted as a criterion of solubility.

Pretreatment with alkali or with subtilisin did not render the ^{14}C-rubber sufficiently soluble to pass through the silica column. However, treatment of the rubber with dilute acid did have the desired effect. Such rubber would pass through a silica column and also through the GPC columns; it could therefore be used for molecular weight measurements (see below). Figure 6 shows the dependence of the 'solubility' of the ^{14}C-rubber on the pH of the media used to treat it.

Overall, the data indicate that the nascent rubber molecules are bound to the WRP on which they are formed. The relatively high acid strength required to disrupt this binding suggests that the putative allylic pyrophosphate end-groups are not involved. It is tempting to speculate that the attachment is via a covalent linkage to the ω-terminal of the newly-formed rubber molecule (see Fig. 5). This might explain why this group has not been detected in *Hevea* rubber. The possibility that polyvalent cations (e.g. Mg^{2+}) may also be involved in this complex, was indicated by other experiments in which the EDTA used to arrest the biosynthesis was omitted. Further, it could be that the well known microgel component of latex (Allen & Bloomfield, 1963) is formed as a consequence of the biosynthetic mechanism.

Although its was appreciated that acid treatment might cause cyclization of the rubber, some measurements of the molecular weight of the treated, labelled rubber were carried out by GPC. Figure 7 shows the molecular weight distribution found for such a sample. The predominant ^{14}C-species has a molecular weight of about 10^5, but significant quantities of molecules of even

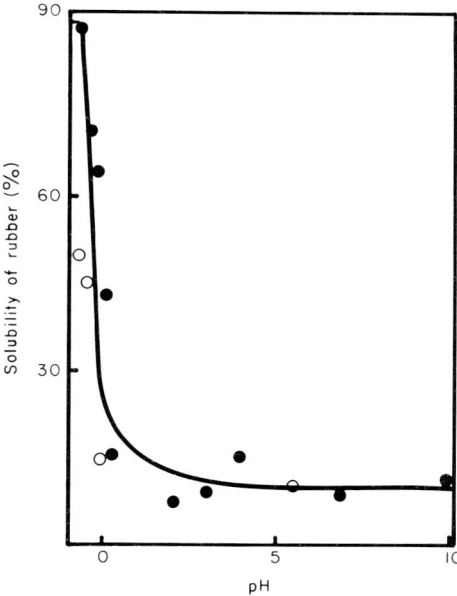

Figure 6. Solubility in tetrahydrofuran of ^{14}C-rubber as a function of pH of pretreatment medium: results for experiments with two different batches of CSS are plotted (open and closed circles).

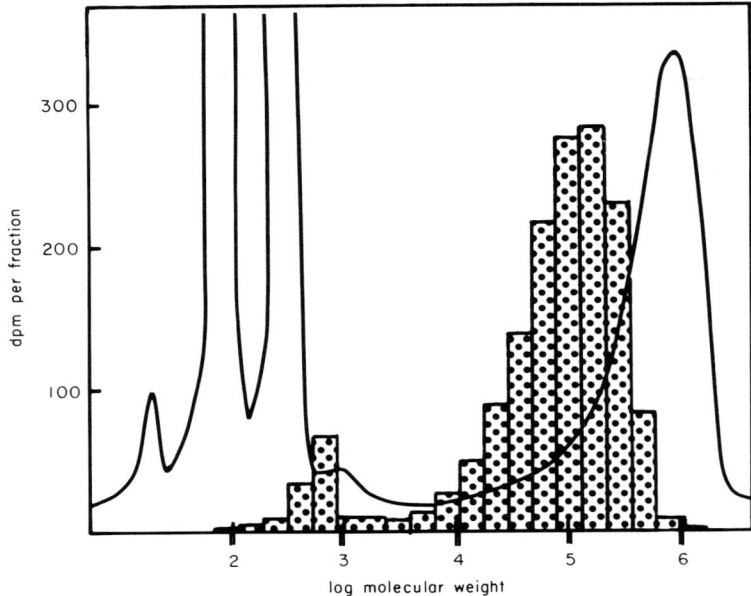

Figure 7. Molecular weight distribution of acid-treated ^{14}C-rubber biosynthesized by WRP isolated from CSS. The distribution for the soluble unlabelled rubber of the WRP is shown by the curve; peaks in low molecular weight region are due to antioxidant, etc. (see Methods).

higher molecular weight have also been formed. The molecular weight of the labelled rubber ($\overline{Mn} = 56{,}000$; $\overline{Mw} = 131{,}000$) is lower than that of the soluble hydrocarbon originally present in the WRP ($\overline{Mn} = 260{,}000$; $\overline{Mw} = 620{,}000$), but no great weight should be attached to this difference because of the arbitrary conditions employed in the incubations. The main value of these experiments is that they indicate that the WRP system might be usefully employed in studies of factors potentially controlling the molecular weight of natural rubber.

CONCLUSIONS

The picture we now have of the biosynthesis of rubber is of polyisoprene molecules originating on the surface of existing small particles. Isopentenyl pyrophosphate is first isomerized to DMAPP or some similar allylic compound, perhaps complexed with a glyco- or lipoprotein, this primary isoprene unit is then built up to the C_{20}-stage (geranylgeranyl pyrophosphate) wholly, or partly, in the *trans*-form. Isoprene units from IPP are then added on to give high molecular weight rubber.

The results recorded above clearly confirm the proposal of Lynen (1969) that IPP isomerase is on the surface of the rubber-particles and the data in Table 3 suggest that the enzyme may also be in a soluble fraction (fraction C). As the rubber transferase assay of Archer & Cockbain (1969) was based on the assumed elongation of existing rubber pyrophosphate molecules its validity must now be questioned, because it may well be that the production of initiator was the rate-limiting step in the biosynthetic sequence.

The separate identity and precise location of the enzyme(s) needed to catalyse the third stage of the reaction are still in doubt. Clearly, the necessary enzymic activity is available on the surface of the rubber particles, but the experiments have yet to be done to *prove* that there is also a soluble form of rubber transferase in the serum phase of *H. brasiliensis* latex, or indeed in other rubber-forming plants. In this laboratory, the addition of starter molecules has always proved more effective in accelerating the uptake of ^{14}C-IPP into rubber in WRP, than any preparation potentially containing such a soluble transferase (Table 4).

The problem of the origin of the first rubber particles also remains unsolved. Energetic considerations preclude the possibility of rubber particles splitting into two. Clearly they must originate within the latex vessel, where several sites have been suggested in the past. These include the Frey–Wyssling organelle (Dickenson, 1969) or the endoplasmic reticulum (Archer & Audley, 1973). It could be that rubber particle biogenesis starts from a molecule of geranylgeranyl pyrophosphate attached to enzyme and possibly other non-rubber molecules, in the form of a micelle, either suspended in the serum of the latex or attached to some surface. As the rubber molecule grows inside the micelle by accretion of IPP, it would form a small sphere surrounded by enzyme, on which other molecules of rubber would be initiated as suggested above.

ACKNOWLEDGEMENTS

The authors wish to thank Mrs C. Beck, K. F. Cheong, Mrs S. M. Pugh, Miss F. Saboowalla, S. Shabani and Dr Valda A. M. Stevens for their valuable assistance. They also thank those who donated chemicals, Dr A. D. Edwards for GPC analyses and the Director of the Rubber-Research Institute of Malaysia for permission to publish the work.

REFERENCES

ALLEN, B. E. & BANTHORPE, D. V., 1981. Partial purification and properties of prenyl transferase from *Pisum sativum*. *Phytochemistry, 20:* 35–40.
ALLEN, K. G., BANTHORPE, D. V., CHARLWOOD, B. V., EKUNDAYO, O. & MANN, J., 1976. Metabolic pools associated with monoterpene biosynthesis in higher plants. *Phytochemistry, 15:* 101–107.
ALLEN, P. W. & BLOOMFIELD, G. F., 1963. Natural rubber hydrocarbon. In L. Bateman (Ed.), *The Chemistry and Physics of Rubber-Like Substances:* 1–17. London: McLaren and Sons Ltd.
ANDERSON, N. G., ANDERSON, N. L., TOLLAKSEN, S. L., HAHN, H., GIERE, F. & EDWARDS, J., 1979. Analytical techniques for cell fractions. *Analytical Biochemistry, 95:* 48–61.
ARCHER, B. L., 1980. Polyisoprene. In E. A. Bell & B. V. Charlwood (Eds), *Encyclopedia of Plant Physiology, 8. Secondary Plant Products:* 309–327. Berlin: Springer-Verlag.
ARCHER, B. L. & AUDLEY, B. G., 1967. Biosynthesis of rubber. In F. F. Nord (Ed.), *Advances in Enzymology, 29:* 221–257. New York: John Wiley.
ARCHER, B. L. & AUDLEY, B. G., 1973. Rubber, Gutta percha and Chicle. In L. P. Miller (Ed.), *Phytochemistry, 2:* 310–343. New York: Van Nostrand Reinhold.
ARCHER, B. L., AUDLEY, B. G. & BEALING, F. J., 1982. Biosynthesis of rubber in *Hevea brasiliensis*. *Plastics and Rubber International, 7:* 109–111.
ARCHER, B. L., AUDLEY, B. G., COCKBAIN, E. G. & McSWEENEY, G. P., 1963. Biosynthesis of rubber. *Biochemical Journal, 89:* 565–574.
ARCHER, B. L. & COCKBAIN, E. G., 1955. The proteins of *Hevea brasiliensis* latex. *Biochemical Journal, 61:* 508–512.
ARCHER, B. L. & COCKBAIN, E. G., 1969. Rubber transferase from *Hevea brasiliensis* latex. In R. B. Clayton (Ed.), *Methods in Enzymology, 15:* 476–480. New York: Academic Press.
BABA, T. & ALLEN, C. M., 1978. Substrate specificity of undecaprenyl pyrophosphate synthetase from *Lactobacillus plantarum*. *Biochemistry, 17:* 5598–5604.

BABA, T. & ALLEN, C. M., 1980. Prenyl transferases from *Micrococcus luteus*. *Archives of Biochemistry and Biophysics*, *200*: 474–484.
BACKHAUS, R. A., 1985. Rubber formation in plants—a mini-review. *Israel Journal of Botany*, *34*: 283–293.
BACKHAUS, R. A. & WALSH, S., 1983. The ontogeny of rubber formation in guayule. *Botanical Gazette*, *144*: 391–400.
BANTHORPE, D. V., CHRISTOU, P. N., PINK, R. P. & WATSON, D. G., 1983. Metabolism of linaloyl, neryl and geranyl pyrophosphates in *Artemesia annua*. *Phytochemistry*, *22*: 2465–2468.
BANTHORPE, D. V., DOONAN, S. & GUTOWSKI, J. A., 1977. Biosynthesis of irregular monoterpenes in extracts from higher plants. *Phytochemistry*, *16*: 85–92.
BARNARD, G. F. & POPJAK, G., 1981. Human liver prenyl transferase and its characterisation. *Biochimica et Biophysica Acta*, *661*: 87–89.
BENEDICT, C. R., 1983. Biosynthesis of Rubber. In J. W. Porter & S. L. Spurgeon (Eds), *The Biosynthesis of Isoprenoid Compounds*, *2*: 355–369. New York: John Wiley.
CARDEMIL, E., VICUNA, J. R., JABALQUINTO, A. M. & CORI, O., 1974. Separation of isoprenoid alcohols and aldehydes by thin-layer chromatography. *Analytical Biochemistry*, *59*: 636–638.
CORNFORTH, R. H. & POPJÁK, G., 1969. Chemical synthesis of substrates of sterol biosynthesis. In R. B. Clayton (Ed.), *Methods in Enzymology*, *15*: 359–390. New York: Academic Press.
DICKENSON, P. B., 1969. Electron microscopical studies of latex vessel system of *Hevea brasiliensis*. *Journal of the Rubber Research Institute of Malaya*, *21*: 543–559.
ESAU, K., 1965. *Plant Anatomy*. New York: John Wiley.
FRY, D. W., COURTLAND WHITE, J. & GOLDMAN, I. D., 1978. Rapid separation of low molecular weight solutes from liposomes without dilution. *Analytical Biochemistry*, *90*: 809–815.
GOMEZ, J. B. & MOIR, G. F. J., 1979. *The Ultracytology of Latex Vessels in Hevea Brasiliensis*. Malaysian Rubber Research and Development Board, Monograph No 4. Kuala Lumpur.
GOSS, R. A., BENEDICT, C. R., KEITHLY, J. H., NESSER, C. L. & STIPANOVIC, R. D., 1984. *cis*-Polyisoprene synthesis in guayule plants (*Parthenium argentatum* Gray) exposed to low, nonfreezing temperatures. *Plant Physiology*, *74*: 534–537.
HEMMING, F. W., 1983. Biosynthesis of dolichols and related compounds. In J. W. Porter & S. L. Spurgeon (Eds), *The Biosynthesis of Isoprenoid Compounds*, *2*: 305–354. New York: John Wiley.
ISHII, K., SAGAMI, H. & OGURA, K., 1986. A novel prenyl transferase from *Parcoccus denitrificans*. *Biochemical Journal*, *233*: 773–777.
KANDUTSCH, A. A., PAULUS, H., LEVIN, E. & BLOCH, K., 1964. Purification of geranylgeranyl pyrophosphate synthetase from *Micrococcus lysodeikticus*. *Journal of Biological Chemistry*, *239*: 2507–2515.
LYNEN, F., 1969. Biochemical problems of rubber synthesis. *Journal of the Rubber Research Institute of Malaya*, *21*: 389–406.
MADHAVAN, S. & BENEDICT, C. R., 1984. Isopentenyl pyrophosphate *cis*-1:4-polyisoprenyl transferase from guayule (*Parthenium argentatum* Gray). *Plant Physiology*, *75*: 908–913.
McMULLEN, A. I., 1959. Nucleotides of *Hevea brasiliensis* latex. *Biochemical Journal*, *72*: 545–549.
McMULLEN, A. I. & McSWEENEY, G. P., 1966. Biosynthesis of rubber. *Biochemical Journal*, *101*: 42–47.
McSWEENEY, G. P., 1968. Ozonolysis of natural rubber. *Journal of Polymer Science, Part A-1*, *6*: 2678–2680.
OSTER, M. O. & WEST, C. A., 1968. Biosynthesis of *trans*-geranylgeranyl pyrophosphate in endosperm of *Echinocystis macrocarpa*. *Archives of Biochemistry and Biophysics*, *127*: 112–123.
POULTER, C. D. & RILLING, H. C., 1983. Prenyl transferases and isomerase. In J. W. Porter & S. L. Spurgeon (Eds), *The Biosynthesis of Isoprenoid Compounds*, *1*: 163–224. New York: John Wiley.
RUDALL, P., 1987. Laticifers in Euphorbiaceae—A conspectus. *Botanical Journal of the Linnean Society*, *94*: 143–163.
TANAKA, Y., 1985. Structural characterisation of naturally occurring *cis*- and *trans*-polyisoprenes. *Proceedings of the International Rubber Conference, Kyoto, Japan, October 1985*.
TANAKA, Y., SATO, H. & KAGEYU, A., 1982. Structural characterisation of polyprenols by ^{13}C-n.m.r. spectroscopy: Signal assignments of polyprenol homologues. *Polymer*, *23*: 1087–1090.
TANAKA, Y., SATO, H. & KAGEYU, A., 1983. Structure and biosynthesis mechanism of natural *cis*-polyisoprene from Golden Rod. *Rubber Chemistry and Technology*, *56*: 299–303.
THRELFALL, D. R., 1980. Polyprenols and terpene quinones and chromanols. In E. A. Bell & B. V. Charlwood (Eds), *Encyclopedia of Plant Physiology*, *8. Secondary Plant Products*: 288–308. Berlin: Springer-Verlag.

Botanical Journal of the Linnean Society (1987) 94: 197–219. With 10 figures

Tumour promoters of the irritant diterpene ester type as risk factors of cancer in man

ERICH HECKER

German Cancer Research Center, Institute of Biochemistry, Im Neuenheimer Feld 280, 6900 Heidelberg, F.R.G.

Received and accepted for publication October 1986

HECKER, E., 1987. **Tumour promoters of the irritant diterpene ester type as risk factors of cancer in man.** It is widely accepted in experimental and epidemiologic oncology that in real human life multifactorial causation of cancer is the rule and classical unifactorial causation the exception. This current and comprehensive concept of the aetiology of cancer has evolved essentially from in-depth investigations of tumour promoters of the irritant DTE type, found occurring in the Euphorbiales (Euphorbiaceae, Thymelaeaceae) in about the last 20 years. Moreover, the irritant DTEs are most powerful tools in experimental cell and cancer research, as illustrated by current results of investigations of their toxicokinetics and toxicodynamics.

Quantitative chemical and toxicologic investigations indicate that life-style oesophageal cancer, prevalent on Curaçao, is the first case of cocarcinogens of the DTE type ('Welensalifactors' from *Croton flavens* L.) shown to be the principal risk factor in a multifactorial human cancer, in conjunction with initiators of the solitary carcinogenic PAH type. The typical tigliane type DTE Welensalifactor F_1 was isolated, also from the tung oil tree (*Aleurites fordii*). It was shown that DTEs of the tigliane, ingenane and daphnane types activate certain human tumour viruses, e.g. the Epstein–Barr virus. The latter virus is known to be associated with nasopharyngeal carcinoma (NPC) prevalent in South China. It is concluded that uncontrolled utilization of Euphorbiaceae-derived materials, e.g. as (ethno-medicinal) drugs, or as technological raw materials, may be responsible for high incidence rates of certain human cancers due to DTE type risk factors of the promoter type.

The identification of well defined cocarcinogens of the DTE type as a new, non-classical group of risk factors in terms of '*per se* essentially non-carcinogenic amplifiers' of carcinogenesis adds a new dimension to the aetiology of cancer in man. Together with the multifactorial concept of carcinogenesis, it may allow a more refined assessment of the cancer risk.

ADDITIONAL KEY WORDS:—Cancer of the oesophagus – cancer of the nasopharynx – cancer prevention – cancer (tumour) aetiology – cocarcinogen – Euphorbiaceae – initiator – promoter – risk assessment of cancer – solitary carcinogen – Thymelaeaceae.

CONTENTS

Introduction	198
Toxicokinetics and toxicodynamics of DTE tumour promoters—a new class of carcinogenic risk factors	201
'Cryptic' irritants and promoters; enzymes involved in metabolizing DTE . . .	202
Structure/activity relations and putative mechanisms of action of DTE . . .	203
Irritant DTE tumour promoters—risk factors of cancer in man	208
Cancer of the oesophagus in Curaçao—local life-style cocarcinogenesis by DTE promoters	209
Cancer of the nasopharynx in South China—another aetiology of cancer involving DTE promoters?	213

0024–4074/87/020197+23 $03.00/0   © 1987 The Linnean Society of London

Increasing refinement in the assessment of the risk of cancer 215
Acknowledgements 216
References. 216

INTRODUCTION

As a universal concept of aetiologies of environmental cancer, *multifactorial carcinogenesis* has become widely accepted in experimental and epidemiological oncology. It visualizes exposure of the host, be it an experimental animal or man, to a multitude of environmental risk factors of cancer and integrates the simple classical case of unifactorial carcinogenesis (e.g. Hecker, 1972, 1984c, 1985a; Higginson, 1979). With regard to the risk of cancer, multifactorial carcinogenesis comes closer to the realities of every day human life than does the classical unifactorial concept. For this conceptual progress in aetiologies of cancer made during the last 20 years, the cocarcinogens of the irritant diterpene ester (DTE) type, widespread in certain Euphorbiales (Euphorbiaceae, Thymelaeaceae), played, and still are playing, a key role (Hecker 1981a; 1984b; Evans & Taylor, 1983).

The multifactorial concept of environmental cancer aetiology adds considerably to scientific and practical complexity. In assessment and management of the cancer risk, in dealing with this complexity and as a heuristic means, a simple but comprehensive system of aetiologies of cancer may be helpful. It may be based upon certain prototype processes of carcinogenesis (e.g. Hecker, 1976a–d, 1981a, b, 1984c, 1985a) which can be defined by typical *experimental models* employing (i) experimental animals with (preferably) non-predisposed target tissue(s), (ii) physicochemically well-defined risk factors of cancer and (iii) specific but simple dose/time protocols as visualized, for example, in Fig. 1.

Typically, any one protocol of carcinogenesis may be considered to start with the stage of 'initiation'. It is elicited by subcarcinogenic dose(s) i of a solitary carcinogen (Fig 1, line 1), but not by single or multiple dosing of a cocarcinogen, e.g. by a promoter of initiation (Fig. 1, line 3). Initiation may be considered the key event of all processes of carcinogenesis at the whole-organism, cell or molecular level. If the stage of 'initiation' is followed by chronic exposure of the host to multiple doses of one and the same solitary carcinogen (e.g. by multiple doses i, Fig. 1, line 2) the corresponding prototype process is called SOLITARY CARCINOGENESIS (Fig. 1, line 2), the classical *unifactorial* case of carcinogenesis mentioned above. *Multifactorial* carcinogeneses are processes of combined exposure of the host to two and more solitary carcinogens (PLURICARCINOGENESIS, not shown) or to a solitary carcinogen followed by a cocarcinogen (COCARCINOGENESIS, Fig. 1, line 4, as initiation/promotion). *Per se* the cocarcinogen is non-carcinogenic (Fig. 1, line 3, stage promotion); but in combination with an initiator, as shown in Fig. 1, it acts as an amplifier of carcinogenesis; in the special process of initiation/promotion, the cocarcinogen is called an initiation- or tumour-promoter. An interpretation at the cell level of cocarcinogenesis verified in the initiation/promotion protocol on mouse skin is given in Fig. 2.

The prototype processes of unifactorial SOLITARY CARCINOGENESIS and of multifactorial COCARCINOGENESIS define solitary- and co-carcinogens (see Fig. 1);

Specific protocol (dose/time)	Effect (tumours)	Prototype processes, stages
■	0	Stage: initiation
■ ■ ■ ■ ■	+	Solitary carcinogenesis
☐ ☐ ☐ ☐	0	Stage: promotion
■ ☐ ☐ ☐ ☐	+	Cocarcinogenesis as inititiation/promotion

■ = Solitarycarcinogen (Initiator)   ☐ = Cocarcinogen (Promoter)

Figure 1. Specific protocols with stages defining prototype processes of uni- and multifactorial carcinogenesis, e.g. solitarycarcinogenesis and cocarcinogenesis (as initiation/promotion, single doses, i, of initiator and p of promoter schematically).

they are considered risk factors of different categories of risk (see below). In human cancer occupational, iatrogenic or life-style aetiologies may be considered *epidemiologic models* reflecting more or less precisely uni- or multi-factorial prototype processes as defined experimentally (see Table 1). (In this context it may be pointed out that terms such as tumour promoter are meaningful only under a well-defined dose/time-effect pattern as verified in experimental models. In the case of epidemiological models, mostly with rather

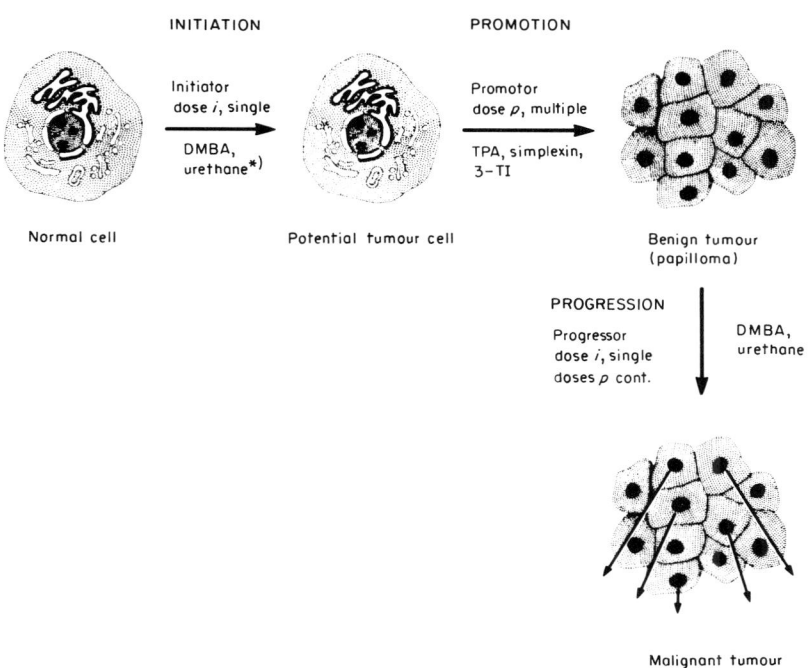

Figure 2. The prototype process of cocarcinogenesis in mouse skin (in the special protocol of initiation/promotion) with its three, operationally defined stages initiation, promotion and progression—one of the most advanced models for toxicokinetic and toxicodynamic investigations in chemical carcinogenesis.

Table 1. Some lifestyle cancers, principal pattern of exposure and prototype processes involved or suspected.

Life style	Carcinoma	Geographic distribution	Principal pattern of exposure	Prototype process and environmental carcinogens involved or suspected.
Excessive sunbathing	Skin (melanoma)	Worldwide	Unifactorial (approved)	Solitary carcinogenesis* (UV light)
Local habit of drinking Welensali tea	Oesophagus	Local (Curaçao)	Multifactorial (suspected)	Cocarcinogenesis of initiation/promotion type (*initiator*: petrol contaminated drinking water, *promoter*: Welensali tea)
Customary utilization of Euphorbiaceae materials	Nasopharynx	Regional (China)	Multifactorial (suspected)	Cocarcinogenesis*, of initiation/promotion type (suspected *initiator*: EB-virus; suspected *promoter*: EBV-activating diterpenes)
Smoking			Multifactorial (suspected)	Partly or totally unspecified
cigarette	Lung	Worldwide		
pipe	Lips	Worldwide		
bhidi	Oral cavity	Regional (India)		
Chewing betel	Oral cavity	Regional (India)		
Warming kangri	Skin	Regional (Himalaya)		

*Genetic predisposition may be involved on the side of the host.

poor dose/time-effect patterns, descriptive terms less specific with regard to the protocol involved, such as cocarcinogen (as defined above), are considered more adequate (Hecker 1984b, c, 1985a).) In life-style oesophageal cancer on the Caribbean island of Curaçao cocarcinogens of the promoter type, as a new class of risk factors of cancer in man, were established for the first time (Table 1 and below) (Hecker et al., 1983, 1984a).

To provide some of the toxicologic background of DTE from Euphorbiaceae and Thymelaeaceae I shall deal at first with some toxicokinetic and toxicodynamic aspects of this new class of risk factors of cancer. This is one main area of research covered at the German Cancer Research Center in Heidelberg, F.R.G.

TOXICOKINETICS AND TOXICODYNAMICS OF DITERPENE ESTER (DTE) TUMOUR PROMOTERS—THE NEW CLASS OF CARCINOGENIC RISK FACTORS

In chemical carcinogenesis, as a mechanistic model the initiation/promotion protocol in mouse skin (Figs 1 & 2) has achieved full validity after the irritant DTE promoters of Euphorbiaceae and Thymelaeaceae origin were established as molecularly uniform, highly active cocarcinogenic entities (Hecker, 1972, 1978; Hecker & Schmidt, 1974). They exert skin irritant and promoting activities at doses comparable to the dose-level of proliferation stimulating hormones or growth factors (e.g. Hecker, 1985b; Hecker & Bresch, 1965). The multitude of different irritant DTE promoters isolated so far from the Euphorbiaceae and Thymelaeaceae, guided by taxonomic leads (Engler, 1964), may be related to only three principal (hypothetic) parent diterpenes, i.e. tigliane, ingenane and daphnane (Fig. 3). In the biologically active entities, the parent diterpenes are polyfunctionalized to carry, besides double bonds and simple oxygen functions, certain ester groups, as shown for example in the three prototype irritants and promoters (Fig. 3) i.e. 12-O-tetradecanoylphorbol-13-acetate (TPA, Hecker, 1978; Hecker & Schmidt, 1974), 3-O-tetradecanoylingenol (3-TI, Sorg & Hecker, 1982; Sorg, Schmidt & Hecker, 1987) and 9,13,14-O-orthodecanoyl-5β-hydroxy-6α,7α-epoxyresiniferonol (simplexin, Zayed, Hafez, Adolf & Hecker, 1977; Roberts et al., 1975). The efficiency of promotion by DTE, as shown in the standardized initiation/promotion protocol for the prototypes TPA, 3-TI and simplexin, is dose-dependant Sorg et al., 1987; Thielman & Hecker, 1969).

Remarkably, the corresponding prototype diterpene parent alcohols phorbol, ingenol and resiniferonol, respectively (Fig. 3), are biologically inactive. They have been used to 'tailor-make' many DTEs by partial syntheses, for investigations of structure/activity relationships and for biochemical investigations (Hecker, 1978, 1985b; Hecker & Schmidt, 1974; Hecker, Adolf et al., 1984; Sorg et al., 1987). Also, remarkably, the scientific validity of the DTE (see Balandrin, Klocke, Wurtele & Bollinger, 1985) has stimulated efforts towards total syntheses of their parent alcohols (phorbol, see Wender, Hillemann & Szymonifka, 1980, ingenol, see Paquette, Nitz, Ross & Springer, 1984).

Most interestingly, the class of defined skin irritant cocarcinogens of the promoter type was extended recently by certain new irritant indole alkaloid and polyacetate promoters (e.g. Fujiki et al., 1984b; Hecker et al., 1984).

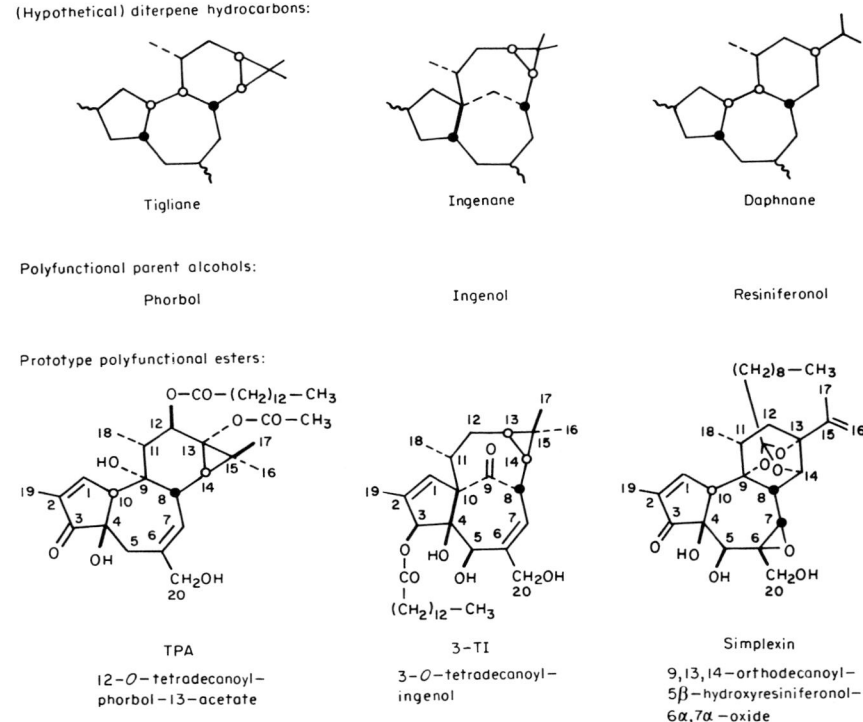

Figure 3. Structures of the prototypes of skin irritant and promoting polyfunctional diterpene esters TPA, 3-TI, and simplexin with hypothetical parent diterpene hydrocarbons, and corresponding polyfunctional parent alcohols.

'Cryptic' irritants and promoters; enzymes involved in metabolizing DTE

Remarkably, in many of the plant species investigated DTEs occur as oligo-20-esters which exhibit low irritant (and promoting) activities (e.g. Hecker, 1978, 1981a; Hecker & Schmidt, 1974; Hecker et al., 1983). Typical oligo-20-esters of the tigliane, ingenane, and daphnane type, their irritant doses 50 and their irritancies are compiled in Table 2 together with the corresponding 20-deacylated DTEs. Selective release of the 20-hydroxyl group in the DTE oligo-20-esters, generally achieved by mild transesterification (e.g. Hecker & Schmidt, 1974; Hecker et al., 1983) causes a significant increase of irritancy (and of promoting activity, not shown). In target tissues, the selective release of the 20-hydroxyl group may take place enzymatically by esterases or lipases as a means of *metabolic activation* (Doege & Hecker, 1984). Therefore, the plant originating low activity DTE oligo-20-esters may be specified as 'cryptic irritants and promoters'. From the toxicokinetic point of view, as some kind of 'pro-drugs', they deserve particular attention as potential risk factors of cancer (see below and Hecker, 1981a, 1984a, c; Hecker, et al., 1983; Hecker, Adolf et al., 1984).

Metabolic investigations of the active forms of DTE promoters, such as the prototypes TPA or 3-TI (see Fig. 3), revealed that hydrolysis of their long chain acyl group is a first step of *metabolic deactivation*, generating metabolites such as

Table 2. Typical forms of polyfunctional irritant DTE promoters: oligo esters of the tigliane, ingenane, and daphnane type carrying 20-ester groups: low activity 'cryptic irritants' and promoters corresponding 20-deacylated esters: relatively high active irritants and promoters; irritant dose 50 on the mouse ear after 24 h, irritancy accordingly $I^{24} = 1/ID_{50}^{24}$*

Parent diterpene	Polyfunctional ester	Irritant activity	
		ID_{50}^{24} [nmoles]	I^{24} [nmoles^{-1}]
Tigliane	TPA-20-tetradecanoate	0.77	1.3
	TPA	0.016	63
Ingenane	3-TI-20-tetradecanoate	0.05	20
	3-TI†	0.011	91
Daphnane	K_7-20-hexadecanoate	0.16	6.3
	Synaptolepis factor K_7	0.02	50
1α-alkyl-daphnane	K_1-20-hexadecanoate	0.019	53
	Synaptolepis factor K_1	0.0015	670

* Standard deviation $\sigma = 1.3$, level of significance $\alpha = 0.05$.
† 3-O-tetradecanoylingenol.

phorbol-13-acetate and phorbol, or ingenol, respectively. In mechanistic terms, therefore, the active forms of DTEs are considered to interact directly with cellular target molecules, i.e. they are 'ultimate irritants and promoters' (Roeser, Sorg & Hecker, 1983; Hecker, 1978, 1979, 1981a). Due to their fascinating biological properties, and guided by important insights into their structure/activity relations, DTEs of the promoter type have proved to be most useful tools for biochemical mechanistic investigations in both experimental cell research and in cancer research.

Structure/activity relations and putative mechanisms of action of DTE

To develop the *toxicodynamic* issue of DTE-promoters their structure/activity relationships were and still are investigated (e.g. Hecker, 1978, 1985b; Sorg *et al.*, 1987). Some of the recent accomplishments and conclusions regarding putative cell membrane associated receptors of DTE promoters shed an interesting light on possible mechanisms of action of these risk factors of cancer (e.g. Hecker, 1985b). Moreover, they also stimulated certain ideas as to the redesign of DTE to obtain medicinally useful anticancer drugs (e.g. Hecker *et al.*, 1984; Fellhauer & Hecker, 1986). Whereas the latter issue is beyond our present topic, its bases are some recent findings regarding structure/activity relations of DTE which will be discussed subsequently.

Role for bioactivity of diterpene moieties and their hydrophilic functionalities

A free primary hydroxyl function in position 20 is the most prominent hydrophilic feature common to all irritant DTE promoters (see Fig. 3), it is essential for their biological activity (e.g. Hecker, 1978; Sorg & Hecker, 1982;

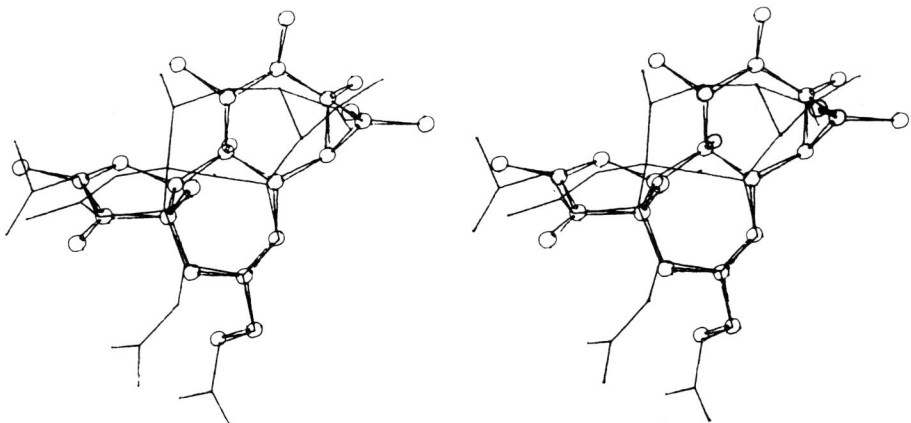

Figure 4. Stereoplot of computer graphs of phorbol ((1), atoms connected by cones) and ingenol-3,5,20-triacetate ((2) atoms connected by red lines) matched in fit 2a involving both C6–(C20)–C7 and C4–O4 (see also text and Table 3).

Thielmann & Hecker, 1969). Therefore it may be postulated that the spatial position of this function, as determined by the different carbon skeletons involved, may play a delicate role in the *hydrophilic interaction* of DTE promoters with cellular membranes and putative receptors (Hecker, 1978). Such hypothesis was tested recently using computer-assisted molecular modelling of DTE.

Pairwise matching by common structural elements of the diterpene moieties, e.g. of the prototypes TPA, 3-TI and simplexin (see Fig. 3), is used to this end. On the graphic display of the computer, three-dimensional graphs were made up from X-ray data obtained previously of (1) phorbol (Hoppe *et al.*, 1969), (2) ingenol-3,5,20-triacetate (Zechmeister *et al.*, 1970) and (3) mezerein (Lotter, Jones & Sturm, 1977). The quality of the fits may be expressed quantitatively by the intermolecular distances of corresponding atoms calculated for the pairs matched i.e. (1)–(2) and (1)–(3), for example see Fig. 4 (1)–(2) (Lotter & Hecker, 1985, 1986). The intermolecular distances of corresponding atoms in the pairs (in Ångstrom) are related to the length of the regular C–C bond as a standard and compared in individual fits (for example, see Table 3).

In this way two hypotheses have been tested consecutively: Firstly, the diterpene moieties of (1), (2) and (3) fit well if, as primary matching group, their common structural element carrying the primary hydroxyl function, i.e. C6–(C20)–C7, is used (e.g. see fit 1, Table 2). Secondly, the diterpene moieties fit even better, if as secondary matching groups, structural elements carying other oxygen functions are used, such as C4–O4 (e.g. fit 2a, Table 3), and alternatively or in addition C9–O9, and C3–O3.

For the pairs (1)–(2) and (1)–(3) both hypotheses were verified (Lotter & Hecker, 1985, 1986). Especially in the case of pair (1)–(2) it was noted that the fit is quite good (i.e. to the extent that the distances of corresponding atoms in the pair are well below the distances of a regular C–C bond, e.g. Table 3), despite the different distribution of the functionalities in the phorbol and the ingenol moiety, respectively.

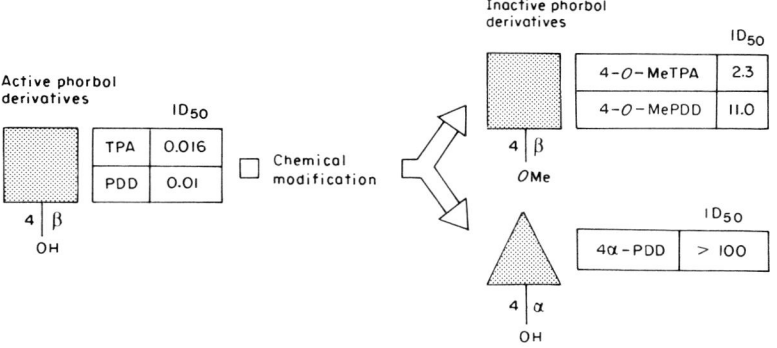

Figure 5. Biologically inactive derivatives 4-*O*-Me-TPA/4-*O*-MePDD, most closely structurally related to their highly active parent DTEs, i.e. TPA and PDD, respectively, and therefore more appropriate as negative control compounds than 4α-TPA/4α-PDD. Parameter for biological activity: irritant dose 50 nmoles/ear, α = 0.05, σ = 1 : 3.

As a consequence, in the case of fit 2a (Table 3), both of the hydrophilic structural elements of the diterpene moieties fitted, i.e. C(6)–C(20)–C(7), and C4–O4 may be considered important for binding to putative cellular target molecules or receptors and, consequently, for bioactivity. That this really is the case was suggested by the results of previous investigations in our laboratory aiming at the 'most appropriate negative control compounds', e.g. for TPA or PDD, in terms of 'non-active DTEs of closest possible structures' as summarized in Fig 5. In this regard 4α-PDD appears less appropriate, since by epimerization in 4-position of PDD the conformation of the entire diterpene entity is changed severely to deviate considerably from that in PDD (or TPA; see Hecker, 1978). Indeed, as the most appropriate negative control compound, the 4-*O*-Me-TPA as proposed has been used for many years for mechanistic investigations of DTE promoters (see Hecker, 1978, 1985a, Hecker, Adolf *et al.*, 1984).

Table 3. Intermolecular distances of corresponding atoms in the pair phorbol–ingenol-3,5,20-triacetate [(1)–(2)] for comparison: length of regular C–C bond 1.53 Å

Structural elements fitted		Further important structural elements	
Atoms	Å	Atoms	Å
Fit 1: matching of C6–(C20)–C7			
C6	0.00	O3	1.87
C7	0.01	O4	0.93
C20	0.01	O9	1.14
Fit 2a: matching of both C6–(C20)–C7 and C4–O4			
C6	0.12	O3	1.62
C7	0.08	O9	0.94
C20	0.10		
C4	0.18		
O4	0.19		

Role for bioactivity of ester moieties

Highly active irritants and promoters, such as TPA, 3-TI and simplexin, carry at least one ester group as a common lipophilic feature. Their (long chain) ester moiety is essential for biological activity, since the metabolite phorbol-13-acetate or the parent alcohols are inactive (Hecker, 1978; Sorg & Hecker, 1982; Sorg et al., 1987; Zayed et al., 1984).

Out of a great many investigations of the role of the ester moieties for biological activity performed in our laboratory, only one of the most remarkable structure/activity relationships observed will be discussed. Briefly, it was shown that TPA-analogues with an unsaturated 12-O-acyl group exhibit irritancies of a similar (high) degree as do the corresponding saturated DTE (Fig. 6, TPA and Ti_8). Surprisingly however, at high degrees of unsaturation in the ester moiety, such as seen in *Euphorbia tirucalli* factor Ti_8 (Fig. 6), the promoting activity becomes negligible (Fürstenberger & Hecker, 1972, 1985, 1986). From this and other examples explored similarly in the ingenol and in the resiniferonol ester series, it may be concluded that generally in mouse skin *irritant (and hyperplasiogenic) activity is (are) necessary, yet incomplete, requirements for promoting activity* (Fürstenberger & Hecker, 1985, 1986; Hecker, 1978). To test this hypothesis, an especially highly unsaturated ester analogue of TPA, 12-O-retinoylphorbol-13-acetate (RPA), was prepared by partial synthesis (Sorg et al., 1982; Tremp & Hecker, 1985). Indeed, and not unexpectedly, RPA is a strong irritant, practically as active as TPA or Ti_8 (Fig. 6). Yet in the standardized initiation/promotion protocol on skin of NMRI-mice it is essentially a non-promoter (Fürstenberger et al., 1981; Marks & Fürstenberger, 1984). Quite apart from mechanistic implications, regarding tumour promotion (see below), findings of this kind may provide certain clues for trials to redesign polyfunctional DTEs, to entirely new principles of potentially useful anticancer drugs (e.g. Fellhauer & Hecker, 1986, Hecker, 1979).

With regard to the mechanism of action of DTEs as risk factors of cancer it was shown that, in conjunction with a preceding single dose of TPA (stage one-(PI)-promotion), followed by multiple doses of RPA, 'stage two (PII) promoter' RPA is the best presently available so-called 'incomplete promoter' (Fürstenberger et al., 1981; Hecker, 1985b; Marks & Fürstenberger, 1984; Slaga et al., 1980). Utilization of RPA in promotional protocols using two stages of promotion (PI, PII) may be useful as a tool to develop further the multistage approach and hence the mechanistic aspects of tumourigenesis and carcinogenesis (e.g. Kinzel et al., 1984; Marks & Fürstenberger, 1984). Such an approach is particularly important, in view of the biological and the biochemical pleiotropism exhibited by irritant promoters of the DTE type, which makes it difficult to pinpoint those biochemical event(s) of promotion causative for selective expression of the transformed phenotype (initiated cell). However, in the near future the two stage approach of promotion may allow to separate the hitherto unknown 'truly promotion specific effects' exhibited by complete promoters, such as TPA, from their putative 'non-specific side effects' such as, for example, irritancy and hyperplasiogenic activity (e.g. Hecker et al., 1984; Hecker, 1985a, b).

Putative cellular receptors of DTE promoters

The results of structure/activity investigations strongly suggest that in target cells TPA and related DTE promoters may interact with membrane-associated,

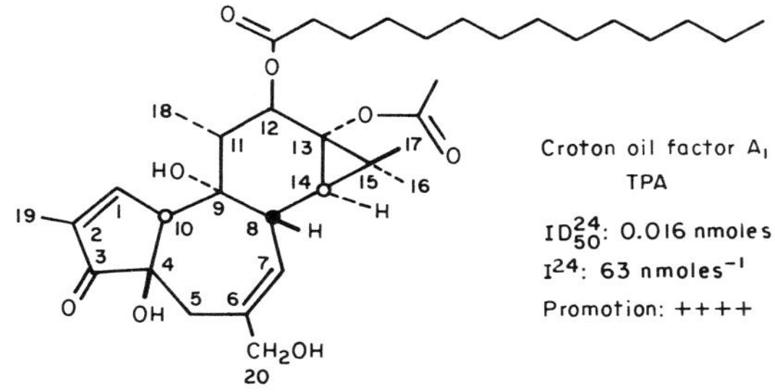

Croton oil factor A_1
TPA
ID_{50}^{24}: 0.016 nmoles
I^{24}: 63 nmoles^{-1}
Promotion: ++++

12-*O*-tetradecanoylphorbol-13-acetate

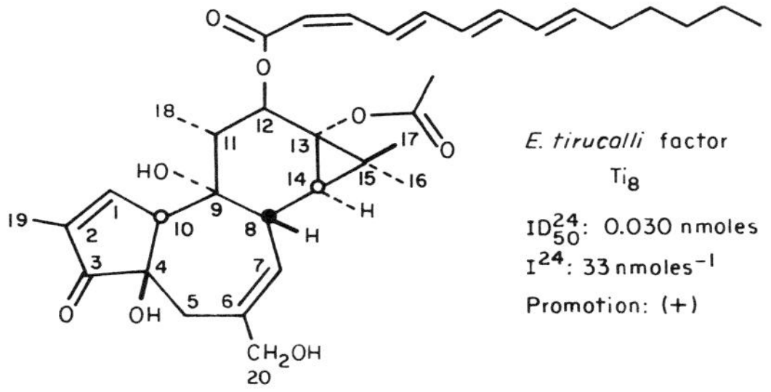

E. tirucalli factor
Ti_8
ID_{50}^{24}: 0.030 nmoles
I^{24}: 33 nmoles^{-1}
Promotion: (+)

12-*O*-tetradeca(tetra-2,4,6,8-en)oylphorbol-13-acetate

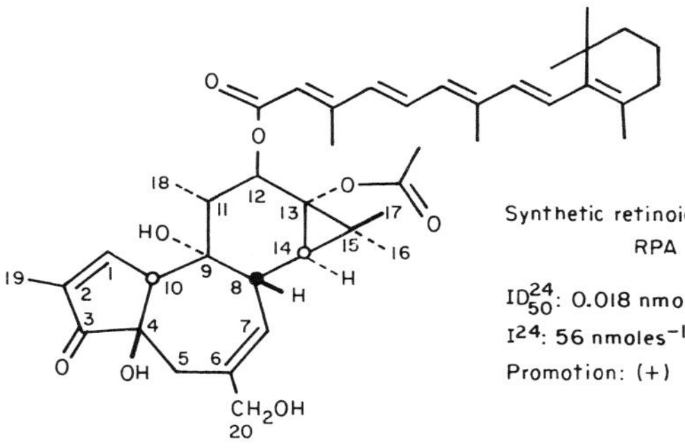

Synthetic retinoic acid ester
RPA
ID_{50}^{24}: 0.018 nmoles
I^{24}: 56 nmoles^{-1}
Promotion: (+)

12-*O*-retinoylphorbol-13-acetate

Figure 6. Irritant and promoting activities of TPA and of the analogous, highly unsaturated DTE Ti_8 and RPA indicate that in mouse skin irritancy is a necessary, yet incomplete requirement for promotion.

specific target molecules, i.e. with 'receptors'. They may trigger single key enzymes and/or enzymatic cascades governing the biological and biochemical effects elicited (Hecker, 1978, 1985a, b; Hecker, Adolf, *et al.*, 1984).

The receptor concept, proposed in 1978 (Hecker, 1978), was greatly strengthened by the finding that, in correlation with their irritant and promoting activities, DTE promoters exhibit agonist type specific (non-covalent) binding to the particulate fraction of mouse epidermis and of other mouse organs or cells (Declos, Nagle & Blumberg, 1980; Hergenhahn & Hecker, 1981; Schmidt *et al.*, 1983; Shoyab, Warren & Todaro, 1981). Use of certain,

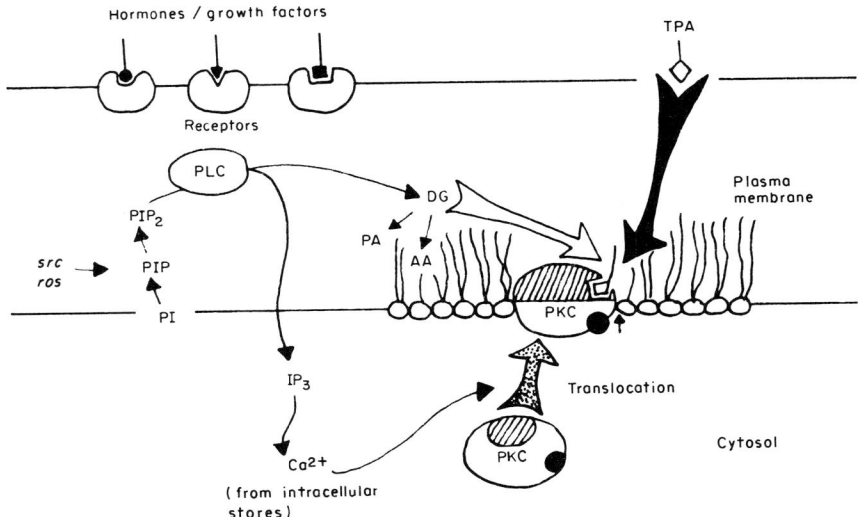

Figure 7. Diagram illustrating membrane-associated receptors of endogenous hormones or growth factors triggering the inositolphospholipid/diacylglycerol second messenger system and putative intervention by *onc* gene products such as *src* or *ros* and/or DTE promoters, such as TPA. PLC = Phospholipase C, PIP_2, PIP, PI, IP_3 = inositolphospholipids, DG = diacylglycerol, PA = phosphatidic acid, AA = arachidonic acid, PKC = protein-kinase C (cytosolic and membrane bound).

specifically bound electron-spin and photoaffinity-labelled phorbol ester analogues indicate a distinct molecular orientation of DTE in membranes and, in their micro-environment, presence of phospholipids essential for activity of the Ca^{2+} dependant proteinkinase C (**PKC**). Activation of PKC was shown for TPA and congeners particularly also in mouse epidermis as one of the target organs of initiation/promotion by DMBA/TPA. Therefore, it may be postulated that DTE promoters interact with the *second messenger system operated by inositolphospholipid/diacylglycerol* which is linked to both the receptors of certain (endocrine) hormones or growth factors and to certain oncogene derived (autocrine) molecular signals (Fig. 7). In this way, DTE promotors mostly tailor-made by partial syntheses, may allow the interlinking of receptor research with parts of processes of carcinogenesis to open up new and fascinating aspects of mechanisms of carcinogenesis at the cell and/or molecular level (Hecker 1985a, b; Hecker, Adolf et al., 1984).

IRRITANT DTE TUMOUR PROMOTERS—RISK FACTORS OF CANCER IN MAN

To provide a scientific background for assessment of the cancer risk, aiming at appropriate measures of primary prevention, experimental analyses of epidemiologic situations of human cancer are indispensable. Soon after the irritant DTE promoters were established, this line of research was taken up by my group at the German Cancer Research Center in Heidelberg. Thus, the methodology developed was employed in investigation of cocarcinogens of the promoter type as possible risk factors of cancer associated with utilization of Euphorbiaceae and Thymelaeaceae material in the human environment

(Hecker, 1981a). Recently, for the first time, the epidemiologic model situation of putative multifactorial life style cocarcinogenesis of the oesophagus prevailing in the Caribbean region (see Table 4) was analysed completely. As principal sources of the multifactorial risk of this cancer on the island of Curaçao, temporary pollution of drinking water by petrol, in combination with the customary life-style associated exposure of the people to promoters of the DTE-type were established (Hecker 1984a–c, Hecker, Lutz, et al., 1983).

Cancer of the oesophagus in Curaçao—local life-style cocarcinogenesis by DTE promoters

On the Caribbean island of Curaçao (Netherlands Antilles) the black and creole population is burdened by an exceedingly high incidence rate of oesophageal cancer (Table 4). Epidemiologic enquiries of the population at risk by the National Cancer Institute of the U.S.A. (NCI) provided evidence for exposure to drinking water contaminated with petrol in the period of 1930–1964. It may suggest exposure to initiating doses of solitary carcinogens of the *polycyclic aromatic hydrocarbon* (PAH) type. Furthermore, by similar enquiries, it was established that, as part of local life style, plant-derived preparations from about 100 different species, including two species of spurge (Euphorbiaceae), are utilized for food, beverages and folk medicines. On the two neighbouring islands (Bonaire, Aruba) and in other parts of the region this habit is uncommon and the incidence rate of oesophageal cancer is average. Consequently extracts of the 23 most frequently used plant species were assayed by the NCI for solitary carcinogenic activity in mice, rats and hamsters. Five species out of the 23 were found active, but not the two Euphorbiaceae species. These findings lead to postulate the aetiology of the hypothesis of 'solitary carcinogenesis' for the local high incidence cancer. To account for a possible combination of exposure of plant-originating solitary carcinogens with those suspected in the petrol-contaminated drinking water, 'pluricarcinogenesis' (see above, text associated with Fig. 1) might be considered alternatively as the process of carcinogenesis involved. However, so far neither of these hypotheses

Table 4. Age-standardized annual incidence rates of oesophageal carcinoma in Curaçao and in some regions of Central America related to 100 000 of the European and World population

Region	Europe		World		References
	♂	♀	♂	♀	
Curaçao*	32.3	19.4	21.2	12.9	de Boer, 1979
Venezuela Falcon, Coro†					Morton, 1974
Puerto Rico	21.5	8.1	14.8	5.4	
Cuba	8.7	3.7	5.7	2.4	Waterhouse et al. (1976)
Jamaica (Kingston)	13.4	6.6	9.1	4.7	

*Data refer to Netherland Antilles (islands of Aruba, Curaçao, Bonaire, St. Maarten, Saba, St. Eustatius) without Aruba and cover the period 1962–1973. Ninety per cent of the population of the Netherland Antilles live in Curaçao. Limited data for Aruba: in the period between 1963 and 1966, five male and two female patients out of a population of about 60 000 suffered from oesophageal carcinoma (Morton, 1968a).
†Approximately 15 oesophageal carcinoma in 28 months out of a population of about 45 000.

has been substantiated, e.g. by rigorous chemical identification of the solitary carcinogen(s) possibly involved, or by verification in corresponding experimental models of the carcinogenic processes postulated (see Hecker, Lutz et al., 1983).

The two species of Euphorbiaceae utilized in Curaçao are the bushes *Croton flavens* L. (Crotoneae), locally known as Welensali, and *Jatropha gossipyfolia* L. (Cluytieae), locally called Belly-ache Bush. The fresh aromatic leaves of Welensali are used to prepare a 'bush tea' as an everyday beverage (Welensali tea). Other habitual utilizations of the leaves and of other parts of the bush, especially of the roots (for chewing), are widespread. Parts of the Belly-ache Bush are used as folk medicinal treatments (Adolf et al., 1984; Morton 1968a, b). Therefore, it was considered not unlikely that a situation of environmental cocarcinogenesis would prevail and an aetiologic 'cocarcinogen hypothesis' was postulated (Hecker, Lutz et al., 1983; Hecker & Weber, 1977; Lutz & Hecker, 1979). Experimental assessment of such hypothesis became the programme of a corresponding research project.

Investigations of *roots and fresh green leaves of the Welensali bush* revealed that their extracts are moderately irritant on the mouse ear. It turned out that they contain 17 and 28 irritant 'Welensalifactors', respectively (Table 5). Their multitude may be subdivided essentially in two activity types, F and F'. Together F- and F'-types represent an estimated minimal content of 1.4 and 0.06% by weight, of the extracts, respectively. The polyfunctional diterpene parent alcohols involved were shown to be the tetracylic *tigliane structures* 16-hydroxy- and 4-deoxy-16-hydroxyphorbol (see Hecker, Lutz et al., 1983). Individually, by HPLC and spectroscopic means, the Welensalifactors were identified as diterpene-12,13-di- (F-types) and -12,13,29-tri-esters (F'-types, i.e. oligo-20-esters, see above), respectively. To represent F and F' types as typical examples (Table 5), data on Welsenalifactor F_1, most abundant in roots and

Table 5. Extracts from roots and leaves of Welensali with irritant activity, total number of Welensalifactors, minimum content determined and typical Welensalifactor F_1 together with corresponding cryptic form F_1-20-decanoate. Irritant activity on the mouse ear as irritant dose 50 and irritancy ($I^{24} = 1/ID_{50}^{24}$), for comparison TPA, ID_{50}^{24} 0.01 µg/ear; I^{24} 100 µg^{-1}.

Typical extracts or factors	ID_{50}^{24} (µg/ear)*	I^{24} (µg^{-1})	Welensalifactors		
			Type of activity	Identified factors (total number)	Minimal content of factors (%)†
Roots, methanol extract (223 g)	6.4	0.15	F,F'	17	1.4
Leaves, ethylacetate fraction (1265 g)	90	0.01	F,F'(F^y)	28	0.06
Welensali-factor F_1	0.022	45	F	—	—
F_1-20-decanoate (F'_1)	0.09	11	F'	—	—

*Standard deviation 1,3, level of significance $\alpha = 0.05$.
†By dry weight of extract.

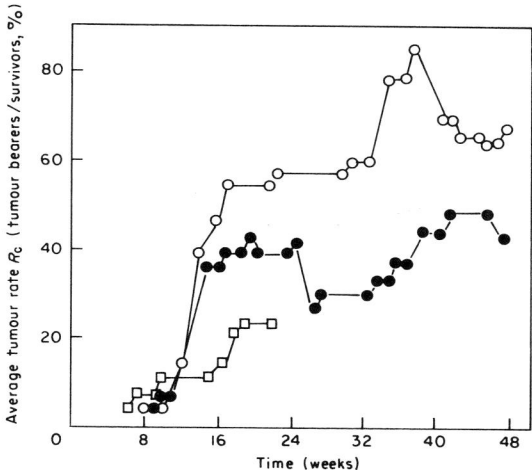

Figure 8. Cocarcinogenic (as promoting) activity of typical Welensalifactor F_1 and its 'cryptic' analogue F_1-20-decanoate measured by tumour rate in the standard assay on back skin of NMRI-mice; TPA (croton oil factor A_1) as positive control. Initiator: DMBA, $i = 100$ nmoles; 28 females/group. Promoter: ○, TPA, $p = 2.5$ nmoles; ●, F_1, $p = 10$ nmoles; □, F_1-20-decanoate, $p = 100$ nmoles.

leaves, and on one of its corresponding oligo-20-esters, F_1-20-decanoate, are specified.

As promoters of initiation in our highly standardized model on the back skin of NMRI-mice, with 7,12-dimethylbenz[a]anthracene (DMBA) as initiator, F-types exhibit strong cocarcinogenic in terms of promoting activity already in small doses (see, for example, p = 10 nmoles of Welensalifactor F_1, Fig. 8). Indeed, qualitatively and quantitatively the irritant (and cocarcinogenic) activity of Welensalifactors of the F-type is comparable to that of the chemically related, well established DTE promoter TPA. As 'cryptic' forms the corresponding F'-types are less active irritants (and promoters) (see, for example, Table 5). In chronic exposures of mouse skin solely to a promotional protocol (see line 3, Fig. 1) the mixture of F- and F'-types obtained from leaves exhibited no solitarycarcinogenic activity (Hecker, Lutz, et al., 1983).

Welensali tea: The plant preparation of *Croton flavens* most frequently consumed on Curaçao contains a mixture of activity type F and activity type F' factors with 16-hydroxyphorbol as the only parent alcohol (see Table 6). Together, they amount to an estimated minimal (relative total) content of 1.6 ppm per one cup of the tea (Table 6, last column). The two mixtures of Welensalifactors activity type F and F', respectively, were separated, and by HPLC it was shown that each comprises three individual Welensalifactors which had been obtained also from the leaves (Table 6, first double column). Of the three Welensalifactors type F the estimated (absolute) minimal content/cup is 0.27 μg (see Table 6). This amount (i.e. 0.41 nmol of F_1), for example, exceeds by more than 10 times the irritant dose 50 on the mouse ear of the typical promoter Welensalifactor F_1 (see Table 7). The corresponding estimated (absolute) minimal content of Welensalifactors type F' in one cup is 0.12 μg (in Table 6). In nanomoles, this amount is practically identical with the irritant dose 50 of

Table 6. The two mixtures of Welensalifactors activity type F and F', obtained from extract of Welensali tea, together with their determined and estimated minimal contents, in one cup of the tea

Welensalifactors F:
$R^2 = OH$; $R^3 = H$
$R^1 = CO(CH_2)_{14}CH_3$: F_1
$R^1 = CO(CH_2)_{12}CH_3$: F_3
$R^1 = CO(CH_2)_{16}CH_3$: F_5

Welensalifactors F':
$R^2 = OH$; $R^3 = Ac$
$R^1 = $ for F'-types
see F_1, F_3

parent alcohol: 16-hydroxyphorbol

Activity type	Composition (Welensali-factors)	Minimal content in one cup of Welensali tea					
		Determined			Estimated		
		Absolute (μg)	Ratio F/F'	Rel. total (ppb)	Absolute (μg)	Ratio F/F'	Rel. total (ppb)
F	F_1 F_3 F_5	0.06*	0.5/1	0.75	0.27	2.2/1	1.6
F'	F_1-20-octanoate F_1-20-decanoate F_3-20-cecanoate	0.12*			0.12*		

*In aqueous extracts such as Welensali tea, hydrophilic contaminations (e.g. proteins) complicate the quantitative determination of the F-types, but not of F'-types.

For estimation of the true minimal content of F-types in the tea an amount of 0.12 of the F'-types was used as an internal standard.

the typical 'cryptic' promoter F_1-20-decanoate (see Table 7). Thus, the estimated (absolute) total of active principles F and F', i.e. a minimal content of 0.39 μg per cup of tea corresponding to 1.6 ppm, may be considered a 'basal dose' regarding irritants and cocarcinogens in the tea. Comparison with the irritant doses 50 by Welensalifactors F_1 and F_1-20-decanoate *on the mouse ear*

Table 7. Comparison of the estimated minimal content of Welensalifactors activity types F and F' in one cup of Welensali tea and irritant dose 50 on the mouse ear of the prototype Welensalifactors F_1 and F_1-20-decanoate.

Mixture of activity types	Estimated minimal content per cup		Prototype Welensali-factors	ID_{50}^{24} (nmoles)	Ratio*
	μg	nmoles			
F	0.27	0.41	F_1	0.033	12
F'	0.12	0.15	F_1-20-decanoate	0.11	1.4

*Example: estimated minimal content of mixture of factors type F over irritant dose 50 of F_1 in nmoles.

(Table 7) suggests: solely by one cup of the tea per day, the dose maintaining chronic irritation in the *human oesophagus* may be exceeded considerably. As a rule, in fact, people on Curaçao consume much more than one cup per day (Morton, 1968b), quite apart from other utilizations of the plant parts of Welensali (and, in addition, of Belly-ache Bush, see above).

Without testing in a corresponding experimental model, prototypes of the solitary and cocarcinogens postulated to be involved in the experimental analysis of the 'Curaçao case' would not have been complete. Therefore, according to the initiation/promotion protocol (see Fig. 1), a single, subcarcinogenic dose of the PAH-type solitarycarcinogen DMBA was administrated as initiator to NMRI mice by stomach tube, followed by repeated intubations of the cocarcinogenic DTE type promoter TPA. In this way tumours of the forestomach were generated (see Hecker, Lutz *et al.*, 1983). Independently and in quite another, different context, similar results were obtained in mice (and also in rats) by others, using analogous protocols but different initiators and promoters. Taking all the experimental data together it appears most likely that in the the 'Curaçao case' limited, previous exposure to *initiators of the solitary carcinogenic PAH-type* through petrol-contaminated drinking water and (simultaneous or) especially *subsequent chronic exposure to promoters of the cocarcinogenic DTE type* by Welensali tea represent the *principal risk factors in this typical life style cancer* (see Table 1).

This conclusion, as drawn from extrapolation of the experimental data obtained to the epidemiologic model of the human cancer, are confirmed by findings in biopsies obtained and registered in the only hospital on Curaçao, which deals with about 90% of all clinical cases on the island: in patients over 50 years of age, without manifested cancer, frequently severe chronic inflammation of the oesophagus is observed. Therefore, for the population at risk, chronic inflammation caused by Welensalifactors, an important element of promotion by diterpene esters also in mouse skin, may be considered *a principal local risk factor of cancer* of the oesophagus in confirmation of the hypothesis postulated (Hecker, Lutz *et al.*, 1983). Corresponding investigations of plant parts of the Belly-ache Bush (*Jatropha gossipyfolia*) are underway in our laboratory; they have shown already that this plant contains irritants of the tigliane type ('Jatropha factors', Adolf, Opferkuch & Hecker, 1984; Biehl & Hecker, 1986).

The results of the experimental analysis presented here may provide the scientific basis for risk assessment by local health authorities in an effort to defray adequate measures of primary prevention. However, as a final step in the scientific evaluation of the 'Curaçao case' it appears desirable to investigate the aetiologic cocarcinogen hypothesis by prospective epidemiological research (see Hecker, 1984a, c; Hecker, Lutz, *et al.*, 1983).

Cancer of the nasopharynx in South China—another aetiology of cancer involving DTE promoters?

Traditionally, the ancient culture of China was committed not only to the *cure of diseases* of all kinds but also to the idea of *prevention*, as reflected in the Chinese saying: "Treat a person before he has fallen ill and not only when he is already sick" (after Zhong, 1984).

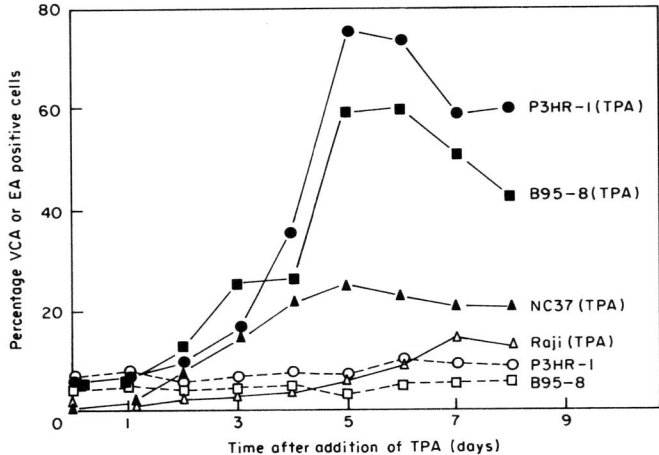

Figure 9. Induction of synthesis of EB-virus capsid antigen (VCA) in the cell lines P3HR-1 and B-95-8, and of early antigen (EA) in the lines NC37 and Raji with and without 3×10^{-8} M TPA.

With this saying in mind, it is of particular interest to note that Welensalifactor F_1, identified as one of the local cocarcinogenic risk factors in oesophageal cancer on Curaçao, is fully identical chemically with the DTE 'hexadecanoyl-hydroxyphorbolacetate' (HHPA) as obtained from *Aleurites fordii*, another species of Euphorbiaceae cultivated as the main source of tung oil. In a similar manner as shown by us previously for TPA and many other promoting DTEs of the tigliane-, ingenane-, and daphnane-type, (Hecker, 1979; zur Hausen et al., 1978; zur Hausen et al., 1979a), also HHPA (or Welensalifactor F_1) was shown to activate the genome of Epstein–Barr virus in immortalized lymphoblastoid cells (Ito et al., 1983a, b).

Induction of Epstein–Barr virus (EBV) synthesis by TPA in certain human lymphoblastoid cell lines (Fig. 9) may be measured by counting the number of cells carrying virus capsid (VCA) or early (EA) antigen; apart from 'producer' cell lines (which produce the virus spontaneously at low rate, such as, for example Fig. 9, P3HR-1 and B 95-8), synthesis of virus related proteins is stimulated by TPA and congeners also in 'non-producer' cell lines (e.g. Fig. 9, NC37 and Raji, see also Table 8). The number of cells carrying the antigen reaches a maximum about five days after exposure to TPA (see Fig. 9). The irritant and promoting activities in mouse skin of selected DTE covering the three principal structural types (see Fig. 3) correlate well with their capabilities to induce in NC37 and Raji cells early antigen of EBV (Table 8). Most recently it was shown that also some of the new indole alkaloid type promoters (see above) induce EBV synthesis in much the same way as TPA (Yamamoto, 1984; Yamamoto et al., 1981). To what extent these findings may be generalized for all structural types of tumour promoters (Takada & zur Hausen, 1984) remains to be seen.

In the rodent *Mastomys natalensis*, closely related to the rat, a synergism for generation of skin papilloma was shown to exist involving activation of endogenous papilloma virus genomes by TPA (Antmann, Volm & Wayss, 1984). Both of the tumour viruses, shown to be activated by DTE promoters,

Table 8. Skin irritant and promoting activities of polyfunctional tigliane, ingenane and daphnane-type diterpenes *in vivo* and their capability to induce synthesis of EB-viral early antigen (EA) in the human lymphoblastoid NC 37 and Raji cells

Polyfunctional diterpenes	Irritant activity*		Promoting activity (relative potency)†	EA induction at 3×10^{-8} M	
	ID_{50}^{24} (nmoles)	I_{50}^{24} (nmoles^{-1})		NC37‡ (% of cells)	Raji§
Tigliane types					
TPA	0.016	6.3	++++	13.2	5.2
4-O-MeTPA	2.3	0.43	(+)	14.1	10.2
PDD	0.01	100	++	10.0	4.9
4α-PDD	>150	<0.01	0	0.2	<0.01
12-Deoxyphorbol-13-decanoate	0.017	59	+++	13.4	5.8
Phorbol	>270	<0.004	0	0.3	<0.01
Daphnane types					
Hippomane factors $M_1 + M_2$	0.02	50	+++	14.9	8.3
Pimelea factors P_2	0.003	333	+++	11.0	7.8
Ingenane types					
Ingenol-3-hexadecanoate	0.086	12	+++	12.8	6.5
Ingenol	>287	<0.093	0	0.1	<0.01

* Irritant dose 50 (ID_{50}^{24}) and irritancy (I_{50}^{24}) on the mouse ear at 24 hours.
† Initiation/promotion standard assay on back skin female NMRI mice with DMBA as initiator, semiquantitative estimation of relative potency (see Sorg *et al.*, 1987).
‡ Spontaneous EA induction 0.1%.
§ Spontaneous EA induction 0.01%.

are related to human cancers: EBV to Burkitt lymphoma and anaplastic nasopharyngeal carcinoma (e.g. Hirayama & Ito, 1981; Ito *et al.*, 1983a, b; zur Hauzen *et al.*, 1979b), and papilloma virus to certain male and female genital cancers containing human papilloma virus (HPV, e.g. Gissmann, 1984). Thus it may be suspected that geographical regions of frequent utilization of Euphorbiaceae derived materials, for example, as ethnomedicinal therapeutics (Hecker, 1981a; Zheng *et al.*, 1983) or as technological products (tung oil, Ito *et al.*, 1983a, b) may be identical with regions of high incidence of NPC. Indeed, in very recent investigations an extended survey for activation of EBV by environmental materials collected in South China, where NPC is prevalent (see Table 1), was accomplished (Zheng, 1986).

INCREASING REFINEMENT IN THE ASSESSMENT OF THE RISK OF CANCER

However complex the multifactorial aetiology of cancer (tumours) may be, one of the most important tasks of basic cancer research always was and still is development of generally valid criteria for *detection and classification of environment carcinogens*, and possibly for grading of the risk imposed by them on human beings (Bohrmann, 1983; Hecker, 1985a). On the background of the progress made in the last years there might be already a general consensus to classify (classical) solitary carcinogens *as category one risk factors of cancer*. In contrast, the (newly established) cocarcinogens of the DTE type, as '*per se* essentially non-carcinogenic amplifiers' of carcinogenesis, which produce cancer only in *appropriate combination* with initiators (or with predisposed tissues), may be classified as *category two risk factors of cancer* (e.g. Hecker, 1985a, 1984a–c, 1981a, and previous publications). By classifying physicochemically well defined environmental risk factors in solitary carcinogens and in cocarcinogens perhaps an acceptable *grading of the risk of cancer* may be achieved, allowing in the future for more refined assessment of cancer risk (Marshall, 1982, Hecker, 1981a). For health authorities this may be particularly helpful to find in practice (for example, remember the saccharin case), a scientifically more appropriate balance between an overprotective legislation aiming at absolute safety and an all permissive absence of restriction, both of which are disadvantageous to the human individual. An incentive progress of this kind, covering both basic and applied areas of cancer research, was given by the detection and chemical, as well as biological, characterization of the DTE promoters occurring in certain Euphorbiales, the Euphorbiaceae and Thymelaeaceae, as documented in many international symposia (see for example, Hecker *et al.*, 1982; Fujiki *et al.*, 1984a).

ACKNOWLEDGEMENTS

The project 'Welensali tea' was generously supported in all its phases by the Wilhelm-und-Maria-Meyenburg-Stiftung, Heidelberg-Leimen. The enthusiastic cooperation in our various projects on DTE by many Dr rer.nat.-students of the Faculty of Pharmacy, University of Heidelberg, and by post-doctoral research fellows with their untiring engagement is gratefully acknowledged.

REFERENCES

ADOLF, W., OPFERKUCH, H. J. & HECKER, E., 1984. Irritant phorbol derivatives from four *Jatropha* species (Euphorbiaceae). *Phytochemistry, 23:* 129–132.

AMTMANN, E., VOLM, M. & WAYSS, K., 1984. Tumour induction in the rodent *Mastomys natalensis* by activation of endogenous papilloma virus genomes. *Nature, 308:* 291–292.

BALANDRIN, M. F., KLOCKE, J. A., WURTELE, E. S. & BOLLINGER, WM. H., 1985. Natural plant chemicals: sources of industrial and medicinal materials. *Science, 228:* 1154–1160.

DE BOER, S. Y., 1979. *Frequentia van Klanker van Mond-Tong-Pharynx-Slokdarm en de Nederlandse Antillen.* Dissertation (Landslaboratorium voor de Volkszondheid, Afdeling Pathologisch anatomisches Laboratorium Curaçao, Niederländische Antillen).

BIEHL, J. & HECKER, E., 1986. Irritant and antineoplastic principles of some species of the genus Jatropha (Euphorbiaceae). *34th Annual Congress on Medicinal Plant Research, Hamburg, Sept. 22–27. Abstracts of Short Lectures and Poster Presentations:* K31: 16. Stuttgart: Thieme-Verlag.

BOHRMANN, J. S., 1983. Identification and assessment of tumor-promoting and cocarcinogenic agents: state-of-art in vitro methods. *Critical Review of Toxicology, 11:* 131–167.

DELCLOS, K. B., NAGLE, D. S. & BLUMBERG, P. M., 1980. Specific binding of phorbol ester tumor promoters to mouse skin. *Cell, 19:* 1025–1032.

DOEGE, T. & HECKER, E., 1984. Distribution and metabolism of the cryptic tumor promoter TPA-t as a possible pro-drug of the irritant cocarcinogenic TPA. III. *International Congress of Inflammation, Paris, 3.–8.9.1984, Abstracts:* 257.

DUNHAM, L. J. & BAILAR, J. C., 1968. World maps of cancer mortality rates and frequency ratios. *Journal of the National Cancer Institute (U.S.) 41:* 155–203.

ENGLER, A., 1964. *Syllabus der Pflanzenfamilien.* Berlin-Nikolassee: Gebr. Bornträger.

EVANS, F. J. & TAYLOR, S. E., 1983. Pro-inflammatory, tumour promoting and anti-tumour diterpenes of the plant families of Euphorbiaceae and Thymelaeaceae *Progress in the Chemistry of Organic Natural Products 44:* 1–99.

FELLHAUER, M., HECKER, E., 1986. Screening of thymelaeaceae species for irritant, cocarcinogenic and antineoplastic activity. *34th Annual Congress on Medicinal Plant Research, Hamburg, Sept. 22–27, Abstracts of Short Lectures and Poster Presentations* P108: 73. Stuttgart: Thieme Verlag.

FUJIKI, H., HECKER, E., MOORE, R. E., SUGIMURA, T. & WEINSTEIN, I. B. (Eds), 1984a. 14th Symposium Princess Takamatsu Cancer Research Fund, Tokyo. *Cellular Interactions of Environmental Promoters.* Tokyo: Japan Science Press; Utrecht: VNU Science Press.

FUJIKI, H., SUGANUMA, M., TAHIRA, T., YOSHIOKA, A., NAKAYASU, M., ENDO, Y., SHUDO,

K., TAKAYAMA, S., MOORE, R. E. & SUGIMURA, T., 1984b. Nakahara Memorial Lecture: New classes of tumour promoters: Teleocidin, Aplysiatoxin and Palytoxin. In: H. Fujiki, E. Hecker, R. E. Moore, T. Sugimura & I. B. Weinstein (Eds), *Cellular Interactions by Environmental Tumour Promoters*: 37–45. Tokyo: Japan Science Society Press; Utrecht: VNU Science Press.

FÜRSTENBERGER, G. & HECKER, E., 1972. Zum Wirkungsmechanismus cocarcinogener Pflanzeninhaltsstoffe. *Planta Medica, 22:* 241–266.

FÜRSTENBERGER, G. & HECKER, E., 1985. On the active principles of Euphorbiaceae XI. The skin irritant and tumor promoting diterpene esters of Euphorbia tirucalli L. originating from South Africa. *Zeitschrift für Naturforschung, 40c:* 631–646.

FÜRSTENBERGER, G. & HECKER, E., 1986. On the active principles of Euphorbiaceae XII. Highly unsaturated irritant diterpene esters from Euphorbia tirucalli L. originating from Madagascar. *Journal of Natural Products (Lloydia), 49:* 386–397.

FÜRSTENBERGER, G., BERRY, D. L., SORG, B. & MARKS, F., 1981. Skin tumor promotion by phorbol esters is a two-stage process. *Proceedings of the Academy of Sciences, U.S.A., 78:* 7722–7726.

GISSMANN, L., 1984. Papilloma viruses and their association with cancer in animals and in man. *Cancer Surveys, 3:* 161–181.

HECKER, E., 1972. Aktuelle Probleme der Krebsentstehung. *Zeitschrift für Krebsforschung, 78:* 99–122.

HECKER, E., 1976a. Definitions and terminology in cancer (tumor) etiology—an analysis aiming at proposals for a current internationally standardized terminology. *International Journal of Cancer, 18:* 122–129.

HECKER, E., 1976b. Definitions and terminology in cancer (tumor) etiology—an analysis aiming at proposals for a current internationally standardized terminology. *Zeitschrift für Krebsforschung, 86:* 219–230.

HECKER, E., 1976c. Definitions and terminology in cancer (tumor) etiology—an analysis aiming at proposals for a current internationally standardized terminology. *Gann, 67:* 471–481.

HECKER, E., 1976d. Definitions and terminology in cancer (tumor) etiology—an analysis aiming at proposals for a current internationally standardized terminology. *Bulletin of the World Health Organisation, 54:* 1–10.

HECKER, E., 1978. Structure–activity relationships in diterpene esters irritant and cocarcinogenic to mouse skin. In T. J. Sloga, A. Sivak & R. K. Boutwell (Eds), *Mechanisms of Tumour Promotion and Cocarcinogenesis*: 11–48. New York: Raven Press.

HECKER, E., 1979. Diterpene ester type modulators of carcinogenesis—new findings in the mechanism of chemical carcinogenesis and the etiology of human tumors. In E. C. Miller, J. A. Miller, T. Horono, E. Sugimura & S. Takayama (Eds): *Naturally Occurring Carcinogens—Mutagens and Modulators of Carcinogenesis:* 263–286. Tokyo: Japan Science Society Press.

HECKER, E., 1981a. Co-carcinogenesis and tumor promoters of the diterpene ester type as possible carcinogenic risk factors. *Journal of Cancer Research and Clinical Oncology, 99:* 103–124.

HECKER, E., 1981b. Prototype processes as experimental models in multifactorial carcinogenesis. *Journal of Cancer Research and Clinical Oncology, 99:* A29.

HECKER, E., 1984a. Co-carcinogens of the tumor promoter type as potential risk factors of cancer in man—a first complete experimental analysis of an etiological model situation and some of its consequences. In M. Börszönyi, N. E. Day, K. Lapis, & H. Yamasaki (Eds), *Models, Mechanisms and Etiology of Tumour Promotion*. IARC Publication No. 56: 441–446. Lyon: IARC.

HECKER, E., 1984b. Concluding remarks and future perspectives. In M. Börszönyi, N. E. Day, K. Lapis, & H. Yamasaki (Eds), *Models, Mechanisms and Etiology of Tumour Promotion*. IARC Publication No. 56: 509–521. Lyon: IARC.

HECKER, E., 1984c. Cocarcinogens of the initiation (or-tumor)-promoter type as environmental risk factors of cancer in man—experimental analysis of an etiologic model situation of life style cancer and current problems of assessment of cancer risk in multifactorial carcinogenesis. *Acta Pharmacologica et Toxicologica, 55 (Suppl. II):* 145–164.

HECKER, E., 1985a. Multifaktorielle und Mehrstufen-Karzinogenese—ätiologische (epidemiologische) und experimentelle Modelle sowie aktuelle Probleme der Bewertung von Krebsrisikofaktoren der Umwelt. In K. E. Appel & A. G. Hildebrandt, (Eds), *bgaschriften, 6, Tumorpromotoren-Erkennung, Wirkungsmechanismen und Bedeutung:* 14–42. Munchen: MMV Medizin Verlag.

HECKER, E., 1985b. Cell membrane associated protein kinase C as receptor of diterpene ester co-carcinogens of the tumor promoter type and the phenotypic expression of tumors. *Arzneimittel-Forschung/Drug Research 35(II), 12a:* 1890–1903.

HECKER, E. & BRESCH, H., 1965. Über die Wirkstoffe des Crotonöls, III. Reindarstellung und Characterisierung eines toxisch, entzündlich und cocarcinogen hochaktiven Wirkstoffs. *Zeitschrift für Naturforschung, 20b:* 216–226.

HECKER, E. & SCHMIDT, R., 1974. Phorbol esters—the irritants and cocarcinogens of *Croton tiglium* L. *Progress in the Chemistry of Organic Natural Products, 31:* 377–467.

HECKER, E. & WEBER, J., 1977. Cocarcinogens from *Croton flavens* L. and the high incidence of esophageal cancer in Curaçao. 7th International Symposium on the Biological Characterization of Human Tumors. Budapest, April 13–15, Abstracts: 52; see also *Experientia (Basel), 34:* 679–682 (1978).

HECKER, E., OSSWALD, H., SCHMIDT, R., OPPFERKUCH, H. J., SORG, B. & YOUSSEF, M., 1984. European Patent No. 0013981 and corresponding patents worldwide.

HECKER, E., FUSENIG, N. E., KUNZ, W., MARKS, F. & THIELMANN, H. W., (Eds), 1982. Cocarcinogenesis and biological effects of tumor promoters. In *Carcinogenesis—a Comprehensive survey, 7.* New York: Raven Press, Inc.

HECKER, E., LUTZ, D., WEBER, J., GOERTTLER, K. & MORTON, J. F., 1983. Multistage tumor development in the human esophagus—the first identification of cocarcinogens of the tumor promoter type as principal carcinogenic risk factors in a local life-style cancer. In *13th International Cancer Congress. Part B. Biology of Cancer (1):* 219–238. New York: Alan R. Liss, Inc.

HECKER, E., ADOLF, W., HERGENHAHN, M., SCHMIDT, R. & SORG, B., 1984. Keynote lecture: Irritant promoters of mouse skin—contributions to etiologies of environmental cancer and to biochemical mechanisms of carcinogenesis by diterpene esters. In H. Fujiki, E. Hecker, R. E. Moore, T. Sugimura & I. B. Weinstein (Eds), *Cellular interactions of environmental promoters:* 3–63. Tokyo: Japan Science Press; Utrecht: VNU Science Press.

HERGENHAHN, M. & HECKER, E., 1981. Specific binding of the tumor promoter TPA in various mouse organs as measured by a 'cold acetone-filter assay'. *Carcinogenesis, 2:* 1277–1281.

HIGGINSON, J., 1979. A global perspective of environmental carcinogenesis. In P. Emmlot & E. Kriek (Eds), *Environmental Carcinogenesis:* 9–22. Amsterdam: Biochemical Press.

HIRAYAMA, T. & ITO, Y., 1981. A new view of the etiology of nasopharyngeal carcinoma. *Preventive Medicine, 10:* 614–622.

HOPPE, W., ZECHMEISTER, K., RÖHRL, M., BRANDL, F., HECKER, E., KREIBICH, G. & BARTSCH, H., 1969. Structure determination of a solvate of phorbol—the diterpene parent of the tumor promoters from Croton oil. *Tetrahedron Letters, 1969:* 667–670.

ITO, Y., OHIGASHI, H., KOSHIMIZU, K. & YI, Z., 1983a. Epstein-Barr-virus-activating principle in the ether extracts of soils collected from under plants which contain active diterpene esters. *Cancer Letters, 19:* 113–117.

ITO, Y., YAMASE, S., TOKUDA, H., KISHISHITA, M., OHIGASHI, H., HIROTA, M. & KOSHIMIZU, K., 1983b. Epstein-Barr virus activation by tung oil, extracts of Aleurides fordii and its diterpene ester 12-O-hexadecanoyl-16-hydroxyphorbol-13-acetate, *Cancer Letters, 18:* 87–95.

KINZEL, V., RICHARDS, J., GOERTTLER, K., LOEHRKE, H., FÜRSTENBERGER, G. & MARKS, F., 1984. Interaction of phorbol derivatives with replicating cells. In M. Börszönyi, N. E. Day, K. Lapis & H. Yamasaki (Eds), *Models, Mechanisms and Etiology of Tumour Production:* 253–264. IARC Scientific Publication No. 56. Lyon: IARC.

LOTTER, H., & HECKER, E., 1985. Wechselwirkung von Tumorpromotoren mit zellulären Rezeptoren—feinstruktureller Vergleich der drei Prototypen irritierender und co-carcinogener Diterpenester aus Euphorbiaceen und Thymelaeaceen, Fresenius. *Zeitschrift für analytische Chemie, 321:* 639–640.

LOTTER, H. & HECKER, E., 1986. Analysis of the structure/activity relationships in diterpene ester irritants and promoters by computergraphic methods. *IX. International Symposium on Medicinal Chemistry, Berlin, Sept. 14–18, Abstracts.*

LOTTER, H., JONES, A. & STURM, M., 1977. Röntgenstrukturanalyse von Mezerein aus Daphne mezereum L. *Zeitschrift für Naturforschung, 32:* 678–682.

LUTZ, D. & HECKER, E., 1979. Esophageal Cancer on Curacao—further new tumor promoters from Croton flavens L. *8th International Symposium on the biological characterization of human tumors, May 8–11, Athens, Greece, Book of Abstracts.*

MARKS, F. & FÜRSTENBERGER, G., 1984. Stages of tumor promotion in skin. In M. Börszönyi, N. E. Day, K. Lapis & H. Yamasaki (Eds), *Models Mechanisms and Etiology of Tumour Production:* 253–264. IARC Scientific Publication No. 56. Lyon: IARC.

MARSHALL, E., 1982. EPA's high risk carcinogen policy. *Science, 218:* 975–978.

MORTON, J. F., 1968a. A survey of medicinal plants of Curacao. *Economic Botany 22:* 87–102.

MORTON, J. F., 1968b. Plants associated with esophageal cancer cases in Curacao. *Cancer Research 28:* 2268–2271.

MORTON, J. F., 1974. Folk remedy plants and esophageal cancer in Coro, Venezuela. *Morris Arboretum Bulletin: 25:* 24–34.

PAQUETTE, L. A., NITZ, T. J., ROSS, R. J. & SPRINGER, J. P., 1984. Ingenane synthetic studies. An expedient approach to highly oxygenated ABC subunits of ingenol via reductive dialkylative annulation and α,β-epoxy ketone photoisomerization. *Journal of the American Chemical Society, 106:* 1446–1454.

ROBERTS, H. B., McCLURE, T. J. RITCHIE, E., TAYLOR, W. C. & FREEMAN, P. W., 1975. The isolation and structure of the toxin of *Pimelea simplex*, responsible for St. George disease of cattle. *Australian Veterinary Journal, 51:* 325–326.

ROESER, H., SORG, B. & HECKER, E., 1983. Irritant and tumor promoting activities of some esters of ingenol and' relationships between structures and activities. International Symposium: Role of Cocarcinogens and Promoters in Human and Experimental Carcinogenesis, Budapest, May 16–18. *Abstracts:* 67. Lyon: IARC.

SCHMIDT, R., ADOLF, W., MARSTON, A., ROESER, H., SORG, B., FUJIKI, H., SUGIMURA, T., MOORE, R. E. & HECKER, E., 1983. Inhibition of specific binding of (^{3}H)phorbol-12,13-dipropionate to an epidermal fraction by certain irritants and irritant promoters of mouse skin. *Carcinogenesis, 4:* 77–81.

SHOYAB, M., WARREN, T. C. & TODARO, G. J., 1981. Tissue and species distribution and developmental variation of specific receptors for biologically active phorbol and ingenol esters. *Carcinogenesis, 2:* 1273–1276.

SLAGA, T. J., FISCHER, S. M., NELSON, K. & GLEASON, G. L., 1980. Studies on the mechanism of skin tumor promotion: evidence for several stages in promotion. *Proceedings of the Academy of Sciences U.S.A. 77:* 3659–3663.

SORG, B. & HECKER, E., 1982. Zür Chemie des Ingenols, II. Ester des Ingenols und des $\Delta^{7\cdot 8}$-Ingenols. *Zeitschrift für Naturforschung, 37b:* 748.

SORG, B., FÜRSTENBERGER, G., BERRY, D. L., HECKER, E. & MARKS, F., 1982. Preparation of retinoic acid esters of phorbol derivatives. *Journal of Lipid Research, 23:* 443–447.

SORG, B., SCHMIDT, R. & HECKER, E. 1987. Structure/activity relationships of polyfunctional diterpenes of the ingenane type. I. Tumor promoting activity of homologous, aliphatic 3-esters of ingenol and of $\Delta^{7\cdot 8}$-isoingenol-3-tetradecanoate. *Carcinogenesis, 8:* 1–4.

TAKADA, K. & ZUR HAUSEN, H., 1984. Induction of Epstein-Barr virus antigens by tumor promoters for epidermal and non-epidermal tissues, *International Journal of Cancer, 33:* 491–496.

THIELMANN, H. W. & HECKER, E., 1969. Beziehungen zwischen der Struktur von Phorbolderivaten und ihren entzündlichen und tumorpromovierenden Eigenschaften, *Fortschritte der Krebsforschung, Molekularbiologie, Wachstumskinetik,* Schattauerverlag Stuttgart-New York: 171–179.

TREMP, G. & HECKER, E., 1985. Stability of the 'second stage' promoter 12-O-retinoylphorbol-13-acetate (RPA). *Cancer Research, 45:* 2390.

WATERHOUSE, J., MUIR, C., CORREA, P. & POWELL, J., 1976. *Cancer Incidence in Five Continents, III: IARC Scientific Publication No. 15.* Lyon: IARC.

WENDER, P. A., HILLEMANN, C. L. & SZYMONIFKA, M. J., 1980. An approach to the tiglianes, daphnanes and ingenanes via the divinylcyclopropane rearrangement. *Tetrahedron Letters, 1980:* 2205–2208.

YAMAMOTO, N., 1984. Interaction of viruses with tumor promoters. *Reviews of Physiology, Biochemistry and Pharmacology, 101:* 111–159.

YAMAMOTO, H., KATSUKI, T., HINOMA, Y., HISHINO, H., MASANO, M., FUJIKI, H. & SUGIMURA, T., 1981. Induction of Epstein-Barr Virus by a new tumor promoter teleocidin, compared to induction by TPA, *International Journal of Cancer, 28:* 125–129.

ZAYED, S., HAFEZ, A., ADOLF, W. & HECKER, E., 1977. New tigliane and daphnane derivatives from *Pimelea prostrata* and *Pimelea simplex. Experientia, 33:* 1554–1555.

ZAYED, S., SORG, B. & HECKER, E., 1984. Structure/activity relations of polyfunctional diterpenes of the tigliane type, VI. Irritant and tumor promoting activities of semi-synthetic mono- and diesters of 12-deoxyphorbol. *Planta Medica, 1:* 65–69.

ZECHMEISTER, K., BRANDL, F., HOPPE, W., HECKER, E., OPFERKUCH, H. J. & ADOLF, W., 1970. Structure Determination of the new tetracyclic diterpene ingenoltriacetate with Triple Product Methods. *Tetrahedron Letters, 35,* 4071–4072.

ZHENG, Y., ZHONG, J. M., MO, Y. K. & MIAO, X. C., 1983. Epstein-Barr virus early antigen induction in Raji cells by Chinese medicinal herbs. *Intervirology, 19:* 201–204.

ZHENG, Y., 1986. Prospective studies on nasopharyngeal carcinoma and Epstein-Barr virus inducers. In G. Wagner & Z. You-hui (Eds), *Cancer of the Liver, the Esophagus and the Nasopharynx:* 162–167. Berlin, Heidelberg, New York, Tokyo: Springer-Verlag.

ZHONG, S. Z., 1984. *Ancient China's Scientists,* Hongkong: Commercial Press Ltd.

ZUR HAUSEN, H., O'NEILL, F. B., FREESE, U. K. & HECKER, E., 1978. Persisting oncogenic herpesvirus induced by the tumor promoter TPA. *Nature, London, 272:* 373–374.

ZUR HAUSEN, H., BORNKAMM, G. W., SCHMIDT, R. & HECKER, E., 1979a. Tumor initiators and promoters in the induction of Epstein-Barr virus. *Proceedings of the National Academy of Sciences, U.S.A. 76:* 782–785.

ZUR HAUSEN, H., SCHULTE-HOLTHAUSEN, H., KLEIN, G., HENLE, W., HENLE, G., CLIFFORD, J. & SANTESSON, L., 1979b. EBV DNA in biopsies of Burkitt tumors and anaplastic carcinomas of the nasopharynx. *Nature, London, 228:* 1076.

The biosynthesis of tigliane and related diterpenoids; an intriguing problem

RICHARD J. SCHMIDT, F.L.S.

Welsh School of Pharmacy, University of Wales Institute of Science and Technology, P.O. Box 13, Cardiff CF1 3XF

Received May 1986, accepted for publication August 1986

SCHMIDT, R. J., 1986. **The biosynthesis of tigliane and related diterpenoids; an intriguing problem.** An hypothesis to explain the relationships between various diterpenoids found in members of the Euphorbiaceae is proposed. The likely nature of the tetraprenyl pyrophosphate precursor is considered, as is the manner in which the precursor may cyclize. It is thought that up to eight possible precursors may exist, being the various *cis/trans* isomers of geranyl-geranyl pyrophosphate. This suggestion arises from an acceptance of the now well-established (but as yet unproven) process of an initial 'head-to-tail' cyclization of the precursor to form the variously cyclized hydrocarbon skeletons of tigliane and related diterpenoids. It is considered that the necessary isomerization of up to three of the prenyl units, as would have to be the case if the tigliane skeleton were produced from the all-*trans* geranyl-geranyl pyrophosphate, is not reconcilable with an initial head-to-tail cyclization of the precursor either on the grounds of spatial considerations or on the evidence of the currently known chemical variation in the whole family of compounds.

ADDITIONAL KEY WORDS:—Euphorbiaceae – isomers – phytochemistry.

CONTENTS

Introduction	221
'Head-to-tail' *vs.* 'concertina-like' cyclization	222
The tetraprenyl pyrophosphate precursor	223
Summary	227
References	228

INTRODUCTION

A wide variety of terpenoids, ranging from mono-, sesqui- and diterpenoids to triterpenoids and steroids are known from members of the Euphorbiaceae. Many occur widely in the plant kingdom; others appear to be of restricted distribution. In particular, the tigliane and related diterpenoids are of special interest because they have a very limited known distribution in nature, being found only in the plant families Euphorbiaceae and Thymelaeaceae (see Evans & Taylor, 1983). These compounds are therefore of chemosystematic value and a better understanding of their biosynthesis can serve only to enhance their value in this respect.

The tigliane and related diterpenoids have apparently failed to attract the attentions of phytochemists with an interest in biosynthesis. Several reasons for this may be suggested, not least of which is the comparatively problematical isolation procedure associated with the compounds and the difficulty in obtaining, or producing, suitably radioactively-labelled compounds for feeding experiments. Thus, to date, no direct attempt to investigate the biosynthetic pathways has been reported in the literature. However, we are now in a position to consider the many such compounds whose structures are known and to formulate hypotheses by which the compounds may be related to one another biosynthetically. Perhaps the most important initial observation that can be made to this end is that the hydrocarbon skeletons are apparently cyclized through a 'head-to-tail' condensation of a tetraprenyl pyrophosphate precursor rather than through the classical 'concertina-like' cyclization that is typical of many diterpenoids, triterpenoids, and steroids (see, for example, Banthorpe & Charlwood, 1980). It is the purpose of this discussion to explore this intriguing observation, with a view to producing a working hypothesis that can trace the biosynthetic relationships between the whole family of compounds and ultimately to provide some new data that will facilitate the classification of the plants from which the compounds are derived.

'HEAD-TO-TAIL' *VS.* 'CONCERTINA-LIKE' CYCLIZATION

The data to be considered are derived from two distinct areas of knowledge. From one area we have a range of chemical structures of diterpenoids from the plant family Euphorbiaceae and elsewhere; from the other we have existing information (and hypotheses) concerning the biosynthesis of terpenoids generally.

We have to assume that the tigliane and related diterpenoids have an isoprenoid origin. Also, by analogy with the terpenoid biosynthesis-mimicking cyclization reactions described by Nishizawa, Takenaka, Nishide & Hayashi (1983), who succeeded in initiating a 'concertina-like' cyclization process in farnesol derivatives using a mercury (II) trifluoromethane sulphonate–amine complex, we can assume that functionalization occurs after cyclization even in 'head-to-tail' cyclization processes. It is convenient, therefore, to ignore functionalization in this discussion.

We also have to assume that a biosynthetic relationship between the various compounds exists notwithstanding the variety of plants (genera and species) from which they were isolated.

We can then recognize three classes of diterpene skeleton—bicyclic, tricyclic and tetracyclic—that may be related to one another. The hydrocarbon skeletons shown in Fig. 1 have been drawn to show their relationships to tigliane and are therefore far from accurate representations of actual molecular shapes. Botanical sources of representative diterpenoids are given in Table 1.

The existence of the monocyclic type is the key to the whole discussion. It can be regarded as either the first product in a 'head-to-tail' cyclization of a tetraprenyl pyrophosphate precursor or as the last product in a 'concertina-like' cyclization sequence that is followed by bond cleavage. Although no member of the monocyclic class of diterpene has yet been reported from the Euphorbiaceae, such compounds are known from other natural sources. Dauben, Thiessen &

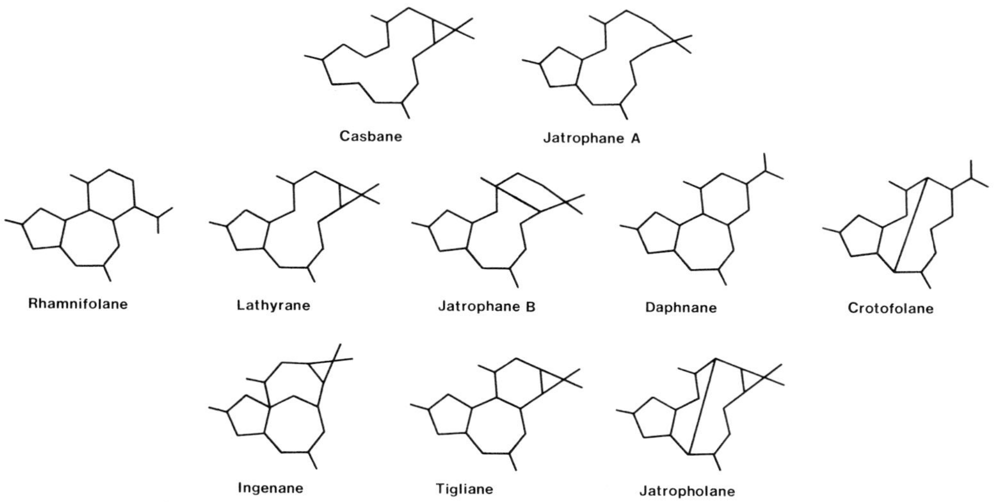

Figure 1. Hydrocarbon skeletons of diterpene classes.

Resnick (1962) provided the first example, cembrene from the oleoresin of *Pinus albicaulis* Engelm. and of "many other pine trees". This was followed very shortly by a report by Kobayashi & Akiyoshi (1962) of the probable structure of thunbergene from "several kinds of pine roots", and another by Roberts & Rowland (1962) describing the duvatriene-1,3-diols from tobacco (*Nicotiana tabacum* L.). The biosynthetic significance of these monocyclic diterpenes was immediately recognized and the concept of 'head-to-tail' cyclization later discussed by Erdtman, Norin, Sumimoto & Morrison (1964) and by Harrison, Scrowston & Lythgoe (1966) in relation to the supposed biosynthetic origins of the verticillane skeleton of verticillol from *Sciadopitys verticillata* Sieb. & Zucc. and the taxane skeleton of the yew alkaloids respectively. Subsequently, Robinson & West (1970a) isolated casbene, a bicyclic 14-membered ring diterpene, from *Ricinus communis* L. seedlings, and also recognized that its biosynthetic derivation from a 'head-to-tail' condensation of a tetraprenyl pyrophosphate precursor was highly plausible. They also showed that geranylgeranyl pyrophosphate serves as a precursor for the formation of their casbene (Robinson & West, 1970b) and that one unique enzyme component appears to carry out the cyclization process.

The concept of 'head-to-tail' cyclization, though by no means proven, thus became respectable. Furthermore, the biosynthetic origins of the tigliane and related diterpenoids can be more easily explained by an initial head-to-tail cyclization than by an initial concertina-like cyclization. The ensuing discussion is based, therefore, on the assumption that head-to-tail cyclization of an appropriate precursor is the first significant event in the biosynthesis of tigliane and related diterpenoids.

THE TETRAPRENYL PYROPHOSPHATE PRECURSOR

What exactly is the nature of the tetraprenyl pyrophosphate precursor? Banthorpe & Charlwood (1980), discussing the biosynthesis of terpenoids, noted

Table 1. Representative diterpenoids and their plant sources.

Diterpene skeleton	Compound	Species	Reference
Casbane	Casbene	*Ricinus communis* L.	Robinson & West (1970a, 1970b)
Jatrophane A	Crotonitenone	*Croton nitens* Sw.	Burke *et al.* (1981)
	Characiols	*Euphorbia characias* L.	Seip & Hecker (1984)
	Euphornin	*Euphorbia maddeni* Boiss.	Sahai *et al.* (1981)
	Euphoscopins	*Euphorbia helioscopia* L.	Shizuri *et al.* (1983)
	Jatrophone	*Jatropha gossypifolia* L.	Kupchan *et al.* (1976)
	Kansuinines	*Euphorbia kansui* Liou	Uemura & Hirata (1975), Uemura *et al.* (1975)
Rhamnifolane	Rhamnifolane derivative	*Croton rhamnifolius* Kunth	Stuart & Barrett (1969)
Lathyrane	Bertyadionol	*Bertya cupressoidea* Airy Shaw	Ghisalberti *et al.* (1973, 1974)
	Euphohelioscopins	*Euphorbia helioscopia* L.	Shizuri *et al.* (1983)
	Ingols	*Euphorbia ingens* E. Meyer	Lotter *et al.* (1979)
	Jokinols	*Euphorbia jolkini* Boiss.	Uemura *et al.* (1976)
	Lathyrols	*Euphorbia lathyris* L.	Zechmeister *et al.* (1970) Narayanan *et al.* (1971)
		Macaranga tanarius Muell. Arg.	Hui *et al.* (1975)
Jatrophane B	Jatrophatrione	*Jatropha macrorhiza* Benth.	Torrance *et al.* (1976)
Daphnane	Excoecariatoxin	*Excoecaria agallocha* L.	Ohigashi *et al.* (1974), Wiriyachitra *et al.* (1985)
		Stillingia sylvatica L.	Adolf & Hecker (1980)
	Hippomane factors	*Hippomane mancinella* L.	Adolf & Hecker (1975)
	Huratoxin	*Hura crepitans* L.	Sakata *et al.* (1971)
	Montanin	*Baliospermum montanum* Muell. Arg.	Ogura *et al.* (1978)
		Cunuria spruceana Baillon	Gunasekera *et al.* (1979)
	Resiniferatoxin	*Euphorbia resinifera* Berg	Adolf *et al.* (1982), Hergenhahn *et al.* (1975)
Crotofolane	Crotofolin A	*Croton corylifolius* Lam.	Chan *et al.* (1975)
Tigliane	Phorbols	*Jatropha* spp.	Adolf *et al.* (1984)
		Baliospermum montanum Muell. Arg.	Ogura *et al.* (1978)
		Ostodes paniculata Blume	Handa *et al.* (1983)
		Vernicia fordii Airy Shaw	Okuda *et al.* (1975), Airy Shaw (1966)
		Croton tiglium L.	Hecker & Schmidt (1974)
		Hippomane mancinella L.	Adolf & Hecker (1975)
		Sapium japonicum Pax & K. Hoffm.	Ohigashi *et al.* (1972)
		Stillingia sylvatica L.	Adolf & Hecker (1980)
		Euphorbia tirucalli L.	Fürstenberger & Hecker (1977)
Jatropholane	Jatropholones	*Jatropha gossypifolia* L.	Purushothaman *et al.* (1979)
Ingenane	Ingenols	*Euphorbia* spp.	Opferkuch & Hecker (1974), Evans & Taylor (1983)
		Elaeophorbia spp.	Evans & Kinghorn (1975)

that geranyl-geranyl pyrophosphate is invariably considered to be the parent compound in the biosynthesis of diterpenes. Robinson & West (1970a, 1970b), and subsequently Adolf & Hecker (1977), who both considered the biosynthesis of diterpenes in the Euphorbiaceae, also assumed that geranyl-geranyl pyrophosphate is the precursor. In fact, there are several reasons why the participation of precursors other than geranyl-geranyl pyrophosphate should be considered in diterpene biosynthesis. These reasons are discussed below.

Since there are three positions in geranyl-geranyl pyrophosphate in which *cis/trans* isomerism may occur, it follows that there are eight possible isomers:

[2E,6E,10E]-geranyl-geranyl pyrophosphate
[2E,6Z,10E]-geranyl-geranyl pyrophosphate
[2E,6E,10Z]-neryl-geranyl pyrophosphate
[2E,6Z,10Z]-neryl-geranyl pyrophosphate

[2Z,6Z,10Z]-neryl-neryl pyrophosphate
[2Z,6E,10Z]-neryl-neryl pyrophosphate
[2Z,6Z,10E]-geranyl-neryl pyrophosphate
[2Z,6E,10E]-geranyl-neryl pyrophosphate

(The trivial nomenclature is used here deliberately to avoid the more cumbersome I.U.P.A.C. nomenclature.)

After identifying the 'head' and the 'tail' of the tetraprenyl pyrophosphate precursor thought to be involved in the biosynthesis of each of the diterpene hydrocarbon skeletons shown in Fig. 1 above, the sequence of prenyl units may be traced and those bonds resulting from intramolecular cyclizations may be identified. Fig. 2 shows the same hydrocarbon skeletons as in Fig. 1, but gives the *cis* ("Z") or *trans* ("E") nature of each prenyl unit. In many cases, no double bond actually exists at the site of the original prenyl. unit(s), but the *cis* or *trans* nature of these units may be determined either from the shape and rigid nature

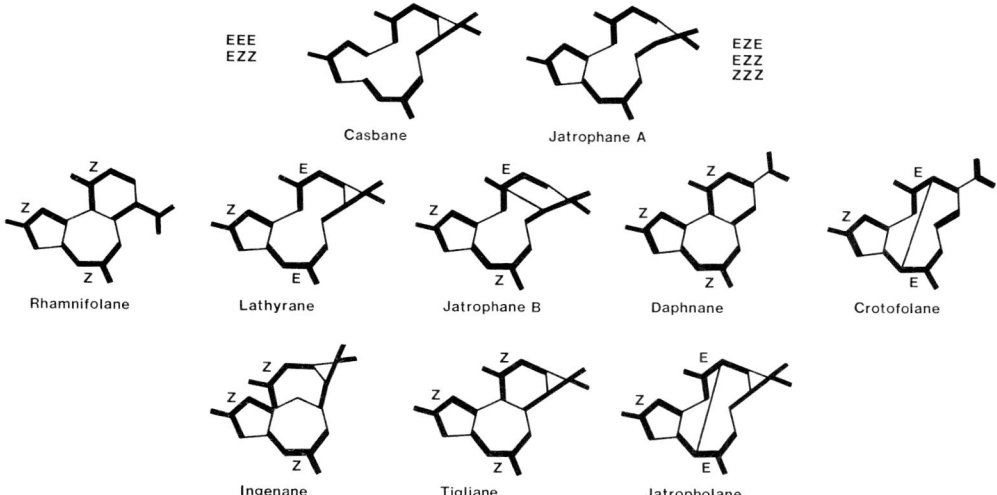

Figure 2. Diterpene hydrocarbon skeletons showing possible *cis–trans* isomerism.

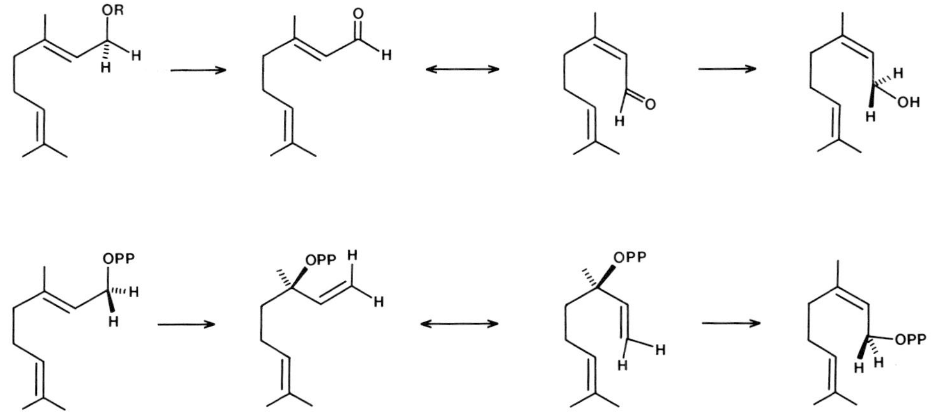

Figure 3. Isomerization mechanisms in monoterpene biosynthesis (see text).

of the ring in which they are found, or from published computer-generated molecular drawings produced from X-ray crystallographic measurements.

Now, since the tigliane and some of the related diterpenoids have more-or-less rigid, highly strained carbon skeletons, their final cyclization almost certainly occurs on an enzyme site. Whether or not cyclization to the final di-, tri- or tetracyclic skeleton occurs at the same enzyme site as the supposed initial head-to-tail cyclization, the simplest substrate has to be the appropriate isomer. Otherwise, in the case of the tigliane skeleton for instance, the enzyme(s) involved would also have to have the ability to carry out isomerization of each of three prenyl units in turn, a sequence of events that is difficult to reconcile with the spatial requirements of an initial head-to-tail condensation of all *trans* (i.e. [2E,6E,10E]-)geranyl-geranyl pyrophosphate. It is far more attractive to propose that the appropriate isomer (i.e. [2Z,6Z,10Z]-neryl-neryl pyrophosphate) is made at one enzyme site whilst the diterpene skeleton is elaborated, and perhaps functionalized, on another. In addition, the two proposed mechanisms (see Cane, 1980) for isomerization of prenyl units in monoterpene biosynthesis—one involving the generation of an $\alpha\beta$-unsaturated aldehydic intermediate, the other involving an intermediate tertiary allylic pyrophosphate—allow only for isomerization at one double bond, that is the one adjacent to the pyrophosphate group (see Fig. 3).

Thus, *de novo* synthesis of the appropriate tetraprenyl pyrophosphate isomer appears to be more likely that the initial utilization of all-*trans* geranyl-geranyl pyrophosphate followed by isomerization to the appropriate isomer.

We should also consider that the four possible isomers of farnesyl pyrophosphate:

[2Z,6Z]-farnesyl pyrophosphate
[2Z,6E]-farnesyl pyrophosphate
[2E,6Z]-farnesyl pyrophosphate
[2E,6E]-farnesyl pyrophosphate

have been utilized in both hypothetical discussions and experimental investigations of the biosynthesis of sesquiterpene hydrocarbon skeletons (see

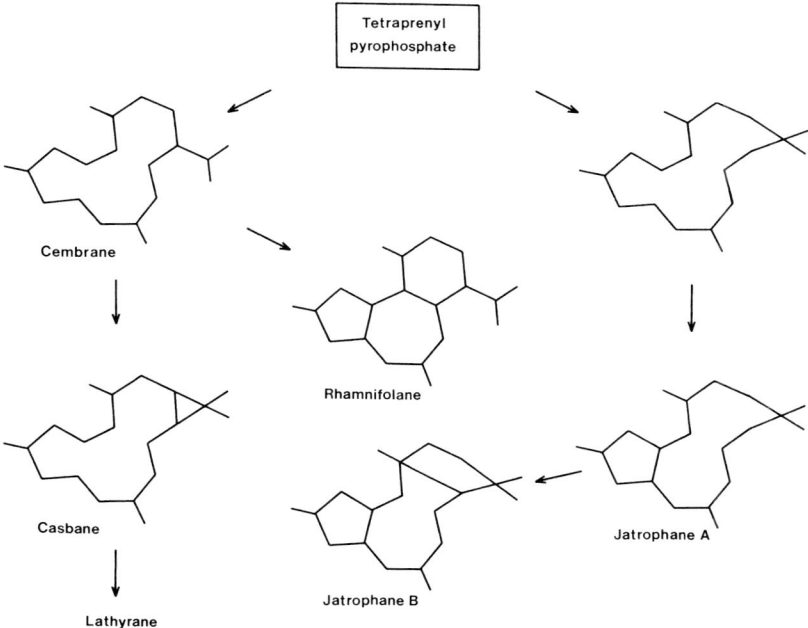

Figure 4. Head-to-tail cyclization of a tetraprenyl pyrophosphate precursor.

Cordell, 1976 and Herz, 1977). It is incongruous that diterpene biosynthesis should proceed through only one precursor.

The final piece of circumstantial evidence presented here in support of the hypothesis that the isomers of geranyl-geranyl pyrophosphate should be considered in diterpene biosynthesis is the phytochemical data, admittedly scanty. Unfortunately, the diterpenoids under discussion here are neither easily detected nor easily isolated; and not one species in the whole of the family Euphorbiaceae has been subjected to an exhaustive search for such diterpenoids for the purpose of studying their biosynthesis. As a group, the best studied, to date, is the genus *Euphorbia*. Using the terminology developed here, several species are known to contain ZZZ-tiglianes or ZZZ-ingenanes together with EZE-lathyranes. It is noteworthy that EZZ-, ZZE- and ZZZ-lathyranes have not been found. At this stage it is easier to envisage the existence of two separate pathways leading to the ZZZ- and EZE- compounds respectively, rather than a single pathway in which EEE-geranyl-geranyl pyrophosphate is converted into an EZE-lathyrane and subsequently a ZZZ-tigliane or ingenane since this pathway involves such major changes in molecular shape between starting and finishing products.

SUMMARY

Thus, the hypothesis discussed here to account for the biosynthesis of the tigliane and related hydrocarbon skeletons may be summarized thus:

1. The first stage is the production of an appropriate tetraprenyl pyrophosphate precursor by *de novo* synthesis, and not necessarily or invariably the all-*trans* geranyl-geranyl pyrophosphate.

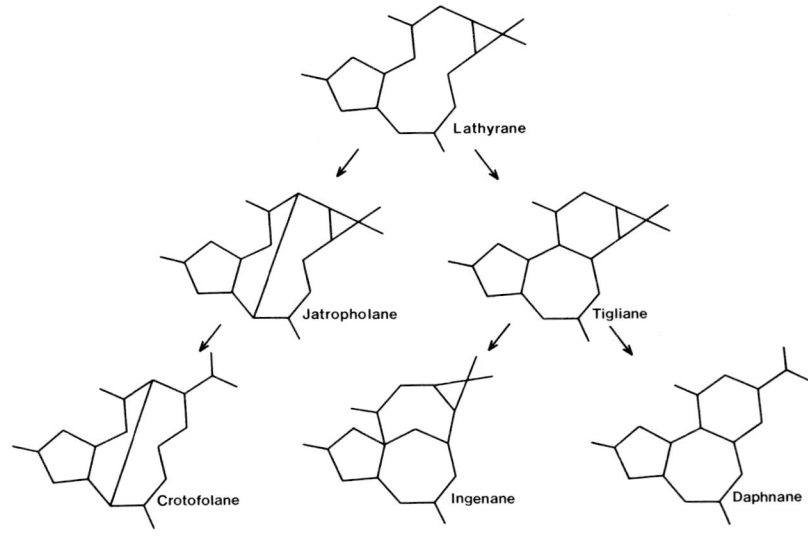

Figure 5. Further transformations of the lathyrane skeleton.

2. The tetraprenyl pyrophosphate precursor is cyclized initially by a head-to-tail cyclization, a process that may in itself generate two families of compounds for each precursor (see Figs 4 & 5).

3. Each of the final diterpene skeletons is produced on a separate enzyme system, with little or no release of 'intermediates'.

REFERENCES

ADOLF, W. & HECKER, E., 1975. On the irritant and cocarcinogenic principles of *Hippomane mancinella*. *Tetrahedron Letters (19)*: 1587–1590.

ADOLF, W. & HECKER, E., 1977. Diterpenoid irritants and cocarcinogens in Euphorbiaceae and Thymelaeaceae: structural relationships in view of their biogenesis. *Israel Journal of Chemistry, 16:* 75–83.

ADOLF, W. & HECKER, E., 1980. New irritant diterpene-esters from roots of *Stillingia sylvatica* L. (Euphorbiaceae). *Tetrahedron Letters (21):* 2887–2890.

ADOLF, W., OPFERKUCH, H. J. & HECKER, E., 1984. Irritant phorbol derivatives from four *Jatropha* species. *Phytochemistry, 23:* 129–132.

ADOLF, W., SORG, B., HERGENHAHN, M. & HECKER, E., 1982. Structure-activity relations of polyfunctional diterpenes of the daphnane type. I. Revised structure for resiniferatoxin and structure-activity relations of resiniferonol and some of its esters. *Journal of Natural Products, 45:* 347–354.

AIRY SHAW, H. K., 1966. Notes on Malaysian and other Asiatic Euphorbiaceae. *Kew Bulletin, 20:* 379–415.

BANTHORPE, D. V. & CHARLWOOD, B. V., 1980. The terpenoids. In E. A. Bell & B. V. Charlwood (Eds), *Encyclopedia of Plant Physiology, vol. 8, Secondary Plant Products:* 185–220. Heidelberg: Springer-Verlag.

BURKE, B. A., CHAN, W. R., PASCO, K. O., BLOUNT, J. F. & MANCHAND, P. S., 1981. The structure of crotonitenone, a novel casbane diterpene from *Croton nitens* Sw. (Euphorbiaceae). *Journal of the Chemical Society Perkin I;* 2666–2669.

CANE, D. E., 1980. The stereochemistry of allylic pyrophosphate metabolism. *Tetrahedron, 36:* 1109–1159.

CHAN, W. R., PRINCE, E. C., MANCHAND, P. S., SPRINGER, J. P. & CLARDLY, J., 1975. The structure of crotofolin A, a diterpene with a new skeleton. *Journal of the American Chemical Society, 97:* 4437–4439.

CORDELL, G. A., 1976. Biosynthesis of sesquiterpenes. *Chemical Reviews, 76:* 425–460.

DAUBEN, W. G., THIESSEN, W. E. & RESNICK, P. R., 1962. Cembrene, a 14-membered ring diterpene hydrocarbon. *Journal of the American Chemical Society, 84:* 2015–2016.

ERDTMAN, H., NORIN, T., SUMIMOTO, M. & MORRISON, A., 1964. Verticillol, a novel type of conifer diterpene. *Tetrahedron Letters (51):* 3879–3886.

EVANS, F. J. & KINGHORN, A. D., 1975. The succulent euphorbias of Nigeria. Part 1. *Lloydia, 38:* 363–365.

EVANS, F. J. & TAYLOR, S. E., 1983. Pro-inflammatory, tumour-promoting and anti-tumour diterpenes of the plant families Euphorbiaceae and Thymelaeaceae. *Progress in the Chemistry of Organic Natural Products, 44:* 1–99.

FÜRSTENBERGER, G. & HECKER, E., 1977. The new diterpene 4-deoxyphorbol and its highly unsaturated irritant esters. *Tetrahedron Letters, (11):* 925–928.

GHISALBERTI, E. L., JEFFRIES, P. R., PAYNE, T. G. & WORTH, G. K., 1973. Structure and stereochemistry of bertyadionol. *Tetrahedron, 29:* 403–412.

GHISALBERTI, E. L., JEFFRIES, P. R., TOIA, R. F. & WORTH, G. K., 1974. Stereochemistry of bertyadionol and related compounds. *Tetrahedron, 30:* 3269–3274.

GUNASEKERA, S. P., CORDELL, G. A. & FARNSWORTH, N. R., 1979. Potential anticancer agents. XIV. Isolation of spruceanol and montanin from *Cunuria spruceana* (Eurphorbiaceae). *Journal of Natural Products, 42:* 658–662.

HANDA, S. S., KINGHORN, A. D., CORDELL, G. A. & FARNSWORTH, N. R., 1983. Plant anticancer agents. XXII. Isolation of a phorbol diester and its $\Delta^{5,6}$-7β-hydroperoxide derivative from *Ostodes paniculata*. *Journal of Natural Products, 46:* 123–126.

HARRISON, J. W., SCROWSTON, R. M. & LYTHGOE, B., 1966. Taxine. Part IV. The constitution of taxine-1. *Journal of the Chemical Society (C)*, 1933–1945.

HECKER, E. & SCHMIDT, R., 1974. Phorbolesters—the irritants and cocarcinogens of *Croton tiglium* L. *Progress in the Chemistry of Organic Natural Products, 31:* 377–467.

HERGENHAHN, M., ADOLF, W. & HECKER, E., 1975. Resiniferatoxin and other esters of novel polyfunctional diterpenes from *Euphorbia resinifera* and *unispina*. *Tetrahedron Letters, (19):* 1595–1598.

HERZ, W., 1977. Biogenetic aspects of sequiterpene lactone chemistry. *Israel Journal of Chemistry, 16:* 32–44.

HUI, W.-H., LI, M.-M. & NG, K.-K., 1975. Terpenoids and steroids from *Macaranga tanarius*. *Phytochemistry, 14:* 816–817.

KOBAYASHI, H. & AKIYOSHI, S., 1962. Thunbergene, a macrocyclic diterpene. *Bulletin of the Chemical Society of Japan, 35:* 1044–1045.

KUPCHAN, S. M., SIGEL, C W., MATZ, M. J., GILMORE, C. J. & BRYAN, R. F., 1976. Structure and stereochemistry of jatrophone, a novel macrocyclic diterpenoid tumor inhibitor. *Journal of the American Chemical Society, 98:* 2295–2300.

LOTTER, H., OPFERKUCH, H. J. & HECKER, E., 1979. Crystal structure and stereochemistry of ingol-3,7,8,12-tetraacetate. *Tetrahedron Letters (1):* 77–78.

NARAYANAN, P., ROHRL, M., ZECHMEISTER, K., ENGEL, D. W., HOPPE, W., HECKER, E. & ADOLF, W., 1971. Structure of 7-hydroxy-lathyrol, a further diterpene from *Euphorbia lathyrus* L. *Tetrahedron Letters (18):* 1325–1328.

NISHIZAWA, M., TAKENAKA, H., NISHIDE, H. & HAYASHI, Y., 1983. A new olefin cyclization agent, mercurcy (II) trifluoromethanesulfonate-amine complex. *Tetrahedron Letters, 24(25):* 2581–2584.

OGURA, M., KOIKE, K., CORDELL, G. A. & FARNSWORTH, N. R., 1978. Potential anticancer agents VIII. Constituents of *Baliospermum montanum* (Euphorbiaceae). *Planta Medica, 33:* 128–143.

OHIGASHI, H., KATSUMATA, H., KAWAZU, K., KOSHIMIZU, K. & MITSUI, T., 1974. A piscicidal constituent of *Excoecaria agallocha*. *Agricultural and Biological Chemistry, 38:* 1093–1095.

OHIGASHI, H., KAWAZU, K., KOSHIMIZU, K. & MITSUI, T., 1972. A piscicidal constituent of *Sapium japonicum*. *Agricultural and Biological Chemistry, 36:* 2529–2537.

OKUDA, T., YOSHIDA, T., KOIKE, S. & TOH, N., 1975. New diterpene esters from *Aleurites fordii* fruits. *Phytochemistry, 14:* 509–515.

OPFERKUCH, H. J. & HECKER, E., 1974. New diterpenoid irritants from *Euphorbia ingens*. *Tetrahedron Letters (3):* 261–264.

PURUSHOTHAMAN, K. K., CHANDRASEKHARAN, S., CAMERON, A. F., CONNOLLY, J. D., LABBÉ, C., MALTZ, A. & RYCROFT, D. S., 1979. Jatropholones A & B, new diterpenoids from the roots of *Jatropha gossypiifolia* (Euphorbiaceae)—crystal structure analysis of jatropholone B. *Tetrahedron Letters (11):* 979–980.

ROBERTS, D. L. & ROWLAND, R. L., 1962. Macrocyclic diterpenes. α- and β-4,8,13-Duvatriene-1,3-diols from tobacco. *Journal of Organic Chemistry, 27:* 3989–3995.

ROBINSON, D. R. & WEST, C. A., 1970a. Biosynthesis of cyclic diterpenes in extracts from seedlings of *Ricinus communis* L. I. Identification of diterpene hydrocarbons formed from mevalonate. *Biochemistry, 9:* 70–79.

ROBINSON, D. R. & WEST, C. A., 1970b. Biosynthesis of cyclic diterpenes in extracts from seedlings of *Ricinus communis* L. II. Conversion of geranylgeranyl pyrophosphate into diterpene hydrocarbons and partial purification of the cyclization enzymes. *Biochemistry, 9:* 80–89.

SAHAI, R., RASTOGI, R. P., JAKUPOVIC, J. & BOHLMANN, F., 1981. A diterpene from *Euphorbia maddeni*. *Phytochemistry, 20:* 1665–1667.

SAKATA, K., KAWAZU, K., MITSUI, T. & MASAKI, N., 1971. Structure and stereochemistry of huratoxin, a piscicidal constituent of *Hura crepitans*. *Tetrahedron Letters (16):* 1141–1144.

SEIP, E. H. & HECKER, E., 1984. Derivatives of characiol, macrocyclic diterpene esters of the jatrophane type from *Euphorbia characias*. *Phytochemistry, 23:* 1689–1694.

SHIZURI, Y., KOSEMURA, S., OHTSUKA, J., TERADA, Y. & YAMAMURA, S., 1983. Structural and

conformational studies on euphohelioscopins A and B and related diterpenes. *Tetrahedron Letters, 24(25):* 2577–2580.
STUART, K. L. & BARRETT, M., 1969. A phorbol derivative from *Croton rhamnifolius*. *Tetrahedron Letters (28):* 2399–2400.
TORRANCE, S. J., WIEDHOPF, R. M., COLE, J. R., ARORA, S. K., BATES, R. B., BEAVERS, W. A. & CUTLER, R. S., 1976. Antitumour agents from *Jatropha macrorhiza* (Euphorbiaceae). II. Isolation and characterization of jatrophatrione. *Journal of Organic Chemistry, 41:* 1855–1857.
UEMURA, D. & HIRATA, Y., 1975. Stereochemistry of kansuinine A. *Tetrahedron Letters (21):* 1701–1702.
UEMURA, D., KATAYAMA, C., UNO, E., SASAKI, K., HIRATA, Y., CHEN, Y.-P. & HSU, H.-Y., 1975. Kansuinine B, a novel multi-oxygenated diterpene from *Euphorbia kansui* Liou. *Tetrahedron Letters (21):* 1730–1706.
UEMURA, D., NOBUHARA, K., NAKAYAMA, Y., SHIZURI, Y. & HIRATA, Y., 1976. The structure of new lathyrane diterpenes, jolkinols A, B, C, and D, from *Euphorbia jolkini* Boiss. *Tetrahedron Letters (50):* 4593–4596.
WIRIYACHITRA, P., HAJIWANGOH, H., BOONTON, P., ADOLF, W., OPFERKUCH, H. J. & HECKER, E., 1985. Investigations of medicinal plants of Euphorbiaceae and Thymelaeaceae occurring and used in Thailand; II. Cryptic irritants of the diterpene ester type from three *Excoecaria* species. *Planta Medica (5):* 368–371.
ZECHMEISTER, K., RÖHRL, M., BRANDL, F., HECHTFISCHER, S., HOPPE, W., HECKER, E., ADOLF, W. & KUBINYI, H., 1970. Röntgenstrukturanalyse eines neuen Makrozyklischen Diterpen-Esters aus der Springwolfsmilch (*Euphorbia lathyris* L.). *Tetrahedron Letters (35):* 3071–3073.

Activity correlations in the phorbol ester series

FRED J. EVANS F.L.S. AND MARY C. EDWARDS

Department of Pharmacognosy, The School of Pharmacy, University of London, 29–39 Brunswick Square, London WC1N 1AX

Received April 1986, accepted for publication October 1986

EVANS, F. J. & EDWARDS, M. C. 1987. **Activity correlations in the phorbol ester series.** The production of inflammation by phorbol esters on mammalian skin correlates on a structural basis with *in vitro* measurements of lymphocyte mitogenesis and mobilization of prostaglandins. All of the pro-inflammatory phorbol esters tested in our laboratory have been shown to activate the enzyme protein kinase C, and such an interaction could in large part explain the induction of an inflammatory response *in vivo*. Certain of these compounds additionally induce aggregation of human and rabbit blood platelets. This activity does not structurally correlate with the induction of inflammation but may correlate with the known tumour-promoting actions of phorbol derivatives. Compounds which induce platelet aggregation stimulate the secretion of a biologically active substance which we have termed 'Factor W'. The production of Factor W occurs into human plasma following platelet stimulation by phorbol tumour-promoting agents. It is an unstable substance, distinct in its aggregating properties from phorbol esters, ADP, 5-hydroxytryptamine, thrombin, platelet aggregating factor and the products of arachidonate oxidative metabolism.

ADDITIONAL KEY WORDS:—Daphnane *ortho* esters – Euphorbiaceae – inflammation – platelets – protein kinase C.

CONTENTS

Introduction	231
In vivo inflammatory effects of phorbol esters	233
In vitro inflammatory models of phorbol ester activity	235
Phorbol ester induced platelet aggregation	236
Correlations of phorbol ester activity	239
Properties and release of a transferable aggregating substance	241
Conclusions	244
References	244

INTRODUCTION

The most widely published toxic hazard associated with plants of the family Euphorbiaceae is their tumour-promoting effect (Hecker, 1975). This is a condition which may develop after chronic exposure to these plants following a subthreshold dose of a carcinogenic agent (Hecker, 1978). Even then this phenomenon has only been conclusively demonstrated on mouse skin, in animals which have been specifically bred to be susceptible to tumour-formation (Kinsella, 1986). The acute effect of the exposure of humans to these plants is

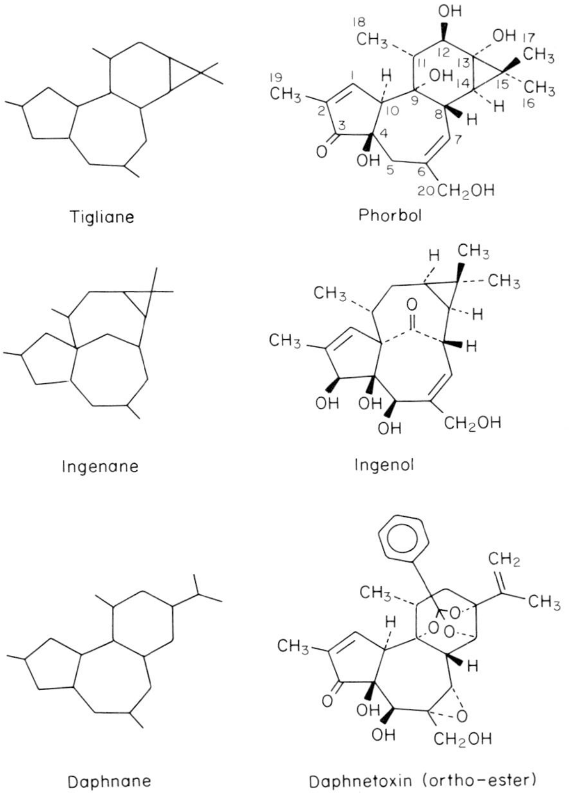

Figure 1. Examples of Euphorbiaceae and Thymelaeaceae diterpenes.

the production of inflammation. From a clinical point of view therefore the most serious and immediate toxic syndrome induced by contact with these plants is the inflammatory condition. This is a direct pro-inflammatory response on skin which is characterized by the classical symptoms (Evans, 1978, and references therein): redness, swelling, pain, thickening of tissues.

Very often in skin burning by *Euphorbia*, open and weeping pustules develop, which at a later stage may lead to the formation of dry flaking skin, scab formation and eventually to necrosis of the tissues (Schmidt & Evans, 1980). When ingested, the classical effects of these plants (Evans, 1986a), can be attributed to their irritant activity. A burning sensation develops on the lips, tongue and the mucous membranes of the mouth, followed sometime later by intestinal pain, vomiting and severe purging (Kinghorn & Evans, 1975). Similarly, when allowed to come into contact with the eyes (Evans & Kinghorn, 1975) a conjunctivitis develops together with a swelling of the eye lids and closure of the eyes due to oedema. This condition can lead to blindness.

The phorbol-12,13-diesters of *Croton tiglium* L. were isolated nearly twenty years ago (Hecker, 1968; Hope *et al.*, 1968) and subsequently a range of related diterpene esters based upon the tigliane, ingenane and daphnane hydrocarbon skeletons were isolated from plant species of the Euphorbiaceae (Fig. 1). Their chemistry and distribution in plants has been regularly reviewed (Hecker, 1968, 1971; Furstenburger & Hecker, 1972; Hecker & Schmidt, 1974; Evans & Soper,

1978; Evans & Taylor, 1983; Evans, 1986b). These compounds were identified as both the tumour-promoting and pro-inflammatory agents responsible for the toxicity of these plants. However, it became clear that whilst the irritant action of these substances was an almost universal property of the group, not all of these substances were tumour-promoting agents. Structure–activity observations (Schmidt & Evans, 1979, 1980), indicated that investigations into biological properties would need to be carefully controlled for meaningful correlations to be made in this series of compounds. Further, *in vitro* results indicated that separate receptors existed for the tumour-promoting and the pro-inflammatory activities of phorbol related esters (Driedger & Blumberg, 1979). The basis for the tumour-promoting actions of the phorbol esters appeared to be their ability to substitute for diacylglycerol in the activation of protein kinase C (Castagna *et al.*, 1982; Gilmore & Martin, 1983). This kinase is now thought to be the phorbol ester receptor site (Kikkawa *et al.*, 1983). Subsequently, during our investigations of the activation of protein kinase C by phorbol derivatives, it became clear that compounds with a variety of different biological effects would activate the enzyme to a greater or lesser extent (Ellis *et al.*, 1985). In fact a correlation in structural terms existed between activation of protein kinase C and the pro-inflammatory activity *in vivo*.

IN VIVO INFLAMMATORY EFFECTS OF PHORBOL ESTERS

The most commonly used test for the assessment of the pro-inflammatory activity of phorbol esters is the determination of erythema on mouse skin. This is an 'all or nothing' assay for the calculation of the dose of challenging agent which will induce a response in 50% of the test animals (Hecker, Immrich, Bresch & Schairer, 1966; Evans & Schmidt, 1979). Using a range of naturally occurring phorbol esters, together with synthetic analogues, structure–activity relationships have been observed (Hecker, 1977, 1978; Evans, 1978; Schmidt & Evans, 1979, 1980; Adolf, Sorg, Hergenhahn & Hecker, 1982) (Table 1). The structure–activity requirements of the related daphnane series of compounds

Table 1. Pro-inflammatory structure activity relationships

Activity requirements
 AB ring *trans* configuration
 Acyl residue at C-12/13 or at C-13

Features increasing potency
 Primary hydroxyl at C-20
 Unsaturation in C-12/13 acyl residue
 Aromatic residue at C-12/13

Features not necessary for activity
 C-12 hydroxyl or acyl group
 Free hydroxyl at C-20
 C-4 tertiary hydroxyl
 Cyclopropane ring D
 C-16 methyl group

R^1 = acyl or acetate
R^2 = OH or acyl
R^3 = H or OH

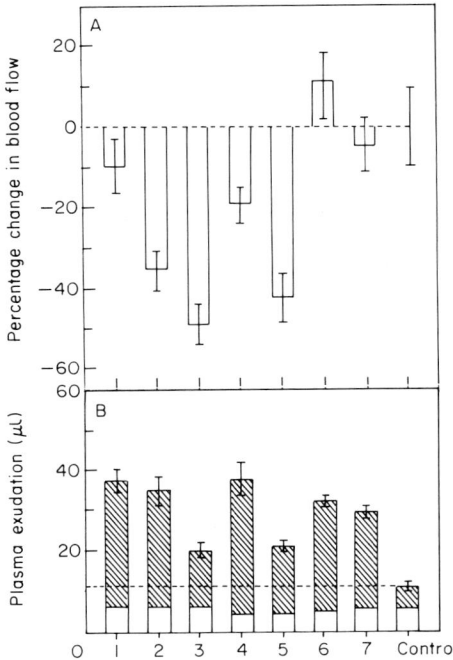

Figure 2. Plasma exudation and local blood flow changes induced by phorbol esters in rabbit skin. A. Local blood changes in response to (1) TPA, (2) 12-deoxyphorbol-phenylacetate-20-acetate, (3) 12-deoxyphorbol phenylacetate (4) 12-deoxyphorbol angelate-20 acetate (5) 12-deoxyphorbolangelate (6) phorbol (7) Resiferonol-*ortho*-phenylacetate (100 ng per site of injection). B. Plasma exudation in response to compounds 1–7 as in A. Hatched areas represent plasma exudation in presence of prostaglandin E_1, 100 ng per site.

differed from those of the phorbol esters and the two groups of compounds might therefore exert their effects by more than one mechanism.

In order to investigate this mechanism more fully, the blood flow changes that occur in the skin together with plasma leakage into surrounding tissues were measured using the rabbit as an experimental animal (Williams, Westwick, Williamson & Evans, 1981). Blood flow changes were measured by ^{133}xenon wash out in skin over that of control sites, and plasma leakage was measured by ^{131}iodine-labelled human serum albumin leakage into tissues from skin blood vessels (Fig. 2). Unexpectedly, a series of phorbol esters did not increase blood flow as would be required for the development of erythema, but in fact induced a vasoconstriction. The daphnane *ortho*-esters were totally inactive. Similarly, no plasma leakage was recorded by either group of compounds. The direct pharmacological effect of phorbol esters on an isolated rabbit aorta was to induce a prolonged contraction. Such an effect upon vessel smooth muscle might explain the necrosis observed during high-dose exposure of skin to phorbol esters (Fig. 3). The induction of erythema appeared not to be due to a direct effect upon skin microvasculature but was more likely to be the result of an indirect action involving the release of endogenous mediators of inflammation. Experiments on mouse skin involving the use of pharmacological antagonists to phorbol ester action (Table 2), seem to confirm that conclusion. Substances which inhibit the development of erythema included calmodulin

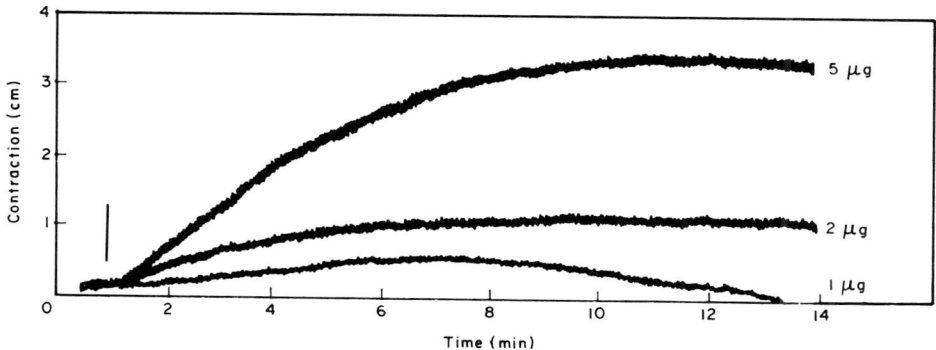

Figure 3. Contraction of isolated rabbit aorta induced by 12-deoxyphorbolphenylacetate-20-acetate.

Ca^{2+} antagonists, phospholipase inhibitors and membrane stabilizing agents (Williamson & Evans, 1981). Furthermore, when experiments involving plasma leakage in rabbit skin were repeated in the presence of added prostaglandin E_1, a vascular permeability developed (Fig. 2B).

IN VITRO INFLAMMATORY MODELS OF PHORBOL ESTER ACTIVITY

Phorbol esters have been shown to mobilize phospholipids, liberate arachidonate and cause the secretion of prostaglandins in cell cultures of various types (Ohuchi & Levine, 1978; Trashjian, Ivey, Declos & Levine, 1978). Mobilization of prostaglandin was demonstrated using human rheumatoid synovial cells obtained following a total synovectomy from rheumatoid arthritic patients (Edwards, Evans, Barett & Gordon, 1985). A range of structurally-related phorbol derivatives, introduced into these cultures in low doses, stimulated the release of prostaglandin E_2 as measured by the use of radio-immunoassay. This action structurally correlated with the production of inflammation on skin and could, like the development of inflammation, be inhibited by corticosteroids (Fig. 4).

A further common feature of the action of phorbol esters in cell culture is their ability to stimulate protein synthesis (Scribner & Slaga, 1973), RNA synthesis (Baird, Melera & Boutwell, 1972) and DNA synthesis (Shinozuka & Ritchie,

Table 2. Antagonists of phorbol ester-induced (0.1 μg) erythema of mouse skin

Compound	Action	Dose (mg)	% Inhibition
Indomethacin	Cyclo-oxygenase inhibitor	1	10
Thioasnisole	Free radical scavenger	0.20	30
Aminopyrine		1	70
Imipramine	Membrane stabilizer	1	80
Promethazine	Calmodulin/Ca^{2+} antagonist	1	75
Trifluoperazine		1	95
Propranolol	Phospholipase A_2 inhibitors	1	70
p-Bromophenacylbromide		1	50
Hydrocortisone	Corticosteroid	0.05	55

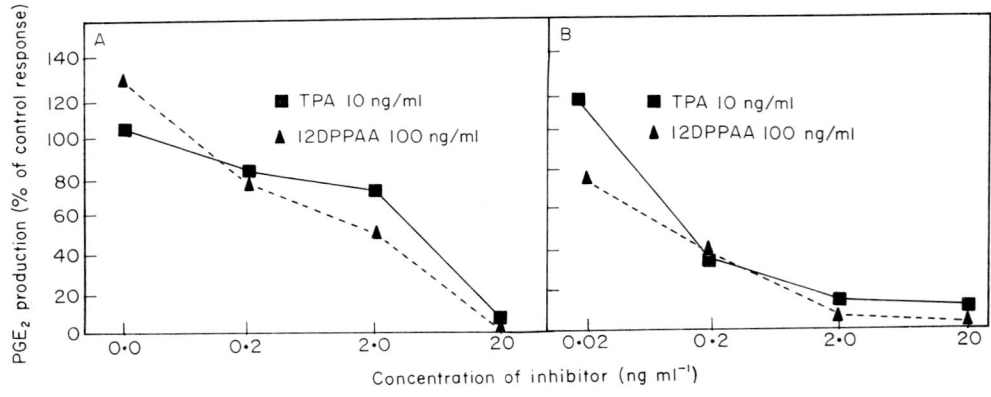

Compound	ED_4 (ng ml^{-1})
TPA	0.28
12-deoxyphorbolphenylacetate	0.38
12-deoxyphorbolphenylacetate-20-acetate (12 DPPAA)	46.0
12-deoxyphorbolangelate	42.0
12-deoxyphorbolangelate-20-acetate	103.0
Phorbol	Not active
Sapintoxin-A	0.26
α-Sapinine	Not active

$ED_4 \equiv$ The does of phorbol ester required to raise the basal level of prostaglandin E_2 production fourfold in human rheumatoid synovial cell culture.

Figure 4. Release of prostaglandin E_2 by phorbol esters from human rheumatoid synovial cells. Inhibition of release by A, indomethacin, and B, dexamethasone.

1967). Phorbol esters behaved as mitogenic agents and stimulated cell growth in very low doses (Frei & Stevens, 1968; Touraine et al., 1977). In a series of 10 phorbol derivatives (Edwards, Nouri, Gordon & Evans, 1983a), seven compounds were shown to induce lymphocyte mitogenesis (Fig. 5). The mitogenic activity of phorbol esters was also inhibited by a similar range of antagonists as had previously been shown to inhibit their pro-inflammatory actions on mammalian skin and their ability to mobilize prostaglandin in cell culture suggesting that these activities of the phorbol esters were biochemically related. Furthermore, lymphocytes stimulated into cell division by phorbol esters appeared to be able to produce receptors to the growth factor interleukin-2. This result was confirmed by Shackelford & Trowbridge (1984) using a different cell culture system.

PHORBOL ESTER INDUCED PLATELET AGGREGATION

Zucker, Troll & Belman (1974) first demonstrated that a pro-inflammatory and tumour-promoting phorbol ester would induce a dose dependent aggregation of human blood platelets. This compound, TPA (tetradecanoyl-phorbol acetate), caused vacuole formation within storage granules and dilation

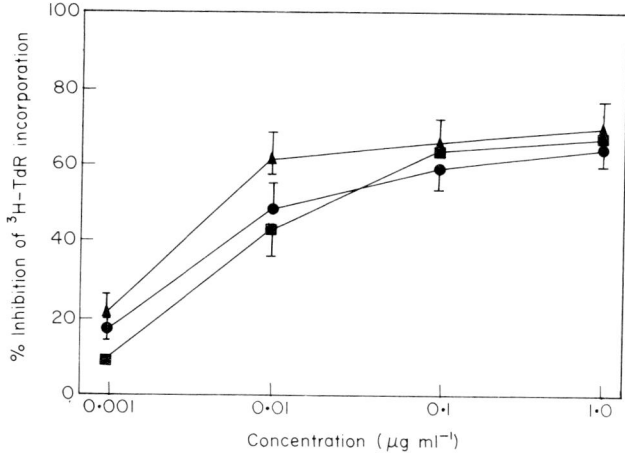

Compound	EC_{40}
Phorbol	Not active
TPA	17
Sapintoxin A	35
Sapintoxin B	560
α-Sapinine	Not active
12-deoxyphorbol phenylacetate	45
12-deoxyphorbolphenylacetate-20-acetate	564

Figure 5. Above: inhibition by dexamethasone of lymphocyte mitogenesis stimulated by $1\mu g\ ml^{-1}$ of: ■, TPA; ●, 12-deoxyphorbolphenylacetate; ▲, 12-deoxyphorbolphenylacetate-20-acetate. Below: mitogenic activity of phorbol esters. EC_{40}, the dose $(ng\ ml^{-1})$ required to increase mitogenesis of hyman lymphocytes by 40%.

of the open canicular system of platelets (Estensen & White, 1974; White & Estensen, 1974). We began our studies on phorbol ester induced platelet aggregation by investigating the aggregating ability of a range of phorbol, deoxyphorbol and hydroxyphorbol esters, together with some daphnane *ortho*-esters (Westwick, Williamson & Evans, 1980; Williamson, Westwick & Evans, 1980; Edwards, Taylor, Williamson & Evans, 1983b). Human blood was collected by forearm venipuncture from donors who had denied receiving any medication within the previous 14 days. Platelet rich plasma (PRP) for use in these *in vitro* experiments was produced by standard centrifugation techniques.

It was found that the daphnane *ortho*-esters did not induce aggregation of human platelets and that the aggregation induced by phorbol esters was rigidly structurally dependant. An AB *trans* configuration, a free C-20 primary hydroxyl group and an ester moiety at C-13 were essential structural features for activity in this system. However, the hydroxyl function at C-4 and the ester function at C-12 were not essential structural features.

The aggregation induced by phorbol esters was shown to be a true aggregation and not merely a non-specific agglutination due to a surfactant type of action (Williamson, Westwick, Kakkar & Evans, 1981). This was confirmed by the fact that EDTA (Fig. 6) would inhibit phorbol ester induced aggregation thus demonstrating the importance of extracellular Ca^{2+} in the reaction and

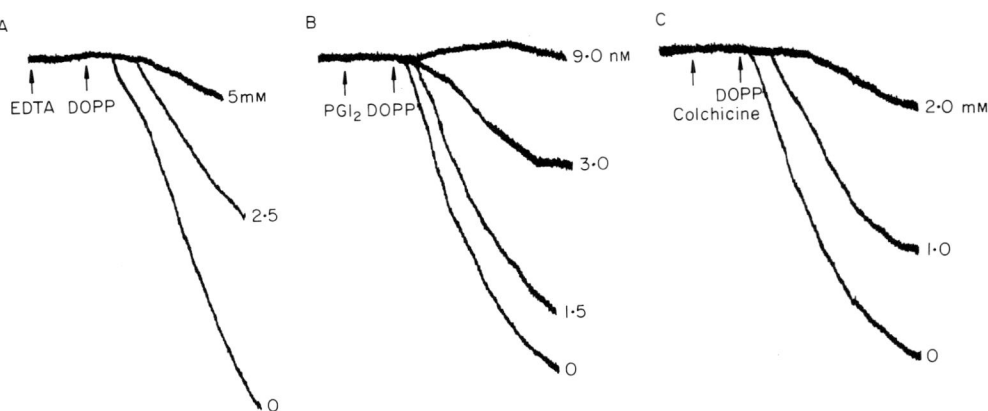

Figure 6. Inhibition of 12-deoxyphorbolphenylacetate-induced (0.89 μm) aggregation of human platelets by A, EDTA; B, prostacyclin; and C, colchicine.

also by the fact that prostacyclin (PGI_2) inhibited the aggregation, demonstrating the significance of intracellular levels of cyclic AMP. Finally, it was demonstrated that this aggregation was inhibited by colchicine, thus confirming the importance of an intact microtubular system for phorbol ester induction of human blood platelet aggregation. Furthermore, this aggregation, unlike that commonly seen by ADP- and thrombin-induced aggregation,

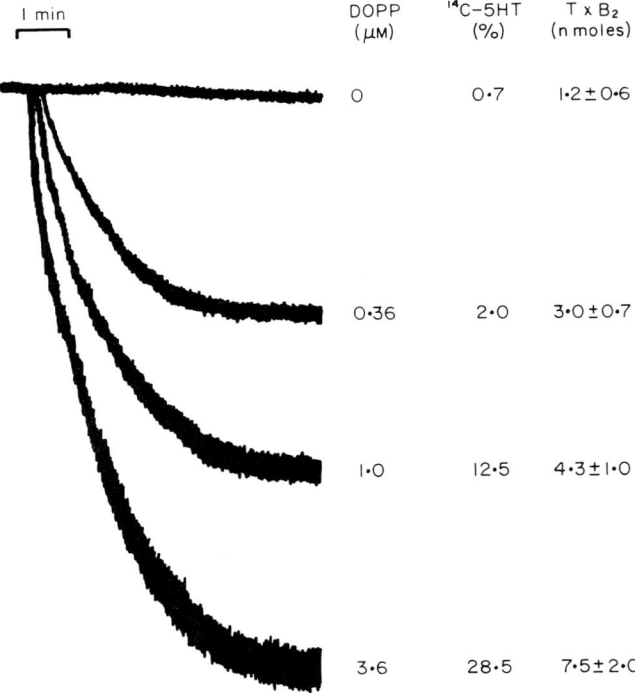

Figure 7. Platelet secretion induced by phorbol esters: stimulation of 5HT release and thromboxane B_2 generation by DOPP.

involved only a small release reaction from dense granules (Fig. 7). This was determined by the measurement of the release of ^{14}C-5HT (5-hydroxytryptamine) into plasma following phorbol ester-induced aggregation. Secretion of the oxidative products of arachidonate metabolism were also shown to play little part in phorbol ester induced aggregation in that only relatively low levels of thromboxane B_2 were detected by radio-immunoassay methods (Fig. 7). Pharmacological experiments demonstrated that the cyclo-oxygenase inhibitor, indomethacin, the inhibitor of cyclo-oxygenase and lipoxygenase, phenidone, the free radical scavenger, thioanisole, and the endoperoxide/thromboxane receptor antagonist, pinane-thromboxane A_2, had no effect upon the aggregation produced. Compounds that would inhibit this aggregation are shown in Table 3. A further distinction between the aggregation induced by phorbol esters and several other platelet agonists such as thrombin was obtained from the measurement of phospholipid turnover subsequent to aggregation. Unlike thrombin, (Mahadevappa & Holub, 1983) phorbol esters did not stimulate phospholipid turnover (Fig. 8) during induction of aggregation.

CORRELATIONS OF PHORBOL ESTER ACTIVITY

Although protein kinase C is generally accepted as being the phorbol ester receptor, correlations of the various activities of phorbol esters enable several observations to be made. A correlation appears to exist in structural terms between activation of protein kinase C and the production of skin inflammation in that all compounds tested which activate that enzyme also act as irritants *in vivo*. Both of these actions of the phorbol esters additionally correlate with the *in vitro* actions of secretion of prostaglandins and the stimulation of lymphocyte mitogenesis. Thus, all the AB ring *trans* compounds available to us from phytochemical studies would induce these four actions either in a cell free enzyme assay, in cell culture or when applied to mouse skin. If the initial biochemical event involved in phorbol ester induction of inflammation is the activation of protein kinase C then a mechanism can be envisaged in which

Table 3. Examples of compounds which effect/do not effect DOPP-induced human platelet aggregation

Compound	Concentration (mM)	Action
Effect DOPP-induced platelet aggregation		
Indomethacin	0.03–0.1	Cyclo-oxygenase inhibitor
Pinane thromboxane A_2	0.001–0.0004	PG endoperoxide/Tx antagonist
Verapamil	0.003–0.3	Calcium flux antagonist
Phenidone	0.5	Cyclo-oxygenase/lipoxygenase inhibitor
CP/CPK	4 mM 40 µml ml^{-1}	ADP scavenger
Imidazole	1–8	Thromboxane synthetase inhibitor
Thioanisole	0.2–2.0	Free radical scavenger
Do not effect DOPP-induced platelet aggregation		
EDTA	1–8	Calcium antagonist
PGI_2	1–5 ng ml^{-1}	Elevates cyclic AMP
Imipramine	0.1–0.35	Membrane stabilizers
Bromophenacyl bromide	0.01–0.30	Phospholipase A_2 inhibitor
Trifluoperazine	0.06–0.30	Calmodulin/Ca^{2+} antagonist

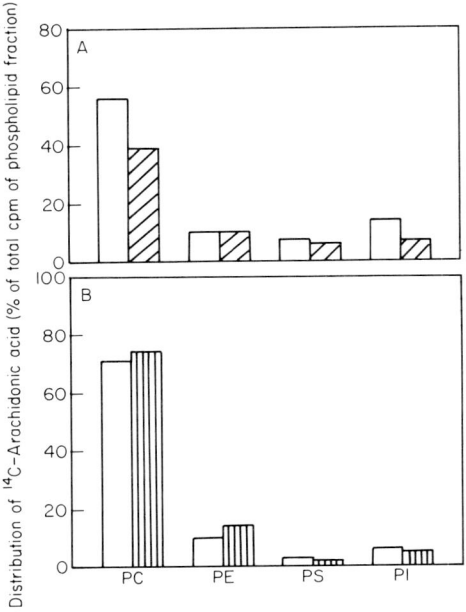

Figure 8. A. Distribution of ^{14}C-arachidonate (^{14}C-AA) in human platelets stimulated by thrombin (results of Bills et al., 1976). Net loss of 25% of ^{14}C-AA from phospholipid fraction on thrombin stimulation (▨, 100 U ml^{-1}); □, controls. B. Distribution of ^{14}C-AA in human platelets stimulated by phorbol ester (unpublished results of Edwards & Evans, 1981). Net gain of 5% of ^{14}C-AA by phospholipid fraction on stimulation with 12-DOPP (▥, 2 μg ml^{-1}) □, controls. PC = Phosphatidylcholine, PE = phosphatidylethanolamine, PS = phosphatidylserine, PI = phosphatidylinositol.

mobilization of intracellular Ca^{2+}, activation of the calmodulin proteins, elevation of cyclic AMP and the mobilization of arachidonate would play a part in the development of the erythema of skin. These effects would overcome, initially, the direct pharmacological activity of the phorbol esters on skin microvasculature. However, on a structural basis the induction of platelet aggregation was found to be non-correlating in this sequence (Fig. 9, Table 4), in that not all of the compounds which activate protein kinase C and induce skin inflammation are platelet-aggregating agents. Only those compounds with a free C-20 primary hydroxyl function were capable of inducing aggregation. This structural feature is also a requirement for their tumour-promoting actions (Hecker, 1975, 1978). These non-correlating activities of the phorbol esters could be due to an extra or separate biochemical property, or perhaps protein kinase C exists in more than one form in the membrane.

PROPERTIES AND RELEASE OF A TRANSFERABLE AGGREGATING SUBSTANCE

Whilst investigating the aggregation of platelets induced by phorbol esters, one characteristic property of these compounds which came to light was their ability to stimulate the secretion of a biologically active factor into human plasma (Williamson et al., 1981). This is a transferable aggregating substance (TAS), which we have termed 'Factor W', the Williamson–Westwick factor. TAS is secreted when platelets are induced to aggregate by 12-deoxyphorbol-

Table 4. Activity correlations of phorbol esters

Activity	Phorbol esters without C-20 acyl, C-20 aldehyde, C-20 methyl, C-5 hydroxyl	All phorbol esters
Pro-inflammatory	+	+
Lymphocyte mitogenesis	+	+
Stimulate prostaglandin secretion	+	+
Aggregate rabbit platelets	+	−
Aggregate human platelets	+	−
Protein kinase C activation	+	+

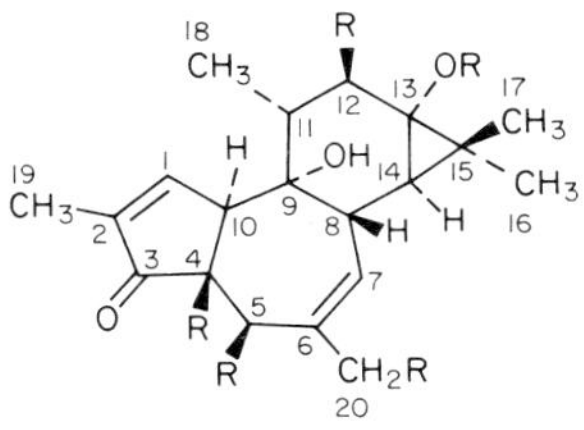

Figure 9. Phorbol esters.

phenylacetate (DOPP) (Fig. 10). The activity of TAS was demonstrated by inducing aggregation of donor platelets with DOPP(0.86 μM) and then transferring 100 μl of plasma (0.09 μm DOPP) to fresh recipient platelets and observing the aggregation induced. The amount of DOPP transferred was insufficient to induce a significant aggregation of recipient platelets. The production of TAS was time dependent (Fig. 11) reaching a maximum within 4 min of the onset of aggregation (Edwards, Westwick & Evans, 1982), with a

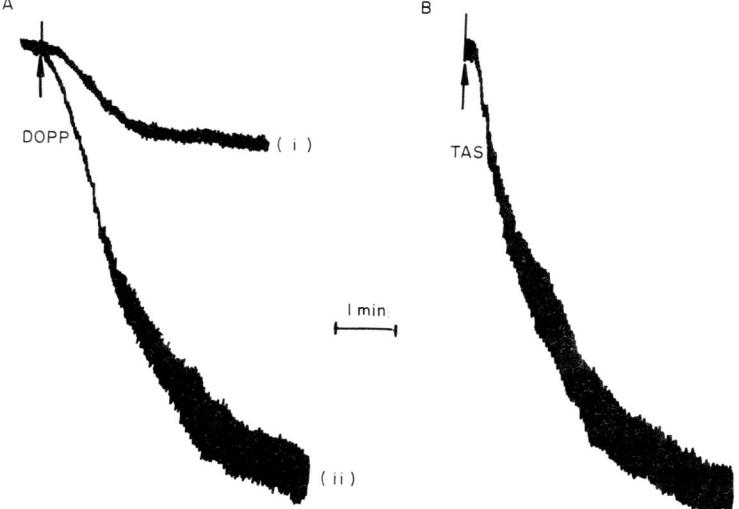

Figure 10. Secretion of TAS by human platelets stimulated by phorbol esters. A. Human platelet aggregation induced by (i) 0.09 μm DOPP, (ii) 0.43 μm DOPP. B. Human platelet aggregation induced by 100 μl of plasma-transferred from A (ii) (TAS) to 400 μl of fresh platelets 4 min after stimulation by DOPP (equivalent to a DOPP concentration of 0.09 μm, A(i))

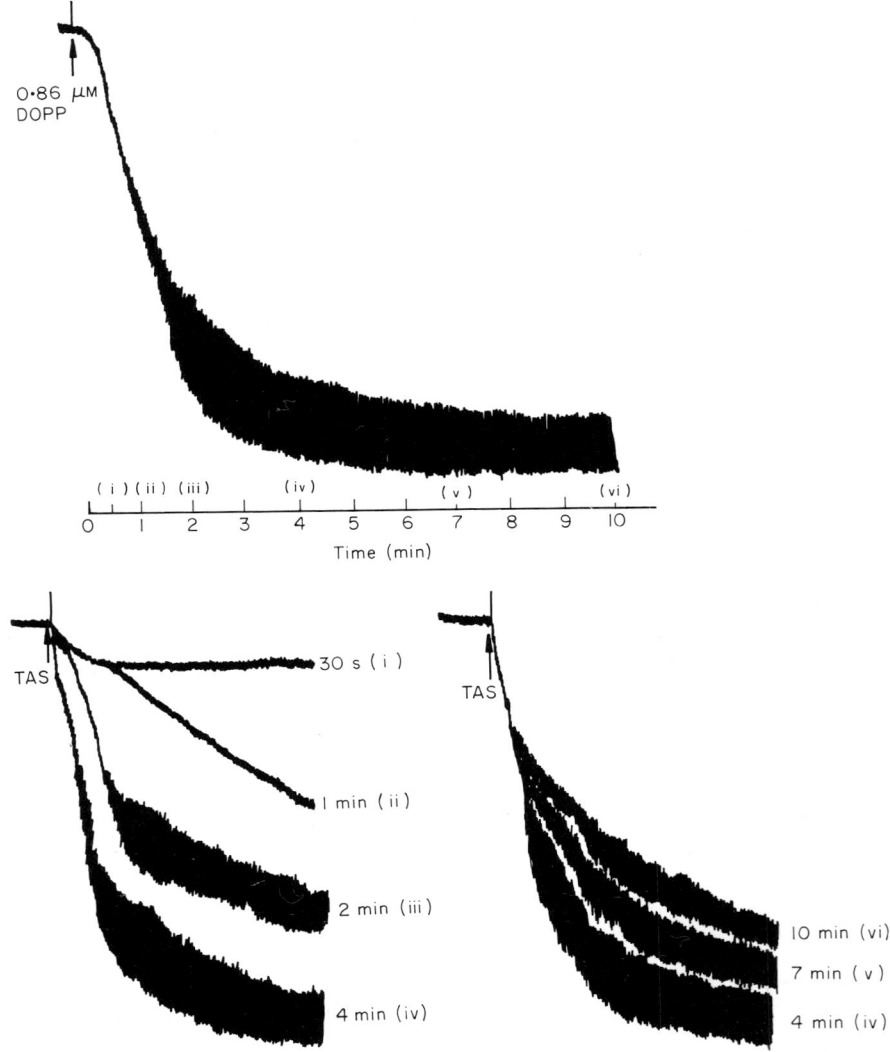

Figure 11. Time-course of the production of TAS by human platelets stimulated by 0.86 μm DOPP.

half-life of 20 min in plasma at 37°C. All activity of TAS containing plasma was lost on storage at $-4°C$ for 24 h. The aggregation induced by TAS was distinct from that induced by DOPP in that platelets which had been stored for 4 h failed to respond to TAS stimulation whilst still responding fully to stimulation by DOPP. A further distinction between the activity of the two agonists was found with a washed platelet system. Washed platelets failed to significantly respond to DOPP unless Ca^{2+} and Mg^{2+} ions were present (Fig. 11), but did respond to TAS without the addition of divalent ions. The secretion of TAS by platelets was found to be species-dependent. Phorbol esters are known to be potent stimulators of rabbit blood platelets (Edwards et al., 1983b) but attempts to demonstrate transfer aggregation into recipient rabbit platelets failed to induce further aggregation.

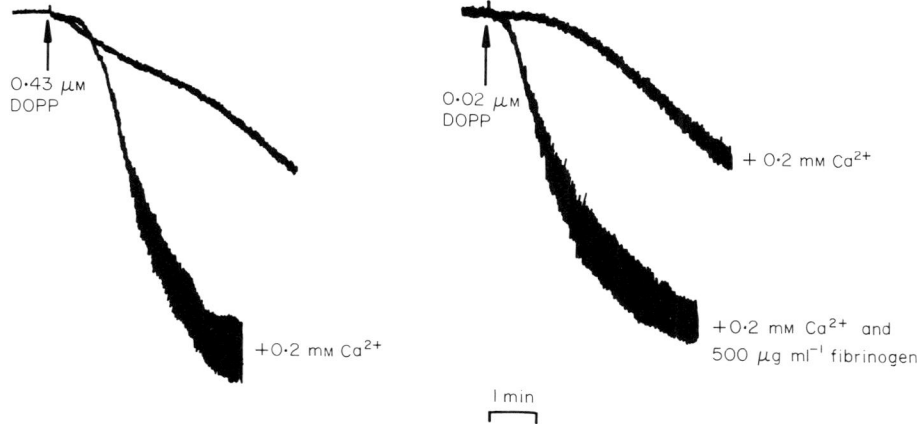

Figure 12. The action of 12-deoxyphorbolphenylacetate (DOPP) on washed human platelets. Resuspension buffer: KCl, 0.0373%; NaCl, 0.671%; CaCl$_2$, 0.0294%; MgCl$_2$, 0.0203%; Tris, 0.375%; glucose, 0.09%, pH 7.4

TAS-induced aggregation of human platelets was a true aggregation in that it was dependent upon the presence of Ca^{2+} ions in plasma and upon intracellular levels of cyclic AMP (Fig. 13). However, unlike the phorbol esters and their interaction with platelets, the aggregation induced by TAS was abolished in the presence of creatine phosphate creatine phosphate kinase, an ADP scaventer. However, TAS was not ADP because β-methylene ATP an ADP receptor antagonist failed to prevent TAS induced aggregation. The possibility remained that TAS was phospholipid-derived and to investigate this possibility, further human platelets were desensitized to platelate aggregating factor (PAF). These desensitized platelets still responded to stimulation by TAS (Fig. 14).

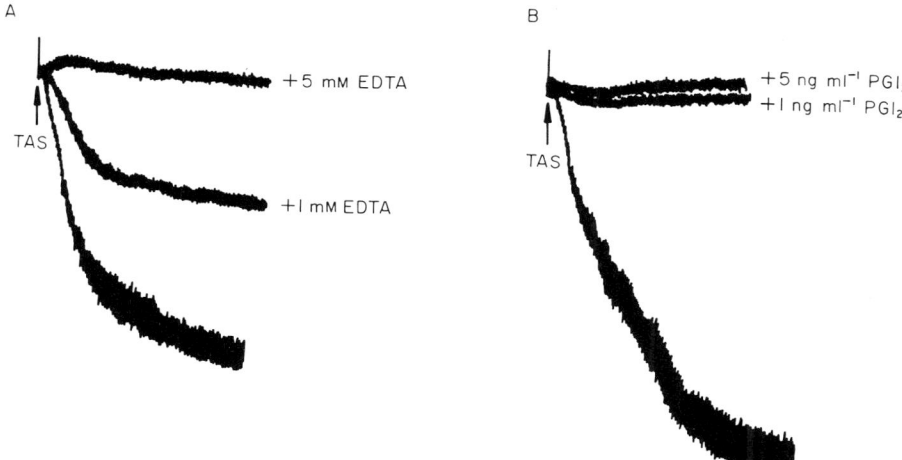

Figure 13. Inhibition of TAS-induced aggregation of platelets. A. Inhibition of TAS-induced aggregation by EDTA (TAS produced by stimulation of platelets by 0.86 μm DOPP). B. Inhibition of TAS-induced aggregation by PGI$_2$ (prostacyclin) (TAS produced by stimulation of platelets by 0.43 μm DOPP).

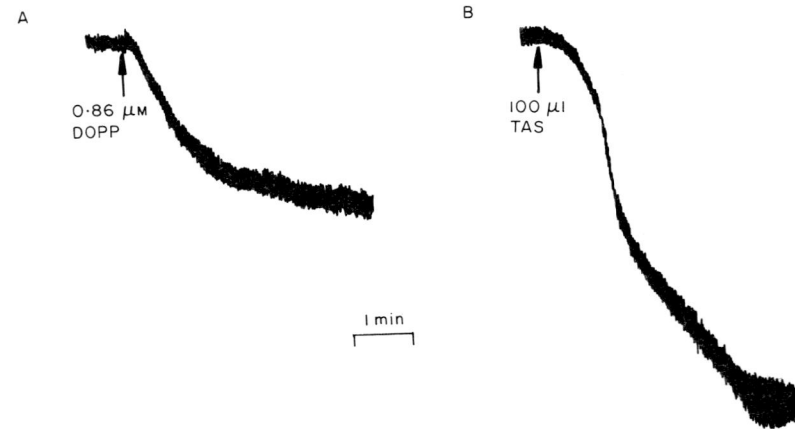

Figure 14. Aggregation of PAF-desensitized platelets by TAS.

The chemical nature of TAS remains to be determined, and preliminary attempts at chromatographic separation on silica gel and on Sephadex resulted in a loss of activity of TAS containing plasma.

CONCLUSIONS

Plant-derived drugs and toxins are of the upmost importance in basic medical science. These novel substances are useful biochemical and pharmacological tools for the elucidation of the biochemistry of diseased states in humans and animals and for our understanding of the mechanisms of drug action. In recent years, the phorbol esters are amongst the most significant compounds that phytochemists have isolated from higher plants in this respect. Their extreme potency, specific biological actions and rigid chemical structure have made them invaluable assets in the study of cancer proliferation and inflammation. These compounds have played a large part in the elucidation of the protein kinase C second messenger system of mammalian membranes, and more recently from correlation and other studies, it would appear that the phorbol esters initiate more than one primary biochemical event. It may be that protein kinase C exists in more than one form in the cell, or perhaps there is more than a single site on this enzyme for binding phorbol esters. Alternatively, a completely separate receptor to phorbol esters may yet be isolated.

REFERENCES

ADOLF, W., SORG, B., HERGENHARN, M. & HECKER, E., 1982. Structure-activity relationships of polyfunctional diterpenes of the daphnane type. *Journal of Natural Products, 45:* 347–354.
BAIRD, W. M., MELERA, P. W. & BOUTWELL, R. K., 1972. Acrylamide gel electrophoresis studies of the incorporation of cytidine into mouse skin RNA at early times after treatment with phorbol esters. *Cancer Research, 32:* 781–787.
BILLS, T. K., SMITH, J. B. & SILVER, M. J., 1976. Metabolism of arachidonic acid by human platelets. *Biochimica Biophysica Acta, 424:* 303–314.
CASTAGNA, M., TAKAI, Y., KAIBUCHI, K., SANO, K., KIKKAWA, U. & NISHIZUKA, Y., 1982. Direct activation of calcium-activated dependent protein kinase by tumour-promoting phorbol esters. *Journal of Biological Chemistry 257:* 7847–7851.

DRIEDGER, P. E. & BLUMBERG, P. H., 1979. Quantitative correlation between *in vitro* and *in vivo* activities of phorbol esters. *Cancer Research, 39:* 714–719.

EDWARDS, M. C., EVANS, F. J., BARRETT, M. L. & GORDON, D., 1985. Structural correlations of phorbol ester stimulation of PGE$_2$ production by human rheumatoid synovial cells. *Inflammation, 9:* 33–38.

EDWARDS, M. C., NOURI, A. M. E., GORDON, D. & EVANS, F. J., 1983a. Tumour-promoting and non-promoting pro-inflammatory phorbol esters act as human lymphocyte mitogens with different sensitivities to inhibition by cyclosporin A. *Molecular Pharmacology 23:* 703–708.

EDWARDS, M. C., TAYLOR, S. E., WILLIAMSON, E. M. & EVANS, F. J., 1983b. New phorbol and deoxyphorbol esters: Isolation and relative potencies in inducing platelet aggregation and erythema of skin. *Acta Pharmacologia et Toxicologia, 53:* 177–187.

EDWARDS, M. C., WESTWICK, J. & EVANS, F. J., 1982. Release of an aggregating substance by human platelets in response to phorbol ester stimulation. *Journal of Pharmacy and Pharmacology* (Suppl.).

ELLIS, C., MORRICE, N., AITKIN, A., PARKER, P. J. & EVANS, F. J., 1985. Tumour-promoting and non-promoting phorbol esters activate protein kinase C. *Journal of Pharmacy and Pharmacology* (Suppl.).

ESTENSEN, R. D. & WHITE, R. J., 1974. Ultrastructural features of the platelet response to phorbol myristate acetate. *American Journal of Pathology, 74:* 441–452.

EVANS, F. J., 1978. The irritant toxins of the blue Euphorbia. *Toxicon, 16:* 51–57.

EVANS, F. J., 1986a. Environmental hazards of plants containing tumour-promoting phorbol esters. In F. J. Evans (Ed.), *Naturally Occurring Phorbol Esters:* 1–32. Boca Raton: CRC Press Inc.

EVANS, F. J., 1986b. The phorbol esters. In F. J. Evans (Ed.), *Naturally Occurring Phorbol Esters:* 171–215. Boca Raton: CRC Press Inc.

EVANS, F. J. & KINGHORN, A. D., 1975. The succulent euphorbias of Nigeria. *Lloydia, 38:* 363–365.

EVANS, F. J. & SCHMIDT, R. J., 1979. An assay procedure for the comparative irritancy testing of esters of the tigliane and daphnane series. *Inflammation, 3:* 215–223.

EVANS, F. J. & SOPER, C. J., 1978. The tigliane, daphnane and ingenane diterpenes, their chemistry, distribution and biological activities. *Lloydia, 41:* 193–233.

EVANS, F. J. & TAYLOR, S. E., 1983. Pro-inflammatory, tumour-promoting and anti-tumour diterpenes of the plant families Euphorbiaceae and Thymelaeaceae. *Fortschritte der Chemie Organischer Naturstoffe, 44:* 1–99.

FREI, J. V. & STEVENS, P., 1968. The correlation of promotion of tumour growth and of induction of hyperplasia in epidermal two stage carcinogenesis. *British Journal of Cancer, 22:* 83–92.

FURSTENBERGER, G. & HECKER, E., 1974. Zum wirkungsmechanismus cocarcinogenes pflanzeninhaltsstoffe. *Planta Medica, 22:* 241–266.

GILMORE, T. & MARTIN, G. S., 1983. Phorbol ester and diacylglycerol induce protein phosphorylation at tyrosine. *Nature, 306:* 487–490.

HECKER, E., 1968. Co-carcinogenic principles from the seed oil of *Croton tiglium* and from other Euphorbiaceae. *Cancer Research, 31:* 2338–2349.

HECKER, E., 1971. Isolation and characterisation of the co-carcinogenic principles of *Croton* oil. In H. Busch (Ed.), *Methods of Cancer Research, 6:* 439–484. London: Academic Press.

HECKER, E., 1975. Co-carcinogens and co-carcinogenesis: with a note on synergistic process in carcinogenesis. In E. Grundman (Ed.), *Handbuch der Allgemeine Pathologie, 6:* 651–675. Berlin: Springer-Verlag.

HECKER, E., 1977. New toxic, irritant and co-carcinogenic diterpene esters from Euphorbiaceae and Thymelaeaceae. *Pure and Applied Chemistry, 49:* 1423–1431.

HECKER, E., 1978. Structure-activity relationships in diterpene esters irritant and co-carcinogenic to mouse skin. In T. J. Slaga, A. Sivak & R. K. Boutwell (Eds), *Carcinogenesis, 2:* 11–38. New York: Raven Press.

HECKER, E., IMRICH, H., BRESCH, J. & SCHAIRER, H. U., 1966. Über die wirkstoffe des Crotonols VI: Entzundungstreste am mauseohr. *Zeitschrift für Krebsforschung, 68:* 366–374.

HECKER, E. & SCHMIDT, R., 1974. Phorbol esters, the irritants and co-carcinogens of *Croton tiglium* L. *Forschritte der Chemie Organischer Naturstoffe, 31:* 377–467.

HOPE, W., BRANDL, F., STRELL, I., ROHRL, M., GASSMAN, I., HECKER, E., BARTSCH, H., KREIBICH, G. & SZCZEPANSKI, CH. V., 1967. X-ray structure analysis of neophorbol. *Angewanate Chemie International Edition, 6:* 809–810.

KIKKAWA, U., TAKAI, Y., TANAKA, Y., MIYAKE, R. & NISHIZUKA, Y., 1983. Protein kinase C as a possible receptor protein of tumour-promoting phorbol esters. *Journal of Biological Chemistry 258:* 11442–11445.

KINGHORN, A. D. & EVANS, F. J., 1975. A biological screen of selected species of the genus *Euphorbia* for skin irritant effects. *Planta Medica, 28:* 325–335.

KINSELLA, R. A., 1986. Multi-stage carcinogenesis and the biological effects of tumour-promoters. In F. J. Evans (Ed.), *Naturally Occurring Phorbol Esters:* 33–62. Boca Raton: CRC Press Inc.

MAHADEVAPPA, V. G. & HOLUB, B. J., 1983. Degradation of different molecular species of phosphatidylinositol in thrombin stimulated human platelets: evidence for preferential degradation of 1-acyl-2-arachidonyl species. *Journal of Biological Chemistry, 258:* 5337–5339.

OHUCHI, K. & LEVINE, L., 1978. Stimulation of prostaglandin production, deacylation of lipids and morphological changes by tumour-promoting phorbol 12,13-diesters and adriamycin in MDCK cells. *Prostaglandins, 15:* 723.

SCHMIDT, R. J. & EVANS, F. J., 1979. Investigations into the skin-irritant properties of resiniferonol orthoesters. *Inflammation, 3:* 273–280.

SCHMIDT, R. J. & EVANS, F. J., 1980. Skin irritant effects of esters of phorbol and related polyols. *Archives of Toxicology, 44:* 279–289.

SCRIBNER, J. D. & SLAGA, T. J., 1973. Multiple effects of dexamethasone on protein synthesis and hyperplasia caused by a tumour-promoter. *Cancer Research, 33:* 542–546.

SCHACKELFORD, D. A. & TROWBRIDGE, I. S., 1984. Induction of expression and phosphorylation of the human IL-2 receptor by a phorbol ester. *Journal of Biological Chemistry, 259:* 11706–11712.

SHINOZUKA, H. & RITCHIE, A. C., 1967. Pretreatment with *Croton* oil, DNA synthesis and carcinogenesis by carcinogen followed by *Croton* oil. *International Journal of Cancer, 20* 77–84.

TRASHJIAN, A. H., IVEY, B. & DECLOS, B. & LEVINE, L., 1978. Stimulation of prostaglandin production in bone by phorbol esters and melittin. *Prostaglandins, 16:* 221–232.

TOURAINE, J. L., HADDEN, J. W., TOURAINE, F., ESTENSEN, R. D. & GOOD, R. A., 1977. Phorbol myristate acetate, a mitogen selective for a T-lymphocyte subpopulation. *Journal of Experimental Medicine 145:* 460–465.

WESTWICK, J., WILLIAMSON, E. M. & EVANS, F. J., 1980. Structure-activity relationships of 12-deoxyphorbol esters on human platelets. *Thrombosis Research, 20:* 683–692.

WHITE, J. G. & ESTENSEN, R. D., 1974. Cytochemical electron microscopic studies of the action of phorbol myristate acetate on platelets. *American Journal of Pathology, 74:* 453–460.

WILLIAMSON, E. M. & EVANS, F. J., 1981. Inhibition of erythema induced by pro-inflammatory esters of 12-deoxyphorbol. *Acta Pharmacologia et Toxicologica, 48:* 47–52.

WILLIAMSON, E. M., WESTWICK, J. & EVANS, F. J., 1980. The effect of daphnane esters on platelet aggregation and erythema of the mouse ear. *Journal of Pharmacy Pharmacology, 32:* 373–374.

WILLIAMSON, E. M., WESTWICK, J., KAKKAR, V. V. & EVANS, F. J., 1981. Studies on the mechanism of action of 12-deoxyphorbolphenylacetate, a potent platelet aggregating tigliane ester. *Biochemical Pharmacology, 30:* 2691–2696.

WILLIAMS, T. J., WESTWICK, J., WILLIAMSON, E. M. & EVANS, F. J., 1981. Vascular changes in rabbit skin induced by pro-inflammatory phorbol and deoxyphorbol esters. *Inflammation, 5:* 29–36.

ZUCKER, M. B., TROLL, W. & BELMAN, S., 1974. The tumour-promoting phorbol ester 12-O-tetradecanoylphorbol-acetate: a potent aggregating agent for blood platelets. *Journal of Cell Biology, 60:* 325–336.

The activation of protein kinase C by daphnane, ingenane and tigliane diterpenoid esters

ALASTAIR AITKEN*

Department of Pharmaceutical Chemistry, School of Pharmacy, University of London, 29/39 Brunswick Square, London, WC1N 1AX

Received May 1986, accepted for publication September 1986

AITKEN, A., 1986. **The activation of protein kinase C by daphnane, ingenane and tigliane diterpenoid esters.** In this review, the mechanism of action of phorbol esters and related diterpenes is described. These compounds have been shown to stimulate a Ca^{2+} and phospholipid dependent protein kinase, termed kinase C. Phorbol esters activate protein kinase C by substituting for the natural effector, the second messenger, diacylglycerol. The various known protein substrates of this enzyme are described. Many of these substrates are involved in regulation of protein synthesis, DNA expression, cell transformation etc. This provides the explanation for the tumour promotion effects of some phorbol esters. Evidence for the biochemical mechanisms of action of phorbol esters that have other biological effects are also described. Recent evidence from our laboratories indicates that phorbol esters with limited biological effects, e.g. inflammatory but not tumour promoting, also act through this protein kinase. These phorbol esters appear to stimulate the phosphorylation of a different range of substrate proteins *in vivo*.

ADDITIONAL KEY WORDS:—phorbol esters – phosphatidylinositol turnover – protein kinase C – protein phosphorylation.

CONTENTS

Introduction	247
Phosphatidylinositol turnover	249
Purification and properties of protein kinase C	250
Protein kinase C substrate specificity	251
Involvement with other hormone-receptor systems	255
The stimulation of protein kinase C by structurally diverse phorbol esters	257
In vivo phosphorylation in response to phorbol esters of distinct biological acitivity	260
Conclusions	260
Acknowledgements	260
References	261

INTRODUCTION

The Euphorbiaceae produce a range of toxic diterpenes belonging to a number of structural types (reviewed Evans & Taylor, 1983). Esters of the diterpenes, tigliane, daphnane and ingenane (Fig. 1) have been shown to possess a range of biological effects including tumour promotion and cell proliferation, activation of blood platelets, lymphocyte mitogenesis, inflammation (erythema

---

* Present address: Laboratory of Protein Structure, National Institute for Medical Research, The Ridgeway, Mill Hill, London NW7 1AA.

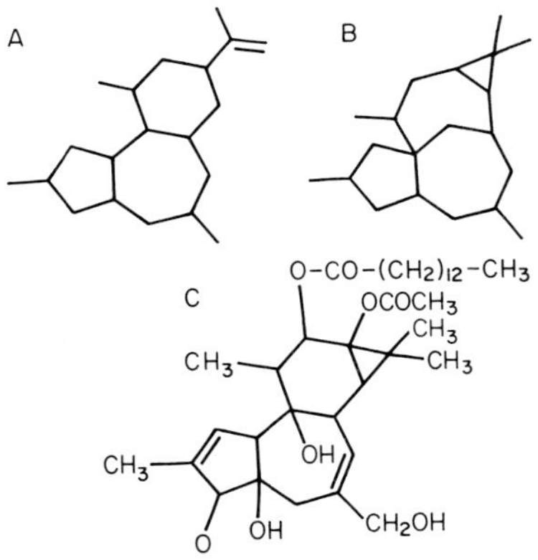

Figure 1. Three titerpene structural types occurring in Euphorbiaceae; A. Daphnane type; B. Ingenane type; C. Tigliane type (TPA is shown).

of the skin), prostaglandin production, and stimulation of degranulation in neutrophils (Blumberg, 1980, 1981).

Phorbol belongs to the tigliane group, and its esters have been found in the genera *Croton*, *Sapium* and *Euphorbia*. One of these, 12-*O*-tetradecanoylphorbol-13-acetate (TPA) is particularly potent and has been used in many studies of biological activity.

Much of what is known about the transmission of the effects of phorbol esters inside cells comes from study of tumour promoting and platelet aggregating phorbol esters. These have been shown to interact with, and activate a recently discovered protein kinase, first described by Nishizuka and collegues (Takai, Kisimoto, Inoue & Nishizuka, 1977; Kuo *et al.*, 1980) termed C kinase or protein kinase C. This is a Ca^{2+}- and phospholipid-dependent protein kinase of apparently ubiquitous tissue distribution. Phorbol esters causing tumour promotion and other biological effects appear to substitute for the natural activator of this protein kinase, diacylglycerol (Castagna *et al.*, 1981; Kikkawa *et al.*, 1983). This is a product of phosphatidylinositol hydrolysis (Fig. 1), an important mechanism by which the effects of many neurotransmitters and hormones are transmitted (Berridge & Irvine, 1984; Aitken, 1986). This activator of protein kinase C, diacylglycerol, may be considered a 'second messenger' acting in an analogous manner to cyclic AMP. Phosphatidylinositol breakdown does not only give rise to diacylglycerol, but the other product, inositol trisphosphate, is implicated as a second messenger for the release of Ca^{2+} from internal stores (Michell, 1984). For a recent detailed treatment of phosphatidylinositol turnover, the reader is referred to the review of Berridge (1984).

The importance of reversible phosphorylation as a major intracellular regulatory mechanism has been fully realized only within the last decade (Cohen, 1982). Not only is this a mechanism for mediating neural and hormonal regulation of enzyme activity in a wide variety of metabolic processes, but the recent discovery of tyrosine-specific protein kinases from growth factor receptors and RNA tumour virus gene products (Hunter & Sefton, 1982) has further emphasized the biological importance of this covalent modification of proteins.

PHOSPHATIDYLINOSITOL TURNOVER
(FIG. 2)

The type of receptor implicated in phosphatidylinositol hydrolysis includes the α_1-adrenergic, H_1-histaminergic, muscarinic-cholinergic and V_1-vasopressin receptors.

On binding to the receptor on the cell surface, the agonist causes hydrolysis of phosphatidylinositol-4,5-bisphosphate (PtdIns4,5P_2) catalysed by a phospholipase C or phosphodiesterase to produce the two second messengers inositol-1,4,5-trisphosphate (Ins1,4,5P_3) and diacylglycerol. The former is responsible for triggering release of Ca^{2+} ions from internal stores. The increased level of Ca^{2+} then activates Ca^{2+}-dependent protein kinases, and many other intracellular events.

In the other branch of the response to the agonist, the 1,2-diacylglycerol activates the Ca^{2+}- and phospholipid-dependent protein kinase C.

As well as functioning as a second messenger in the stimulation of protein kinase C activity, diacylglycerol may also be a precursor for release of arachidonic acid which is required not only for synthesis of prostaglandins but also leukotrienes and thromboxane (Lapetina, 1982). There is evidence for

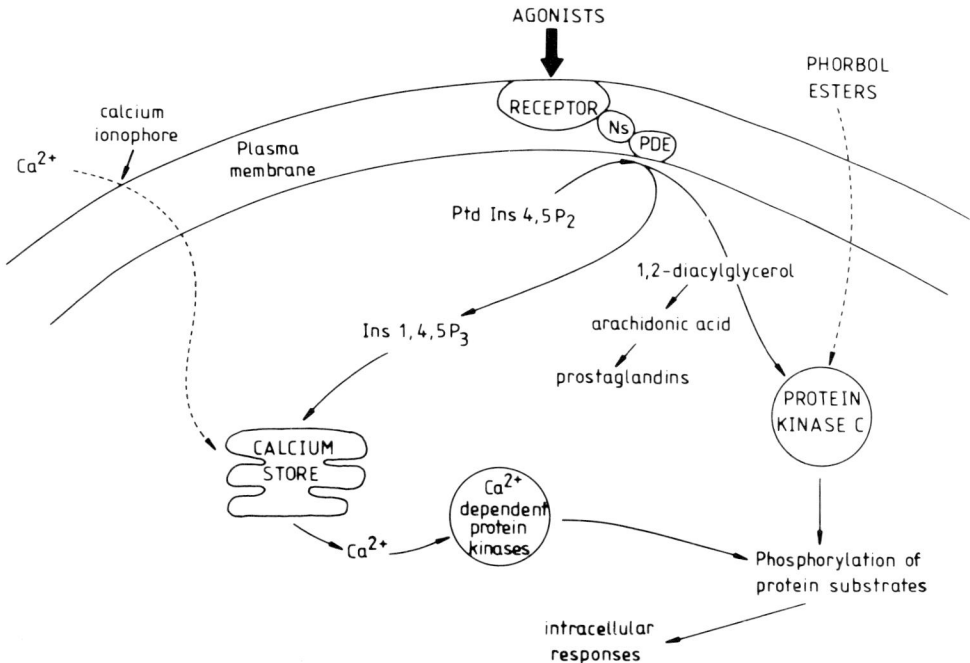

Figure 2. Phosphoinositide hydrolysis. On binding to its receptor, the agonist stimulates the activity of a phosphodiesterase (PDE). There is recent evidence that a guanine nucleotide binding regulatory protein may be involved and is designated N_s to indicate possible analogy with the regulatory protein in the adenylate cyclase receptor. The products from phosphatidylinositol-4,5-bisphosphate (PtdIns,4,5P_2) are inositol-1,4,5-trisphosphate (Ins,1,4,5P_3) which releases calcium from the endoplasmic reticulum and 1,2-diacylglycerol which activates protein kinase C. The direct stimulation of protein kinase C by phorbol esters is illustrated. Calcium ionophores may increase intracellular calcium levels directly, leading to synergistic effects of two artificial stimuli. Other details are given in the text.

preferential breakdown of those phosphoinositides that carry arachidonic acid in the 2 position, the most common fatty acid at this position (Mahedevappa & Holub, 1983).

In the normal induction of phosphatidylinositol breakdown by agonists to produce diacylglycerol and inositoltrisphosphate there exists inside the cell degradative pathways that would remove these two internal signals rapidly when the external signal, the agonist, is removed. Diacylglycerol is converted either to phosphatidic acid (by a diacylglycerol kinase) or to a monoacylglycerol.

This is an important point to consider when seeking a basis for the tumour promoting and other effects of phorbol esters. These biological effects may be due to the fact that the activation of protein kinase C by phorbol esters is of longer duration than would be activation by diacylglycerol. Evidence that the relative importance of each branch of the phosphatidylinositol hydrolysis pathway may vary with time has been obtained in a comparative study of the effects of 12-O-tetradecanoylphorbol-13-acetate (TPA) on platelet myosin phosphorylation (see below). Calcium may be more involved in the initiation of some of the effects while diacylglycerol may be more important in the maintenance of the response to agonists.

A possible biochemical mechanism by which RNA tumour viruses may cause cell proliferation has been discovered through their effect on metabolites of PtdIns4,5,P_2 hydrolysis. This 'tyrosine-specific' protein kinase activity, the oncogene product, can also cause phosphorylation of PtdIns and PtdIns4P as well as phosphorylation of 1,2-diacylglycerol (Michell, 1984). The related tyrosine kinase activity in growth hormone receptors for insulin and epidermal growth factor also has this ability. These enzymes act by modifying the activity of cellular lipid kinases. The phosphorylation of the phosphatidylinositol intermediates would have the effect of increasing the pool of PtdIns4,5P_2 and lead to a greater intensity of effect of agonists that stimulate PtdIns4,5P_2 hydrolysis. This type of synergistic effect has been observed where a combination of growth hormone (insulin or epidermal growth factor) *and* PtdIns4,5P_2 agonist has led to much greater activation of cell proliferation than one type of agonist alone. Platelet derived growth factor (PDGF) and epidermal growth factor (EGF) have been shown to possess the ability to stimulate both tyrosine kinase activity and to enhance PtdIns4,5P_2 breakdown.

In most systems so far studied, the stimulation of phosphatidylinositol metabolism is independent of an increase in the intracellular level of calcium. However, in a few cases, e.g. in neutrophils (Cockcroft, 1981), stimulation of phosphatidylinositol turnover does appear to be dependent on calcium. There is evidence that besides formation of inositol-1,4,5-trisphosphate (releasing Ca^{2+} from internal stores), phosphatidylinositol metabolism may also be responsible for increasing the permeability of the plasma membrane to calcium.

PURIFICATION AND PROPERTIES OF PROTEIN KINASE C

This enzyme was first purified from rat brain by Kikkawa *et al.* (1982) in the group of Nishizuka. Protein kinase C has recently been purified to homogeneity from bovine brain by Parker, Stabel & Waterfield (1984) who have shown it to be a monomeric protein of M_r 79000 in fairly close agreement with the value obtained by most other groups. The purified enzyme is dependent on Ca^{2+} (with a K_a in the micromolar range) and is dependent on phospholipid (particularly phosphatidylserine). Five other phospholipids (phosphatidyl-inositol, -ethanolamine, -choline, phosphatidic acid and sphingomyelin) were shown to be ineffective in stimulating protein kinase C phosphorylation of proteins (Iwasa & Hosey, 1984). Indeed phosphatidyl choline and sphingomyelin inhibited the enzyme (Kaibuchi, Takai & Nishizuka, 1981). Although it is not a calmodulin-dependent enzyme, protein kinase C is inhibited by the phenothiazine and other type of inhibitors of Ca^{2+}-calmodulin regulated enzymes (Mori et al., 1980; Feinstein & Hadjian, 1982). The purified enzyme from bovine brain exhibits two apparent K_a values for phosphatidylserine of approximately 0.6–2.0 and 35–80 μg ml^{-1}. Single K_a values of 6–20 μg ml^{-1} were reported for the enzyme from other tissues. In the presence of 1,2-diacylglycerol (diolein) the K_a for Ca^{2+} is lowered from 10 μM to 1 μM (Kaibuchi et al., 1981).

The enzyme from bovine and rat brain has been shown to bind close to one mole ^{3}H-phorbol dibutyrate, furnishing proof that protein kinase C is the phorbol ester receptor. The bovine and rat brain preparations were shown to bind this phorbol ester with K_a values of 15 and 8 nM respectively. Diacylglycerol has also been shown to competitively inhibit this phorbol ester binding (Sharkey, Leach & Blumberg, 1984).

The presence of diacylglycerol or phorbol dibutyrate lowered the K_a values for phosphatidylserine two- to three-fold in bovine brain protein kinase C.

Protein kinase C appears to be present in all tissues, from mammals to insects (Kuo et al., 1980) and is present in comparatively large amounts in brain, where it is partially localized in synaptosomes.

There is evidence that, on activation by either diacylglycerol or phorbol esters, protein kinase C, which is mainly associated with the cytoplasm, becomes attached to the inside of the plasma membrane (Kraft & Anderson, 1983). Although this active form of the enzyme is now membrane-associated, it is still capable of phosphorylating specific intracellular proteins.

PROTEIN KINASE C SUBSTRATE SPECIFICITY

Protein kinase C is an enzyme that specifically phosphorylates serine and threonine residues in proteins. In common with many other classes of protein kinase, for example tyrosine-specific, and cyclic-AMP and cyclic-GMP dependent enzymes, protein kinase C has been shown to have the ability to autophosphorylate. In this case, there is no known physiological significance of the reaction. There would seem to be a clear distinction between protein kinases that phosphorylate serine and threonine and those that are specific for tyrosine residues. The latter class(es) of protein kinase have been shown to be an intrinsic part of growth hormone receptors. RNA tumour virus gene products have also been shown to be tyrosine specific protein kinases that are involved in cellular transformation (Hunter & Sefton, 1982; Bishop, 1983). If protein kinase C could be shown to affect the state of phosphorylation of substrates for these tyrosine-specific protein kinases this could provide a molecular mechanism for at least some of the effects of phorbol esters—this has in fact been seen in a number of systems. A combination of 'tyrosine' kinase and PtdIns4,5P$_2$ hydrolysis stimulus leads to synergistic effects in cell proliferation (Michell, 1983; Berridge, 1984).

Protein kinase C has been shown to phosphorylate growth hormone receptors (Hunter, Ling & Cooper, 1984). The best studied example is epidermal growth factor (EGF) receptor which is phosphorylated by protein kinase C on a threonine residue nine amino acids from the cytoplasmic end of a proposed transmembrane domain (Table 1).

The *in vivo* phosphorylation of EGF receptor mediated by protein kinase C has been shown in A431 cells to lead to a reduction in the ability of EGF to bind to its receptor and to reduce the tyrosine protein kinase activity. Both TPA and the indole alkaloid tumour promoter, teleocidin, stimulated this phosphorylation. EGF has been shown to increase the hydrolysis of PtdIns4,5P$_2$. The role of this EGF receptor phosphorylation by kinase C may therefore be a feedback mechanism to modulate the response of the EGF receptor. Gilmore & Martin (1983) have shown that an $M_r = 42\,000$ polypeptide that is phosphorylated at a tyrosine residue in avian sarcoma virus transformed cells and in cells stimulated with growth hormones (EGF, PDGF and multiplication-stimulating activity), also becomes phosphorylated at tyrosine residues in cells treated with TPA. Exogenously added 1-oleoyl-2-acetyl-glycerol also stimulates the phosphorylation of the $M_r = 42\,000$ protein on tyrosine. The most probable explanation was that this tyrosine phosphorylation was not mediated directly by stimulation of protein kinase C activity but that C-kinase (with a specificity for serine and threonine phosphorylation) was stimulating tyrosine specific protein kinases(s).

In 3T3-L 1 fibroblasts, the state of phosphorylation of an $M_r = 80\,000$ protein is increased in response to phorbol esters and the growth factors PDGF and EGF (Blackshear, Wen, Glynn & Witters, 1986). These authors have evidence for a role in cholinergic neurotransmission in neuronal cells, for this protein. This substrate is distinct from protein kinase C itself, which has a similar molecular weight. We have also shown that this protein is phosphorylated on the same peptides by kinase C, and by stimulation with the polypeptide mitogen, embryonal carcinoma-derived growth factor, ECDGF (Mahadevan, Aitken, Heath & Foulkes, in press).

The large number of basic amino acids surrounding the side of phosphorylation is immediately apparent in the EGF receptor. Other sites of phosphorylation in the basic proteins histone H1, H2B and myelin basic protein (all phosphoserine residues) are preceded by basic amino acid residues (see Table 1). The primary structure requirements of protein kinase C would therefore appear to be similar to that of many other protein kinases, particularly cyclic AMP and cyclic GMP dependent protein kinases. There may in fact be some overlap in structural specificity of cyclic AMP-dependent and C-kinase.

The same serine residue in ATP-citrate lyase is phosphorylated by both protein kinases (Table 1). The state of phosphorylation of this site is increased by insulin.

We have shown that protein kinase C phosphorylates acetyl CoA carboxylase on a different site to that phosphorylated by cyclic-AMP dependent protein kinase (Hardie *et al.*, 1986) and the phosphorylation does not result in activation of the enzyme. Tyrosine hydroxylase phosphorylation by protein kinase C does result in activation of this enzyme. This phosphorylation is on the same site labelled with cyclic-AMP dependent protein kinase. Administration of

Table 1. Amino acid sequences surrounding the sites of phosphorylation of some substrates of protein kinase C

Substrate	Sequence	References
Rat mammary gland ATP citrate lyase	Lys or Arg Thr-Ala-Ser(P)-Phe-Ser-Glu-Ser-Arg	Hardie et al., 1986
Epidermal growth factor	Arg-Arg-Arg-His-Ile-Val-Arg-Lys-Arg-Thr(P)-Leu-Arg-Arg	Hunter et al., 1984
Calf thymus histone H1	Ala-Lys-Arg-Lys-Ala-Ser^{38}(P)-Gly-Pro-Val-Ser	Takai et al., 1977
Calf thymus histone H2B	Gly-Lys-Lys-Arg-Lys-Arg-Ser^{32}(P)-Arg-Lys-Gly-Ser^{36}*(P)-Tyr	Takai et al., 1977
Bovine myelin basic protein	Arg-Phe-Ser^{115}(P)-Trp	Turner et al., 1984
Ribosomal protein S6	Arg-Arg-Leu-Ser-Ser(P)-Leu-Arg	Gabrielli et al., 1984

insulin has been shown to lead to the stimulation of the phosphorylation of an $M_r = 47\,000$ protein in rat hippocampus (Akers & Routtenberg, 1984). This substrate, protein F1, is probably identical to the central nervous system B-50 protein ($M_r = 48\,000$) which is a substrate for protein kinase C (Aloyo, Zwiers & Gispen, 1983). This protein, and the following ribosomal protein, are examples of substrates whose state of phosphorylation is increased in response to phorbol esters, although it is not clear yet whether the phosphorylation is mediated directly by protein kinase C or by indirect activation of an additional protein kinase.

TPA can stimulate phosphorylation of ribosomal protein S6 in hepatoma cells, on serine residues (Trevillyan, Kulkarni & Byus, 1984). This H35 rat hepatoma cell line has been shown to increase ornithine decarboxylase activity and polyamine synthesis in response to tumour promoting phorbol esters. The growth hormones insulin and insulin-like growth factor have also been shown to enhance phosphorylation in protein S6. A function for ribosomal protein S6 phosphorylation has been suggested in transition into the G1 phase of the cell cycle and in altering the affinity of the ribosome for certain classes of mRNA.

Wettenhall and colleagues (Gabrielli et al., 1984) have shown that within a palindromic sequence (Table 1) the serine residues are phosphorylated by cyclic-AMP-dependent protein kinase, a protease-activated kinase and protein kinase C. Cyclic-AMP-dependent kinase preferentially phosphorylates the first serine while the latter kinases appear to prefer the other site.

Phosphorylation of vinculin (a cytoskeletal protein) in fibroblasts and Swiss 3T3 cell cultures is increased in response to TPA and phorboldibutyrate (Werth & Pastan, 1984). The latter, which is a less potent phorbol ester, resulted in a smaller increase in phosphorylation (inactive phorbol esters showed no increase in phosphorylation). Significant increases in phosphorylation were seen in three peptides, on serine and threonine residues. This protein is also a substrate for the transforming protein of Rous sarcoma virus, when it is phosphorylated on a tyrosine residue. Vinculin has been postulated as having had a role in the morphological changes seen in Rous sarcoma virus transformation. This work establishes a link between the morphological changes induced by phorbol esters and those seen following viral transformation.

TPA administration induces terminal differentiation into macrophages in normal bone marrow promyelocytes and in myeloid leukaemic cells. A large increase in the phosphorylation of two cystosolic proteins with $M_r = 17\,000$ and $27\,000$ in the promyelocytic cell line GK-60 has been noted in response to biologically active phorbol esters (Feuerstein & Cooper, 1984). Increase in the intracellular level of Ca^{2+} and activation of an Na^+/H^+ exchange carrier are the main ionic events occurring that are responsible for the onset of cell proliferation (Moolenaar, Tsien, van der Saag & der Laat, 1983). Protein kinase C appears to phosphorylate and activate this neutral Na^+/H^+ exchanger (Rosoff, Stein & Cantley, 1984). Induction of a murine pre-lymphocyte cell line differentiation is accomplished by activation of this Na^+ uptake system. In 3T3 cells, administration of phorbol esters has been shown to increase internal pH (Burns & Rozengurt, 1983).

In blood platelets, in response to stimulation by 12-O-tetradecanoyl phorbol-13-acetate (TPA) there is a marked increase in phosphorylation of two proteins with M_r values of 20 000 and 40 000. The $M_r = 40 000$ protein appears to be a specific substrate for protein kinase C and the state of phosphorylation of this protein is also increased by thrombin on an identical peptide (Lyons & Atherton, 1979). The protein of $M_r = 20 000$ has been shown to be identical to myosin light chain (Naka, Nishikawa, Adelstein & Hidaka, 1983). The major phosphorylation site has been shown by peptide mapping to be different from that phosphorylated by the Ca^{2+} calmodulin dependent myosin light chain kinase. The latter is the site phosphorylated when platelets are stimulated with thrombin and when purified human platelet myosin is phosphorylated by myosin light chain kinase (Naka et al., 1983).

The phosphorylation of both the $M_r = 20 000$ (platelet myosin light chain) and $M_r = 40 000$ proteins therefore appears to be important, although their exact role(s) in the regulation of platelet function is not yet known. A role for the $M_r = 40 000$ protein phosphorylation in serotonin release has been suggested by the group of Nishizuka (Sano et al., 1983).

It is interesting to note that phosphorylation of the myosin light chain was much faster in response to thrombin than TPA (Naka et al., 1983). This suggests that the effects of thrombin on the stimulation of $PtdIns4,5P_2$ hydrolysis to inositol trisphosphate (which leads to activation of Ca^{2+}-calmodulin dependent myosin light chain kinase), is faster than the protein kinase C effects. The human platelet myosin light chain response to TPA (Naka et al., 1983) showed an additional minor in vivo site of phosphorylation that appeared to be identical to that phosphorylated in response to the thrombin activation of platelets (the Ca^{2+}-calmodulin activated-myosin light chain kinase site). This indicates a possible effect of protein kinase C on $PtdIns,4,5P_2$ hydrolysis which would subsequently raise Ca^{2+} levels to activate the former Ca^{2+}-calmodulin dependent protein kinase.

The effects of the two, second messenger signals that result from phosphatidylinositol hydrolysis can be mimicked in platelets by the addition of Ca^{2+} ionophores acting synergistically with activators of protein kinase C (Michell, 1983). Addition of the calcium ionophore A23187 together with the 12-O-tetradecanoyl phorbol-13-acetate also produces a greater response in a number of systems including lymphocytes (Mastro & Smith, 1983), in insulin secreting pancreatic islet cells (Malaisse et al., 1983), and in release of acetylcholine from guinea pig ileum (Tanaka, Taniyama & Kusunoki, 1984). However, this synergism between effectors of protein kinase C and Ca^{2+} levels has been best studied in platelets, where addition of the calcium ionophore together with 1-oleoyl-2-acetylglycerol results in a full secretory response.

INVOLVEMENT WITH OTHER HORMONE-RECEPTOR SYSTEMS

Protein kinase C phosphorylates both rat liver (Cuidad et al., 1984) and skeletal muscle (Ahmad, Lee, DePaoli-Roach & Roach, 1984) glycogen synthase in vitro. Phosphate is incorporated into at least two sites. This phosphorylation is accompanied by a decrease in the activity of the enzyme, in keeping with the well-studied effects of phosphorylation on glycogen synthase by a number of other protein kinases, including the β-adrenergic, second messenger-stimulated cyclic-AMP-dependent protein kinase. Ca^{2+}-dependent protein kinases also inactivate glycogen synthase, but phosphorylation on a separate site by another protein kinase (GSK5) has no direct effect on the activity (Cohen, 1982). In a similar study by Imazu, Strickland, Chrisman & Exton (1984), no decrease in activity of glycogen synthase was observed in either liver or muscle enzymes in response to protein kinase C.

In rat hepatocytes in response to phorbol esters, Roach & Goldman (1983) have shown that the state of phosphorylation in sites on glycogen synthase associated with protein kinase C is increased while those associated with other Ca^{2+}-dependent protein kinases (including phosphorylase kinase) were unaffected. This would indicate that one part of the phosphatidylinositol hydrolysis pathway, namely activation of protein kinase C, is involved, while the other branch, stimulation of release of Ca^{2+} from internal stores, is unaffected. Cuidad et al. (1984) studied glycogen synthase phosphorylation in rat hepatocytes in response to adrenaline, vasopressin, and glucagon. They reported changes in the phosphorylation located on two different peptides and concluded that regulation of liver glycogen synthase by multi-site phosphorylation involved other protein kinases in addition to cyclic nucleotide and Ca^{2+} dependent enzymes. Regulation of protein phosphatase activity may also be involved. The precise role of protein kinase C therefore remains to be elucidated.

Vasopressin, angiotensin II and α-adrenergic agonists cause increased phosphorylation of 10 proteins in hepatocytes, three of which were phosphorylated in response to TPA alone while the phosphorylation state of the others was increased by the ionophore A23187 which would exert its effect through increased intracellular calcium levels (Garrison, Johnson & Campanile, 1984).

In turkey erythrocytes, the β-adrenergic receptor has been shown to be phosphorylated in response to TPA (Kelleher, Pessin, Ruoho & Johnson, 1984). This results in a time dependent desensitization of isoprenaline (β-adrenergic agonist) stimulated adenylate cyclase activity. The effects appear to be directed to the uncoupling of receptor interactions with N_s (the stimulatory guanine nucleotide regulatory protein of adenylate cyclase). The authors did not show that the protein kinase C directly phosphorylated the β-adrenergic receptors. Compared to many other cell types, the agonist-induced desensitization of adenylate cyclase in turkey erythrocytes has some differences and it is not yet known if this is a general mechanism. However, it is interesting to note the effects of phorbol esters (through protein kinase C?) on the adenylate cyclase N_s protein, since it has been reported that the effects of the agonists that stimulate PtdIns4,5P$_2$ hydrolysis may also be mediated by a guanine nucleotide binding regulatory protein (Gomperts, 1983). This receptor subunit may have some similarity to the GTP-binding protein, P21, the product of ras oncogenes (Cockcroft & Gomperts, 1985). This forges a further link between phosphatidylinositol turnover, protein kinase C and cell transformation.

Protein kinase C phosphorylates a number of cardiac sarcolemma proteins from chicken heart, including phospholamban (Iwasa & Hosey, 1984). This protein, which is also phosphorylated by a Ca^{2+}-calmodulin dependent protein kinase, regulates the sacroplasmic reticulum Ca^{2+}-ATPase, and provides evidence for the involvement of protein kinase C in the regulation of cardiac contraction.

THE STIMULATION OF PROTEIN KINASE C BY STRUCTURALLY DIVERSE PHORBOL ESTERS

Recent studies by Miyake *et al.* (1984) have shown that the daphnane derivative, merzerein, also stimulates protein kinase C. Leach & Blumberg (1985) have recently shown that in addition to TPA and merzerein, the semisynthetic diterpene derivatives phorbol 12-retinoate 13-acetate, 4-O-methylphorbol 12-myristate 13-acetate and 12-deoxyphorbol 13-isobutyrate activated protein kinase C. The concentration of these four derivatives required for maximal stimulation was 3 μM (compared to 100 nM for TPA).

We have studied the ability of a wide range of tigliane and ingenane diterpenoids to stimulate the activity of protein kinase C (Ellis *et al.*, 1985).

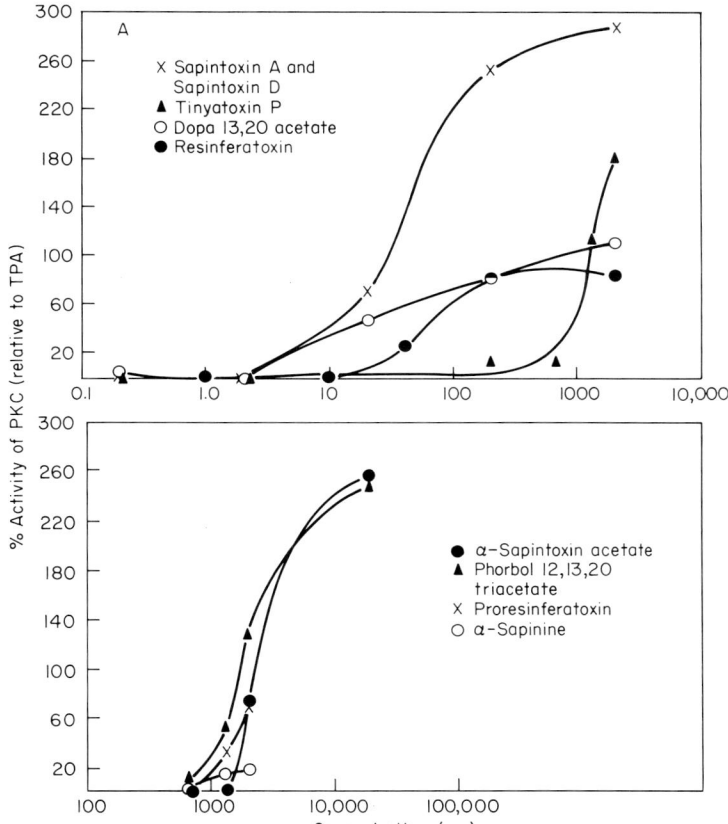

Figure 3. Activation of protein kinase C by phorbol derivatives. Protein kinase C was assayed at 30°C for 10 min in 50 mM, Tris–HCl, pH 7.5, using histone IIIS as substrate in a total volume of 45 μl containing (i) 12.5 mM $MgCl_2$, (ii) 1.5 mM $CaCl_2$, (iii) 0.625 mg ml^{-1} histone, (iv) 0.03 mg ml^{-1} phosphatidylserine, (v) 5 μl enzyme, (vi) 5 μl phorbol derivative (60 pmol to 2 μmol). The reaction was initiated by the addition of 5 μl of (γ^{32}-P-ATP). The reaction was terminated by spotting the reaction mixture onto Whatman phosphocellulose paper, followed by washing six times in water, rinsing in acetone and drying. The effect on protein kinase C activity was in each case compared to that of TPA. The calcium concentration in the assay was controlled by addition of EGTA to obtain maximal stimulation by the diterpenoid derivatives. Fig. 3A shows the activation of the enzyme by biologically active compounds and Fig. 3B shows the effect of biologically inactive derivatives on protein kinase C.

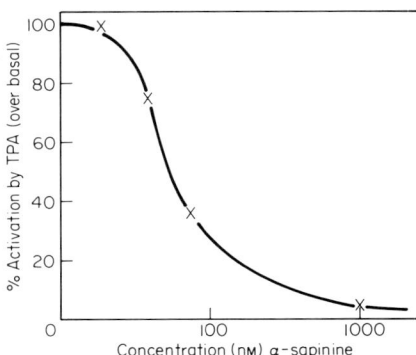

Figure 4. Competitive binding of inactive derivatives to protein kinase C. After addition of components (i) to (v) in the assay (legend to Fig. 3), 2 µl of a non-activating derivative (α-sapinine, sonicated in Tris buffer/acetone 1:1) was added. The mix was incubated for 5 min to allow binding. The active compound (TPA, 2 µl) was added and 5 min allowed for equilibration. The assay was initiated as before by the addition of γ-^{32}P-ATP. The concentration of TPA was kept constant (75 nM) and the concentration of the α-sapinine was varied from 18.75 to 75 nM.

Compounds that are biologically active, stimulated this protein kinase with K_a values of 20–50 nM, (Fig. 3a). In contrast, the biologically inactive compounds, including α-sapinine and phorbol-12,13,20-triacetate stimulated protein kinase C only at high concentrations (Fig. 3b). Typical K_a values in this latter case are approximately two orders of magnitude higher, i.e. 1–2 µM.

Compounds that do not activate protein kinase C or do so only at very high concentrations, still bind competitively with active daphnane and tigliane esters (Fig. 4).

Figure 5. Phosphorylation of proteins *in vivo* in response to phorbol esters of district biological activity.

Harvested GH$_3$ cells (approx. 10^7 cells per 9 ml culture) were washed three times in balanced salt solution (135 mM NaCl, 4.5 nM KCl, 1.5 mM CaCl$_2$, 0.5 mM MgCl$_2$, 5.6 mM glucose, 10 mM HEPES (pH 7.4) 0.1% w/v BSA, 1 mM Na pyruvate) and resuspended in a final solution of 1 ml to give a final concentration of 2×10^7 cells per ml.

Cells were radiolabelled by incubation for 15–60 min with ^{32}P-orthophosphate at 37°C. Five minutes prior to the termination of the labelling period, 125 µl aliquots of the cell suspension were removed and added to a 1:1 suspension of the phorbol ester in 20 mM Tris : acetone to produce a final volume of 200 µl. After incubation at 37°C, cells were pelleted by centrifugation at 200 **g** for 3 min and washed three times in ice cold 0.03 M Na Phosphate, pH 7.6. The suspension was rapidly frozen in dry ice, and thawed in 0.1 M NaF, 0.01 EDTA, 25 mM HEPES (pH 7.5), 0.5 mM β-mercaptoethanol.

After sonication, the suspension was centrifuged at 15 000 **g** for 20 min at 4°C, and the resulting supernatant recentrifuged at 100 000 **g** for 1 h at 4°C to yield 'cytosolic' and 'membrane' fractions. The samples were subjected to 10% SDS-polyacrylamide gel electrophoresis.

The gels were stained with Coomassie blue and destained. After drying, gels were autoradiographed using Kodak X-Omat S X-ray film with an intensifying screen at $-70°$C for 2–7 days.

Fig. 5A shows the autoradiograph obtained from SDS polyacrylamide gel electrophoresis of the cytosolic proteins from cells incubated with the compounds indicated. Fig. 5B shows a densitometer scan of the autoradiographs. The stimulation of phosphorylation (shaded area) over the control level (lower trace in each panel) is indicated. TPA, tetradecanoyl phorbol 13-acetate; DOPPA, 12-deoxyphorbol-13-phenylacetate-20-acetate; SAP A, sapintoxin A (2-*N*-methylaminobenzoyl-4-deoxyphorbol-13-acetate).

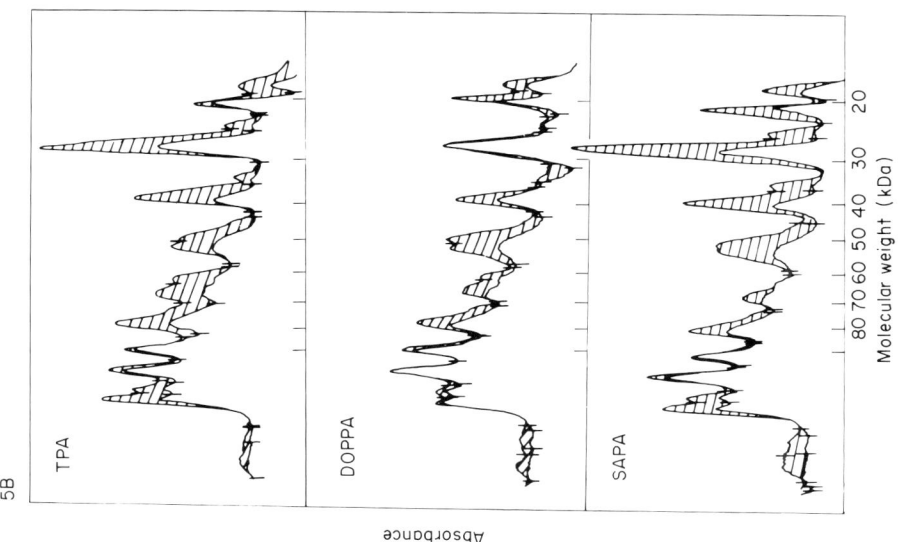
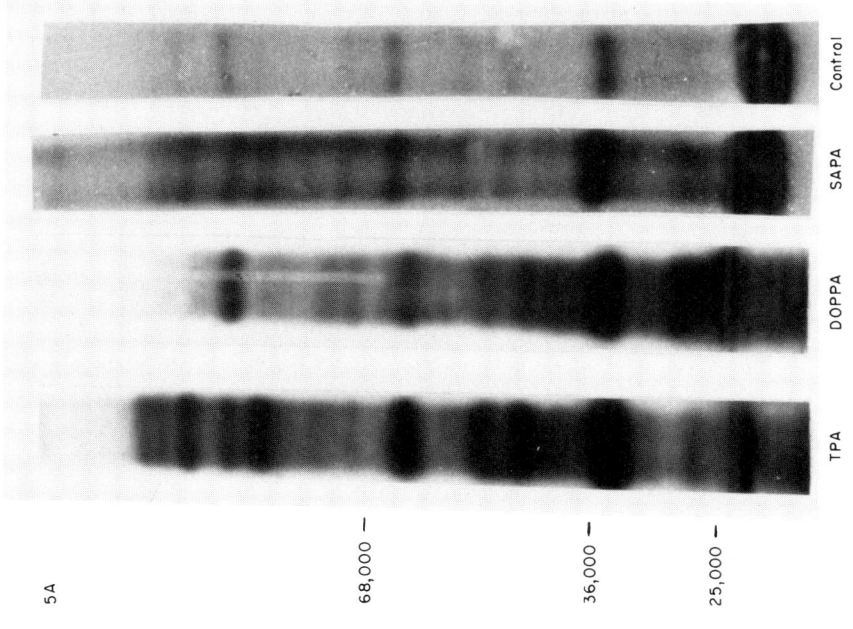

IN VIVO PHOSPHORYLATION IN RESPONSE TO PHORBOL ESTERS OF DISTINCT BIOLOGICAL ACTIVITY

The action of a range of tigliane and ingenane diterpenoids on the *in vivo* phosphorylation of the proteins of GH_3 cells have been studied (Brooks *et al.*, 1987). These phorbol derivatives that have been used have a different spectrum of biological activity (Evans & Taylor, 1983; Edwards, Evans, Barrett & Gordon, 1985). For example, 12-deoxyphorbol-13-phenylacetate-20-acetate (DOPPA) has pro-inflammatory properties, but is not tumour promoting (Edwards *et al.*, 1983a, b; Williamson, Westwick, Kakkar & Evans, 1981).

The GH_3 cell line is a human pituitary cell line that uses phosphatidylinositol turnover to mediate the effects of thyrotropin releasing hormone (Drummond & Raeburn, 1984). We have shown that *in vivo* administration of such phorbol esters can lead to the stimulation of phosphorylation of distinct substrate proteins (Fig. 5). The compounds used in this study were TPA and sapintoxin A (SAPA) that have a wide range of biological effects, and DOPPA, with a much more limited ability.

We are now studying the structure and function of these phosphoproteins whose state of phosphorylation is differentially stimulated by the above derivatives. For example, the protein(s) of $M_r = 29\,000$ that is phosphorylated in the presence of TPA, but not DOPPA, may be specifically involved in mediating the tumour-promoting effects of TPA since the latter compound is not tumour-promoting.

The results suggest that those phorbol esters may act *in vivo* by stimulating protein kinase with distinct substrate specificities. This would explain the ability of certain derivatives to evoke a limited range of biological responses (Ellis *et al.*, 1985). Whether these 'receptors' for phorbol esters are different forms of protein kinase C or distinct protein kinases is the subject of continuing investigation. The recent discovery of three closely related genes for protein kinase C (Parker *et al.*, 1986) has lent weight to the first hypothesis.

CONCLUSION

The activation of protein kinase C by phorbol esters has been described. There is conclusive evidence that this enzyme is the phorbol ester 'receptor'—certainly for the involvement in the effects on platelet aggregation and tumour promotion. There is not complete structure/function correlation between all the biologically active phorbol esters. Some derivatives, with a different spectrum of biological action, stimulate the phosphorylation of a distinct range of substrate proteins. The phosphorylation and alteration in activity of these particular substrates would be implicated in mediating the individual biological effects.

ACKNOWLEDGEMENTS

The work in our laboratory is funded by the Cancer Research Campaign and the S.E.R.C. The generous supply of GH_3 cells by Dr A. H. Drummond is greatly appreciated.

REFERENCES

AHMAD, Z., LEE, F.-T., DEPAOLI-ROACH, A. & ROACH, P. J., 1984. Phosphorylation of glycogen synthase by the Ca^{2+} and phospholipid-activated protein kinase (protein kinase C). *Journal of Biological Chemistry*, 259: 8743–8747.

AITKEN, A., 1986. The biochemical mechanism of action of phorbol esters. In F. J. Evans (Ed.), *Naturally Occurring Phorbol Esters*. Boca Raton: C.R.C. Press.

AKERS, R. F. & ROUTTENBERG, A., 1984. Brain protein phosphorylation *in vitro*: selective substrate action of insulin. *Life Sciences*, 35: 809–813.

ALOYO, V. J., ZWIERS, H. & GISPEN, W. H., 1983. Phosphorylation of B-50 protein by calcium-activated, phospholipid-dependent protein kinase and B-50 protein kinase. *Journal of Neurochemistry*, 41: 649–653.

BERRIDGE, M. J., 1984. Inositol trisphosphate and diacylglycerol as second messengers. *Biochemical Journal*, 220: 345–360.

BERRIDGE, M. J. & IRVINE, R. F., 1984. Inositol trisphosphate, a novel second messenger in cellular signal transduction. *Nature (London)*, 312: 315–321.

BISHOP, J. M., 1983. Cellular oncogenes and retroviruses. *Annual Review of Biochemistry*, 52: 301–354.

BLACKSHEAR, P. J., WEN, L., GLYNN, B. P. & WITTERS, L. A., 1986. Protein kinase C-stimulated phosphorylation *in vitro* of a M_r 80000 protein phosphorylated in response of phorbol esters and growth factors in intact fibroblasts. *Journal of Biological Chemistry*, 261: 1459–1469.

BLUMBERG, P. M., 1980. *In vitro* studies on the mode of action of the phorbol esters. Potent tumour promoters: Part 1. *Critical Reviews of Toxicology*, 8: 153–197.

BLUMBERG, P. M., 1981. *In vitro* studies on the mode of action of the phorbol esters. Potent tumour promoters: Part 2. *Critical Reviews of Toxicology*, 8: 199–234.

BROOKS, S. F., ELLIS, C., EVANS, F. J., EVANS, A. T., MORRICE, N. & AITKEN, A., in press. Tumour-promoting and non-promoting phorbol esters stimulate the phosphorylation of different proteins *in vivo*. *Nature (London)*.

BURNS, C. P. & ROZENGURT, E., 1983. Serum, platelet-derived growth factor, vasopressin, and phorbol esters increase intracellular pH in Swiss 3T3 cells. *Biochemical and Biophysical Research Communications*, 116: 931–938.

CASTAGNA, M., TAKAI, Y., KAIBUCHI, K., SANO, K., KIKKAWA, U. & NISHIZUKA, V., 1982. Direct activation of calcium-activated, phospholipid-dependent protein kinase by tumour promoting phorbol esters. *Journal of Biological Chemistry*, 257: 7847–7851.

COCKCROFT, S., 1981. Does phosphatidylinositol breakdown control the Ca^{2+}-gating mechanism? *Trends in Pharmacological Science*, 2: 340–342.

COCKCROFT, S. & GOMPERTS, B. D., 1985. Role of guanine nucleotide binding protein in the activation of polyphosphoinositide phosphodiesterase. *Nature (London)*, 314: 534–536.

COHEN, P., 1982. The role of protein phosphorylation in neural and hormonal control of cellular activity. *Nature (London)*, 296: 613–620.

COUSSENS, L., PARKER, P. J., RHEE, L., YANG-FENG, T. L., CHEN, E., WATERFIELD, M. D., FRANCKE, U. & ULLRICH, A., 1986. Multiple distinct forms of bovine and human protein kinase C suggest diversity in cellular signaling pathways. *Science*, 233: 859–866.

CUIDAD, C., CAMICI, M., AHMAD, Z., WANG, Y., DEPAOLI-ROACH, A. & ROACH, P. J., 1984. Control of glycogen synthase phosphorylation in isolated rat hepatocytes by epinephrine, vasopressin and glucagon. *European Journal of Biochemistry*, 142: 511–520.

DRUMMOND, A. H. & RAEBURN, C. A., 1984. The interaction of lithium with thyrotropin-releasing hormone-stimulated lipid metabolism in GH_3 pituitary tumour cells. *Biochemical Journal*, 224: 129–136.

EDWARDS, M. C., EVANS, F. J., BARRETT, M. L. & GORDON, D., 1985. Structural correlations of phorbol ester induced stimulation of PGE_2 production by human rheumatoid synovial cells. *Inflammation*, 9: 33–38.

EDWARDS, M. C., NOURI, A. M. E., GORDON, D. & EVANS, F. J., 1983a. Tumour-promoting and nonpromoting proinflammatory esters act as human lymphocyte mitogens with different sensitivities to inhibition by cyclosporin A. *Molecular Pharmacology*, 23: 703–708.

EDWARDS, M. C., TAYLOR, S. E., WILLIAMSON, E. M. & EVANS, F. J., 1983b. New phorbol and deoxyphorbol esters: Isolation and relative protencies in inducing platelet aggregation and erythema of skin. *Acta Pharmacologia et Toxicologia*, 53: 177–187.

ELLIS, C., MORRICE, N., AITKEN, A., PARKER, P. J. & EVANS, F. J., 1985. Activation of protein kinase C by tumour-promoting and non-promoting phorbol esters. *Journal of Pharmacy and Pharmacology*, 37: 23.

EVANS, F. J. & TAYLOR, S. E., 1983. Pro-inflammatory tumour promoting and anti-tumour diterpenes of the plant families Euphorbiaceae and Thymeleaceae. In W. Herz, H. Grisebach & G. W. Kirby (Eds), *Progress in the Chemistry of Natural Products*:1–99. New York: Springer Verlag.

FEINSTEIN, M. B. & HADJIAN, R. A., 1982. Effects of the calmodulin antagonist trifluoperazine on stimulus induced calcium mobilisation aggregation, secretion and protein phosphorylation in platelets. *Molecular Pharmacology*, 21: 422–431.

FEUERSTEIN, N. & COOPER, H. L., 1984. Rapid phosphorylation-dephosphorylation of specific proteins induced by phorbol ester in HL-60 cells, further characterisation of the phosphorylation of 17-kilodalton and 27-kilodalton proteins in myeloid leukemic cells and human monocytes. *Journal of Biological Chemistry*, 259: 2782–2788.

GABRIELLI, B., WETTENHALL, R. E. H., KEMP, B. E., QUINN, M. & BIZONOVA, L., 1984. Phosphorylation of ribosomal protein S6 and a peptide analogue of S6 by a protease activated kinase isolated from rat liver. *FEBS Letters, 175:* 219–226.

GARRISON, J. C., JOHNSON, D. E. & CAMPANILE, C. P., 1984. Evidence for the role of phosphorylase kinase, protein kinase C and other Ca^{2+} sensitive protein kinases in the response of hepatocytes to angiotensin II and vasopressin. *Journal of Biological Chemistry, 259:* 3283–3292.

GILMORE, T. & MARTIN, G. S., 1983. Phorbol ester and diacylglycerol induce protein phosphorylation at tyrosine. *Nature (London), 306:* 487–490.

GOMPERTS, B. D., 1983. Involvement of guanine nucleotide-binding protein in the gating of Ca^{2+} by receptors. *Nature (London), 306:* 64–66.

HARDIE, D. G., CARLING, D., FERRARI, S., GUY, P. S. & AITKEN, A., 1986. Characterisation of the phosphorylation of rat mammary ATP-citrate lyase and acetyle-CoA carboxylase by Ca^{2+} and calmodulin-dependent multiprotein kinase and Ca^{2+} and phosphospholipid dependent protein kinase. *European Journal of Biochemistry, 157:* 553–561.

HUNTER, T., LING, N. & COOPER, J. A., 1984. Protein kinase C phosphorylation of the EGF receptor at a threonine residue close to the cytoplasmic face of the plasma membrane. *Nature (London), 311:* 480–483.

HUNTER, T. & SEFTON, B. M., 1982. Protein kinases and viral transformation. In P. Cohen & S. Van Heyningen, (Eds), *Molecular Action of Toxins and Viruses:* 337–366. Amsterdam: Elsevier Biomedical Press.

IMAZU, M., STRICKLAND, W. G., CHRISMAN, T. D. & EXTON, J. H., 1984. Phosphorylation and inactivation of liver glycogen synthase by liver protein kinases. *Journal of Biological Chemistry, 259:* 1813–1821.

IWASA, Y. & HOSEY, M. M., 1984. Phosphorylation of cardiac sarcolemma proteins by the calcium-activated phospholipid-dependent protein kinase. *Journal of Biological Chemistry, 259:* 534–540.

KAIBUCHI, K., TAKAI, Y. & NISHIZUKA, Y., 1981. Cooperative roles of various membrane phospholipids in the activation of calcium-acitivated phospholipid dependent protein kinase. *Journal of Biological Chemistry, 256:* 7146–7149.

KELLEHER, D. J., PESSIN, J. E., RUOHO, A. E. & JOHNSON, G. L., 1984. Phorbol ester induces desensitisation of adenylate cyclase and phosphorylation of the β-adrenergic receptor in turkey erythrocytes. *Proceeding of the National Academy of Sciences USA, 81:* 4316–4320.

KIKKAWA, U., TAKAI, Y., MINAKUCHI, R., INOHARA, S. & NISHIZUKA, Y., 1982. Calcium-activated, phospholipid-dependent protein kinase from rat brain. Subcellular distribution, purification and properties. *Journal of Biological Chemistry, 257:* 13341–13348.

KIKKAWA, U., TAKAI, Y., TANAKA, Y., MIYAKE, R. & NISHIZUKA, Y., 1983. Protein kinase C as a possible receptor protein of tumour promoting phorbol esters. *Journal of Biological Chemistry, 258:* 11442–11445.

KRAFT, A. S. & ANDERSON, W. B., 1983. Phorbol esters increase the amount of Ca^{2+}, phospholipid dependent protein kinase associated with plasma membrane. *Nature (London), 301:* 621–623.

KUO, J. F., ANDERSON, R. G. G., WISE, B. C., MACKERLOVA, L., SALMONSSON, I., BRACKETT, N. L., KATOH, N., SHOJI, M. & WRENN, R. W., 1980. Calcium dependent protein kinase. Widespread occurrence in various tissues and phyla of the animal kingdom and comparison of effects of phospholipid, calmodulin and trifluoperazine. *Proceedings of the National Academy of Sciences, U.S.A., 77:* 7039–7043.

LAPETINA, E. G., 1982. Regulation of arachidonic acid production: role of phospholipases C and A_2. *Trends in Pharmacological Science, 3:* 115–118.

LEACH, K. L. & BLUMBERG, P. M., 1985. Modulation of protein kinase C activity and ^{3}H-phorbol-12,13-dibutyrate binding by various tumour promoters in mouse brain cytosol. *Cancer Research, 45:* 1958–1963.

LYONS, R. M. & ATHERTON, R. M., 1979. Characterisation of a platelet protein phosphorylated during the thrombin-induced release reaction. *Biochemistry, 18:* 544–552.

MAHADEVAPPA, V. G. & HOLUB, B. J., 1983. Degradation of different molecular species of phosphatidylinositol in thrombin-stimulated human platelets. Evidence for preferential degradation of 1-acyl-2-arachidonyl species. *Journal of Biological Chemistry, 258:* 5337–5339.

MAHADEVAN, L. C., AITKEN, A., HEATH, J. & FOULKES, J. G., in press. Embryonal carcinoma-derived growth factor activates protein kinase C *in vivo* and *in vitro*. *EMBO Journal*.

MALAISSE, W. J., LEBRUN, P., HERCHUELZ, A., SENER, A. & MALAISSE-LAGAIE, F., 1983. Synergistic effect of a tumour-promoting phorbol ester and a hypoglycemic sulphonylurea upon insulin release. *Endocrinology, 113:* 1870–1877.

MASTRO, A. M. & SMITH, M. C., 1983. Calcium-dependent activation of lymphocytes by ionophore, A23187 and a phorbol ester tumour promoter. *Journal of Cellular Physiology, 116:* 51–56.

MIYAKE, R., TANAKA, Y., TSUDA, T., KAIBUCHI, K., KIKKAWA, U. & NISHIZUKA, Y., 1984. Activation of protein kinase C by the non-phorbol tumour promoter, mezerein. *Biochemical and Biophysical Research Communications, 121:* 649–656.

MICHELL, R. H., 1983. Ca^{2+} and protein kinase C: two synergistic cellular signals. *Trends in Biochemical Science, 8:* 263–265.

MICHELL, R. H., 1984. Oncogenes and inositol lipids. *Nature (London), 308:* 770.
MOOLENAAR, W. H., TSIEN, R. Y., VAN DER SAAG, P. T. & DE LAAT, S. W., 1983. Na$^+$/H$^+$ exchange and cytoplasmic pH in the action of growth factors in human fibroblasts. *Nature (London), 304:* 645–648.
MORI, T., TAKAI, Y., MINAKUCHI, R., YU, B. & NISHIZUKA, Y., 1980. Inhibitory action of chlorpromazine, dibucaine and other phospholipid interacting drugs on calcium activated phospholipid dependent protein kinase. *Journal of Biological Chemistry, 255:* 8378–8380.
NAKA, M., NISHIKAWA, M., ADELSTEIN, R. S. & HIDAKA, H., 1983. Phorbol ester-induced activation of human platelets is associated with protein kinase C phosphorylation of myosin light chains. *Nature (London), 306:* 490–492.
PARKER, P. J., STABEL, S. & WATERFIELD, M. D., 1984. Purification to homogeneity of protein kinase C from bovine brain-identity with the phorbol ester receptor. *EMBO Journal, 3:* 953–959.
PARKER, P. J., COUSSENS, L., TOTTY, N., RHEE, L., YOUNG, S., CHEN, E., STABEL, S., WATERFIELD, M. D. & ULLRICH, A., 1986. The complete primary structure of protein kinase C—the major phorbol ester receptor. *Science, 233:* 853–859.
SANO, K., KAIBUCHI, K., HOSHIJIMA, M., YAMANISHI, J., KIKKAWA, U., TAKAI, Y. & NISHIZUKA, Y., 1983. Phospholipid turnover as transmembrane signalling for protein phosphorylation and platelet activation. In B. Connor Johnson (Ed.), *Posttranslational Covalent Modifications of Proteins:* 195–223. New York: Academic Press.
SHARKEY, N. A., LEACH, K. L. & BLUMBERG, P. M., 1984. Competitive inhibition by diacylglycerol of specific phorbol ester binding. *Proceedings of the National Academy Sciences, U.S.A., 81:* 607–610.
ROACH, P. J. & GOLDMAN, M., 1983. Modification of glycogen synthase activity in isolated rat hepatocytes by tumour-promoting phorbol esters: Evidence for differential regulation of glycogen synthase and phosphorylase. *Proceedings of the National Academy of Sciences, U.S.A., 80:* 7170–7172.
ROSOFF, P. M., STEIN, L. F. & CANTLEY, L. C., 1984. Phorbol esters induce differentiation in a pre-B-lymphocyte cell line by enhancing Na$^+$/H$^+$ exchange. *Journal of Biological Chemistry, 259:* 7056–7060.
TAKAI, Y., KISHIMOTO, A., INOUE, Y. & NISHIZUKU, Y., 1977. Studies on a cyclic-nucleotide-independent protein kinase and its proenzyme in mammalian tissue I. Purification and characterisation of an active enzyme from bovine cerebellum. *Journal of Biological Chemistry, 252:* 7610–7616.
TANAKA, C., TANIYAMA, K. & KUSUNOKI, M., 1984. A phorbol ester and A23187 act synergistically to release acetylcholine from the guinea pig ileum. *FEBS Letters, 175:* 165–169.
TREVILLYAN, J. M., KULKARNI, R. K. & BYUS, C. V., 1984. Tumour-promoting phorbol esters stimulate the phosphorylation of ribosomal protein S6 in quiescent Reuber H35 hepatoma cells. *Journal of Biological Chemistry, 259:* 897–902.
TURNER, R. S., CHOU, J. C-H., MAZZEI, J. G., DEMBURE, P. & KUO, J. F., 1984. Phospholipid-sensitive Ca^{2+}-dependent protein kinase preferentially phosphorylates serine-115 of bovine myelin basic protein. *Journal of Neurochemistry, 43:* 1257–1264.
WERTH, D. K. & PASTAN, I., 1984. Vinculin phosphorylation in response to calcium and phorbol esters in intact cells. *Journal of Biological Chemistry, 259:* 5264–5270.
WILLIAMSON, E. M., WESTWICK, J., KAKKAR, V. V. & EVANS, F. J., 1981. Studies on the mechanism of action of 12-deoxyphorbolphenylacetate, a potent platelet aggregating tigliane ester. *Biochemical Pharmacology, 30:* 2961–2966.

Botanical Journal of the Linnean Society (1987), 94: 265–282. With 4 figures

A review of the evidence from *in vitro* and *in vivo* studies for a role for phorbol ester tumour promoters from the Euphorbiales in the selection and clonal expansion of specific cell populations

ANNE R. KINSELLA

Paterson Institute for Cancer Research, Christie Hospital and Holt Radium Institute, Manchester M20 9BX

Received September 1986, accepted for publication October 1986

KINSELLA, A. R., 1987. **review of the evidence from *in vitro* and *in vivo* studies for a role for phorbol ester tumour promoters from the Euphorbiales in the selection and clonal expansion of specific cell populations.** The identification of the active principle of croton oil as 12-*O*-tetradecanoylphorbol-13-acetate (TPA) and the advent of *in vitro* studies, facilitated the identification of a plethora of biochemical effects, certain of which could be correlated with biological and differentiation end points. Evidence is presented for the ability of TPA to select and expand specific cell populations in two different cell culture systems. The roles of differentiation state and second messenger systems will be discussed in relation to models of multi-step carcinogenesis.

ADDITIONAL KEY WORDS:—Clonal expansion – Chromosomal events – Differentiation – Multi-step carcinogenesis – Phorbol esters – 12-*O*-tetradecanoylphorbol-13-acetate (TPA) – Tumour promotion.

CONTENTS

Introduction	265
In vitro studies	267
Elimination of metabolic co-operation	268
Susceptibility of skin fibroblasts from individuals with a genetic predisposition to cancer to TPA-induced transformation	268
Selection and clonal expansion of a specific keratinocyte cell populations	271
Oncogenes and TPA responsiveness	275
TPA-induced chromosomal aberrations	276
TPA and the phosphatidylinositol second messenger system	277
Conclusions	278
Acknowledgements	278
References	278

INTRODUCTION

One of the most extensively studied systems in terms of the multi-step nature of carcinogenesis is the so-called two-stage mouse skin system first defined by

Berenblum (1941). Berenblum discovered that a group of naturally occurring substances found in croton oil, the seed oil of the leafy Euphorbiaceous shrub *Croton tiglium* L., had little or no carcinogenic activity when applied alone topically to mouse skin, but when applied repeatedly to mouse skin following a low dose of carcinogen, such as benzopyrene or dimethylbenzanthracene (DMBA), greatly increased the number of skin tumours that occurred, and shortened the lag-time for tumour appearance. The most abundant and active tumour promoting principle of croton oil was identified more than twenty years later as a diester, 12-O-tetradecanoylphorbol-13-acetate (TPA), of the parent tetracyclic diterpene alcohol phorobol (Hecker, 1968; Van Duuren & Sivalc, 1968). Thus, 'initiation' is achieved by a single treatment with a sub-carcinogenic dose of a carcinogenic agent, whilst 'promotion' is a long-term process comprising repeated promoter treatments (Friedewald & Rous, 1944; Boutwell, 1974). The observation that the promoter treatment could be delayed for several months and still be effective (Berenblum, 1941; Boutwell, 1974) lead to the assumption that latent tumour cells were present in the epidermis as a result of the treatment with a single small dose of carcinogen, and that the induction process (initiation) and the growth stimulating process (promotion) depended on two very different classes of stimuli. However at the present time, although it is quite clear that tumour promoters differ strikingly in their basic biological properties from initiating agents (Table 1) and have been used extensively *in vivo* and more recently in *in vitro* studies, there is no single, simple mechanism which explains their role in the carcinogenic process.

Most tumour promoters are irritants that produce epidermal hyperplasia (Berenblum, 1944; Saffiotti & Shubik, 1963; Frei & Stephens, 1968), but not all the irritants or hyperplasiogenic agents are promoters. The graded tumour promoting activities of a series of phorbol esters (TPA and related esters) can be correlated with their ability to induce sustained stimulation of DNA, RNA and protein synthesis in mouse epidermis associated with epidermal hyperplasia (Baird, Sedgwick & Boutwell, 1971; Raick, 1973). Indeed, although *in vivo* studies have revealed differences between the different strains of mice in the extent of their hyperplasiogenic response to TPA, a consistent correlation has been observed between epidermal hyperplasia and tumour promotion (Sisskin, Gray & Barrett, 1982; DiGiovanni *et al.*, 1984). The subdivision of tumour

Table 1. Comparison of the Biological properties of initiating and promoting agents (Weinstein *et al.*, 1978; Kinsella, 1986)

Initiating agents	Promoting agents
1. Carcinogenic by themselves 'solitary carcinogens'	1. Not carcinogenic alone
2. Must be given before a promoting agent	2. Must be given after an initiating agent
3. A single exposure is sufficient	3. Require prolonged and repeated exposure
4. Action is irreversible and additive	4. Action is reversible up to a certain stage
5. Yield electrophiles that bind covalently to macromolecules	5. No evidence of co-valent binding of macro molecules
6. Mutagenic	6. Not mutagenic
7. Do not alter gene expression at low initiating doses	7. Alter gene expression
8. DNA binding are resultant repair are detectable responses	8. Plethora of biological and biochemical responses

promotion itself into two clearly defined stages (Boutwell, 1964; Furstenberger et al., 1981; Slaga et al., 1982) and the development of TPA analogues such as 12-O-retinoylphobol-13-acetate (RPA) (Furstenberger, Berry, Sorg & Marks, 1981) has lead to further refinement of these theories as will be discussed later.

IN VITRO STUDIES

The advent of extensive *in vitro* studies showed TPA and related phorbol esters to modulate specific enzyme activities (Colburn, Lau & Head, 1975; O'Brien, Simsiman & Boutwell, 1975; Wigler & Weinstein, 1976; Yuspa et al., 1976 and Table 2). TPA was shown to mimic *in vitro* in cellular transformation assay systems the observation made *in vivo* in mouse skin (Mondal, Brankow & Heidelberger, 1976; Kinsella, pers. comm). TPA was confirmed to be non-mutagenic in *in-vitro* mutagenesis assay systems (Kinsella 1981). In fact, the action of TPA was shown to resemble that of hormones or growth factors in that it exerted its highly pleiotropic gene regulatory effects (Table 2) at nanomolar concentrations via highly specific phorbol ester receptors (Delclos, Nagle & Blumberg, 1980; Dunphy, Kochenburger, Castagna & Blumberg, 1981). However it is probably the *in vitro* properties of elimination of metabolic co-operation (Murray & Fitzgerald, 1979; Yotti, Chang & Trosko 1979; Kinsella, 1981), transformation of human skin fibroblasts from patients with an inherited predisposition to cancer (Kopelovich, Bias & Helson, 1979; Gainer, Schor & Kinsella, 1984), selection and clonal expansion of specific keratinocyte cell populations (Yuspa, Ben, Hennings & Lichti, 1982; Kinsella, unpublished observations) and the induced chromosomal changes (Kinsella & Radman, 1978; Nagasawa & Little, 1979; Fusenig & Dzarlieva, 1982; Kinsella, Gainer & Butler, 1983; Gainer & Kinsella, 1983; Gainer et al., 1984) which have made the greatest overall contribution to defining a role for tumour promoters in a general model for multi-step carcinogenesis.

Table 2. *Evidence that tumour promoters act as reversible depressors or gene regulators*

1. Increase in plasminogen activator activity	Wigler & Weinstein (1976)
2. Increase in ornithine decarboxylase activity	O'Brien et al. (1975), Yuspa et al. (1976)
3. Depression of *iso*proteronol-stimulated cyclic-AMP	Belman & Troll (1974), Mufson et al. (1977)
4. Depression of histidase activity	Colburn et al. (1975)
5. Increase in prostaglandin synthesis	Bresnick et al. (1979)
6. Induction of the expression of viral genes	Zur Hausen et al. (1978), Amtmann & Sauer (1982)
7. Induction of specific proteins	Laszlo et al. (1981), Cabral et al. (1981), Bazill et al. (1984)
8. Modulation of differentiation	Diamond et al. (1978), Cohen et al. (1977), Ishii et al. (1978), Huberman & Callahan (1979), Huberman et al. (1979), Reiners & Slaga (1983), Kulesz-Martin et al. (1980)
9. Enhanced transcription	Murdoch et al. (1985), Angel et al. (1985, 1986)
10. Modulation of expression of known c-*oncs*	Greenberg & Ziff (1984)
11. Induction of gene amplification	Varshavsky (1981), Barsoum & Varshavsky (1983)

ELIMINATION OF METABOLIC CO-OPERATION

The observation that low concentrations of the tumour promoter TPA and related phorbol ester promoters could eliminate cell–cell communication or metabolic co-operation between adjacent cells at high density (Murray & Fitzgerald, 1979; Yotti et al., 1979; Kinsella, 1981), lead very swiftly to the suggestion that TPA could release cells from the constraints of their normal environment. Thus, TPA-induced elimination of metabolic co-operation may reflect or coincide with the induction of a protease activity that releases cells from contact inhibition of growth (Kinsella, 1981).

This discovery helped in turn with the interpretation of data obtained from studies with TPA using skin fibroblasts from individuals who exhibit a familial predisposition to cancer, such as retinoblastoma (RB) and familial polyposis coli (FCP) (Kopelovich, 1979; Gainer et al., 1984).

SUSCEPTIBILITY OF SKIN FIBROBLASTS FROM INDIVIDUALS GENETICALLY PREDISPOSED TO CANCER TO TPA-INDUCED TRANSFORMATION

In cancers with such a clearly defined genetic basis it can be assumed that skin fibroblasts (as representative somatic cells) might carry a mutation analogous to the initiation event of classical two-stage carcinogenesis (Gainer et al., 1984). Retinoblastoma (RB) is an intraocular tumour of early childhood the tendency to which may be inherited (40% of cases) in an autosomal dominant fashion (Jensen & Miller, 1971; Knudson, 1971). Familial polyposis coli (FPC) also exhibits a dominant mode of inheritance characterized by the presence of numerous colorectal adenomatous polyps, one or a few of which might give rise to colon adenocarcinoma at a younger age and higher frequency than occurs in the normal population (Muto, Bussey & Morson, 1975). Thus, in the genetic forms of the diseases cited above, at least one mutation is inherited or occurs germinally and it may be supposed that another event(s) must occur in the somatic cells for tumour formation or transformation to occur. Skin fibroblasts from such individuals might therefore be more susceptible to transformation by a tumour promoter alone.

In order to examine this possibility, the effects of the tumour promoter TPA on skin fibroblasts from normal individuals were compared with those from patients with hereditary RB and FCP, with regard to the ease with which properties of the transformed phenotype were induced. Preliminary investigation of the cell strains to be studied for existing properties, which would be consistent with a transformed phenotype such as (1) ability to grow and clone well in low serum (Ts'O, 1982), (2) altered migration patterns in collagen gels (Schor et al., 1982) and (3) tumourigenicity in nude mice, showed none of the RB or FPC cell strains to exhibit a strong predisposition to the transformed phenotype. Exposure of two hereditary RB cell lines to TPA at each medium change and passage, followed by analysis for anchorage independence and altered migration patterns in collagen gels, showed TPA to have no influence on the induction of the transformed phenotype. Transformation of human cells is notoriously difficult (Kakunaga, 1978) and these results might have been dismissed as inconclusive if it had not been for the observations made with the FPC fibroblast cell strains. Treatment of cell strains FPC6 and FPC7 with TPA gave rise to anchorage independent colonies at passages 2, and 2 and 8, respectively (Gainer et al., 1984; Table 3). Analysis of the same FPC cell strains for changes in their migration pattern in collagen gels during the transformation experiment showed the migration patterns to fluctuate between normal and abnormal regardless of treatment. Also, the abnormal migration patterns, which can be taken as an indication of the transformed phenotype, failed to correlate with the ability of certain cells in the population to exhibit anchorage independent growth in methyl cellulose medium (Table 4). Consistent with this, none of the isolated anchorage independent clones isolated from methyl cellulose back into normal culture medium and then assayed for their migration patterns in collagen gels (data not shown) or injected into nude mice (Table 5), exhibited altered migration patterns or gave rise to tumours. Thus, it seems that treatment of certain susceptible strains of FPC skin fibroblasts, with the tumour promoter TPA only, brings out partial, reversible transformation (Gainer et al., 1984). This is not so surprising in the light of the recent data and theories which propose that three stages are required for transformation (Hennings et al., 1983). In this context the susceptible FPC skin fibroblasts can be considered analogous

Table 3. Anchorage independent colonies formed by skin fibroblasts from familial polyposis coli (FPC) patients following treatment with TPA

Cell strain	Passage No.	No. of anchorage independent colonies/plate	
		Control	0.1 ug ml^{-1} TPA
FPC5a	0	0	0
	2	0	0
	4	0	0
	6	0	0
	8	0	0
	10	0	0
FPC6c	0	0	0
	2	0	6
	4	0	0
	6	0	0
	8	0	0
	10	0	0
FPC7c	0	0	0
	2	0	15
	4	0	0
	6	0	0
	8	0	1
	10	0	0

Three 25 cm^2 flasks per treatment were seeded at a density of 1×10^5 cell/flask 5 ml HAMS F10 growth medium. The monolayers were medium changed after 1 week and the treatments repeated for chronic administrations. After a fortnight the cells were subcultured and re-seeded at 1×10^5 cells/25 cm^2 flask with the appropriate treatments and if appropriate cells taken to assay for anchorage independence, migration patterns in collagen gels and tumourigenicity in nude mice (Gainer et al., 1984).

Table 4. Migration patterns of FPC strains in collagen gels following treatment with TPA during experiment to assay for transformation

		Migration pattern*	
Cell strain	Passage No.	Control	$0.1\ \mu g/ml^{-1}$ TPA
FPC 5a	2	N	N
	4	N	A^{++}
	7	A^+	N
FPC 6c	2	N	N
	4	A^{++}	N
	7	N	A^{++}
	9	A^{++}	—
FPC 7c	2	N	N
	4	A^+	N^{++}
	7	N	N
	9	A^+	A^+

* N denotes a normal migration pattern. A denotes an abnormal migration pattern and + denotes the degree of deviation from the normal pattern. The normal pattern of migration occurred when relatively more cells migrated into the gel at the lower cell density (10^4 cells/plate) than at the higher density (10^5 cells/plate). The abnormal pattern occurs when more cells migrate into the gel at the higher density.

Table 5. Tumorigenicity in nude mice of FPC cell strains following treatment with TPA

Cell strain	No. of mice injected	Tumours
FF_1 (untreated)	5	0
FPC 6c (untreated)	8	0
FPC 7c (untreated)	8	0
FPC 6c ($0.1\ \mu g\ ml^{-1}$ TPA)	6	0
FPC 7c ($0.1\ \mu g\ ml^{-1}$ TPA)	6	0
FPC 7I$^+$ (anchorage independent clone maintained + TPA)	6	0
FPC 7I$^-$ (anchorage independent clone maintained − TPA)	7	0

1×10^6 cells in 0.2 ml sterile saline were injected per mouse and the mice examined weekly for appearance of subcutaneous tumours.

to 'initiated' mouse epidermal cells, where treatment of initiated mouse skin with TPA readily brings about the induction of benign papillomas and only rarely complete carcinomas. Hennings et al. (1983) proposed that a second "mutational event" was required in mouse skin to convert the papillomas to carcinomas, which is probably an accurate assumption for the TPA-induced anchorage-independent FPC cell clones described in Table 3. It would appear from these data and those of Reddy & Fialkow (1983), on the clonal nature of the initiation–promotion versus initiation–initiation-induced carcinomas, that TPA is responsible for cell selection and clonal expansion, very rarely being responsible for taking the process to completion.

SELECTION AND CLONAL EXPANSION OF A SPECIFIC MOUSE KERATINOCYTE CELL POPULATION

Nowhere is the ability of TPA to select and allow the expansion of a certain cell population better demonstrated than in the mouse keratinocyte system first described by Kulesz-Martin et al. (1980) and confirmed in this laboratory. The normal pattern of epithelial differentiation characterized by limited proliferation, keratin synthesis, desmosome formation and terminal differentiation (Kulesz-Martin, Koehler, Hennings & Yuspa, 1980) is markedly altered by reduction of the ionic calcium (Ca^{2+}) concentration from the 1.2–1.4 mM levels of normal culture medium to less than 0.1 mM. Under these conditions, desmosome formation and stratification are prevented and the keratinocytes continue to proliferate to form a uniform layer of polygonal 'basal' cells. Treatment of the keratinocyte cultures on day 2 with an initiating dose of carcinogen such as dimethylbenz[a]anthracene (DMBA), followed by maintenance in low Ca^{2+} (<0.1 mM) (Fig. 1) for 4–9 weeks, confers on a proportion of the population a resistance to the terminal differentiation that normally accompanies a switch to high Ca^{2+}. Instead of terminally differentiating and sloughing from the dish within 72 h, the resistance population exhibits an altered morphology (compare Figs 1 and 2) and continues to proliferate indefinitely. Addition of TPA to the basal cell population results in a heterogenous response (Yuspa et al. 1982, and confirmed by this laboratory) with the induction of rounded cells which undergo TPA-induced terminal differentiation (Fig. 3), whilst the remaining basal cell population is resistant to the effects. These TPA resistant cells are also resistant

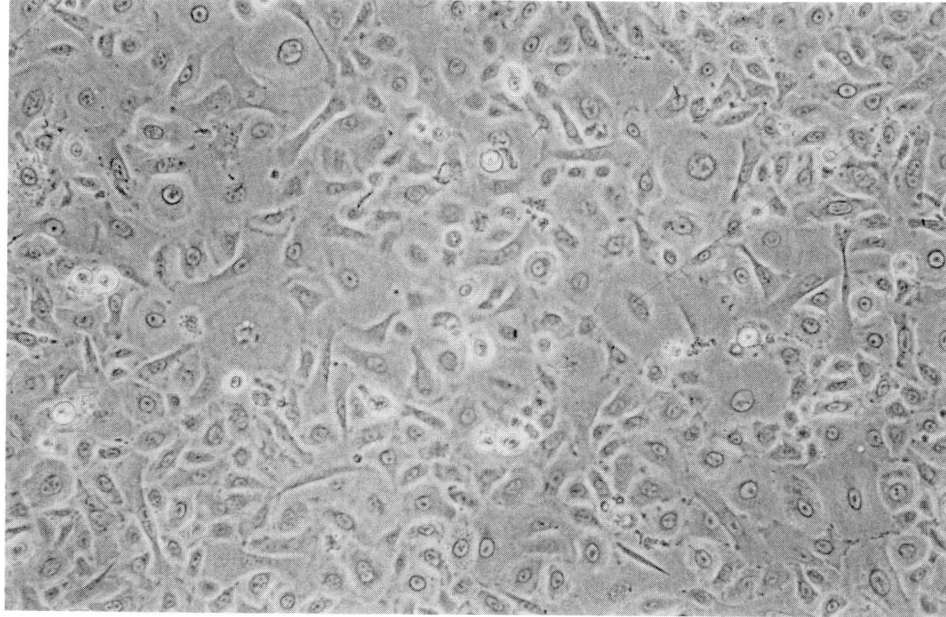

Figure 1. Balb/C mouse keratinocyte culture, treated with DMBA day 2 and maintained in low Ca^{2+} (0.04 mM) for 4 weeks.

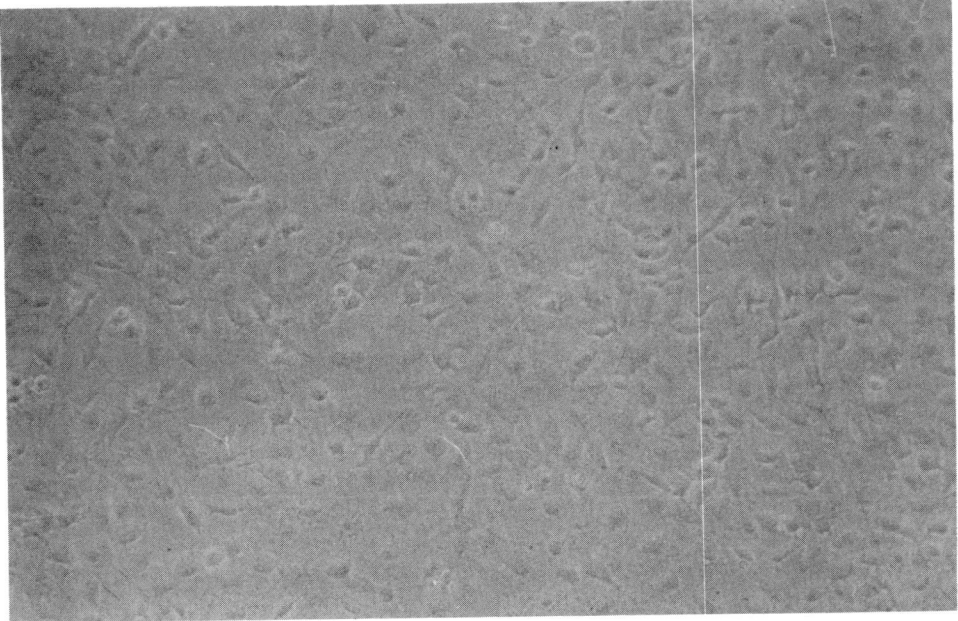

Figure 2. Balb/C mouse keratinocyte culture treated with DMBA day 2 after switch to high Ca^{2+} (1.2 mM) week 9.

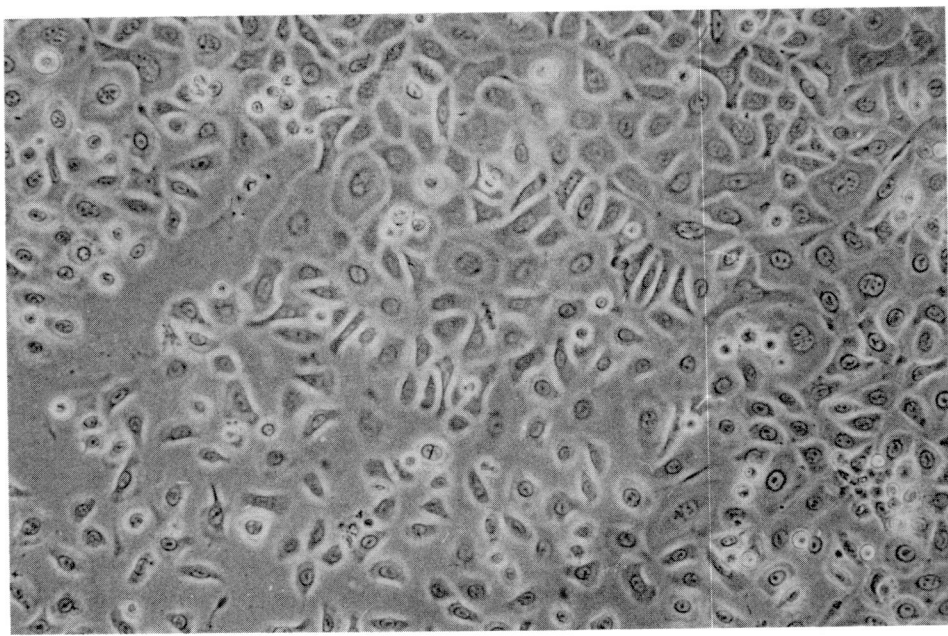

Figure 3. Balb/C mouse keratinocyte culture treated with TPA day 2 and continuously thereafter. Maintained in low Ca^{2+} medium for 4 weeks.

to the terminal differentiation induced by a switch to a high Ca^{2+} and to differentiation induced by suspension in methocellulose or cornified envelope formation (Parkinson Grabham & Emmerson, 1983) but do not exhibit anchorage independence or tumorigenicity in nude mice. That the cell population being examined were indeed keratinocytes, and not any of the other cell populations present in the epidermis, was confirmed by labelling with fluorescein tagged anti-human callous antibody (Fig. 4 A & B). Evidence that the low Ca^{2+} population that resisted the switch to high Ca^{2+} was indeed a basal cell population was obtained when the cell population which switched to high Ca^{2+} failed to stain with the fluoroscein-tagged RSKE 60 anti-keratin antidody specific for the suprabasal cell population (data not shown), confirming that carcinogen-altered cells are arrested at an early stage in their differentiation.

In summary, these data suggest that the primary effect of initiation is to alter the normal differentiation commitment of the basal cells or stem cells. Promoters presumably then act on both the 'initiated' and the surrounding cells, as in the keratinocyte system cited above and reported previously by Yuspa and co-workers, to redistribute cell populations as in regenerative hyperplasia (Furstenberger et al., 1981).

Tumour promotion in combined *in vivo/in vitro* studies has been shown to parallel the hyperplasiogenic response to chronic TPA administration (Sisskin et al., 1982; DiGiovanni, Pritchett, Dencina & Diamond, 1984). However, the degree of the hyperplasia following single TPA exposures does not show such a correlation, suggesting that hyperplasia might only be a component of the period following the first week of promotion. Recent studies with neonatal mice (Fustenberger et al., 1985) which are resistant to the first stage of promotion, shows them also to be resistant to the hyperplasiogenic response to TPA, re-establishing the necessity for hyperplasia in the early events of promotion. A fact that is supported by the observation from Slaga's group (Klein-Szanto, Major & Slaga, 1980) that steroids (such as fluocinolone acetonide) which are potent inhibitors of epidermal hyperplasia are potent inhibitors of both stage I and stage II promotion. What is masked, therefore, by the hyperplasiogenic response of stage I promoters that distinguishes them from stage II promoters and from non-promoting hyperplasiogenic agents?

The differential resistance to TPA-induced keratinocyte terminal differentiation described above and by other workers for mouse keratinocytes, for human keratinocytes and bronchial epithelial cells (Yuspa et al., 1982; Parkinson et al., 1983; Willey, Moser, Lechner & Harris, 1984) immediately provides the basis for a selective growth advantage. The observation that TPA selects for a specific 'initiated' keratinocyte population with an altered pattern of differentiation has its support and parallels in much earlier studies in mouse skin. Raick (1973, 1974) showed a single application of TPA to mouse skin to induce an ordered sequence of ultrastructural changes in cells of all layers of the epidermis, including the induction of dark cells. Dark basal cells are found in large number in embryonic skin, but rarely in adult skin and are thought to be primitive stem cells. In fact, TPA was shown to be five times more efficient at dark cell induction than the weakly promoting diterpene ester mezerein, a potent second stage promoter (Klein-Szanto et al., 1980). Moreover, the synthetic steroid fluocinolone acetonide in addition to inhibiting epidermal

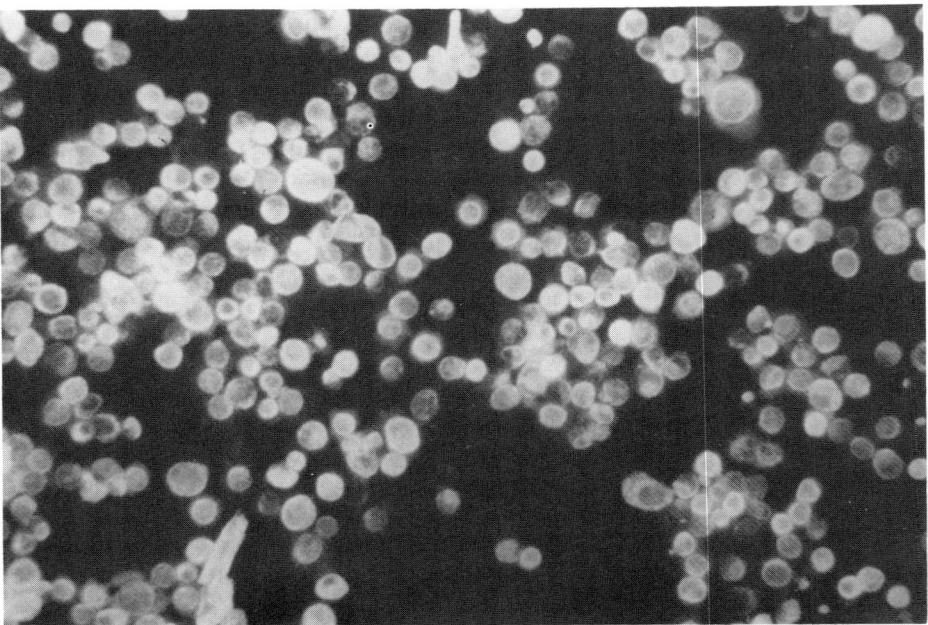

Figure 4. Above: untreated control epidermal culture, spontaneously resistant to the switch to high Ca^{2+} medium fails to label with fluorescein-tagged anti-keratin antibody. Below: Balb/C mouse keratinocyte culture (Fig. 2) resistant to high Ca^{2+} (1.2mM) induced terminal differentiation week 9, labels positively with fluorescein-tagged anti-keratin antibody.

hyperplasia, also inhibited dark cell formation in this system, as did a synthetic antiprotease another potent inhibitor of first-stage promotion. Dark cells also occur in large numbers in papillomas and carcinomas (Raick, 1974) and cell lines from carcinomas, papillomas and virally infected keratinocyte and bronchial epithelial cultures are all resistant to TPA-induced terminal differentiation (Parkinson et al., 1983; Yuspa, Kilkenny, Stanley & Lichti, 1985; Yoakum et al., 1985). No such differential response to terminal differentiation signals is observed with non-promoting hyperplasiogenic agents (Parkinson & Emmerson (1984) whilst the release of TGF-β which accompanies wound healing hyperplasia might be expected to produce the same selective advantages as TPA if the data of the effects of TGF-β on human bronchial epithelia are correct (Willey et al., 1984).

So, how do the differentiation modulating and hyperplasiogenic properties of TPA combine to provide for its role as a highly potent tumour promoter? TPA enhances the number of mouse keratinocyte colonies resistant to normal terminal differentiation signals (Yuspa & Morgan, 1981; and my unpubl. obs.). Keratinocyte subpopulations have been identified in which TPA fails to induce differentiation changes, as if TPA selects for cells at a particular step of their differentiation pathway. Unlike normal keratinocytes 'initiated' cells would appear to be resistant to TPA-induced terminal differentiation and expand into the space left by the suprabasally differentiating normal keratinocytes (Steinert & Yuspa, 1978; Yuspa et al., 1980, 1982; Parkinson & Emmerson, 1984). This is readily explained if the 'initiated' cells are retained within the stem cell populations responsible for long term regeneration of the epidermis as described by Morris, Fisher & Slaga (1985) and indicated by the lack of this laboratory's TPA-resistant keratinocytes to exhibit suprabasal characteristics on staining with the RSKE-60 antibody.

Thus to summarize, the initiation event in keratinocyte systems and probably in the FPC system, is a genetic alteration in the programme of terminal differentiation. Promotion with agents such as TPA provides the proper conditions (differential resistance) within the tissue to allow these altered cells to escape the constraints of a restrictive environment (as evidenced by studies on elimination of metabolic co-operation). Implicit in such a hypothesis is that an initiated cell possesses all the characteristics required to form a papilloma, evidence for which is provided by mouse skin painting experiments, where the promotion stage has been delayed (Hennings & Yuspa, 1985). That the promotion event is truly selective, and not simply mitogenic as reported for a variety of cell culture systems, is supported by the observation of a dissociation of the mitogenic responsiveness and promotion sensitivity in the JB6 mouse epithelial cell line which is induced to irreversible anchorage independence by the tumour promoter TPA (Colburn, Wendel & Abruzzo, 1981).

ONCOGENES AND TPA RESPONSIVENESS

Prior to malignancy, papillomas may first acquire the ability to grow autonomously in the differentiating zone of the epidermis. Activated c-Ha-*ras* oncogenes have been identified in chemically-induced benign papillomas as well as carcinomas (Balmain & Pragnell, 1983; Balmain, Ramsden, Bowden & Smith, 1984), suggesting that c-Ha-*ras* activation might be an early event in skin carcinogenesis. Also infection of mouse keratinocytes with oncogenic retrovirus blocks keratinocytes in an early phorbol ester responsive stage of terminal differentiation (Yuspa, Vass & Scolnick, 1983; Yuspa et al., 1985). Transfection of human broncal epithelial cells with the v-Ha-*ras* oncogene results in cell lines that are immortal, aneuploid and are also resistant to inducers of squamous cell terminal differentiation (Yoakum et al., 1985). Recently, in fibroblast systems, *ras* transfection has been shown to confer on the transfectants a TPA responsive phenotype which appears as foci of morphologically transformed cells against a monolayer of normal cells (Dotto, Parada & Weinberg, 1985). Thus, TPA appears to allow the *ras* transfectants to overcome the growth inhibitory effect imposed on them by the surrounding monolayer of normal cells just as initiated cells *in vivo* are induced to grow out into papillomas by continued TPA application. Interestingly, the second stage promoters RPA and mezerein were just as efficient in this system suggesting that *ras* gene activaton bypasses the requirements for first stage promotion. Recent *in vivo* experiments confirm that exposure of scratched mouse skin to Harvey sarcoma virus confers promoting sensitivity (Brown et al., 1986).

TPA-INDUCED CHROMOSOMAL ABERRATIONS

However, as stated previously TPA in mouse skin readily brings about the induction of papillomas, and in *in vitro* keratinocyte systems probably brings about the induction and expansion of an equivalent population, but very rarely takes the process to completion (13%) (Reddy & Fialkow, 1983). Hennings et al. (1983) proposed that a second mutational event was required. If the term 'mutational event' proposed for the conversion of papillomas to carcinomas is used in its broadest sense, then it could be that chromosomal rearrangements are involved. Kinsella & Radman (1978) proposed that one of the rate-limiting steps leading to neoplastic transformation might be the expression or segregation of a recessive chromosomal lesion leading to an increase in the relative representation of a mutated or affected chromosome segment. Evidence for the chromosomal effects of TPA are numerous (Kinsella & Radman, 1978; Nagasawa & Little, 1979; Emerit & Cerrutti, 1981; Kinsella et al., 1983; Gainer & Kinsella, 1983; Fusenig & Dzarlieva 1983). Chromosome analysis performed

Table 6. Summary of chromosome abrerrations and ploidy changes induced by TPA during the course of transformation experiments

Cell strain (treatment)	No. of Metaphases	Mean breaks/cell	Mean % ploidy
FF_1 (control)	124	0.056 (0.00–0.08)*	4.8
FF_1 (0.1 µg ml^{-1} TPA	154	0.97 (0.07–0.12)	12.3
FPC 5a (control)	138	0.036 (0.03–0.03)	5.8
FPC 5a (0.1 µg m^{-1} TPA)	119	0.92 (0.06–0.28)	10.1
FPC 6c (control)	161	0.068 (0.02–0.12)	8.1
FPC 6c (0.1µg ml^{-1} TPA)	156	0.089 (0.04–0.17)	14.1
FPC 7c (control)	171	0.064 (0.06–0.09)	5.3
FPC 7c (0.1 µg ml^{-1} TPA)	175	0.103 (0.00–0.24)	13.1

*Denotes range of values.

Table 7. Chromosome aberrations and ploidy changes induced by TPA in normal human skin fibroblasts

Cell strain	Passage No.	No. of Metaphases	Breaks/cell*	Polyploidy
FF_1 (control)	11	50	0.08	2
	13	21	0.05	14
	15	50	0.04	4
FF 1 (0.1 µg ml^{-1} TPA)	11	50	0.08	4
	13	14	0.07	14
	15	40	0.1Q	20
	17	50	0.12Q	14

* Q denotes quadriradial chromosomes.

during the course of the FPC transformation experiments showed TPA to increase the frequency of tetraploidy/polyploidy and quadriradial chromosomes (Table 6). Quadriradial chromosomes always occurred in the polyploid cells probably due to an increase in pairing time brought about by TPAs effect on ploidy. Similar observations had also been made by the author's group in mormal human skin fibroblasts (Table 7) (Kinsella et al. 1983). Ploidy increase and mitotic recombination, as evidence by quadriradial chromosomes, were two of the mechanisms proposed to increase the relative representation of a mutated or affected chromosome segment (Kinsella & Radman, 1978). Thus, this low level ability of TPA to generate a mutation-like event could give rise to the low level malignant conversion observed *in vivo*. This step normally being independent of the action of a tumour promoter.

TPA AND THE PHOSPHATIDYLINOSITOL SECOND MESSENGER SYSTEM

Finally, the membrane receptor specified for TPA has been identified as protein kinase C (Castagna et al., 1982; Niedel, Kuhn & Vandenbark, 1983). Nishizuka and co-workers (Kikkawa et al., 1983) suggested that protein kinase C was activated by phosphatidylinositol turnover leading to diacylglycerol formation. TPA can substitute for unsaturated diacylglycerol to increase the affinity of protein kinase C for calcium and phosphatidylserine thereby activating the enzyme. Similarly, the interaction of certain hormones or growth factors with their receptors and stimulation of phosphatidylinositol turnover, activates protein kinase C (Nishizuka, 1984). In turn, oncogenes are known to code for various aspects of this key pathway in the regulation of cell proliferation and differentiation (Muller et al., 1983; Berridge & Irvine, 1984). TPA has been shown to regulate the expression of certain of these oncogenes in relation to growth state (Greenberg & Ziff, 1984). Therefore, it is attractive to suggest that the uncontrolled production of an active form of protein kinase C, whether or not it is the product of a cellular proto-oncogene (Moelling et al., 1984), may 'promote' carcinogenesis. The pleiotropic effects of the phorbol ester tumour promoters may in fact be due to the location of protein kinase C at the crossover in pathways of hormone action and cell proliferation involving calcium, inositol phospholipids, arachidonic acid and prostaglandins. Finally, Heyworth, Whetton, Kinsella & Houslay, (1984) in studies of TPA inhibition of adenylate cyclase, proposed that TPA mediated its effect on adenylate cyclase via protein kinase C, and that the site at which protein kinase C exerted its effect was the guanyl nucleotide regulatory protein. Interestingly, the *ras* proto-oncogene products have both structural (Hurley *et al.*, 1984) and biochemical similarities (McGrath, Capon, Goeddel & Levinson, 1984) to the guanyl nucleotide regulatory proteins, which might explain the specific growth response of *ras*-transformed cells to tumour promoters mentioned previously (Dotto *et al.*, 1985) and the promotion sensitivity of Harvey sarcoma virus infected mouse skin (Brown *et al.*, 1986).

CONCLUSIONS

The data presented and reviewed above, all lend support to the model that an initiating carcinogen creates a small number of initiated cells. The tumour promoter then acts to induce the clonal expansion of each initiated cell (i.e. an initiated cell is one which experiences a selective growth advantage in the presence of a promoter) in one or more of which a second genetic event will occur which is usually promoter independent. The identification of protein kinase C as the specific TPA receptor (see also Blumberg & Aitken, this volume) suggests that TPA may intervene in the inositol lipid second messenger system to regulate cell proliferation and differentiation (Berridge & Irvine, 1984), facilitating the selection and clonal expansion of the specific cell populations referred to above. However, the main, and not uncontroversial, question remains—what genes determine sensitivity to tumour promotion, and what functions do they specify (Dotto *et al.*, 1985; Brown *et al.*, 1986; Colburn, 1986).

ACKNOWLEDGEMENTS

This work was funded by the Cancer Research Campaign, which is gratefully acknowledged.

REFERENCES

AMTMANN, E. & SAUER, G., 1982. Activation of non-expressed bovine papilloma virus genomes by tumour promoters. *Nature, 296:* 675–677.
ANGEL, P., RAHMSDORF, H. J., POTING, A., LUCKE-HULE C. & HERRLICH, P., 1985. 12-O-tetradecanoylphorbol-13-acetate (TPA)-induced gene sequences in primary diploid fibroblasts and their expression in SV40-transformed fibroblasts. *Journal of Cellular Biochemistry, 29:* 351–360.
ANGEL, P., POTING, A., MALLICK, U., RAHMSDORF, H. J., SCHORPP, M. & HERRLICH P., 1986. Induction of metaleiothionein and other mRNA species by carcinogens and tumour promoters in primary human skin fibroblasts. *Molecular and Cell Biology, 6:* 1760–1766.
BAIRD, W. M., SEDGEWICK, J. A. & BOUTWELL R. K., 1971. Effects of phorbol and four diesters of phorbol on the incorporation of tritiated precursors into DNA, RNA and protein in mouse epidermis. *Cancer Research, 31:* 1434–1439.
BALMAIN, A & PRAGNELL, I. B., 1983. Mouse skin carcinomas induced *in vivo* by chemical carcinogens have a transforming Harvey-*ras* oncogene. *Nature, 303:* 72–74.
BALMAIN, A., RAMSDEN, M., BOWDEN, G. T. & SMITH, J., 1984. Activation of the mouse cellular Harvey-ras gene in chemically-induced benign skin papillomas. *Nature, 307:* 658–660.
BARSOUM, J. & VARSHAVSKY, A., 1983. Mitogenic hormones and tumour promoters greatly enhance the incidence of colony-forming cells bearing amplified dihydrofolate reductase genes. *Proceedings of the National Academy of Sciences, U.S.A., 80:* 5330–5334.
BAZILL, G, deWYNTER, E., FUJIKI, H. & KINSELLA, A., 1984. A comparison of the effects of tumour promoters 12-O-tetradecanoylphorbol-13-acetate and teleocidin on gene expression in human skin cell fibroblasts. *Gann, 75:* 672–680.

BERENBLUM, I., 1941. The mechanisms of carcinogenesis. A study of the significance of co-carcinogenic action and related phenomena. *Cancer Research, 1:* 807–814.
BERENBLUM, I., 1944. Invitation and carcinogenesis. *Archives of Pathology, 38:* 233–244.
BELMAN, S. & TROLL, W., 1974. Phorbol-12-myristate-13-acetate effect on cyclic adenosine 3'–5' monophosphate levels in mouse skin and inhibition of phorbol myrinstate acetate-promoted tumourigenesis. *Cancer Research, 34:* 3446–3455.
BERRIDGE, M. J. & IRVINE, R. F., 1984. Inositol triphosphate, a novel second messenger in cellular signal transduction. *Nature, 312:* 315–321.
BOUTWELL, R. K., 1964. Some biological aspects of skin carcinogenesis. *Progress in Experimental Tumour Research, 4:* 207–250.
BOUTWELL, R. K., 1974. The function and mechanisms of promoters of carcinogenesis. *CRC Critical Reviews of Toxicology, 2:* 419–443.
BROWN, K., QUINTANILLA, M., RAMSDEN, M., KERR, I. B., YOUNG, S. & BALMAIN A., 1986. v-ras genes from Harvey and Balb murine sarcoma viruses can act as initiators of two-stage mouse skin carcinogenesis. *Cell, 46:* 447–456.
BRESNIK, E., MEVNIER, R. & LAMDEU, M., 1979. Epidermal prostaglandins after topical application of a tumorpromoter. *Cancer Letters, 7:* 121–124.
CABRAL, F., GOTTESMAN, M. M. & YUSPA, S. H., 1981. Induction of specific protein synthesis by phorbol esters in mouse epidermal cultures. *Cancer Research, 41:* 2025–2031.
CASTAGNA, M., TAKAI, Y., KAITUCHI, K., SANO, K., NIKKAWA, U. & KISHIZUKA, Y., 1982. Direct activation of calcium-activated, phospholipid dependent protein kinase by tumour promoting phorbol esters. *Journal of Biological Chemistry, 257:* 7847–7851.
COHEN, R., PACIFICI, M., RUBINSTEIN, N., BEIHL, J. & HOLTZER H., 1977. Effect of a tumour promoter on myogenesis. *Nature, 226:* 538–540.
COLBURN, N. H., LAU, S. & HEAD, R., 1975. Decrease of epidermal histidase activity by tumour promoting phorbol esters. *Cancer Research, 35:* 3154–3159.
COLBURN, N. H., WENDEL, E. J. & ABRUZZO, G., 1981. Dissociation of mitogenesis and late-stage promotion of tumour cell phenotype by phorbol esters: mitogenic resistant variants are sensitive to promotion. *Proceedings of the National Academy Sciences, U.S.A., 78:* 6192–6916.
COLBURN, N. H., 1986. The genetics of tumour promotion. In J. C. Barrett (Ed.), *Mechanisms of Environmental Carcinogenesis, 1.* Boca Raton: CRC Press.
DELCLOS, K. B., NAGLE, D. S. & BLUMBERG, P. M., 1980. Specific binding of phorbol ester tumour promoters to mouse skin. *Cell, 19:* 1025–1032.
DIAMOND, L., O'BRIEN, T. & ROVERA, G., 1978. Tumour promoters inhibit terminal cell differentiation in culture. In T. J. Slaga, A. Sivak & R. K. Boutwell (Eds), *Carcinogenesis, 2, Mechanisms of Tumour Promotion and Co-carcinogenesis:* 335–341. New York: Raven Press.
DIGIOVANNI, J., PRITCHETT, W. P., DECINA, P. C. & DIAMOND, L., 1984. DBA/2 mice are as sensitive as Sencar mice tumour promotion by 12-O-tetradecanoyl phorbol-13-acetate. *Carcinogenesis, 5:* 1493–1498.
DOTTO, G. P., PARADA, L. F. & WEINBERG, R. A., 1985. Specific growth response of *ras*-transformed embryo fibroblasts to tumour promoters. *Nature, 318:* 472–475.
DUNPHY, W. G., KOCHENBURGER, R. J., CASTAGNA, M. & BLUMBERG, P. M., 1981. Kinetics and subcellular localisation of specific [^{3}H] phorbol 12, 13-dibutyrate binding by mouse brain. *Cancer Research, 41:* 2640–2647.
EMERIT, I. & CERUTTI, P. A., 1981. Tumour promoter phorbol-12-myristate-13-acetate induces chromosomal damage via indirect action. *Nature, 293:* 144–146.
FREI, J. V. & STEPHENS, P., 1968. The correlation of promotion of tumour growth and the induction of hyperplasia in epidermal two-stage carcinogenesis. *British Journal of Cancer, 22:* 83–92.
FRIEDEWALD, W. F. & ROUS, P., 1944. The initiating and promoting elements in tumour production. *Journal of Experimental Medicine, 80:* 101–126.
FURSTENBERGER, G., BERRY, D. L., SORG, G. & MARKS, F., 1981. Skin tumour promotion by phorbol esters is a two-stage process. *Proceedings of the National Academy of Sciences, U.S.A., 78:* 7722–7726.
FUSTENBERGER, G. S., SCHWEIZER, J. & MARKS F., 1985. Development of phorbol ester responsiveness in neonatal mouse epidermis: correlation between hyperplastic response and sensitivity to first-stage promotion. *Carcinogenesis, 6:* 289–294.
FUSENIG, N. E. & DZARLIEVA., 1982. Phenotypic and chromosomal alterations in cell structures as indicators of tumour promoting activity. In E. Hecker, N. E. Fusenig, W. Dunz, F. Marks & H. W. Thielmann, (Eds), *Carcinogenesis, 7:* 201–216. New York: Raven Press.
GAINER, H. St. C. & KINSELLA, A. R., 1983. Analysis of spontaneous, carcinogen induced and promoter-induced chromosomal instability in patients with hereditary retinoblastoma. *International Journal of Cancer, 32:* 449–453.
GAINER, H. St. C., SCHOR, S. & KINSELLA, A. R., 1984. Susceptibility of skin fibroblasts from individual genetically predisposed to cancer, to transformation by the tumour promoter 12-O-tetradecanoyl-phorbol-13-acetate. *International Journal of Cancer, 34:* 349–357.
GREENBERG, M. E. & ZIFF, E. B., 1984. Stimulation of 3T3 cell induces transcription of c-fos proto-oncogene. *Nature, 311:* 433–437.

HECKER, E., 1968. Co-carcinogenic principles from the seed oil of *Croton tiglium* and other Euphorbiaceae. *Cancer Research, 28:* 2338–2349.

HENNINGS, H., SHONES, R., WENK, M. L., SPANGLER, E. F., TARONE, R. & YUSPA, S. H., 1983. Malignant conversion of mouse skin tumours is increased by tumour initiators and unaffected by tumour promoters. *Nature, 304:* 67–69.

HENNINGS, H. & YUSPA, S. H., 1985. Two-stage tumour promotion in mouse skin: An alternative interpretation. *Journal of the National Cancer Institute, 74:* 735–740.

HEYWORTH, C. M., WHETTON, A. D., KINSELLA, A. R. & HOUSLAY, M. D., 1984. The phorbol ester TPA inhibits glucagon-stimulated adenylate cyclase activity. *FEBS Letters, 170:* 38–42.

HUBERMAN, E., HERCHMAN, C. & LANGENBACH, R., 1979. Stimulation of differentiated functions in human melanoma cells by tumour promoting agents and dimethyl sulphoxide. *Cancer Research, 39:* 2618–2624.

HUBERMAN, E. & CALLAHAN, M. F., 1979. Induction of terminal differentiation in human promyelocytic leukaemia cells by tumour promoting agents. *Proceedings of the National Academy of Sciences, U.S.A., 76:* 1293–1297.

HURLEY, J. B., SIMON, M. I. TEPLOW, D. B., ROBINSHAW, J. D. & GILMAN, A. G., 1984. Homologies between signal transducing G. proteins and *ras* gene products. *Science, 206:* 860–862.

ISHII, D. B., FIBACH, E., YAMASAKI, H. & LUCINSTEIN, J. B., 1978. Tumour promoters inhibit morphological differentiation in cultured mouse neuroblastoma cells. *Science, 200:* 556–561.

JENSEN, R. D. & MILLER, R. W., 1971. Retinoblastoma: epidemiologic characteristics. *New England Journal of Medicine, 285:* 307–311.

KAKUNAGA, T., 1978. Neoplastic transformation of human diploid fibroblast cells by chemical carcinogens. *Proceedings of the National Academy of Sciences, U.S.A., 75:* 1334–1338.

KIKKAWA, Y., TAKAI, Y., TANAKA, Y., MIYAKI, R. & NISHIZUKA, U., 1983. Protein kinase c as a potential receptor of tumour promoting phorbol ester. *Journal of Biological Chemistry, 258:* 11442–11445.

KINSELLA, A. R. & RADMAN, M., 1978. Tumour promoter induces sister chromatid exchanges: relevance to mechanisms of carcinogenesis. *Proceedings of the National Academy of Sciences, U.S.A., 75:* 6149–6153.

KINSELLA, A. R., 1981. Investigation of the effects of the phorbol ester TPA on carcinogen-induced foward mutagenesis to 6-thioguanine resistance in V79 Chinese hamster cells. *Carcinogenesis, 2:* 43–47.

KINSELLA, A. R., GAINER, H. St. C. & BUTLER, J., 1983. Investigation of a possible role for superoxide anion production in tumour promotion. *Carcinogenesis, 4:* 717–719.

KINSELLA, A. R., 1986. Multi-stage carcinogenesis and the biological effects of tumour promoters. In F. J. Evans (Ed.), *Naturally Occuring Phorbol Esters.* Boca Raton, U.S.A.: CRC Press.

KLEIN-SZANTO, A. J. P., MAJOR, S. M. & SLAGA, T. J., 1980. Induction of dark keratinocytes by 12-O-tetradecanoylphorbol-13-acetate and mezerein as an indicator of tumour promoting efficiency. *Carcinogenesis, 1:* 399–405.

KNUDSON, A. G., 1971. Mutation and cancer: statistical study of retinoblastoma. *Proceedings of the National Academy of Sciences, U.S.A., 68:* 820–823.

KOPELOVICH, L., BIAS, N. E. & HELSON, L., 1979. Tumour promoter alone induces malignant transformation of fibroblasts from humans genetically pre-disposed to cancer. *Nature, 282:* 619–621.

KULESZ-MARTIN, M. F., KOEHLER, B., HENNINGS, H., & YUSPA, S. H., 1980. Quantitative assay for carcinogen-altered differentiation in mouse epidermal cells. *Carcinogenesis, 1:* 995–1006.

LASZLO, A., RADKE, K., CHIN, S. & BISSELL, M. J., 1981. Tumour promoters alter gene expression and protein phosphorylation in avian cells in culture. *Proceedings of the National Academy of Sciences, U.S.A., 78:* 6241–6245.

McGRATH, J. P., CAPON, D. J., GOEDDEL, D. V. & LEVINSON, A. D., 1984. Comparative biochemical properties of normal and activated human *ras* p21 protein. *Nature, 310:* 644–649.

MOELLING, K., HEIMANN, B., BEINLING P., RAPP, U. R. & SANDER, T., 1984. Serine and threonine-specific protein kinase activities of purified gag-mil and gag-raf proteins. *Nature, 312:* 558–561.

MONDAL, S., BRANKOW, D. W. & HEIDELBERGER, C., 1976. Two-stage chemical carcinogenesis in cultures of $C_3H10T\frac{1}{2}$ cells. *Cancer Research, 36:* 2254–2260.

MORRIS, R. J. FISHER, S. M. & SLAGA, T. J., 1985. Evidence that the centrally and peripherally located cells in the murine epidermal proliferative unit are 2 distinct cell populations. *Journal of Investigational Dermatology, 84:* 277–283.

MUFSON, R. A., SIMSIMAN, R. C & BOUTWELL, R. K. 1977. The effect of phorbol ester tumour promoters on the basal and catecholamine stimulated levels of cyclic adenosine 3′-5′-monophosphate in mouse skin and epidermis *in vivo*, *Cancer Research 37:* 665–669.

MULLER, R., TREMBLAY, J. M., ADAMSON, E. D. & VERMA I. M., 1983. Tissue and cell specific expression of two human c-onc genes. *Nature, 304:* 454–456.

MURDOCH, G. H., WATERMAN, M., EVANS, R. M. & ROSENFIELD, M. G., 1985. Molecular mechanisms of phorbolesterthyatropin releasing hormone and growth factor stimulation of prolactin gene transcription. *Journal of Biological Chemistry, 260:* 11852–11858.

MURRAY, W. W. & FITZGERALD, D. J., 1979. Tumour promoters inhibit metabolic co-operation in co-cultures of epidermal and 3T3 cells. *Biochemical and Biophysical Research Communications, 91:* 395–401.

MUTO, T., BUSSEY, H. J. R. & MORSON, B. C., 1975. The evolution of cancer of the colon and rectum. *Cancer Research, 36:* 2251–2270.

NAGASAWA, H. & LITTLE, J. B., 1979. Effect of tumour promoters, protease inhibitors and repair processes on X-ray-induced sister chromatid exchanges in mouse cells. *Proceedings of the National Academy of Sciences, U.S.A., 76:* 1943–1947.

NIEDEL, J. E., KUHN L. J. & VANDENBARK G. R., 1983. Phorbol diester co-purifies with protein kinase C. *Proceedings of the National Academy of Sciences, U.S.A., 80:* 36–40.

NISHIZUKA, Y., 1984. The role of protein kinase C in cell surface signal transduction and tumour promotion. *Nature, 308:* 693–698.

O'BRIEN, T. G., SIMSIMAN, R. C. & BOUTWELL, R. K., 1975. Induction of polyamine biosynthetic enzymes in mouse epidermis by tumour promoting agents. *Cancer Research, 35:* 1662–1670.

PARKINSON, E. K., GRABHAM, P. & EMMERSON A., 1983. A subpopulation of human keratinocytes which is resistant to the induction of terminal differentiation-related changes by phorbol-12-myristate, 13-acetate; evidence for an increase in the resistant population following transformation. *Carcinogenesis, 4:* 857–861.

PARKINSON, E. K. & EMMERSON, A., 1984. Non-promoting hyperplasiogenic agents do not mimic the effects of phorbol, 12-myristate, 13-acetate on terminal differentiation of normal and transformed human keratinocytes. *Carcinogenesis, 5:* 687–690.

RAICK, A. N., 1973. Ultrastructural, histological and biochemical alterations produced by 12-O-tetradecanoylphorbol-13-acetate on mouse epidermis and their relevance to skin tumour promotion. *Cancer Research, 33:* 269–286.

RAICK, A. N., 1974. Cell differentiation and tumour promoting action in skin carcinogenesis. *Cancer Research, 34:* 2915–2925.

REDDY, A. L. & FIALKOW, P. J., 1983. Papillomas induced by initiation-promotion differ from those induced by carcinogen alone. *Nature, 304:* 69–72.

REINERS, J. J. & SLAGA, T. J., 1983. Effects of tumour promoters on the rate and commitment to terminal differentiation of subpopulations of murine keratinocytes. *Cell, 32:* 247–255.

SAFFIOTTI, U. & SHUBIK, P., 1963. Studies on the promoting action of skin carcinogenesis. *National Cancer Institute Monograph, 10:* 489–492.

SCHOR, S. L., SCHOR, A. M., WINN, B. & RUSHTON, G., 1982. The use of three dimensional collagen gels for the study of tumour cell invasion *in vitro:* experimental parameters influencing cell migration into the gel matrix. *International Journal of Cancer, 29:* 57–62.

SISSKIN, E. E., GRAY, T. & BARRETT, J. C., 1982. Correlation between sensitivity to tumour promotion and sustained epidermal hyperplasia of mice and rats treated with 12-O-tetradecanoylphorbol-13-1-acetate. *Carcinogenesis, 3:* 403–407.

SLAGA, T. J., FISCHER, S. M., WEEKS, C. E., NELSON, K., MAMRACK, M. & KLEIN-SZANTO, A. J. P., 1982. Specificity and mechanisms of promoter inhibitors in multi-stage promotion. In E. Hecker N. E. Fusenig, W. Kunz, F. Marks, & H. W. Thielmann, (Eds), 19–34. New York: Raven Press.

TS'O, P. O. P., 1982. Progress in the study of *in vitro* neoplastic transformation In C. Nicolini (Ed.), *Chemical Carcinogenesis:* 347–362. New York: Plenum Press.

VAN DUUREN, B. L. & SIVAK, A., 1968. Tumour promoting agents from *Croton tiglium* and their mode of action. *Cancer Research, 28:* 2349–0000

VARSHAVSKY, A., 1981. Phorbol ester dramatically increase the incidence of methotrexate-resistant mouse cells: possible mechanisms and relevance to tumour promotion. *Cell, 25:* 561–572.

WEINSTEIN, I. B., WIGLER, M., FISHER, P. B., SISSKIN, E. & PIETROPAOLO, C., 1978. Cell culture studies on the biological effects of tumour promoters. In T. J. Slaga, A. Sivak & R. K. Boutwell (Eds), *Carcinogenesis, 2, Mechanisms of Tumour Promotion and co-carcinogenesis:* 313–333. New York: Raven Press.

WIGLER, M. & WEINSTEIN, I. B., 1976. Tumour promoter induces plasminogen activator. *Nature, 259:* 232–233.

WILLEY, J. C., MOSER, C. E., LECHNER, J. J. & HARRIS, C. C., 1984. Differential effects of 12-O-tetradecanoylphorbol-13-acetate on cultured normal and neoplastic human bronchial epithelial cells. *Cancer Research, 44:* 5124–5126.

YOAKUM, G. H., LECHNER, J. F., GABRIELSON, E. W., KORBA, B. E., MALAN-SHIBLEY, L., WILLEY, J. C., VALERIO, M. G., SHAMSUDDIN, A. M., TRUMP, B. F. & HARRIS, C. C., 1985. Transformation of human bronchial epithelial cells transfected by Harvey *ras* oncogene. *Science, 227:* 1175–1179.

YOTTI, L. P., CHANG, C. C. & TROSKO, J. E., 1979. Elimination of metabolic co-operation in Chinese hamster cells by a tumour promoter. *Science, 206:* 1089–1091.

YUSPA, S. H., LICHTI, U., BEN, T., PATTERSON, E., HENNINGS, H., SLAGA, T. J., COLBURN, N. H. & KELSEY, W., 1976. Phorbol ester tumour promoters stimulate DNA synthesis and ornithine decarboxylase activity in mouse epidermal cell cultures. *Nature, 262:* 402–404.

YUSPA, S. H. & MORGAN, D. L., 1981. Mouse skin cells resistant to terminal differentiation associated with initiation of carcinogenesis. *Nature, 293:* 72–74.

YUSPA, S. H., BEN, T., HENNINGS, H. & LICHTI, U., 1982. Divergent responses in epidermal basal cells exposed to the tumour promoter 12-O-tetradecanoylphorbol-13-acetate. *Cancer Research, 42:* 2344–2349.

YUSPA, S. J., VASS, W. & SCOLNICK, E., 1983. Altered growth and differentiation of cultured mouse epidermal cells infected with oncogenic retrovirus: contrasting effects of virus and chemicals. *Cancer Research, 423:* 6021–6030.

YUSPA, S. H., KILKENNY, A. E., STANLEY, J. & LICHTI, U., 1985. Harvey and Kirsten sarcoma viruses block keratinocytes in an early phorbol ester responsive stage of terminal differentiation. *Nature*, *314:* 459–462.

ZUR HAUSEN, H. Z., O'NEILL, F. J. & FREESE, U. K., 1978. Persisting oncogenic herpes virus induced by a tumour promoter RPA. *Nature, 272:* 373–375.

Phorbol esters as probes of the modulatory site on protein kinase C—an overview

PETER M. BLUMBERG, TERUO NAKADATE, BARBOUR S. WARREN, MARIE DELL'AQUILA, TAKASHI SAKO, GABRIELLA PASTI AND NANCY A. SHARKEY

Molecular Mechanisms of Tumor Promotion Section, Laboratory of Cellular Carcinogenesis and Tumor Promotion, National Cancer Institute, Bethesda, MD 20892, U.S.A.

Received and accepted for publication October 1987

BLUMBERG, P. M., NAKADATE, T., WARREN, B. S., DELL'AQUILA, M., SAKO, T., PASTI, G. & SHARKEY, N. A., 1987. **Phorbol esters as probes of the modulatory site on protein kinase C—an overview.** The phorbol esters, diterpene derivatives produced by members of the family Euphorbiaceae, are of major scientific importance as tumour promoters. Their primary site of action is protein kinase C, an enzyme involved in the message transduction pathway for a large class of hormones which cause enhanced phosphatidylinositol turnover. Heterogeneity in the response of biological systems to the phorbol esters may be explained in part by heterogeneity in the phospholipids associated with protein kinase C, in its subcellular localization, and by proteolytic processing, as well as by isoforms of the enzyme. Development of other classes of activators of protein kinase C through computer modelling and natural product isolation suggests possible new approaches for intervention in the pathway.

ADDITIONAL KEY WORDS:—Euphorbiaceae.

Considerable insight into biochemistry and cellular physiology has developed from studies on the mechanism of biologically active natural products. Croton oil, the seed oil of *Croton tiglium* L., has been used for hundreds of years in human and veterinary medicine as a violent cathartic and as a counter-irritant. It became the focus of intense research interest after it was identified as one of the most potent of a class of compounds known as tumour promoters (Boutwell, 1974). Tumour promoters are agents which, although not themselves carcinogenic, enhance the development of tumours induced by previous exposure to sub-effective doses of a carcinogen. The active constituents in croton oil were identified by Hecker and co-workers as 12,13-diesters of phorbol, a novel tetracyclic diterpene (Hecker, 1968) (Fig. 1).

In addition to their tumour promoting activity, the phorbol esters were found to have profound effects on a wide variety of biological systems (Blumberg, 1981; Slaga, 1984). Of particular potential medical importance, the phorbol esters induce terminal differentiation in a number of leukaemic cell lines (Abrahm & Rovera, 1980). Based on these biological results, it seemed likely

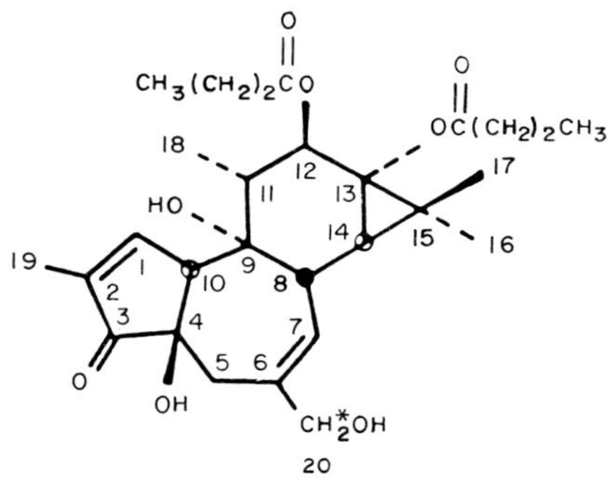

Figure 1. Structure of phorbol 12,13-dibutyrate. The asterisk indicates the position at which tritium is incorporated in the compound for binding studies.

that the phorbol esters were acting to modulate some general cellular regulatory mechanism.

Efforts to identify the site of phorbol ester action directly through binding studies were frustrated by the high non-specific binding of phorbol 12-myristate 13-acetate, the most potent phorbol ester and the derivative studied most extensively *in vitro*. A contribution of this laboratory was the introduction of a different derivative, phorbol 12,13-dibutyrate, which we predicted would possess the optimal ratio of binding activity to lipophilicity (Driedger & Blumberg, 1980). Using that derivative, we and subsequently a large number of other laboratories were able to demonstrate specific phorbol ester binding in cultured cells and in particulate tissue preparations (Blumberg et al., 1984a). Characterization of this binding activity, which we referred to as the phorbol ester receptor, revealed that it was present at the highest level in brain, at intermediate levels in spleen and neutrophils, and at decreasing levels in other tissues. In the brain, if an average molecular weight were assumed for the receptor, then it would account for 0.2% of the total particulate protein. The receptor was highly conserved over evolution, being present in nematodes, fruit flies and sea urchins, was sensitive to low concentrations of Ca^{2+} and was associated with specific phospholipids.

These properties of the phorbol ester receptor markedly resembled those of a Ca^{2+}- and phospholipid-dependent enzyme from brain, protein kinase C, which was being characterized at this same time by Nishizuka and co-workers (Takai, Kishimoto & Nishizuka, 1982). The possible relationship between phorbol ester binding activity and the kinase was further strengthened by the demonstration that, under conditions of limiting Ca^{2+} and phospholipid, the phorbol esters stimulated protein kinase C enzymatic activity (Castagna *et al.*, 1982). A troubling discrepancy, however, remained the apparent difference in subcellular distribution between the kinase and binding activities: binding activity was particulate whereas the kinase, although partially membrane-bound, was largely cytosolic. Resolution of this discrepancy came from recognition that protein kinase C is an apo-enzyme, requiring Ca^{2+} and phospholipids for enzymatic activity. In the presence of these co-factors, phorbol ester binding activity could be demonstrated in cytosol and co-fractionated with protein kinase C during purification, ultimately to homogeneity (Leach, James & Blumberg, 1983; Niedel, Kuhn & Vandenbark, 1983; Kikkawa *et al.*, 1983).

Based on these findings, it is now generally accepted that the quantitatively major phorbol ester binding activity is protein kinase C.

As discussed in greater detail elsewhere, identification of the major phorbol ester receptor as protein kinase C was of great interest because protein kinase C had been postulated to mediate one of the two arms of a major message transduction pathway (Nishizuka, 1984, 1986). For a large class of hormones and other cellular effectors, receptor occupancy activates an intracellular phospholipase C, breaking down phosphatidylinositol 4,5-bisphosphate to generate diacylglycerol and inositol 1,4,5-triphosphate. The diacylglycerol, like the phorbol esters, stimulates protein kinase C; the inositol 1,4,5-triphosphate leads to elevated intracellular calcium and activation of Ca^{2+}-dependent protein kinases. The finding that the major phorbol ester receptor was protein kinase C meant that the phorbol esters could be a powerful tool for defining the role of the protein kinase C pathway in cellular physiology.

However, a critical issue is whether the quantitatively major phorbol ester binding site, protein kinase C, is really the functionally major receptor, accounting for most phorbol ester responses. Three types of experiments support the conclusion that it is.

The first approach is pharmacological (Blumberg et al., 1984a). Comparison of structure-activity relations for phorbol ester binding and for biological response shows quite good agreement. However, because of the amount of effort involved, only a rather limited number of responses have been examined. In addition, several natural products structurally dissimilar to the phorbol esters have been identified which inhibit phorbol ester binding. These include the indole alkaloids teleocidin, isolated from *Streptomyces coelicolor*, and lyngbyatoxin, isolated from *Lyngbya majuscula*, as well as the polyacetate aplysiatoxin, isolated from the sea hare (Fig. 2). The indole alkaloids, which have been studied most extensively, are potent mouse skin tumour promoters and induce similar responses to the phorbol esters in each of the numerous systems in which they have been examined (Sugimura & Fujiki, 1983). Aplysiatoxin induces similar responses in the lesser number of systems in which

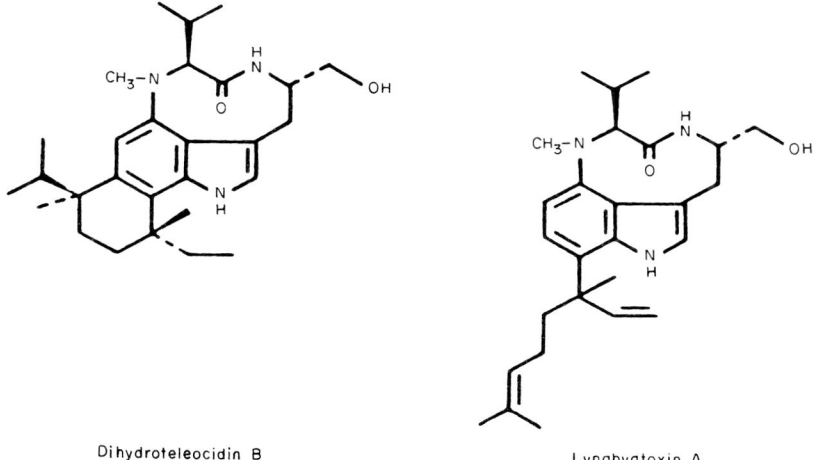

Figure 2. Structure of the indole alkaloids dihydroteleocidin B and lyngbyatoxin.

its activity has been studied. Given the structural diversity of these compounds, it seems unlikely that they would share secondary sites of interaction in the cell.

The second approach has been to show that loss of protein kinase C leads to loss of biological responsiveness. Unfortunately, although a number of laboratories have succeeded in isolating unresponsive variants, in no case have the variants been shown to be deficient in phorbol ester binding. In several systems, however, chronic phorbol ester treatment leads to down regulation of phorbol ester binding and loss of protein kinase C enzymatic activity. Concomitantly, the mitogenic response of the cells to phorbol esters, but not to other mitogens, was lost (Rozengurt, Rodriguez-Pena & Sinnett-Smith, 1985). To test whether the loss of protein kinase C and of mitogenic response are causally related, we microinjected purified protein kinase C back into down-regulated Swiss 3T3 cells. Microinjection restored the mitogenic response to the phorbol esters (Pasti et al., 1986).

The third approach has been biochemical, to show that a response induced by the phorbol esters in intact cells can be mimicked in broken cell preparations by addition of purified protein kinase C. Particularly elegant examples of this approach include the analysis of the phosphorylation of a 42 kD substrate in platelets (Nishizuka, 1984) and of the epidermal growth factor receptors in A431 cells (Hunter, Ling & Cooper, 1984). As part of a collaborative effort with the laboratory of Dr A. I. Tauber, Boston University School of Medicine, we have investigated the mechanism of activation of the oxidative burst in human neutrophils in response to the phorbol esters (Cox et al., 1985). This process is of particular interest in tumour promotion because of considerable indirect evidence for some role of active oxygen in the promotion process (Cerutti, 1985). In the intact neutrophil, phorbol esters activate the NADPH oxidase at doses consistent with the high affinity of the phorbol ester receptor in these cells. *In vitro*, the NADPH oxidase can be activated in preparations containing a plasma membrane fraction, a cytosolic fraction, phosphatidylserine, phorbol esters, calcium, ATP and NADPH. The requirement for the cytosolic fraction can be replaced by purified protein kinase C.

Despite the weight of evidence that protein kinase C is involved in many of the phorbol ester responses, considerable other work demonstrates heterogeneity in phorbol ester binding and response (Blumberg et al., 1984b). At the biological level, this evidence includes the following: First, different phorbol esters may induce different spectra of biological responses. For example, whereas phorbol 12-myristate 13-acetate and mezerein are similar in potency for inducing inflammation and acute hyperplasia in mouse skin, mezerein is very weak as a complete tumour promoter. Secondly, the same phorbol ester may show marked differences for inducing different end-points in the same system. In G-292 osteosarcoma cells, for example, mezerein was 25-fold more potent for inhibiting epidermal growth factor binding than for inducing prostaglandin E2 production. Thirdly, although the phorbol esters typically give normal dose–response curves, in a number of systems the dose–response curves are biphasic, with both the ascending and descending phases at concentrations appropriate for specific phorbol ester effects. Finally, in a few systems the specific effects of one phorbol ester are antagonized by a second, more hydrophilic phorbol ester.

Here again, analysis of natural products promises to be of considerable importance for dissecting biochemical mechanisms. Bryostatin 1, a macrocyclic lactone from the marine bryozoan *Bugula neritina*, was isolated by G. R. Pettit as a potential anti-neoplastic agent (Pettit et al., 1982). Although structurally very different from the phorbol esters, bryostatin 1 was shown to activate protein kinase C and inhibit phorbol ester binding. It activated the oxidative burst in neutrophils (Berkow & Kraft, 1985) and stimulated mitogenesis in Swiss 3T3 cells (J. A. Smith, L. Smith & Pettit, 1985). These results suggested that bryostatin was simply another structural class of protein kinase C activator. Quite excitingly, the bryostatins were found to be inactive in inducing differentiation of HL-60 promyelocytic leukemia cells, which is a well characterized phorbol ester response (Kraft, J. B. Smith & Berkow, 1986). Moreover, they blocked the response to the phorbol esters when added in combination with them. Inhibition of phorbol ester activity is also found in two other differentiating systems examined, keratinocytes and Friend erythroleukaemia cells (unpublished results, collaborative experiments of this laboratory with G. R. Pettit).

Binding analysis supports the conclusion of receptor heterogeneity. Scatchard plots of [^{3}H]PDBu binding to mouse skin particulate preparations yielded curved plots, consistent with three classes of sites (Dunn & Blumberg, 1983). This heterogeneity in binding does not simply reflect an artifact of the particulate preparations. Heterogeneous binding was also found for intact, primary mouse keratinocytes, where the degree of heterogeneity depended on the differentiation state of the cells. Moreover, in both the particulate preparations and the intact cells the different apparent binding sites varied not only in absolute binding affinity but also in structure-activity relations; 12-deoxyphorbol 13-isobutyrate, an irritant but non-promoting phorbol ester isolated from *Euphorbia resinifera* Berg., failed to inhibit radioactive PDBu binding to one of the major binding sites in both systems.

This heterogeneity in phorbol ester binding could reflect the existence of additional targets for the phorbol esters other than protein kinase C. Alternatively, it could represent generation of heterogeneity from protein kinase C itself. One attractive mechanism for generating such heterogeneity is association of protein kinase C with different phospholipids. Since the active form of the protein both for phorbol ester binding and for enzymatic activity is a protein–phospholipid complex, association of the same protein with different phospholipids would yield different active complexes. Indeed, the binding affinity of [^{3}H]PDBu for protein kinase C was 30-fold greater for the complex of enzyme with phosphatidylserine than it was for the complex with a phospholipid mixture corresponding to that of human red blood cells (Konig, DiNitto & Blumberg, 1985a). Similarly, the structure–activity relations for phorbol ester binding depended on the lipids in the complex. Although the shift in binding affinities between the two lipid environments was similar for many of the phorbol ester derivatives, several, including phorbol 12,13-distearate, phorbol 12-retinoate 13-acetate, and ingenol 3-tetradecanoate, showed markedly smaller shifts. Photoaffinity labelling of the phorbol ester receptor in brain demonstrated selective labelling of phosphatidylserine and phosphatidylethanolamine (Delclos, et al., 1983). Because of problems of non-specific binding, other systems have not yet been examined to determine whether a different spectrum of lipids would be labelled.

A second attractive mechanism for the generation of heterogeneity is through variation in subcellular localization. Anderson and co-workers first demonstrated that treatment of cells with phorbol esters caused apparent translocation of the cytosolic protein kinase C to the particulate fraction (Kraft, Anderson, Cooper & Sando, 1982; Kraft & Anderson, 1983). Since protein kinase C interacts with membrane-bound phorbol esters or diglycerides (Sharkey & Blumberg, 1985a), this apparent translocation probably represents a difference in the rate of release of the kinase from membranes rather than the rate of transfer to membranes. Differences in the distribution of the phorbol esters between different cellular membranes should be reflected in corresponding differences in the localization of protein kinase C. The difference in localization of protein kinase C in turn should be reflected in differential accessibility to substrates. Precedent for the important role of subcellular localization on functional activity is afforded by the tyrosine kinase coded by the src oncogene. Mutation to abolish its membrane binding, although it does not affect its *in vitro* kinase activity, reduces its transforming ability (Krueger et al., 1984).

Proteolysis of protein kinase C is a third mechanism. The regulatory domain, which is thought to account for membrane binding, can be cleaved from the enzyme by mild proteolysis, generating a fragment which is now active in the absence of calcium and phospholipids and would be located in the cytosol (Nishizuka, 1984). Activation of the intact enzyme by phorbal esters sensitizes it to proteolytic cleavage *in vitro*; *in vivo*, generation of the catalytic fragment in response to phorbol ester treatment has been confirmed in both neutrophils (Melloni et al., 1986) and platelets (Tapley & Murray, 1985). Formation of the catalytic fragment would alter responses in two ways. First, the substrate selectivity of the catalytic fragment differs from that of the intact kinase (unpubl. obs.). Secondly, by virtue of its cytosolic location the catalytic fragment should have access to a different profile of substrates than does the membrane-bound kinase.

In the course of preparing the catalytic fragment, we observed that phorbol ester binding activity was stable under conditions of proteolysis that degraded both the catalytic fragment as well as the intact kinase. This binding fragment in cells would represent an uncoupled or pseudo-receptor. Its binding characteristics have not as yet been clearly defined. Potentially, under conditions of limiting phorbol ester or endogenous diglyceride, the pseudo-receptor could antagonize responses to these agent by reducing available activator.

Modification of protein kinase C cannot account for all of the heterogeneity in response to phorbol esters and related analogues. Resiniferatoxin, isolated from *Euphorbia resinifera*, is structurally related to the active phorbol esters but differs in that it is substituted at C20 with homovanillic acid. Although for typical phorbol esters substitution at C20 is strongly deactivating, resiniferatoxin is at least 100-fold more irritant than phorbol 12-myristate 13-acetate. In contrast, both for phorbol ester binding and for biological activity for inducing typical phorbol ester responses other than irritancy, resiniferatoxin is several hundred-fold less potent than phorbol 12-myristate 13-acetate. The magnitude of this descrepancy between irritancy and binding strongly argues that resiniferatoxin interacts through a different mechanism.

The diglycerides are postulated endogenous activators of protein kinase C. In an effort to determine whether the diglycerides act as phorbol ester analogues, we examined their ability to block phorbol ester binding to protein kinase C reconstituted with phosphatidylserine *in vitro*. Diglycerides reduced phorbol ester binding affinity with no effect on the number of binding sites (Sharkey, Leach & Blumberg, 1984). This behaviour is characteristic of competitive inhibitors. On the other hand, since we had previously demonstrated that the lipid environment affects the receptor binding characteristics, the alternative interpretation was that the diglyceride was simply altering the lipid environment.

Three tests help to distinguish between these alternatives. All supported the competitive mechanism. First, the change in phorbol ester binding affinity as a function of inhibitor concentration quantitatively fitted the predicted relationship for a competitive inhibitor (Sharkey *et al.*, 1984). Second, as expected for a competitive inhibitor, diglyceride did not affect the rate of release of the phorbol ester from the receptor (Sharkey & Blumberg, 1985b). Thirdly, the stoichiometry of binding was 1:1 (Konig, DiNitto & Blumberg, 1985b).

Comparison of absolute potencies indicated that the phorbol esters were all more potent than similarly substituted diglycerides, but the magnitude of the difference depended markedly on the specific derivative (Sharkey & Blumberg, 1985a, 1986). Thus, phorbol 12,13-dilaurate was 20-fold more potent than glycerol 1,2-dilaurate, whereas phorbol 12-myristate 13-acetate was 30 000-fold more potent. The greater potency of a toxin relative to the natural analogue is not unusual. A classic example is alpha-bungarotoxin, which has a vastly higher affinity for the nicotinic cholinergic receptor than does the natural agonist acetylcholine (Oswald & Freeman, 1981). The rationale is that a regulatory ligand should not bind too tightly, because tight binding limits the rate at which the ligand can be released. For a toxin, in contrast, greater potency is an advantage because it reduces the amount required to achieve a toxic dose.

The ability of diglycerides to inhibit phorbol ester binding competitively suggested that treatments which elevate cellular diglyceride levels should mimic phorbol ester effects. This prediction has been confirmed using four approaches. First, diglycerides can be added exogenously to the culture medium. In order to be active, the diglycerides must be less hydrophobic than natural diglycerides so that they can transfer through the aqueous medium to the cells. Suitable derivatives have proven to be glycerol 1-oleate 2-acetate (Nishizuka, 1984), glycerol 1,2-dioctanoate (Ebeling *et al.*, 1985), and glycerol 1-myristate 2-acetate (Sharkey & Blumberg, 1986). Secondly, addition of bacterial phospholipase C will degrade membrane phospholipids to diglycerides (Jeng, Lichti, Strickland & Blumberg, 1985). Thirdly, triglyceride synthesis can be blocked by bromooctanoic acid, leading to formation of diglyceride (Mayorek & Bar-Tena, 1985). In keratinocytes, bromooctanoic acid treatment partially inhibits epidermal growth factor binding and elevates ornithine decarboxylase, both typical phorbol ester responses. Finally, loss of endogenously generated diglyceride through metabolism by diglyceride kinase can be blocked by diglyceride kinase inhibitors (de Chaffoy de Courcelles, Roevens & Van Belle, 1985).

One consequence of the identification of structurally dissimilar natural products functioning as protein kinase C analogues was that it provided evidence for the involvement of protein kinase C in the mechanism of phorbol ester action. An even more important potential impact is that it permits computer modelling to identify structural features in common which represent the critical sites of interaction with the enzyme. Because of the structural complexity of the phorbol nucleus, which possesses eight asymmetric carbons, *de novo* synthesis of phorbol derivatives altered in the carbon backbone is a formidable synthetic task. Identification of the relevant structural features may permit development of simple ring structures which still possess the relevant functionalities.

The comparable activity of phorbol 12,13-diesters and ingenol 3-monoesters strongly argues against the suggestion (Nishizuka, 1984) that the 12,13-diester of phorbol represents the relevant overlap with the 1,2-diester of diglycerides. In the highly active ingenol 3-monoesters, the hydroxyl groups at 12 and 13 are missing. Comparison between the phorbol esters and the indole alkaloids suggests, rather, correspondence between the 4, 9, and 20 hydroxyl groups of phorbol and the N-13, N-1, and 0-24 atoms of teleocidin (Wender *et al.*, 1986). Diacylglycerol likewise gives a good fit to these oxygens, supporting the biochemical data that the phorbol esters and diglycerides interact at the receptor in a similar fashion.

Based on their computer modelling, Wender and co-workers synthesized several simplified ring systems isosteric with the proposed pharmacophore. These compounds competitively inhibited phorbol ester binding, albeit with somewhat lower potency than did diglycerides. The compounds also induced in intact cells several responses characteristic of the phorbol esters such as inhibition of epidermal growth factor binding in 3T3 cells or stimulation of phosphorylation of a 42 kD protein in platelets. In some other systems, toxicity intervened before specific responses were observed.

The studies on the phorbol esters well exemplify the impact of natural product isolation on our knowledge of biochemistry and cellular physiology. The phorbol esters have been invaluable tools in defining the central role of protein kinase C in cellular message transduction. They and the bryostatins are beginning to clarify the basis of heterogeneity in this system. This knowledge in turn is vital for therapeutic intervention in the pathway without unacceptable side effects. With the natural structures as a guide, new structural classes can be synthesized to further explore structure-activity relations in an attempt to identify antagonists, compounds which block activation of the kinase by endogenous activators.

ACKNOWLEDGEMENTS

We thank Stuart H. Yuspa for critical reading of this manuscript.

REFERENCES

ABRAHM, J. & ROVERA, G., 1980. The effect of tumor-promoting phorbol diesters on terminal differentiation on cells in culture. *Molecular Cell Biochemistry, 31:* 165–175.
BERKOW, R. L. & KRAFT, A. S., 1985. Bryostatin, a non-phorbol macrocyclic lactone, activates intact human polymorphonuclear leukocytes and binds to the phorbol ester receptor. *Biochemical and Biophysical Research Communications, 131:* 1109–1116.
BLUMBERG, P. M., 1981. *In vitro* studies on the mode of action of the phorbol esters, potent tumour promoters. *Critical Reviews in Toxicology, 8:* 153–234.
BLUMBERG, P. M., DUNN, J. A., JAKEN, S., JENG, A. Y., LEACH, R. L., SHARKEY, N. A. & YEH, E., 1984a. Specific receptors for phorbol ester tumor promoters and their involvement in biological

responses. In Slaga, T. J. (Ed.), *Mechanisms of Tumor Promotion, 3, Tumor Promotion and Carcinogene* 143–184. Boca Raton, Florida: CRC Press.

BLUMBERG, P. M., JAKEN, S., KONIG, B., SHARKEY, N. A., LEACH, K. L., JENG, A. E., 1984b. Mechanism of action of the phorbol ester tumor promoters: specific receptors fc ligands. *Biochemical Pharmacology, 33:* 933–940.

BOUTWELL, R. K., 1974. The function and mechanism of promoters of carcinogenesis. *Criti Toxicology, 2:* 419–443.

CASTAGNA, M., TAKAI, Y., KAIBUCHI, K., SANO, K., KIKKAWA, U. & NISHIZUKA, Y., 1982. Direct activation of calcium-activated, phospholipid-dependent protein kinase by tumor-promoting phorbol esters. *Journal of Biological Chemistry, 257:* 7847–7851.

CERRUTTI, P. A., 1985. Prooxidant states and tumor promotion. *Science, 227:* 375–381.

COX, J. A., JENG, A. Y., SHARKEY, N. A., BLUMBERG, P. M. & TAUBER, A. I., 1985. Activation of the human neutrophil nicotinamide adenine dinucleotide phosphate (NADPH)-oxidase by protein kinase C. *Journal of Clinical Investigation, 76:* 1932–1938.

DE CHAFFOY DE COURCELLES, D., ROEVENS, P. & VAN BELLE, H., 1985. R 59 022, a diacylglycerol kinase inhibitor. Its effects on diacylglycerol and thrombin-induced C kinase activation in the intact platelet. *Journal of Biological Chemistry, 260:* 15762–15770.

DECLOS, K. B., YEB, E. & BLUMBERG, P. M., 1983. Specific labeling of mouse brain membrane phospholipids with [20-^{3}H]phorbol 12-p-azidobenzoate 13-benzoate, a photolabile phorbol ester. *Proceedings of the National Academy of Sciences of the United States of America, 80:* 3054–3058.

DRIEDGER, P. E. & BLUMBERG, P. M., 1980. Specific binding of phorbol ester tumor promoters. *Proceedings of the National Academy of Sciences of the United States of America, 77:* 567–571.

DUNN, J. A. & BLUMBERG, P. M., 1983. Specific binding of [20-^{3}H] 12-deoxyphorbol 13-isobutyrate to phorbol ester receptor subclasses in mouse skin particulate preparations. *Cancer, 43:* 4632–4637.

DUNN, J. A., JENG, A. Y., YUSPA, S. H. & BLUMBERG, P. M., 1985. Heterogeneity of [^{3}H]phorbol 12,13-dibutyrate binding in primary mouse keratinocytes at different stages of maturation. *Cancer Research, 45:* 5540–5546.

EBELING, J. G., VANDENBARK, G. R., KUHN, L. J., GANONG, B. R., BELL, R. M. & NIEDEL, J. E., 1985. Diacylglycerols mimic phorbol diester induction of leukemic cell differentiation. *Proceedings of the National Academy of Sciences of the United States of America, 82:* 815–819.

HECKER, E., 1968. Cocarcinogenic principles from the seed oil of *Croton tiglium* and from other Euphorbiaceae. *Cancer Research, 28:* 2338–2349.

HUNTER, T., LING, N. & COOPER, J. A., 1984. Protein kinase C phosphorylation of the EGF receptor at a threonine residue close to the cytoplasmic face of the plasma membrane. *Nature, 311:* 480–483.

JENG, A. Y., LICHTI, U., STRICKLAND, J. E. & BLUMBERG, P. M., 1985. Similar effects of phospholipase C and phorbol ester tumor promoters on primary mouse epidermal cells. *Cancer Research, 45:* 5714–5721.

KIKKAWA, U., TAKAI, Y., TANAKA, Y., MIYAKE, R. & NISHIZUKA, Y., 1983. Protein kinase C as a possible receptor protein of tumor-promoting phorbol esters. *Journal of Biological Chemistry, 258:* 11442–11445.

KONIG, B., DINITTO, P. A. & BLUMBERG, P. M., 1985a. Phospholipid and Ca^{++} dependency of phorbol ester receptors. *Journal of Cell Biochemistry, 27:* 255–265.

KONIG, B., DINITTO, P. A. & BLUMBERG, P. M., 1985b. Stoichiometric binding of diacylglycerol to the phorbol ester receptor. *Journal of Cell Biochemistry, 29:* 37–44.

KRAFT, A. S. & ANDERSON, W. B., 1983. Phorbol esters increase the amount of Ca^{++}, phospholipid-dependent protein kinase associated with plasma membrane. *Nature (London), 301:* 621–623.

KRAFT, A. S., ANDERSON, W. B., COOPER, H. L. & SANDO, J. J., 1982. Decrease in cytosolic calcium/phospholipid-dependent protein kinase activity following phorbol ester treatment of EL4 thymoma cells. *Journal of Biological Chemistry, 257:* 13193–13196.

KRAFT, A. S., SMITH, J. B. & BERKOW, R. L., 1986. Bryostatin, an activator of the calcium phospholipid-dependent protein kinase, blocks phorbol ester-induced differentiation of human promyelocytic leukemia cells HL-60. *Proceedings of the National Academy of Sciences of the United States of America, 83:* 1334–1338.

KRUEGER, J. G., GARBER, E. A., CHIN, S. S. M., HANAFUSA, H. & GOLDBERG, A. R., 1984. Size-variant pp60src proteins of recovered avian sarcoma viruses interact with adhesion plaques as peripheral membrane proteins: effects on cell transformation. *Molecular Cell Biology, 4:* 454–467.

LEACH, K. L., JAMES, M. L. & BLUMBERG, P. M., 1983. Characterization of a specific phorbol ester aporeceptor in mouse brain cytosol. *Proceedings of the National Academy of Sciences of the United States of America, 80:* 4208–4212.

MAYOREK, N. & BAR-TANA, J., 1985. Inhibition of diacylglycerol acyltransferase by 2-bromooctanoate in cultured rat hepatocytes. *Journal of Biological Chemistry, 260:* 6528–6532.

MELLONI, E., PONTREMOLI, S., MICHETTI, M., SACCO, O., SPARATORE, B. & HORECKER, B. L., 1986. The involvement of calpain in the activation of protein kinase C in neutrophils stimulated by phorbol myristic acid. *Journal of Biological Chemistry, 261:* 4101–4105.

NIEDEL, J. E., KUHN, L. J. & VANDENBARK, G. R., 1983. Phorbol diester receptor copurifies with protein kinase C. *Proceedings of the National Academy of Sciences of the United States of America, 80:* 36–40.

NISHIZUKA, Y., 1984. The role of protein kinase C in cell surface signal transduction and tumor promotion. *Nature, (London), 308:* 693–698.

NISHIZUKA, Y., 1986. Studies and perspectives of protein kinase C. *Science, 233:* 305–312.

OSWALD, R. E. & FREEMAN, J. A., 1981. Alpha-bungarotoxin binding and central nervous system nicotinic acetylcholine receptors. *Neuroscience, 6:* 1–14.

PASTI, G., LACAL, J.-C., WARREN, B. S., AARONSON, S. A. & BLUMBERG, P. M., 1986. Loss of mouse fibroblast cell response to phorbol esters restored by microinjected protein kinase C. *Nature (London), 324:* 375–377.

PETTIT, G. R., HERALD, C. L., DOUBEK, D. L., HERALD, D. L., ARNOLD, E. & CLARDY, J., 1982. Isolation and structure of bryostatin 1. *Journal of the American Chemical Society, 104:* 6846–6848.

ROZENGURT, E., RODRIGUEZ-PENA, A. & SINNETT-SMITH, J., 1985. Signalling mitogenesis in 3T3 cells: role of Ca^{++}-sensitive, phospholipid-dependent protein kinase. In Evered, D, Nugent, J. & Whelan, J. (Eds), *Biology & Medicine:* 205–223. London: Pitman Publishing Ltd.

SHARKEY, N. A. & BLUMBERG, P. M., 1985a. Highly lipophilic phorbol esters as inhibitors of specific [^{3}H]phorbol 12,13-dibutyrate binding. *Cancer Research, 45:* 19–24.

SHARKEY, N. A. & BLUMBERG, P. M., 1985b. Kinetic evidence that 1,2-diolein inhibits phorbol ester binding to protein kinase C via a competitive mechanism. *Biochemical and Biophysical Research Communications, 133:* 1051–1056.

SHARKEY, N. A. & BLUMBERG, P. M., 1986. Comparison of the activity of phorbol 12-myristate 13-acetate and the diglyceride glycerol 1-myristate 2-acetate. *Carcinogenesis, 7:* 677–679.

SHARKEY, N. A., LEACH, K. L. & BLUMBERG, P. M., 1984. Competitive inhibition by diacylglycerol of specific phorbol ester binding. *Proceedings of the National Academy of Sciences of the United States of America, 81:* 607–610.

SLAGA, T. J. (Ed.), 1984. *Mechanisms of Tumor Promotion.* Boca Raton: CRC Press.

SMITH, J. A., SMITH, L. & PETTIT, G. R., 1985. Bryostatins: potent, new mitogens that mimic phorbol ester tumor promoters. *Biochemical and Biophysical Research Communications, 132:* 939–945.

SUGIMURA, T. & FUJIKI, H., 1983. Effect of tumor promoters in relation to the cell membrane. In Makita, A., Tsuiki, S., Fujiki, H. & Warren, L. (Eds), *Gann Monograph on Cancer Research, 29:* 3–15. Tokyo: University of Tokyo Press.

TAKAI, Y., KISHIMOTO, A. & NISHIZUKA, Y., 1982. Calcium and phospholipid turnover as transmembrane signaling for protein phosphorylation. In Cheung, W. Y. (Ed.), *Calcium and Cell Function, 2,* 385–412. New York: Academic Press.

TAPLEY, P. M. & MURRAY, A. W., 1985. Evidence that treatment of platelets with phorbol ester causes proteolytic activation of Ca^{++}-activated, phospholipid-dependent protein kinase. *European Journal of Biochemistry, 151:* 419–423.

WENDER, P. A., KOEHLER, K. F., SHARKEY, N. A., DELL'AQUILA, M. L. & BLUMBERG, P. M., 1986. Analysis of the phorbol ester pharmacophore on protein kinase C as a guide to the rational design of new classes of analogs. *Proceedings of the National Academy of Sciences of the United States of America, 83:* 4214–4218.

Botanical Journal of the Linnean Society (1987) 94: 293–326. With 10 figures

The chemical constituents and economic plants of the Euphorbiaceae

ABDEL-FATTAH M. RIZK

Chemistry Department, Faculty of Science, Qatar University, P.O. Box 2713, Doha, Qatar

Received November 1986, accepted for publication November 1986

RIZK, A.-F. M., 1987. **The chemical constituents and economic plants of the Euphorbiaceae.** A chemical review of the different classes of compounds which have been isolated from the Euphorbiaceae (other than the diterpenoids) is given. This includes triterpenoids and related compounds (sterols, alcohols and hydrocarbons), phenolic compounds (flavonoids, lignans, coumarins, tannins, phenanthrenes, quinones, phenolic acids, etc.), alkaloids, cyanogenic glucosides and glucosinolates. A summary of the industrial and medicinal uses of members of the Euphorbiaceae is provided.

ADDITIONAL KEY WORDS:—alcohols – alkaloids – anthraquinones – arrow poison – coumarins – cyanogenic glucosides – fish poison – flavonoids – glucosinolates – lignans – medicinal plants – naphthaquinones – oils – paper – petroleum plants – phenanthrenes – phenolic acids – rubber – steroids – tannins – triterpenoids – ubiquinones – waxes – woods.

CONTENTS

Introduction	293
Chemical constituents	294
Terpenoids and related substances	294
Fatty acids	295
Phenolic substances	298
Alkaloids	306
Economic plants	313
Industrial plants and/or their products	313
Medicinal plants	317
References	321

INTRODUCTION

The plants of the Euphorbiaceae contain acrid, milky or colourless juice. Chemical data are available for several genera, especially *Euphorbia*, where more than 120 species have been investigated. A survey of this data showed that the triterpenoids, followed by flavonoids and alkaloids are the main classes of substances of interest to phytochemists. However, the presence of other substances, e.g. coumarins, cyanogenic glucosides and tannins are also reported.

The family contains several species of considerable economic importance used as foodstuffs, as medicinal plants and in industry.

CHEMICAL CONSTITUENTS

Terpenoids and related substances

More than 55 triterpenoids (tetra- and pentacyclic) have been identified from the Euphorbiaceae. A review has been given recently by Rizk & El-Missiry (1986) who discussed the distribution of triterpenoids in this family. Table 1 summarizes the triterpenoids identified in plants of the Euphorbiaceae; see also Fig. 1. They have been isolated from the latex as well as from the different parts (bark, cortex, flowers, leaves, roots and stems). Some of the triterpenes (e.g. α- and β-amyrin) have been found either free or as their esters (acetates). Some also occurred as glycosides, e.g. geniculatin (a triterpenoid saponin) isolated from *Euphorbia geniculata* Orteg. (Tripathi & Tiwari, 1980). Another triterpenoid glycoside (gypsogenic acid derivative) was found in some *Euphorbia* species (Soboleva, 1979).

The major constituents of the latex of many *Euphorbia* species are triterpenes and their esters. Monoesters and diesters are also found in the latex of *Euphorbia* species but they are only minor components (Warnaar, 1981).

The stem bark of *Croton oblongifolius* Roxb. contains, besides other components, a group of six compounds having the pimarane skeleton, *viz*. oblongifoliol, deoxyoblongifoliol, oblongifolic acid, *ent*-iso-pimara-7,15-diene, *ent*-isopimara-7,15-dien-10-aldehyde and 19-hydroxy-*ent*-isopimaradiene (Aiyar & Seshadri, 1971).

The analysis of the essential oils in several *Croton* species was reported by Craveiro *et al.* (1978, 1981) and the following compounds were identified:

MONOTERPENOIDS: car-3-ene, α-phellandrene, myrcene, α-pinene, β-pinene, sabinene, *v*-terpinene, α-terpinoline, α-thujene, *p*-cymene, borneol, camphor, 1,8-cineole, geranial, linalool, neral, α-terpineol, terpinen-4-ol, ascaridol and isoscaridol;

PHENYLPROPANOIDS: t-anethole, estragol, eugenol, methyl eugenol, *n*-propyl catechol and elemicin;

SESQUITERPENOIDS: aromadendrene, α-bergamoptene, *v*-cadinene, β-caryophyllene, α-copaene, β-elemene, *v*-elemene, δ-elemene, α-farnesene, β-farnesene, germacrene B, β-guayene, α-humulene and *v*-muurolene.

The essential oil of *Euphorbia monostyla* Prokh. contains the following compounds: thymol, α-pinene, α-thujene, camphene, α-phellandrene, myrcene, α-cardinene, β-cardinene, *p*-cymene, linalool, citronellol and geraniol (Baslas, 1982).

β-Sitosterol has been identified in the sterol fraction of the different species of the Euphorbiaceae studied. Other sterols also occur, but in relatively small amounts, e.g. stigmasterol, campesterol, dihydrobrassicasterol, 28-isofucosterol, Δ^7-isofucosterol and cholesterol (Rizk & El-Missiry, 1986). 7-Oxo-, 7-α-hydroxy- and 7-β-hydroxy- derivatives of campesterol, stigmasterol and sitosterol were isolated from the roots of *Euphorbia fischeriana* Steud. (= *E. pallasii* Turcz.), a drug used for its anti-tumour properties in Chinese traditional medicine (Schroeder, Rohmer, Beck & Anton, 1980).

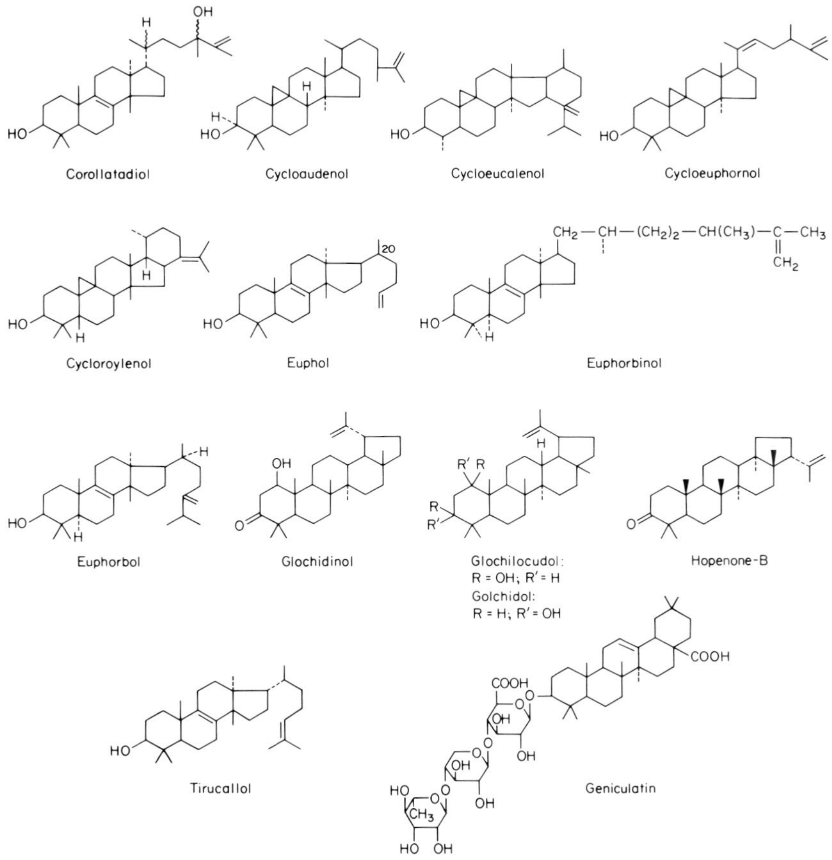

Figure 1. Some triterpenoids of the Euphorbiaceae.

In the genus *Euphorbia*, squalene oxide leads both to 4,4-dimethylleuphoids and 4,4-dimethylsteroids, but it is not clear from the literature whether both or only one of these is metabolized further to 4-desmethyl compounds. The existence of 4,4-dimethylleuphoids and a 24-methylene group, as in the case of euphorbol in *E. triangularis* Desf., *E. ingens* E. Mey and *E. resinifera* Berg, shows that the C-24 alkylation can actually occur after cyclization to the euphoidal structure (Sekula & Nes, 1980).

Long-chain fatty alcohols (particularly *n*-octacosanol and *n*-hexacosanol) and hydrocarbons have been identified from the different genera especially *Euphorbia* spp. (Rizk & El-Missiry, 1986). The latter species have been reported to yield a considerable amount of hydrocarbons and alcohols, and a number of them have been suggested as potential hydrocarbon-producing crops (Calvin, 1980, 1982).

Fatty acids

The fatty acid composition and characterization has been reported for relatively few species of the Euphorbiaceae. In addition to the usual saturated and unsaturated fatty acids, several others were obtained from the seeds oils.

Table 1. Triterpenoids isolated from the Euphorbiaceae

Triterpenes	Species	References
1. Alnusenone	*Euphorbia watanabei* Mark.	Takemoto & Ishiguro (1966)
2. α-Amyrin	*Euphorbia* spp., *Macaranga tanarius* Muell.-Arg.	Rizk & El-Missery (1986), Hui et al. (1975)
3. β-Amyrin	*Euphorbia* spp., *Phyllanthus acidus* Skeel.	Rizk & El-Missery (1986)
4. β-Amyrenone	*Macaranga tanarius* Muell.-Arg.	Hui et al. (1975)
5. δ-Amyrenone	*Euphorbia esula* L., *E. parviflora* L.	Manners & Davis (1984), Rizk & El-Missery (1986)
6. Betulin	*Euphorbia* spp., *Sarcococa pruniformis* Lindl.	Chopra et al. (1969), Rizk & El-Missery (1986)
7. Betulinic acid	*Phyllanthus discoides* Muell., *P. reticulatus* Poir.	Rizk & El-Missery (1986)
8. Butyrospermol	*Euphorbia* spp., *Hura crepitans* L.	Piatak & Reimann (1972)
9. Corollatadiol	*Euphorbia corollata* L.	Rizk & El-Missery (1986)
10. Cycloartenol	*Euphorbia* spp.	Anjaneyulu et al. (1985), Rizk & El-Missiry (1986)
11. Cycloart-25-ene-3-β-24-diols	*Euphorbia* spp.	
12. Cycloaudenol, and 3-epi-cycloaudenol	*Euphorbia caudicifolia* Mans.	Govardhan et al. (1984)
13. Cycloeucalenol	*Euphorbia royleana* Boiss, *E. paralias* L.	Rizk & El-Missiry (1986)
14. Cycloeuphornol	*Euphorbia tirucalli* L.	Afza et al. (1979b)
15. Cycloroylenol	*Euphorbia royleana* Boiss.	Bhat et al. (1982)
16. Dammara-20,24-diene-3-β-ol	*Antidesma bunias* Spreng.	Sainsbury (1970)
17. Euphadienol	*Euphorbia condylocarpa* Bieb.	Roschin & Kir'yalov (1970)
18. Euphol (α- & β-)	*Euphorbia* spp.	Rizk & El-Missiry (1986)
19. Euphorbinol	*Euphorbia tirucalli* L.	Afza et al. (1979a)
20. Euphorbol and *iso*-euphorbol	*Euphorbia* spp.	Rizk & El-Missiry (1986)
21. Friedelan-3-α-ol	*Bischofia trifoliata* (Roxb.)–Hook. fil., *Euphorbia* spp., *Glochidion macrophyllum* Müll.-Arg.	
22. Friedelin	*Bischofia trifoliata* (Roxb.) Hk. fil, *Euphorbia hirta* L., *Glochidion macrophyllum*, *Phyllanthus reticulatus* Poir, *Sapium sebiferum* Roxb.	
23. Fridelinol (Friedelan-3-β-ol)	*Bischofia trifoliata*, *Euphorbia* spp.	Rizk & El-Missiry (1986)
24. *epi*-Fridelinol	*Antidesma bunias* Spreng., *Baccaurea sepida* Muell., *Bischofia trifoliata* (Roxb.) Hook fil, *Bridelia* spp., *Euphorbia* spp., *Sapium discolor* Muell.-Arg.	
25. Germanicol (and 3-*epi*-germanicol)	*Euphorbia* spp.	

26. Glut-5-(6)-en-3-ol	*Euphorbia cyparissias* L.	Starrat (1966)
27. Glut-5-(6)-3n-3-one	*Euphorbia cyparissias* L.	
28. Glochidiol	*Glochidion* spp.	
29. Glochidol	*Glochidion* spp.	Rizk & El-Missiry (1986)
30. Glochidone	*Bridelia moonii* Thw., *Glochidion* spp.	
31. Glochidonol	*Glochidion* spp., *Phyllanthus reticulatus* Poir.	
32. Glocilocudol	*Glochidion multiloculare* Voigt	
33. Hopenol-B	*Euphorbia supina* Rafin. ex Boiss.	Matsunaga & Morita (1983)
34. Hopenone-B	*Euphorbia cyparissias* L.	Starrat (1969)
35. Lanostadienol	*Euphorbia balsamifera* Aiton	Rizk & El-Missiry (1986)
36. Lanostenol	*Euphorbia* spp.	Rizk & El-Missiry (1986)
37. Lanosterol	*Euphorbia* spp.	Rizk & El-Missiry (1986), Rizk et al. (1980b)
38. Lup-20-en-3-β-16-β-diol	*Bayeria* spp.	
39. Lup-20(29)-en-1-β-3-β-diol	*Glochidion* spp.	
40. Lup-20(29)-en-3-α-23-diol	*Glochidion thomsoni* Hook. fil.	Rizk & El-Missiry (1986)
41. Lupenone	*Euphorbia* spp.	
42. Lupeol	*Euphorbia* spp., *Glochidion* spp., *Phyllanthus* spp.	
43. 3-*epi*-Lupeol	*Euphorbia maddeni* Boiss., *Glochidion hohenackeri* Bedd.	
44. Maculatol	*Euphorbia maculata* L.	Takemoto & Inagaki (1958)
45. 24-Methylenecycloartenol	*Euphorbia* spp., *Hura crepitans* L.	Rizk et al. (1974, 1980b), Rizk & El-Missiry (1986)
46. Moretenol	*Aleurites moluccana* L., *Sapium sebiferum*	Rizk & El-Missiry (1986)
47. Moretenone	*Aleurites moluccana* L., *Euphorbia* spp., *Sapium sebiferum*	Rizk & El-Missiry (1986)
48. Nerifoliol	*Euphorbia nerifolia* L.	Rao & Row (1966)
49. Obtusifoldienol	*Euphorbia echinus* Hook., *E. obtusifolia* Poir.	
50. Obtusifoliol	*Euphorbia obtusifolia* Poir.	
51. Oleanolic acid	*Euphorbia paralias* L.	Rizk & El-Missiry (1986)
52. Phyllanthol	*Phyllanthus acidus* Skeel, *P. engleri*	
53. Taraxasterol (and *pseudo*-taraxasterol)	*Euphorbia* spp.	
54. Taraxerol	*Bridelia micrantha* Baill., *Croton sparsiflorus* Morong., *Euphorbia* spp.	Rizk & Rimpler (1977), Rizk & El-Missiry (1986)
55. Taraxerone	*Bridelia micrantha* Baill., *Euphorbia* spp., *Trevia nudiflora* L.	Rizk & El-Missiry (1986), Chopra et al. (1969)
56. Tirucallol	*Euphorbia* spp.	Rizk & El-Missiry (1986)
57. Ursolic acid	*Euphorbia paralias* L.	Rizk et al. (1974)
58. Uvaol	*Euphorbia paralias* L.	Rizk et al. (1974)

Castor oil contains 80–90% of the glycerol ester of ricinoleic acid (Vickery & Vickery, 1979):

$$CH_3(CH_2)_5CHOHCH_2CH=CH(CH_2)_7COOH \quad \textit{Ricinoleic acid}$$

α-Eleostearic acid has been identified from the seed oil of *Trewia nudiflora* L., (Sarkar & Chakrabarty, 1956a):

$$CH_3(CH_2)_3(CH=CH)_3(CH_2)_7COOH \quad \textit{α-Eleostearic acid}$$

A new conjugated fatty acid identified as deca-2,4-trienoic acid along with its four isomers have been detected in the latex of *Euphorbia pulcherrima* Klotzsch (Baslas, 1982).

Euphorbia lagascae Spreng. seed contains a unique epoxy-bearing oil (58–62%, 12,13-epoxyoleic acid) (Krewson & Scott, 1966). The fatty acids of the triacylglycerol fraction of the latex of *Hevea brasiliensis* Muell. Arg. consists of 97% of a C_{18} furanoid fatty acid, 10,13-epoxy-11-methyloctadeca-10,12-dienoic. A dioxy fatty acid, 10,13-dioxy-11-methyloctadecanoic acid was also isolated from the free fatty acid fraction of the same plant (Li Ken Jie & Sinha, 1981).

Although the terpenes of the latex of *Euphorbia* species have been extensively studied, much less attention has been given to the saponifiable part of the terpene esters (Warnaar, 1981). Triterpene esters in the latex of *E. pulcherrima* appeared to be esterified mainly with conjugated decatrienoic acid (Warnaar, 1981). Diterpenes isolated from the latices of *E. tirucalli* L. (Fuerstenberger & Hecker, 1971a, b; Kinghorn, 1979) and *E. lathyris* L. (Adolf & Hecker, 1971) were also esterified with conjugated fatty acids. These acids have eight to 14 carbon atoms in the chain which has two to five double bonds conjugated with carboxylic groups. The fatty acids esterified with the triterpene alcohols of *E. lathyris* were analysed by Warnaar (1981); 77% of the fatty acids were conjugated, the main components of which were decadienoic and decatrienoic acids.

Phenolic substances

Flavonoids

The family Euphorbiaceae is rich in flavonoids, particularly flavones and flavonols, which have been identified from several genera. They occur both as *O*- and *C*-glycosides and as methyl ethers. Flavanones also occur, but in relatively few plants (Table 2 & Fig. 2). The flavonoids were detected in different parts of the plant other than the roots. The two common flavonols kaempferol and quercetin (and their glycosides) are most widely distributed in the different genera of the family.

Robidanol (3,7,3′,4′5′-pentahydroxyflavene) was identified in *Euphorbia palustris* L. and *E. stepposa* Zoz (Bondarenko *et al.*, 1971). Robidanol-3-gallate, also occurred in *E. palustris* (Rizk & El-Missiry, 1986). A chalcone identified as 2′,4′,6′4-tetrahydroxychalcone was isolated from the stem wood of *Excoecaria agallocha* L. (Prakash, Khan, Khan & Zaman, 1983).

Anthocyanins (e.g. cyanidin, delphinidin, and pelargonidin glucosides) were identified in several *Euphorbia species* (Yoshitama, Ishii & Yasudo, 1980; Baslas, 1982; Rizk, 1986).

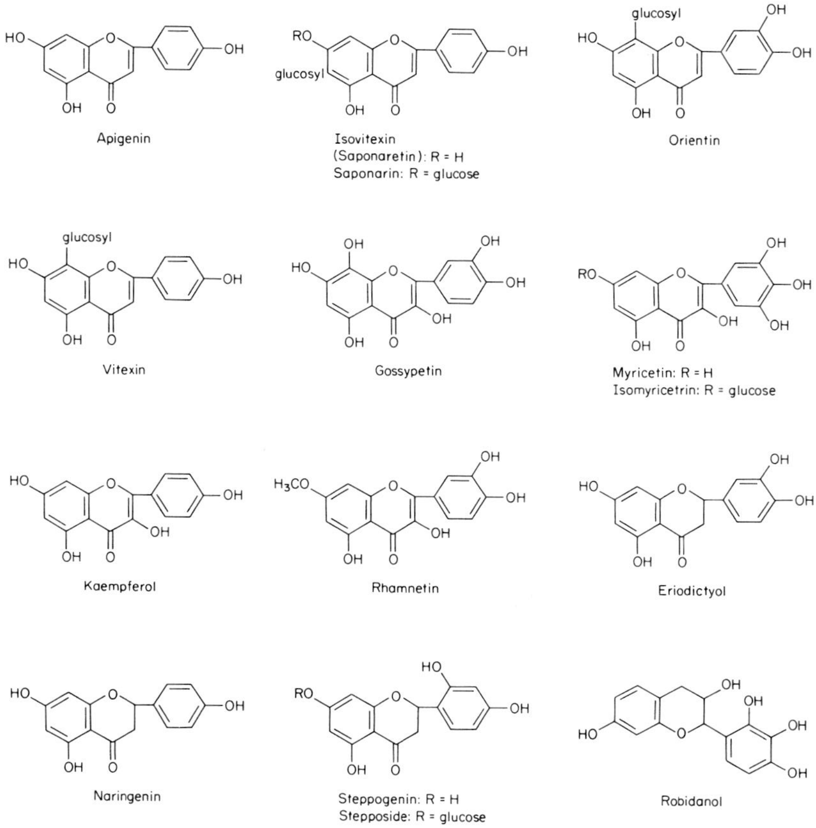

Figure 2. Some flavonoids of the Euphorbiaceae.

Coumarins

These have been isolated from relatively few plants. The seeds of *Euphorbia lathyris* have been reported to contain two bicoumarins (euphorbetin and isoeuphorbetin) (Table 3 & Fig. 3). Three bergenin derivatives were isolated from the bark of *Mallotus japonicus* Muell.-Arg. and identified as 11-*O*-galloylbergenin, 4-*O*-galloylbergenin and 11-*O*-galloyldemethylbergenin (Yoshida, Seno, Takamo & Okuda, 1982).

Euphorbia royleana Boiss. contains three benzocoumarins identified as 7-hydroxy-3,4-benzocoumarin, 7-methoxy-3,4-benzocoumarin and 2,7-dihydroxy-3,4-benzocoumarin (Baslas, 1982). Table 3 summarizes the other coumarins isolated from the Euphorbiaceae.

Lignans

Lignans have been, so far, identified in only two genera, *viz. Jatropha* and *Phyllanthus*. The leaves of *Phyllanthus niruri* L., contain phyllanthin and hypophyllanthin (Rizk & El-Missiry, 1986) (Fig. 4). *Jatropha gossypifolia* L. (stem, root and the seeds) yielded 2-piperonylidene-3-veratryl-3-*R*-*v*-butyrolactone (I) (Chatterjee, Das, Pascard & Prange, 1981) and gadain

Table 2. Flavonoids of the Euphorbiaceae

Flavonoids	Species	References
I. Flavones		
1. Apigenin	*Jatropha curcas* L., *J. gossypifolia* L.	Rizk & El-Missiry (1986)
2. Apigenin 7-*O*-glucoside (Cosmosiin)	*Euphorbia* spp.	Ismail *et al.* (1977), Rizk *et al.* (1982), Rizk & El-Missiry (1986)
3. 5,4′-Dihydroxy-3,7,8-trimethoxyflavone	*Beyeria leschenaultii* Baill., *Ricinocarpos muricatus* Muell.-Arg.	Jefferies (1979), Rizk & El-Missiry (1986)
4. Isoorientin	*Croton zambezicus* Muell.	Wagner *et al.* (1970)
5. Isovitexin (Apigenin-6-C-glucoside, Saponaretin	*Croton zambezicus* Muell., *Jatropha* spp., *Hevea brasiliensis* Muell.-Arg.	Rizk & El-Missiry (1986)
6. Orientin	*Croton zambezicus* Muell.	Wagner *et al.*, (1970)
7. 5,7,3′,4′-Tetrahydroxy-3,8-dimethoxyflavone	*Ricinocarpos muricatus*	Jefferies (1979), Rizk & El-Missiry (1986)
8. 5,7,3′-Trihydroxy-3,8,4′,5′-tetramethoxyflavone	*Beyeria brevifolia* Baill.	Jefferies (1979), Rizk & El-Missiry (1986)
9. Vicenin-1	*Croton zambezicus* Muell.	Wagner *et al.* (1970)
10. Vitexin	*Croton zambezicus* Muell., *Hevea brasiliensis* Muell.-Arg., *Jatropha* spp.	Rizk & El-Missiry (1986)
II. Flavonols		
11. Gossypetin hexamethyl ether	*Ricinocarpos stylosus* Diels	
12. Isomyricitrin	*Euphorbia* spp.	
13. Isoquercitrin	*Euphorbia segueriana* Neck., *E. semivillosa* Prokh.	
14. Isorhamnetin	*Croton oblongifolius* Roxb, *Euphorbia kaleniczkii* Czern. ex Trautv.	Rizk & El-Missiry (1986)
15. Isorhamnetin-3-rutinoside (Narcissin)	*Mercurialis annua* L.	
16. Kaempferol	*Euphorbia* spp., *Phyllanthus emblica*	
17. Kaempferol 3-*O*-galactoside (Trifolin)	*Euphorbia myrsinitis* L.	
18. Kaempferol 3-*O*-glucoside (Astragalin)	*Euphorbia* spp., *Phyllanthus emblica*	Rizk *et al* (1980a, b), Rizk & El-Missiry (1986)
19. Kaempferol 4′-*O*-glucoside	*Euphorbia maddeni* Boiss.	Sahai *et al.* (1981)
20. Kaempferol 3-*O*-glucuronide	*Euphorbia* spp.	
21. Kaempferol 3-*O*-rhamnoside	*Euphorbia lunulata* Bunge, *E. myrsinites* L.	
22. Kaempferol 3-*O*-rutinoside	*Euphorbia* spp., *Mercurialis perennis* L.	
23. Myricetin	*Euphorbia* spp., *Ricinocarpos stylosus*	
24. Myricetin hexamethyl ether	*Ricinocarpos stylosus*	Rizk & El-Missiry (1986)

25. Quercetin	*Croton oblongifolius* Roxb., *Euphorbia* spp. *Jatropha heynii* Bal., *Ricinus communis* L.	Rizk *et al.* (1976, 1982), Rizk & El-Missiry (1986)
26. Quercetin 3-*O*-arabinoside	*Euphorbia paralias* L., *E. stricta* L.	Rizk *et al.* (1979), Rizk & El-Missiry (1986)
27. Quercetin-3-*O*-galactoside (Hyperoside, Hyperin)	*Croton oblongifolius* Roxb., *Euphorbia* spp., *Jatropha heynii* Bal., *Ricinus communis* L.	Rizk *et al.* (1980), Rizk & El-Missiry (1986)
28. Quercetin-3-digalactoside (Heliosin)	*Euphorbia helioscopia* L.	Chen *et al.* (1979), Rizk & El-Missiry (1986)
29. Quercetin-3,5-di-galactoside (Tithymalin)	*Euphorbia helioscopia* L., *E. verrucosa* Lam.	Pohl *et al.* (1975)
30. Quercetin-3-*O*-galactoside-2″-gallate	*Euphorbia dulcis* L., *E. platiphyllos* L.	
31. Quercetin-3-*O*-galactoside-6″-gallate		
32. Quercetin-3-*O*-glucuronide	*Euphorbia cyparissias* L., *E. lathyris* L.	Rizk & El-Missiry (1986)
33. Quercetin-3-*O*-glucoside	*Euphorbia* spp., *Sapium sebiferum*	
34. Quercetin-3-*O*-rhamnoside (Quercitrin)	*Aleuritis cordata* Stand., *Euphorbia* spp.	Ismail *et al.* (1977), Rizk *et al.* (1982), Rizk & El-Missiry (1986)
35. Quercetin-3-*O*-rhamnoglucoside (Quercetin-3-*O*-rutinoside, Rutin)	*Aleurites cordata* Stand., *Croton sparsiflorus* Motong., *Euphorbia* spp., *Manihot utilissima* Pohl., *Mercurialis* spp., *Ricinus communis* L.	Rizk *et al.* (1982), Rizk & El-Missiry (1986)
36. Quercetin-3′-*O*-xyloside	*Euphorbia paralias* L.	Rizk *et al.* (1976)
37. Rhamnetin	*Euphorbia* spp.	Rizk & El-Missiry (1986)
38. Rhamnetin-3-*O*-arabinoside	*Euphorbia amygdaloides* L.	Mueller & Pohl (1970)
39. Rhamnetin-3-*O*-diarabinoside	*Euphorbia amygdaloides* L.	Mueller & Pohl (1970)
40. Rhamnetin-3-*O*-galactoside	*Euphorbia peplus* L., *E. prostrata* Aiton	Ismail *et al.* (1977), Rizk *et al.* (1980a, b)
41. Rhamnetin-3-*O*-rhamnoside	*Euphorbia amygdaloides* L.	Mueller & Pohl (1970)
42. Rhamnetin-3-*O*-dirhamnoside	*Euphorbia amygdaloides* L.	Mueller & Pohl (1970)

III. Flavanones

43. 5-methyleriodictyol 7-*O*-xyloarabinoside	*Sapium sebiferum* Roxb.	Bhat *et al.* (1981)
44. Naringenin	*Mobea caudata* Peth.	
45. Naringenin-3-*p*-coumaryl-glucoside	*Mobea caudata*	
46. Naringenin-3,6-dicoumaroyl-glucoside	*Mobea caudata*	Barros *et al.* (1982)
47. Naringenin-7-*O*-glucoside	*Euphorbia condylocarpa* Bieb.	
48. Steppogenin	*Euphorbia palustris* L., *E. stepposa* Zoz.	Rizk & El-Missiry (1986)
49. Stepposide	*Euphorbia* spp.	

Table 3. Some coumarins of the Euphorbiaceae

Coumarins	Species	References
1. Bergenin	*Fluegga microcarpa* Bl., *Mallotus japonicus* Muell.-Arg.	Ahmad *et al.* (1972), Shigematsu *et al* (1983)
2. Daphnetin	*Euphorbia dracunculoides* Lam.	Chawla *et al.* (1980)
3. 3,7-Dihydroxycoumarin	*Euphorbia terracina* L.	Mahmoud & Abdel Salem (1979)
4. 6,7-Dihydroxycoumarin	*Euphorbia lunulata* Bunge.	Rizk & El-Missiry (1986)
5. 6,7-Dihydroxy-8-methoxy (or 6,8-Dihydroxy-7-methoxy) coumarin	*Ricinus communis* L.	Khafagy *et al.* (1979)
6. 3,4-Dimethoxy-6,8-dihydroxycoumarin	*Ricinus communis* L.	Khafagy *et al.* (1979)
7. Esculin	*Euphorbia acanthothamnos* Heldr. & Sart.	Kallimanis & Philianos (1980)
8. Esculetin	*Euphorbia* spp.	Rizk & El-Missiry (1986)
9. Euphorbetin (and Isoeuphorbetin)	*Euphorbia lathyris* L.	Dutta *et al.* (1972, 1975)
10. Fraxetin	*Jatropha glandulifera*	Parthasarathy & Saradhi (1984)
11. 3-Hydroxy-7-methoxycoumarin	*Euphorbia paralias* L.	Mahmoud & Abdel Salam (1979)
12. Maoyancaosu	*Euphorbia lunulata* Bunge.	Shang *et al.* (1979)
13. Scopoletin	*Euphorbia acanthothamnos*, *Nealchornea yapurensis* Huber.	Gunasekera *et al.* (1980), Rizk & El-Missiry (1986)
14. Scopolin	*Euphorbia acanthothamnos*	Kallimanis & Philianos (1980)

Euphorbetin

Isoeupphorbetin

Maoyancaosu

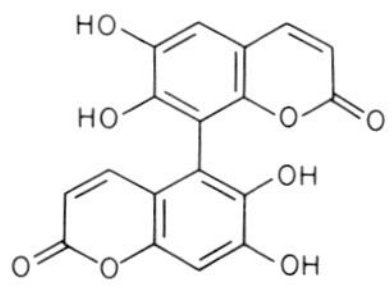

Figure 3. Some coumarins of the Euphorbiaceae.

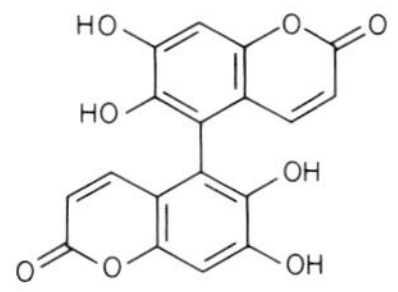

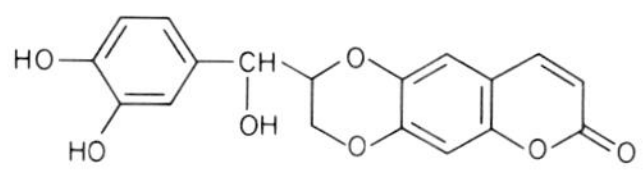

(Banerji, Das, Chatterjee & Shoolery, 1984) A coumarino-lignan (II), involving a dioxan type link between a coumarin and a phenyl propane precursor, similar to the one found in the flavano-lignans, silymarin and hyndocarpin, was isolated from the roots of *Jatropha glandulifera* Roxb. (Parthasarathy & Saradhi, 1984).

Tannins

Hydrolysable tannins have been detected in several species of the Euphorbiaceae, e.g. *Euphorbia maculata* L., *Gleditsia japonica* Miq., *Mallotus japonicus* Muell.-Arg., *Phyllanthus emblica* L. and *Sapium sebiferum* Roxb. (Rizk & El-Missiry, 1986). The bark of *Phyllanthus sebiferum* contains 16.64% tannins. The main tannin of *Triadica sebifera* Small (= *Sapium sebiferum* Roxb.) is geraniin (Fig. 5). The latter compound and mallotusinic acid are the main tannins of *Mallotus japonicus* (Okuda, Mori & Hatano, 1980a; Okuda, Yoshida & Hatano, 1980b; Okuda & Kaoru, 1981). The leaf of *M. japonicus* contained a third tannin identified as mallotinic acid (Okuda & Seno, 1981). Corilagin (a hydrolysate product of geraniin) was identified in the bark of the same plant (Yoshida, Seno, Takamo & Okuda, 1982).

Phyllanthin

Hypophyllanthin

I

Gadain

II

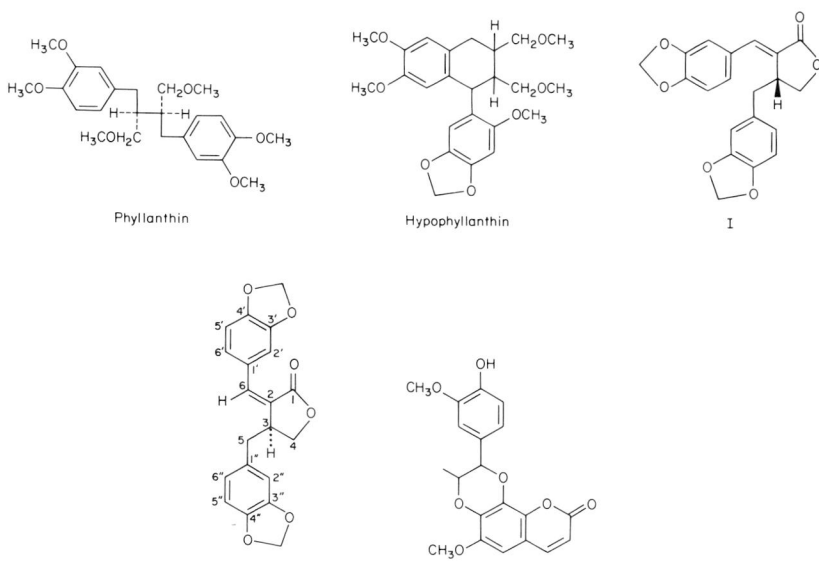

Figure 4. Some lignans of the Euphorbiaceae.

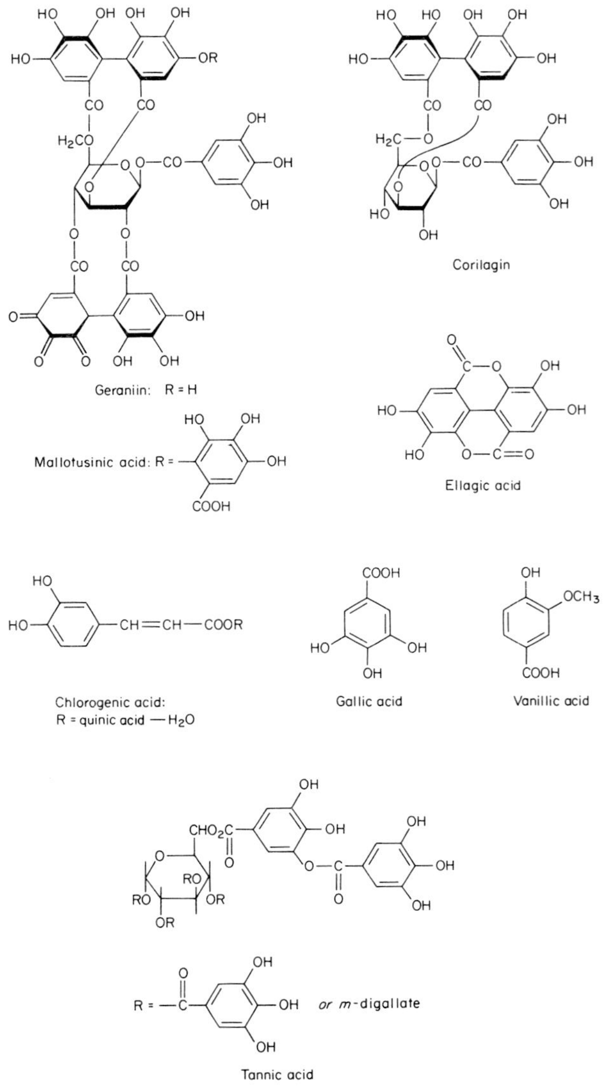

Figure 5. Some phenolic compounds of the Euphorbiaceae.

Ellagic acid (a constituent of some tannins) has been identified in several species, e.g. some species of *Euphorbia* (Rizk, Rimpler & Ismail, 1977; Rizk & El-Missiry, 1986), *Mallotus japonicus* (Yoshida *et al.*, 1982) and *Sapium sebiferum* (Kouno, Saishoji, Sugiyama & Kawano, 1983). Tri-*O*-methylellagic acid was also detected in several species, e.g. *Acalypha indica* L. (Talapatra, Goswami & Talapatra, 1981) and *Mallotus japonicus* (Yoshida *et al.*, 1982). 2,3-Dimethylellagic acid and 3,3-di-*O*-methylellagic acid (and its acetate) have been isolated from *Euphorbia royleana* Boiss. (Nazir, Ahmad, Bhatty & Karimullah, 1966) and *Euphorbia wallichii* Hook. fil. (Baslas, 1982), respectively.
Gallic and tannic acids were identified from *Euphorbia hirta* L. (Rizk, 1986)

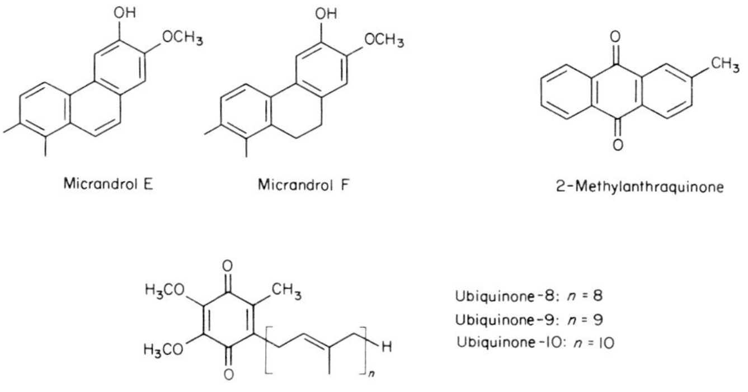

Figure 6. Some phenanthrenes and quinones of the Euphorbiaceae.

and phyllembin (ethyl gallate) from *Emblica officinalis* Gaertn. (syn. *Phyllanthus emblica* L.) (Chopra, Chopra & Varma, 1969).

Phenanthrenes and quinones

Two micrandrols, E (6-hydroxy-7-methoxy-1,2-dimethylphenanthrene) and F (6-hydroxy-7-methoxy-1,2-dimethyl-9,10-dihydrophenanthrene) have been identified from the trunk wood of *Sagotia racemosa* Baill. (Fig. 6) (de Alvarenga, Gottlieb & Magalhães, 1976). The trunk of *Jatropha glandulifera* Roxb. contains 3,3-dimethyl-acryl shikonin and acetylshikonin (Ballantine, 1969). 2-Methylanthraquinone was isolated from both *Acalypha indica* L. (Talapatra et al., 1981) and *Euphorbia pulcherrima* (Baslas, 1982). Ubiquinones (-8, -9 and -10) were identified from *Hevea brasiliensis* (Law, Threlfall & Whistance, 1970).

Phenolic acids

In addition to ellagic, gallic and tannic acids, several other phenolic acids were identified in plants of this family. Examples of these acids are: vanillic and veratric acids from *Euphorbia resinifera* (Boe, Wisnes, Nordal & Bernatek, 1969); *p*-coumaric acid from *Euphorbia acanthothamnos* Heldr. & Sart. (Kallimanis & Philianos, 1980); chlorogenic and neochlorogenic acids from *Ricinus communis* L. and *Mercurialis perennis* L. (Soboleva, 1980); and *m*-hydroxybenzoic acid from *Euphorbia royleana* (Baslas, 1982).

Other phenolic compounds

Four rottlerin-like phloroglucinol derivatives were isolated from the fruits of *Mallotus japonicus*; three of which identified as 3-(3,3-dimethylallyl)-5-(3-acetyl-2,4 - dihydroxy -5- methyl -6- methoxybenzyl) - phloroacetophenone, 3-(3-methyl-2-hydroxybut-3-enyl)-5-(3-acetyl-2,4- dihydroxy-5-methyl-6-methoxybenzyl)-phloroacetophenone (Shigematsu, Kouno & Kawano, 1983) and 2,6-dihydroxy-3-methyl-4-methoxyacetophenone (Kouno, Shigematsu, Iwagam & Kawano, 1985). The structures of these compounds resemble rottlerin from *M. phillipinensis* Muell.-Arg. (Shigematsu *et al.*, 1983).

The root bark of the Chinese tallow tree, *Sapium sebiferum* Roxb. contains the

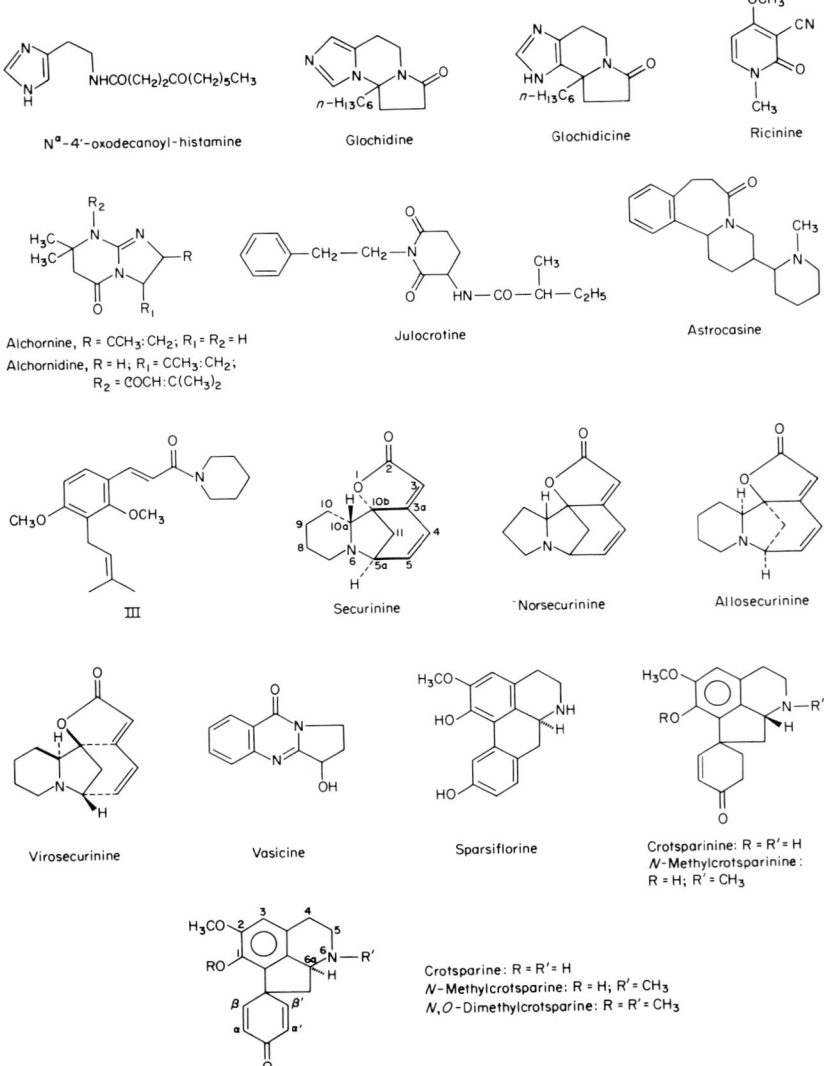

Figure 7. Some alkaloids of the Euphorbiaceae.

following two compounds: xanthoxylin (2-hydroxy-4,6-dimethoxyacetophenone), and a xylosylglucoside of xanthoxylin identified as 2-acetyl-3,5-dimethoxyphenyl-O-β-xylopyranosyl-$(1\rightarrow 6)$-β-D-glucopyranoside (Kouno et al., 1983).

Vanillin (p-hydroxy-m-methoxybenzaldehyde) occurred in *Croton eluteria* Benn. (Claus, 1961).

Alkaloids

Several classes of alkaloids occur in certain genera of the family Euphorbiaceae and in particular *Croton*, *Phyllanthus* and *Securinega* species

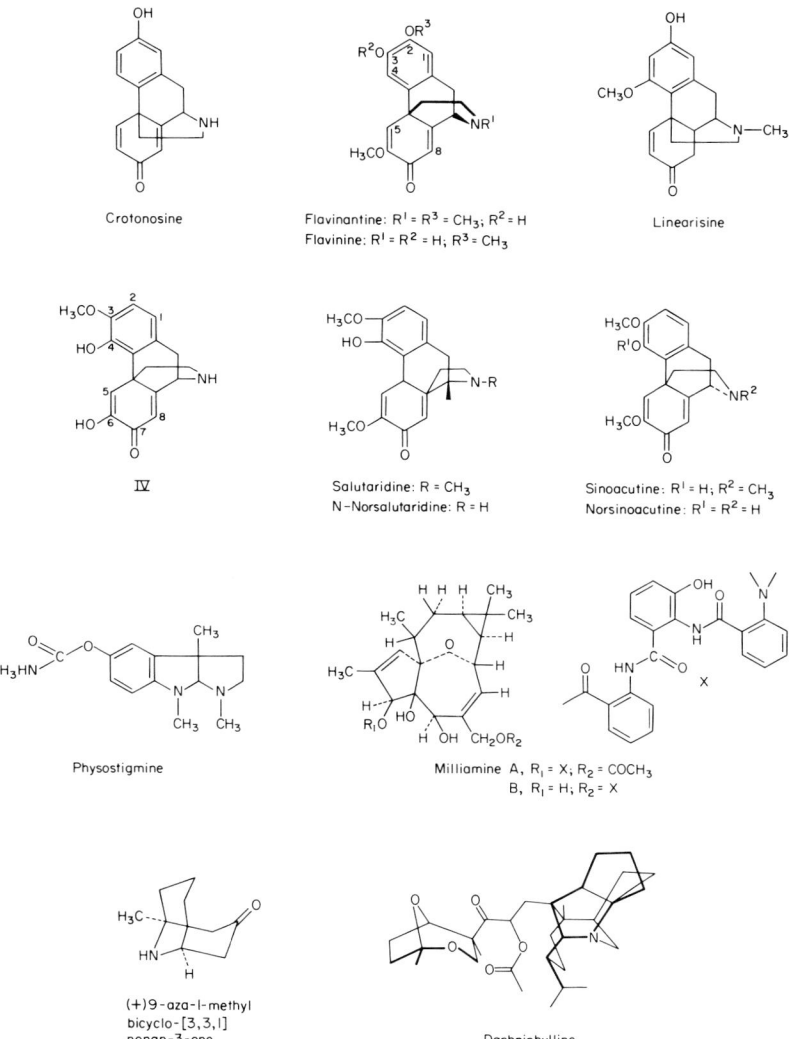

Figure 7 continued.

(Fig. 7). Imidazole alkaloids have been detected in only the genus *Glochidion*. *Croton* species contain several types of alkaloids, viz. isoquinolines (aporphines, e.g. sparsiflorine; proaporphines, e.g. crotsparine; and dihydroproaporphines, e.g. crotsparinine), morphinandienones (e.g. crotonosine), quinolines (e.g. vasicine), a pyrrolidine alkaloid (4-hydroxyhygrinic acid) and others (e.g. taspine, an alkaloid of unusual structure). Pyrimidine and guanidine alkaloids were isolated from only one species, viz. *Alchornea javanensis* Muell.-Arg.; while purine nuclei alkaloids were identified in only *Jatropha basiacantha* Pax & K. Hoffm. Several quinolizidine alkaloids were identified in the family, but mostly from *Phyllanths* and *Securinega* species. The presence of other classes in plants of the Euphorbiaceae was also reported, e.g. pyridine, piperidine and indole alkaloids (Table 4). Glycoalkaloids were reported in *Euphorbia dracunculoides*

Table 4. Alkaloids of the Euphorbiaceae

Alkaloids	Species	References
I. Imidazole alkaloids		
1. Nα-Cinnamoylhistamine	*Glochidion philippicum* (Cav. C. B. Rob. (= *G. philippense* Benth.)	Johns & Lamberton (1967), Manske (1968)
2. Nα-4′-Oxodecanoylhistamine		
3. Glochidicine	*Glochidion philippicum* (Cav.) C. B. Rob. (= *G. philippense* Benth.)	Johns & Lamberton (1967)
4. Glochidine		
II. Pyrimidine alkaloids		
5. Alchornidine	*Alchornea javanensis* Muell.-Arg.	Hart et al. (1969, 1970)
6. Alchornine (2-Isopropenyl-7,7-dimethyl-5-oxo-2,3,5,6,7,8-hexahydroimidazo[1,2-a]-pyrimidine		
III. Pyrrolidine alkaloids		
7. 4-Hydroxy-hygrinic acid	*Croton* spp.	Ponsinet & Ourisson (1965)
IV. Pyridine alkaloids		
8. Ricinine	*Ricinus communis* L.	Rizk & El-Missiry (1986)
9. Nudiflorine	*Trewia* spp.	Ponsinet & Ourisson (1965)
V. Piperidine alkaloids		
10. Astrocasine	*Astrocasia phyllanthoids* Robinson & Millsp.	Manske (1968)
11. 2,4-Dimethoxy-3-ψ-ψ-dimethylallyl-*trans*-cinnamoylpiperidide (III)	*Excoecharia agallocha* L.	Prakash et al. (1983)
12. Julocrotine	*Julocroton* spp.	Manske (1968), Yamaguchi (1970)
VI. Quinolizidine alkaloids		
13. Allosecurine	*Phyllanthus discoides* Muell.-Arg., *Securinega suffruticosa* Rehd.	Bevan et al. (1964)
14. Dihydrosecurine	*Phyllanthus* spp., *Securinega suffruticosa* Rehd.	Ponsinet & Ourisson (1965), Rizk & El-Missiry (1986)
15. Norsecurinine	*Fluegga* spp., *Phyllanthus* spp., *Securinega virosa* Pax. Hofm.	Ponsinet & Ourisson (1965), Yamaguchi (1970)
16. Phyllantidine	*Phyllanthus discoides*	Parello & Munavalli (1965)
17. Phyllantine (Metoxysecurinine)	*Phyllanthus discoides*	Parello & Munavalli (1965)

18. Phyllochrysine	*Phyllanthus* spp.	Ponsinet & Ourisson (1965)
19. Securinine	*Phyllanthus* spp., *Securinega suffruticosa* Rehd.	Ponsinet & Ourisson (1965), Yamaguchi (1970), Rizk & El-Missiry (1986)
20. Suffruticosine	*Phyllanthus* spp.	Ponsinet & Ourisson (1965)
21. Virosecurinine	*Fluegga* spp., *Securinega virosa* Pax. & Hofm.	Ponsinet & Ourisson (1965), Yamaguchi (1970)

VII. Quinazolone alkaloids

22. Tumuriquirensine	*Croton* spp.	Ponsinet & Ourisson (1965)
23. Vasicine	*Croton* spp.	Ponsinet & Ourisson (1965)

VIII. Isoquinoline alkaloids
  (a) *Aporphines*

24. Glaucine	*Croton draconoides* Muell.-Arg.	Bettolo & Scarpati (1979)
25. Orotonosine	*Croton* spp.	Ponsinet & Ourisson (1965)
26. Pronuciforine	*Croton* spp.	Ponsinet & Ourisson (1965), Bhakuni et al. (1970)
27. Sparsiflorine		
28. Thaliporphine	*Croton draconoides* Müll.-Arg.	Bettolo & Scarpati (1979)

  (b) *Dihydroproaporphines*

29. Crotsparinine (Jacularine)	*Croton linearis* Jacq, *C. sparsiflorus* Morong.	Bhakuni & Dhar (1969), Bhakuni et al. (1970)
30. *N*-Methylcrotsparinine	*Croton sparsiflorus* Morong.	Bhakuni et al. (1970)

  (c) *Proaporphines*

31. Crotsparine (Crotoflorine)	*Croton sparsiflorus* Morong. (*C. bonplandium* Bail.)	Bhakuni & Dhar (1968), Bhakuni et al. (1970)
32. *N,O*-Dimethylcrotsparine	*Croton sparsiflorus*	Bhakuni et al. (1970)
33. *N*-Methylcrotsparine		

IX. Morphinandienone alkaloids

34. Crotonosine	*Croton linearis* Jacq.	Yamaguchi (1970)
35. Flavinantine	*Croton flavens* L.	Stuart et al. (1969)
36. Flavinine	*Croton linearis* Jacq.	Yamaguchi (1970)
37. Linearisine	*Croton bonplandianum* Bail.	Tiwari et al. (1981)
38. 3-Methoxy-4,6-dihydroxy-morphinandien-7-one (IV)		
39. *N*-Norsalutaridine	*Croton salutaris* Casar.	Barnes & Soeiro (1981)
40. Norsinoacutine	*Croton* spp.	Manske (1968), Stuart et al. (1969), Tiwari et al. (1981)

Table 4. continued.

Alkaloids	Species	References
41. Salutaridine	*Croton salutaris* Casar.	Manske (1968), Barnes & Soeiro (1981)
42. Salutarine	*Croton salutaris* Casar.	Barnes & Soeiro (1981)
X. Indole alkaloids		
44. Physostigmine	*Hippomane* spp.	Ponsinet & Ourisson (1965)
45. Yohimbine	*Alchornea* spp.	Ponsinet & Ourisson (1965)
XI. Guanidine alkaloids		
46. N^1,N^1-Diisopentenyl-guanidine	*Alchornea javanensis* Muell.-Arg.	Hart et al. (1970)
47. N^1,N^2,N^3-Triisopentenyl-guanidine		
XII. Diterpenoid alkaloids		
48. Milliamine A	*Euphorbia milli* Desmoul.	Uemura & Hirata (1971)
49. Milliamine B		
XIII. Other alkaloids		
50. 9-Aza-1-methyl-bicyclo-[3,3,1]-nonan-3-one	*Euphorbia atoto* Forst.	Hart et al. (1967)
51. Daphnicadine	*Daphniphyllum calycinum* Benth.	Manske (1968
52. Daphnicaline		
53. Daphnicamine		
54. Daphniphylline	*Daphniphyllum macropodium* Miq.	Manske (1968)
55. Flueggeine	*Fluegga microcarpa* Blume (*F. virosa* Baill.)	Paris et al. (1955), Manske (1968)
56. Neodaphniphylline	*Daphniphyllum macropodium* Miq.	Manske (1968)
57. Purine nuclei alkaloids	*Jatropha basicantha* Pax & K. Hoffm.	Rizk & El-Missiry (1986)
58. Taspine	*Croton* spp.	Bettolo & Scarpati (1979)
59. Xanthorhamin	*Euphorbia hirta* L.	Rizk (1986)
60. Yuzurimine	*Daphniphyllum macropodium* Miq.	Manske (1968)

Table 5. Cyanogenic Glucosides of the Euphorbiaceae

Cyanogenic glucosides	Species	References
1. Acalyphin	*Acalypha indica* L.	Nahrstedt *et al.* (1982)
2. Linamarin (Phaseolunatin = manihotoxin)	*Cnidoscolus texanus* Muell.-Arg., *Hevea brasiliensis* Muell.-Arg., *Manihot* spp.	Valen (1978), Rizk & El-Missiry (1986)
3. Lotaustralin	*Manihot carthaginensis* Muell.-Arg.	Valen (1978)
4. Taxiphyllin (Phyllanthin)	*Phyllanthus gastroemii* Muell.-Arg.	Valen (1978)
5. Triglochinin	*Andrachne colchia* Fisch & Mey., *Securinega suffruticosa* (Pall.) Rehdr.	Valen (1978), Rizk & El-Missiry (1986)

Lam. (Singh & Srivastava, 1966) and *Euphorbia hirta* L. (Rizk & El-Missiry, 1986).

Cyanogenic glucosides

Though several species of the Euphorbiaceae have been reported as cyanophoric, yet relatively few cyanogenic glucosides have been identified (Table 5 & Fig. 8). Cyanogensis in the family have been surveyed by Valen (1978). Cyanophoric taxa occur frequently in the Euphorbiaceae. Both subfamilies, i.e. Phyllanthoideae and Euphorbioideae, comprise taxa which are able to produce hydrocyanic acid (Hegnauer, 1966; Valen, 1978). A number of species belonging to the following genera have been reported as cyanogenic: *Andrachne, Beyeria, Bridelia, Cnidoscolus, Colliguaja, Elateriospermum, Euphorbia, Gymnanthes, Hevea, Jatropha, Manihot, Mercurialis, Phyllanthus, Poranthera, Securinega* (Valen, 1978) and *Stillingia* (Lewis & Elvin-Lewis, 1977).

Glucosinolates

Few glucosinolates have been identified from only limited species of the Euphorbiaceae (Fig. 9). Examples of the glucosinolates detected are

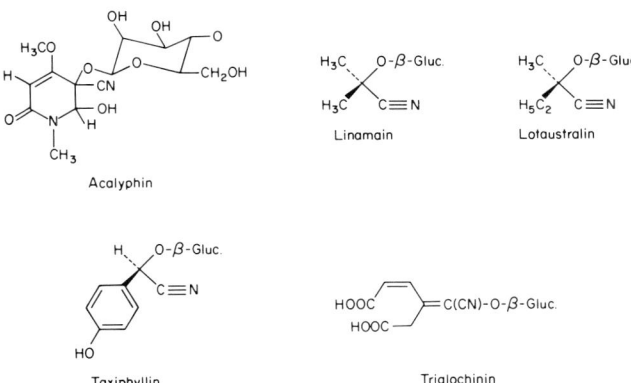

Figure 8. Cyanogenic glucosides of the Euphorbiaceae.

Glucocleomin

Cleomin

Glucocochlearin

Glucoputranjivin

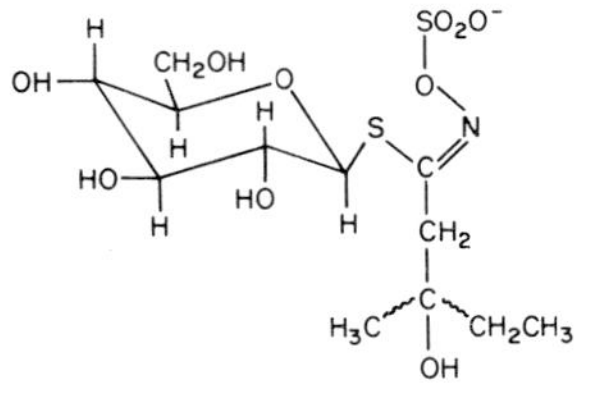

Figure 9. Some glucosinolates of the Euphorbiaceae.

glucoputranjivin, glucocochlearin, glucojaputin and glucocleomin (from kernels of *Putranjiva roxburghii* Wall.) and benzylisothiocyanate (an enzymatic hydrolysate of a glucosinolate from the latex of *Jatropha multifida* L.) (Kjaer & Friis, 1962; Rizk & El-Missiry, 1986).

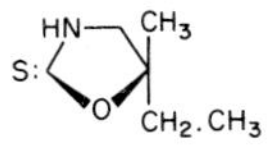

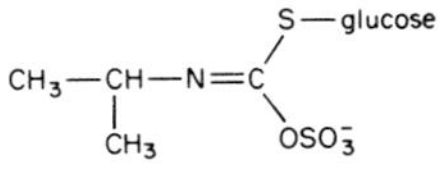

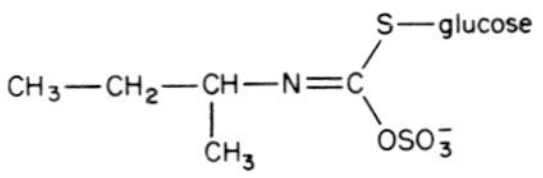

Miscellaneous compounds

A number of other compounds have been isolated from plants of the Euphorbiaceae (Fig. 10). Examples of these substances are: quinic and shikimic acids from *Euphorbia* species (Rizk & El-Missiry, 1986); amides, e.g. acalyphamide (an amide of tyramine and acid; $C_{31}H_{63}COOH$) and succinimide from *Acalypha indica* L. (Talapatra *et al.*, 1981); 1-methyl-6-hydroxy-1,2,3,4-tetrahydroxyisoquinoline-3-carboxylic acid from *Euphorbia myrsinites* L. (Mueller & Schuette, 1968); calcium 5,5-dimethyl-2-oxo-5,6-dihydro-2H-pyran-3,4-dicarboxylate from *Euphorbia biglandulosa* Desf. (Falsone & Spur, 1979); 1-tridecene-3,5,7,9,11-pentyne and *trans*-dehydromatricaria ester from *Ricinus communis* (Schulte, Reisch & Bornfleth, 1964); D-(+)-α-hydroxyglutaric acid, myoinositol, L(−)-inositol from *Euphorbia resinifera* (Boe *et al.*, 1969); cardiac glycosides from *Mallotus japonicus* (Shigematsu *et al.*, 1983); laiodiplodin, a potent antileukemic macrolide, from the stems of *Euphorbia splendens* Bojer (*E. milii* Boiss.) (Lee *et al.*, 1982); melissic acid, $CH_3(CH_2)_{28}COOH$, from *Euphorbia hirta* (Rizk, 1986); and diacetylmaderine from *Phyllanthus maderaspatensis* L. (Chopra, Chopra & Varma, 1969).

Laiodiplodin

Quinic acid

Shikimic acid

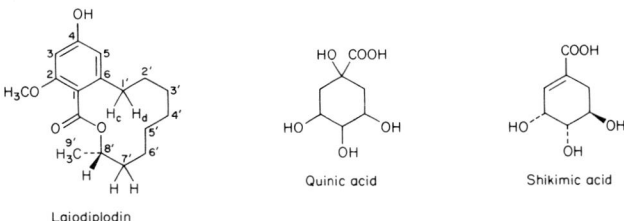

Figure 10. Some miscellaneous compounds of the Euphorbiaceae.

ECONOMIC PLANTS

Several plants of the Euphorbiaceae are of considerable economic importance, and products obtained from this family include castor oil (*Ricinus*), tung oil (*Aleurites*), cassava, tapioca (*Manihot*) and rubber (*Hevea*). By no means have all of these been shown to contain irritant carcinogens (Kinghorn, 1979). Many species, and in particular *Euphorbia* and *Croton* species, are reported to cause poisoning of human beings and livestock (Watt & Breyer-Brandwijk, 1962; Kingsbury, 1964). Poisoning is caused by the acrid and irritant substances (recently identified as diterpenoid esters). Ethnobotanical records (Watt & Breyher-Brandwijk, 1962) indicate that the toxicological dangers presented by *Euphorbia* species have been recognized as far as the succulent species are concerned.

Many of these toxic plants are ornamental and household plants, including several *Euphorbia* species (e.g. *E. milii*, *E. tirucalli* L., *E. lactea* Roxb.), *Acalypha wilksiana* Muell.-Arg., *Ricinus communis*, *Jatropha multifida* L., *Sapium sebiferum* and *Syndadenium grantii* hook. fil.

Industrial plants and/or their products

Food plants

Manihot: Next to the sweet potato, the cassava (*Manihot esculenta* Crantz) is the most important of the tropical root crops and furnishes the basic food for millions of people. Cassava also called manioc, mandioc or yuca is one of the most wholesome foods. The tubers are eaten raw or cooked. Cassava bread has a high food value and replaces wheat bread in some diets (Hill, 1952). The tubers of *M. esculenta* contain about 77.5–88.5% starch (Chopra *et al.*, 1969) but less than 1% protein (Vickery & Vickery, 1979). The fresh tubers contain a useful amount of vitamin C (27 mg per 100 g) and Ca (25 mg per 100 g) (Vickery & Vickery, 1979). Though sweet varieties of cassava contain linamarin, this cyanogenic glucoside occurs only in the outer peel and is removed before the tubers are eaten. Despite its lack of protein, cassava is a useful staple food, as the mature tubers can be left in the ground for up to two years. West African villagers like to keep a plot of cassava in case other crops fail (Vickery & Vickery, 1979).

Cassava starch, Para or Brazilian arrowroot, is the starch obtained from the thickened roots of the butter cassava (*Manihot utilissima* Pohl) and the sweet cassava (*M. palmata* Muell.-Arg.). These plants are reported to produce large roots weighing as much as 5 kg (Hill, 1952).

Cassava is one of the principal crops in much of Africa, India and Brazil. The latter is the largest world producer of Cassava (Vickery & Vickery, 1979). In India it is grown as a subsidiary food crop and as an excellent raw material for sago manufacture (Chopra *et al.*, 1969).

Other plants: The otaheite gooseberry (*Phyllanthus acidus*) yields yellow cherrylike fruits (Hill, 1952). The dried bark of *Croton eluteria* Benn. is used in flavouring liquors and in scenting tobacco. It contains 1–1.5% volatile oil containing eugenol, limonene, cascarillin, tannin and vanillin (Claus, 1961). The nuts of *Cnidoscolus marcgravii* Pohl., a tree identical with *Jatropha oligandra*

Muell.-Arg. are used by the natives in Rio de Janeiro as food and the oil is pressed from them for cooking (Bondar, 1942).

Fatty oils and waxes

Castor oil: The seeds of *Ricinus communis* L., are the source of castor oil, the yield being up to 54%. Castor oil contains 1% tocopherol, 21.6% non-sterol fraction of the unsaponifiable matter and 0.9% castor germ oil. The main constituents of the oil are glycerin esters of ricinoleic acid (Rizk, 1986).

Castor oil has been put to many industrial uses, in addition to its traditional use as a soothing local application and as a successful local application to get rid of warts. It is useful as a plasticizer in the lacquer industry, mixed with oil derivatives as a lubricant and mixed with alcohol as a brake fluid. It is also used in making Turkey red oil, a moistener used in the textile industry (Watt & Breyer-Brandwijk, 1962). The oil is water resistant and therefore is used for coating fabrics and for protective coverings for aeroplanes, insulation, food containers, gums, etc. It is also used in paints, varnishes, soap, inks, plastics and as an illuminant.

Ricinus zanzibarinus G. H. Popova yields 53.7–65.0% oil resembling castor oil and is usable for the same purposes (Watt & Breyer-Brandwijk, 1962).

Tung oil: Tung oil, sometimes called Chinawood oil is the fixed oil obtained from the seeds of both *Aleurites fordii* Hemsl. and *A. montana* E. H. Wilson. The oil is not edible but is a quick-drying solvent for paints and varnishes. It is a good preservative and very resistant to weathering and is therefore valuable for outside paints. It is also used in the manufacture of linoleum, oilcloth, water proofing fabrics, brake linings, leather dressings, soap, inks, insulating compounds and fibreboard. The quick-drying properties of tung oil are due to its content of eleostearic acid (Hill, 1952; Vickery & Vickery, 1979).

Candlenut oil, also known as lumbang oil, is obtained from the hard-shelled seeds or roots of *Aleurites moluccana* Willd. and has similar uses as tung oil. The oil is also used as a preservative of the hulls of vessels (Hill, 1952).

Aleurites species are a profitable source of income, and have also solved the problem of eroded and waste land as they grow on soils unsuitable for other types of agriculture (Hill, 1952).

The seeds of *Trewia nudiflora* L. yield an oil similar to tung oil (Sarkar & Chakrabarty, 1956a).

Tallow: The seeds of *Sapium sebiferum* represent the source of Chinese vegetable tallow (Hill, 1952). An early report suggested its use as a possible substitute of animal tallows in the manufacture of cotton (Puntambeker & Krishna, 1932). The outer layer of the seeds of *Stillingia sebifera* Trillot contained 54.5% tallow (Kaufmann & King, 1939).

Candilla wax: Candilla wax is obtained from *Euphorbia antisyphilitica* Zucc., *E. cerifera* Alcocet, *Pedilanthus aphyllus* Boiss. and *P. pavonis* Boiss. (Hill, 1952; Daugherty et al., 1953). It is generally used as extender wax mixed with others.

Other oils: The evaluation of the seed oils of several other Euphorbiaceous plants has been reported. Some of the oils have been reported as equal or superior to the common oils, e.g. the seed oils of *Euphorbia heterophylla* L. and *E. marginata* Pursh have been stated as equal to, or superior to, linseed as a drying oil (Earle et al., 1960). Seeds of *Jatropha gossypifolia* L. were reported as a common oilseed crop as compared with palm oil (Husain *et al.*, 1982). The nut-kernels of *Cnidoscolus margravii* contained 28.8% total fat and may serve for the manufacture of paints, soap and other articles where soybean oil is used (Bondar, 1942). *Chrozophora plicata* A. Juss. yields lineolate rich seed oil (Hassan *et al.*, 1980). The seed oil content of some other species are shown in Table 6.

Rubber plants

Several rubber plants belong to the family Euphorbiaceae. The *Hevea* or Para rubber tree (*Hevea brasiliensis*) is the major source of the rubber produced throughout the world. Elastica, caoutchouc or India rubber is the prepared latex or milk of *Hevea brasilensis* and probably of other species of *Hevea* (Claus, 1961).

Ceara or manicobe rubber is obtained from *Manihot glaziovii* Muell.-Arg., while caura rubber of Venezuela is obtained from species of *Micandra*. Species of *Cnidoscolus* yield chilte rubber (Hill, 1952).

Many *Euphorbia* species either yield, or have been reported as a possible source of, rubber, e.g. *E. intisy* (Hill, 1952). The latex obtained from *E. caducifolia* Haines gives marginally better natural rubber than does *Calotropis procera* (Banerji & Millns, 1981). The rubber percentages of some other *Euphorbia* lattices are as follows: *E. abyssinica* J. F. Gmel., 16.7%; *E. candelabrum* Trem. ex Klotzsch, 20%; *E. dregeana* E. Mey., 17.6%; and *E. mauritanica* L., 15.81% (Watt & Breyer-Brandwijk, 1962).

Wood and paper

Few Euphorbiaceous trees have been reported to provide timber suitable for making furniture and for structural work. Examples of these plants are: the sandbox tree (*Hura crepitans* L.) which produces timber suitable for furniture (Vickery & Vickery, 1979); trees of *Bridelia ferruginea* Benth., *Oldfeldia africana* Benth. & Hook. fil. and *Uapaca* species provide timber for structural work (Vickery & Vickery, 1979); *Androstachy johnsonii* Prain produces a hard timber, durable and resistant to termites (Watt & Breyer-Brandwijk, 1962); and *Ricinodendron affricanus* Muell.-Arg. produces a valuable timber (Watt & Breyer-Brandwijk, 1962).

Table 6. The seed-oil content of some euphorbiaceous plants

Species	Oil (%)	References
1. *Antidesma diandrum* Spreng.	10.2	Sarkar & Chakrabarty (1965b)
2. *Bischofia javanica* Blume	21.4	
3. *Euphorbia calycina* N. E. Br.	20.8	Watt & Breyer-Brandwijk (1962)
4. *E. dracunculoides* Lam.	25.0	Singh & Srivastava (1966)
5. *E. gregaria* Marl.	40.8	Watt & Breyer-Brandwijk (1962)
6. *E. helioscopia* L.	32.6	
7. *Hevea brasiliensis* (A. Juss.) Muell.-Arg.	23–32	Hill (1952)
8. *Jatropha* spp.	32.0	Sarkar & Chakrabarty (1956b)
9. *Ricinodendron africannum* Muell.-Arg.	15.5	
10. *R. rautanenii* Schinz.	30–40 (edible oil)	Watt & Breyer-Brandwijk (1962)

Possibilities of obtaining pulp and paper from many plants of the Euphorbiaceae are reported, e.g. *Hura crepitans* (Rojas, 1964), *Macaranga tanorius* Muell.-Arg. (Hui, Li & Ng, 1975), *Ricinodendron heudelotti* Pax (Vickery & Vickery, 1979) and *Sapium sebiferum* (Tutiya & Kato, 1940).

Ricinus communis L. yields a fibre suitable for paper making of superior quality, if made with an admixture of bamboo pulp (Watt & Breyer-Brandwijk, 1962).

The mucilages of *Manihot* are used for paper making (Oguri & Shinohara, 1957).

Petroleum plants

Some crops have been recently developed for the production of fuels and materials in addition to food and fibre. *Euphorbia* species, and in particular *E. lathyris*, have been suggested as crops for the production of a whole-plant oil low in isoprene content. Research on several latex-rich species showed that low molecular weight hydrocarbons suitable for cracking to liquid fuels occur in reasonable percentages (10% or more of total dry weight) in many species of the Euphorbiaceae. The Gopher plant (*Euphorbia lathyris*), ranked among the highest species analysed, contains up to 10% reduced hydrocarbons (acetone extractable material) (Rizk, 1986).

Euphorbia lathyris is one of the plants which pass through summer drought conditions, overwinter and, with warming temperatures in spring, commence growth. The plant has been estimated to yield approximately ten barrels of hydrocarbon material per acre in a 7-month growing period on semiarid land (Calvin, 1979). The product is a mixture of hydrocarbons consisting mainly of open-chain and cyclic terpenoid isoprenes (Nemethy, Otvos & Calvin, 1981a). According to Calvin (1982), 95% of the oil from *E. lathyris* can be cracked to useful products, making it very valuable petrochemical feedstock from which gasoline itself can be made. The evaluation of this species as one of the few potential hydrocarbon-producing crops has been reported by several authors (e.g. Calvin, 1979, 1980, 1982, 1983; Calvin, Nemethy, Redenbaugh & Otvos, 1982; Hinman, Hoffman, McLaughlin & Peoples, 1980; Nemethy *et al.*, 1981a, b; Peoples *et al.*, 1981).

Fish poison

The latex of several genera of the Euphorbiaceae, and in particular species of *Euphorbia*, have been used extensively by fishermen in different countries as a fish poison of high biological activity (Watt & Breyer-Branwijk, 1962; Novack, Crea & Falsone, 1980). The rhizomes of *Euphorbia biglandulosa* Desf., for example, are pounded in order to free the latex and thrown into stagnant waters or rivers with a lower water level. The rapidly dissolving poison then kills the fish by an apparent paralysis. However, though the sap causes intensive irritation of the human skin, surprisingly, no intoxications of the people who handle the latex and eat the fish were described (Novack, Crea & Falsone, 1980). Examples of the Euphorbiaceous plants used as fish poison are *Aleurites montana* E. H. Wilson, *Antidesma venosum* E. Mey., *Croton sylvaticus* Hochst. (Watt & Breyer-Brandwijk, 1962), *Elaeophorbia drupifera* Stapf (Vickery & Vickery, 1979) and *Euphorbia* species (Baslas, 1982).

Arrow poison

In the past, arrow poisons have been very important to the people of Africa and other countries, as they were often the only way of killing large animals, especially in poaching valuable game animals such as leopards and lions (Vickery & Vickery, 1979). Examples of the plants used as arrow poisons are *Euphorbia* species (Watt & Breyer-Brandwijk, 1962) and *Hippomane mancinella* L. (Ayensu, 1981).

Other uses

The main toxic principle of the seeds of *Ricinus communis* is the phytotoxin ricin (a protein of the globulin type) converted by acid hydrolysis into non-toxic compounds which are of nutritional value. The seed cake (after removal of the oil) is used as a fertilizer. The oil cake of *Aleurites* species is a good fertilizer, but cannot be used as a feed (Hill, 1952). *Androstachys johnsonii* Prain has been used as an animal feed (Watt & Breyer-Brandwijk, 1962).

Medicinal plants

Since antiquity curative effects have been attributed to many plants of the Euphorbiaceae; for example *Euphorbia fischeriana* Steud. (= *E. pallasii* Turcz.) has been used in traditional Chinese medicine for more than 2000 years as antitumour drug (Schroeder et al., 1980). Plants of this family have been used to treat cancer, tumours and warts from at least the time of Hippocrates (circa 400 B.C.), and references to their used have appeared in the literature of many countries (Kupchan et al., 1976). Certain species have been described in homoeopathic pharmacopoeias, e.g. *Acalypha indica* (Nahrstedt, Kant & Wray, 1982).

In India, members of the following genera are reported as medicinal plants: *Acalypha, Aleurites, Andrachne, Antidesma, Aporosa, Baliospermum, Bischofia, Breynia, Bridelia, Chrozophora, Cicca, Cleistanthus, Croton, Euphorbia, Excoecaria, Fluegga, Glochidion, Hippomane, Homonoia, Hura, Jatropha, Macaranga, Mallotus, Manihot, Phyllanthus, Putranjiva, Ricinus, Sapium, Sebastiania, Tragia* and *Trewia* (Chopra et al., 1983; Kirtikar & Basu, 1984). Species of Euphorbiaceae have been used by local populations of many countries in folk medicines as remedies against several diseases. Table 7 summarizes the medicinal uses.

The plants of this family are known to be extremely toxic; causing inflammations of the skin and mucous membranes, conjuctivitis and even blindness. Several species are toxic to livestock and allelopathic to desirable forage plants. Examples of the poisonous plants are *Aleurites fordii* Hemsl., *Croton* species, *Euphorbia* species, *Excoecaria venenifera* Pax, *Hippomane mancinella, Hura crepitans, Jatropha* species, *Pedilanthus tithymaloides* Poit., *Phyllanthus* species, *Sapium* species, *Stillingia treculeana* I. M. Johnston and *Syndenium grantii* (Watt & Breyer-Brandwijk, 1962; Kingsbury, 1964; Lewis & Elvin-Lewis, 1977).

The attempts over 150 years to establish the cathartic principles of *Croton tiglium* L. were given a new impetus after the discovery of the tumour-promoting (carcinogenic) activity of *Croton* oil (Kinghorm, 1979). These efforts culminated with the isolation of the irritant cocarcinogenic factors of the oil as esters of

Table 7. Medicinal uses of some euphorbiaceous plants

Used as/Treatment of	Species	References
1. Abdominal pains	*Antidesma venosum* E. Mey., *Briedilia scleroneuroides* Pax	Watt & Breyer-Brandwijk (1962)
2. Abortion	*Alchornea cordifolia* (Schum. & Thonn.) Stapf, *Bridelia micrantha* Hochst. (Baill.) (anti-abortive)	Ayensu (1978)
3. Anthelmintic (Vermifuge)	*Acalypha indica* L., *Alchornea cordifolia*, *Andrachne ovalis* Muell.-Arg., *Croton macrostachys* A. Rich., *Euphorbia* spp., *Jatropha curcas* L., *Mallotus* spp.	Caius & Mhaskar (1923), Watt & Breyer-Brandwijk (1962), Lewis & Elvin-Lewis (1977), Ayensu (1978, 1981), Biswas & Chopra (1982)
4. Antiasthmatic	*Croton* spp., *Euphorbia* spp., *Jatropha gossypifolia* L., *Ricinus communis* L.	Claus (1961), Watt & Breyer-Brandwijk (1962), Ayensu (1978, 1981), Baslas (1982), Boulos (1983), Duke & Ayensu (1985)
5. Antibiotic (mainly anti-bacterial)	*Acalypha wilkesiana* Muell.-Arg., *Cnidoscolus urens* Arthur, *Croton sellowii* Baill., *Euphorbia* spp., *Flueggea virosa* Baill. (Willd.) Baill., *Sauropus rostratus* Miq.	Watanabe et al. (1956), Watt & Breyer-Brandwijk (1962), Lewis & Elvin-Lewis (1977), Adesina et al. (1980), Duke & Ayensu (1985)
6. Anticancer	*Croton* spp., *Euphorbia* spp.	Kupchan et al. (1976), Lee et al. (1982), Baslas (1982)
7. Antihistaminic	*Euphorbia hirta* L.	Duke & Ayensu (1985)
8. Antipyretic (antifebrile)	*Euphorbia nerifolia* L.	Watt & Breyer-Brandwijk (1962), Hallet & Parks (1953)
9. Antispasmodic (spasmolytic)	*Euphorbia* spp.	Baslas (1982), Duke & Ayensu (1985)
10. Aphrodisiac	*Mallotus* spp., *Phyllanthus* spp., *Richeria grandis* Vahl.	Ayensu (1978), Duke & Ayensu (1985)
11. Bronchitis	*Euphorbia* spp.	Ayensu (1978), Chen et al. (1979), Baslas (1982)
12. Chest complaints	*Croton flavens* L., *Erythrococca rigidifolia* Pax, *Euphorbia* spp., *Flueggea virosa* Baill., *Mildbraedia fallax* Hutch., *Monadenium invenustum* N. E. Br., *Phyllanthus muellerians* (O. Ktze.) Exell.	Watt & Breyer-Brandwijk (1962), Ayensu (1978, 1981), Boulos (1983), Duke & Ayensu (1985)
13. Choleretic	*Euphorbia* spp.	Baslas (1982)
14. Conjunctivitis	*Bridelia micrantha*, *Croton tiglium* L., *Phyllanthus* spp.	Ayensu (1978), Duke & Ayensu (1985)
15. Contraceptive	*Mallotus philippinensis* Muell.-Arg.	Duke & Ayensu (1985)
16. Cough	*Alchornea cordifolia*, *Elaeophorbia drupifera* (Thonn.), *Phyllanthus englei* Pax	Lewis & Elvin-Lewis (1977), Ayensu (1978)
17. Dermatitis	*Croton cortesianus* H. B. & K., *Glochidion* spp.	Lewis & Elvin-Lewis (1977), Duke & Ayensu (1985)
18. Diabetes	*Bridelia ferruginea* Benth., *Croton* spp., *Euphorbia hirta* L., *Phyllanthus* spp.	Goñalons & Fontana (1926), Watt & Breyer-Brandwijk (1962), Ayensu (1978, 1981), Baslas & Agarwal (1980)
19. Diaphoretic	*Euphorbia genistoides* Berg.	Watt & Breyer-Brandwijk (1962)

20. Diarrhoea	*Acalypha australis* L., *Alchornea cordifolia*, *Bridelia ferruginea*, *Euphorbia* spp., *Jatropha* spp, *Phyllanthus* spp., *ricinus communis* L., *Ricinodendron africanum* Muell.-Arg., *Sauropus rostratus* Miq.	Watt & Breyer-Brandwijk (1962), Ayensu (1978, 1981), Baslas & Agarwal (1980), Boulos (1983)
21. Diuretic	*Acalypha erxardii* Gagnep., *Alchornea cordifolia*, *Bischovia javanica* Bl. *Bridelia ferruginea*, *Euphorbia* spp., *Hippomane mancinella* L., *Homonoia riparia* L., *Jatropha curcas* L., *Kirganelia reticulata* (Poir.) Baill, *Mercurialis annua* L., *Manihot esculenta* Crantz, *Phyllanthus* spp., *Sapium sebiferum* (L.) Roxb.	Woerd (1941), Watt & Breyer-Brandwijk (1962), Chopra et al. (1969), Lewis & Elvin-Lewis (1977), Baslas (1982), Biswas & Chopra (1982), Boulos (1983), Kouno et al. (1983), Duke & Ayensu (1985)
22. Dysentry	*Acalypha australis* L., *Alchornea cordifolia*, *Breynia* spp., *Bridelia ferruginea*, *Caloxylon* spp., *Euphorbia* spp., *Glochidion* spp., *Mallotus oppositifolius* (Geisel.) Muell.-Arg.	Watt & Breyer-Brandwijk (1962), Ayensu (1978, 1981), Duke & Ayensu (1985)
23. Dyspepsia	*Euphorbia antiquorum* L., *Jatropha curcas* l.	Ayensu (1978), Duke & Ayensu (1985)
24. Elephantiansis	*Glochidion puberum* (L.) Hutch., *Hura crepitans* L.	Chopra et al. (1969), Duke & Ayensu (1985)
25. Emetic	*Acalypha indica*, *Alchornea cordifolia*, *Euphorbia* spp, *Hura crepitans*, *Jatropha curcas*, *Mareya micrantha* Muell.-Arg., *Mildraedia fallax* Hutch., *Pedilanthus tithymaloides* Poit.	Watt & Breyer-Brandwijk (1962), Ayensu (1978, 1981), Duke & Ayensu (1985)
26. Emmenagogue	*Ricinus communis*	Watt & Breyer-Brandwijk (1962)
27. Enema	*Jatropha curcas*, *Mallotus oppositifolius* Muell.-Arg., *Phyllanthus muellerianus* (Kuntze) Exell.	Ayensu (1978)
28. Expectorant	*Acalypha indica*	Lewis & Elvin-Lewis (1977)
29. Febrifuge	*Sarococca pruniformis* Lindl.	Chopra et al. (1969)
30. Fevers	*Acalypha australis*, *Alchornea cordifolia*, *Breynia* spp., *Bridelia ferruginea*, *Croton linearis*, *Euphorbia* spp. *Jatropha* spp., *Phyllanthus* spp., *Sauropus rostratus*, Miq., *Saussurea lappa* C. B. Clarke, *Tetrorchidium didymostemon* (Baill.) Pox & K. Hoffm., *Trigonostemon longifolius* Baill.	Serra (1944), Chopra et al. (1969), Lewis & Elvin-Lewis (1977), Ayensu (1978, 1981), Duke & Ayensu (1985)
31. Galactogogue	*Euphorbia* spp.	Watt & Breyer-Brandwijk (1962)
32. Heart diseases	*Euphorbia hirta* L.	Ayensu (1978)
33. Haemorrhage	*Acalypha australis*, *Alchornea cordifolia*, *Breynia* spp.	Ayensu (1978), Duke & Ayensu (1985)
34. Hepatitis	*Euphorbia splendens* Bojer (*E. millii* Desmoulins), *Mallotus apelta* (Lour.) Muell.-Arg., *Phyllanthus urinaria* L.	Lee et al. (1982), Duke & Ayensu (1985)
35. Hypnotic	*Euphorbia* spp.	Baslas (1982)
36. Hypotensive	*Aleurites jordii*, *Euphorbia maddeni*	Claus (1961), Sahai et al. (1981)
37. Influenza	*Clutia abyssinica* Spach., *Glochidion purpureum* (L.) Hutch.	Watt & Breyer-Brandwijk (1962), Duke & Ayensu (1985)
38. Insecticides	*Croton* spp., *Hura crepitans*, *Mallotus rapandus* (Willd.) Muell.-Arg.	Lewis & Elvin-Lewis (1977), Duke & Ayensu (1985)
39. Jaundice	*Euphorbia humifusa* Willd., *Jatropha curcas*, *Ricinus communis*	Ayensu (1978), Boulos (1983), Duke & Ayensu (1985)
40. Lactogogue	*Euphorbia thymifolia* L., *Jatropha curcas*	Ayensu (1978, 1981)
41. Leprosy	*Hura crepitans*, *Mareya micrantha*, *Ricinus communis*	Chopra et al. (1969), Ayensu (1978), Boulos (1983)
42. Malaria	*Alchornea cordifolia*, *Croton* spp, *Glochidion puberum* Hutch.:	Watt & Breyer-Brandwijk (1962), Ayensu (1978), Duke & Ayensu (1985)

Table 7. continued

Used as/Treatment of	Species	References
43. Menstruation	*Euphorbia atoto* Forst.	Lewis & Elvin-Lewis (1977)
44. Nephritis	*Euphorbia* spp., *Phyllanthus urinaria* L.	Duke & Ayensu (1985)
45. Neuralgia	*Euphorbia tirucalli*	Baslas (1982)
46. Oedema	*Bridelia ferruginea*	Ayensu (1978)
47. Ophthalmic diseases	*Acalypha* spp., *Alchornea cordifolia*, *Euphorbia* spp., *Jatropha curcas*, *Phyllanthus muellerianus*	Ayensu (1978), Boulos (1983), Duke & Ayensu (1985)
48. Piles	*Alchornea cordifolia*, *Aleurnus lucidus* (Sw.) Rohm., *Homonia riparia* Lour., *Mallotus oppositifolius* (Geisel.) Muell.-Arg.	Ayensu (1978, 1981), Biswas & Chopra (1982)
49. Purgative (Cathartic, Laxative)	*Acalypha* spp., *Alchornea cordifolia*, *Bridelia* spp., *Claoxylon polot* Merr., *Clutia abyssinica* Spach., *Croton* spp., *Elaeophorbia drupifera* Stapf., *Euphorbia* spp., *Hippomane mancinella* L., *Homonia riparia* Lour., *Hura crepitans*, *Jatropha* spp. *Mallotus philippinensis* Muell.-Arg., *Maprouna africana* Muell.-Arg., *Mareya micrantha* (Benth.), Muell.-Arg., *Mercurialis annua* L., *Midbraedia fallax* Hutch., *Phyllanthus* spp., *Pseudolachnostylis* spp., *Ricinus communis*, *Sapium sebiferum*, *Tetrochidium didymostemon* (Baill.) Pax & K. Hoffm.	Claus (1961), Watt & Breyer-Brandwijk (1962), Chopra et al. (1969), Ayensu (1978, 1981), Biswas & Chopra (1982), Baslas (1982), Lewis & Elvin-Lewis (1977), Kouno et al. (1983), Duke & Ayensu (1985)
50. Rheumatism	*Alchornea cordifolia*, *Bridelia ferruginea*, *Croton* spp., *Euphorbia* spp., *Jatropha* spp., *Mareya micrantha*, *Mercurialis annua*, *Ricinus communis*, *Sarococca pruniformis* Lindl., *Speranskia tuberculata* (Bunge) Baill.	Ballantine (1969), Chopra et al. (1969), Ayensu (1978, 1981), Baslas (1982), Boulos (1983), Duke & Ayensu (1985)
51. Scabies	*Bridelia micrantha* (Hochst.) Baill.	Ayensu (1978)
52. Schistosomicidal	*Euphorbia pekinensis* Rupr.	Duke & Ayensu (1985)
53. Scorpion-bite	*Pedilanthus tithymaloides* Poit.	Chopra et al. (1969)
54. Smallpox	*Kirganelia reticulata* (Poir.) Baill. (= *Phyllanthus reticulatus* Poir.)	Chopra et al. (1969)
55. Snake-bite	*Acalypha* spp., *Claytia similis* Muell.-Arg., *Elaeophorbia druptifera*, *Erythrococca rigidifolia* Pax, *Euphorbia* spp., *Manihot esculenta* Crantz, *Mareya micrantha*, *Sapium sebiferum*	Watt & Breyer-Brandwijk (1962), Lewis & Elvin-Lewis (1977), Ayensu (1978, 1981), Duke & Ayensu (1985)
56. Toothache	*Alchornea* spp., *Bischofia javanica*, *Breynia* spp., *Croton alamosanum* N. E. Rose, *Euphorbia* spp., *Ricinus communis*	Lewis & Elvin-Lewis (1977), Ayensu (1978, 1981)
57. Tonic	*Croton eluteria* (L.) Sw., *Euphorbia* spp.	Ayensu (1978, 1981)
58. Veneral diseases (Gonorrhoea and syphilis)	*Alchornea cordifolia*, *Bridelia ferruginea*, *Breynia* spp., *Euphorbia* spp., *Hippomane mancinella*, *Homonia riparia*, *Jatropha curcas*, *Kirganelia reticulata* Baill., *Mallotus* spp, *Mareya micrantha*, *Monadenium lugardae* N. E. Br., *Phyllanthus* spp., *Ricinodendron africanum* Muell.-Arg.	Watt & Breyer-Brandwijk (1962), Chopra et al. (1969), Ayensu (1978, 1981), Biswas & Chopra (1982), Duke & Ayensu (1985)
59. Urinary diseases	*Croton humilis* L.	Lewis & Elvin-Lewis (1977)

tetracyclic diterpenoid phorbol (Hecker, 1968). After the structure elucidation of phorbol, many related diterpene esters have been identified from the family Euphorbiaceae, especially from *Euphorbia* species (Evans & Soper, 1978; Kinghorn, 1979). Certain of these compounds are tumour promoters while others have anti-tumour action. However, all of them are extremely potent direct primary irritants on mammalian skin (Rizk, 1986). Investigation of *Euphorbia esula* L. and *Croton tiglium* L. which have been used widely in folk medicine for treating cancers led to the isolation of two diterpenoid esters which showed antileukemic activity (Kupchan, 1976). On the other hand there are several genera which have been reported either responsible for irritant dermatitis (e.g. *Cnidoscolus, Dalechampia, Jatropha* and *Tragia*) or causing allergic reactions (e.g. *Codiaeum, Croton, Euphorbia, Hippomane, Hura* and *Phyllanthus*) (Lewis & Elvin-Lewis, 1977).

REFERENCES

ADESINA, S. K., OGANTIMEIN, B. J. & AKINWUSHI, D. D., 1980. Phytochemical and biological evaluation of the leaves of *Acalypha wikesiana* (red acalypha, Euphorbiaceae). *Quarterly Journal of Crude Drug Research, 18:* 45–48.

ADOLF, W. & HECKER, E., 1971. Further new diterpene esters from the irritant and co-carcinogenic seed oil and latex of the caper spurge. *Experientia, 27:* 1393–1394.

AFZA, N., MALIK, A. & SIDDIQUI, S., 1979a. A new triterpenoid from *Euphorbia tirucelli*. *Pakistan Journal of Scientific and Industrial Research, 22:* 124–127.

AFZA, N., MALIK, A. & SIDDIQUI, S., 1979b. Isolation and structure of cycloeuphornol, a new triterpene from *Euphorbia tirucalli*. *Pakistan Journal of Scientific and Industrial Research, 22:* 173–176.

AHMAD, S. A., KAPOOR, S. K. & ZAMAN, A., 1972. Bergenin in *Fluegga microcarpa*. *Phytochemistry, 11:* 452.

AIYAR, V. N. & SESHADRI, T. R., 1971. Chemical components of *Croton oblongifolius* Roxb. Part V. *Indian Journal of Chemistry 9:* 613.

DE ALVARENGA, M. A., GOTTLIEB, O. R. & MAGALHÃES, M. T., 1976. Methyl-phenanthrenes from *Sagotia racemosa*. *Phytochemistry, 15:* 844–845.

ANJANEYULU, V., RAO, G. S. & CONNOLLY, J. D., 1985. Occurrence of 24-epimers of cycloart-25-ene-3, 24-diols in the stem of *Euphorbia trigona*. *Phytochemistry, 24:* 1610–1612.

AYENSU, E. S., 1978. *Medicinal Plants of West Africa*. Algonac, Michigan, U.S.A.: Reference Publications, Inc.

AYENSU, E. S., 1981. *Medicinal Plants of West Indies*. Algonac, Michigan, U.S.A.: Reference Publications Inc.

BALLANTINE, J. A., 1969. The isolation of two esters of the naphthaquinone alcohol, shikonin, from the shrub *Jatropha glandulifera*. *Phytochemistry, 8:* 1587–1590.

BANERJI, M. S. & MILLNS, W., 1981. *Rubber India, 33(1):* 13–18, (via *Chemical Abstracts, 96:* 105549).

BANERJI, J., DAS, B., CHATTERJEE, A. & SHOOLERY, J. N., 1984. Gadain, a lignan from *Jatropha gossypifolia*, *Phytochemistry, 23:* 2323–2327.

BARNES, R. A. & SOEIRO, O. M., 1981. The alkaloids of *Croton salutaris*. *Phytochemistry, 20:* 543–544.

BARROS, D. A. D., DE ALVARENGA, M. A., GOTTLIEB, O. R. & GOTTLIEB, H. E., 1982. Naringenin coumaroylglucosides from *Mobea caudata*, *Phytochemistry, 21:* 2107–2109.

BASLAS, P. K. & AGARWAL, R., 1980. Isolation and characterization of different constituents of *Euphorbia hirta* Linn. *Current Science, 49:* 311–312.

BASLAS, R. K., 1982. Phytochemical studies of the plants of the genera *Euphorbia*. Part II. *Herba Hungarica, 21:* 115–126.

BENTLEY, R. & TRIMEN, H., 1880. *Medicinal Plants, 4.* London: J. A. Churchill.

BETTOLO, R. M. & SCARPATI, M. L., 1979. Alkaloids of *Croton draconoides*. *Phytochemistry, 18:* 520.

BEVAN, C. W. L., PATEL, M. B. & REES, A. H., 1964. Minor alkaloids of *Phyllanthus discoides*. *Chemistry and Industry,* 2054.

BHAKUNI, D. S. & DHAR, M. M., 1968. Crotsparine, a new proaporphine alkaloid from *Croton sparsiflorus* Morong. *Experientia, 24:* 10–11.

BHAKUNI, D. S. & DHAR, M. M., 1969. Crotsparinine, a dihydroaporphine alkaloid from *Croton sparsiflorus*. *Experientia, 25:* 354.

BHAKUNI, D. S., SATISH, S. & DHAR, M. M., 1970. The alkaloids of *Croton sparsiflorus*. *Phytochemistry, 9:* 2573–2580.

BHAT, S. K., DIXIT, V. & SINGH, K. V., 1981. A flavanone glycoside from seeds of *Sapium sebiferum*. *Phytochemistry, 20:* 2442.

BHAT, V. S., JOSHI, V. S. & NANAVATI, D. D., 1982. Cycloroylenol, a cyclopropane containing euphoid from *Euphorbia royleana*. *Tetrahedrom Letters*, 5207–5210.

BHUSHAN, D., RAM, A. & PRAKASH, O., 1951. Mandara seed and its oil. *Journal and Proceedings of the Oil Technologists Association, Kanpur, India*, 7: 68–73; (via *Chemical Abstracts*, 48: 9720).

BISWAS, K. & CHOPRA, R. N., 1982. *Common Medicinal Plants of Darjeeling and the Sikkim Himalayas*. Delhi: Periodical Experts Book Agency.

BOE, J. E., WINSNES, R., NORDAL, A. & BERNATEK, E., 1969. New constituents of *Euphorbia resinifera*. *Acta Chemica Scandinavica*, 23: 3609.

BONDARENKO, O. M., CHAGOVETS, R. K., LITVINENKO, V. I., OBOLENTSEVA, G. V., SILA, V. I. & KIGEL, T. B., 1971. *Euphorbia palustris* and *Euphorbia stepposa* flavonoids and their pharmacological properties. *Farmatsevticheskii Zhurnal* (Kiev), 26(6): 46–48; (via *Chemical Abstracts*, 76: 121697).

BONDAR, G., 1942. Penão, a new source of oil (in the province) of Bahia; *Cnidoscolus marcgravii* Pohl. *Revista de quimica industrial (Rio de Janeiro)*, 11(128): 15–18; (via *Chemical Abstracts*, 37: 4267).

BOULOS, L., 1983. *Medicinal Plants of North Africa*. Algonac, Michigan: Reference Publications, Inc.

CAIUS, J. F. & MHASKAR, K. S., 1923. The correlation between the chemical composition of anthelmintics and their therapeutic value in connection with the hookworm inquiry in the Madras Presidency. XXI. Miscellaneous anthelmintics. *Indian Journal of Medical Research*, 11: 353–370.

CALVIN, M., 1979. Petroleum plantations. In R. R. Hautala, R. B. King & C. Kutal (Eds), *Solar Energy: Chemical Conversion and Storage*. U.S.A.: The Humana Press.

CALVIN, M., 1980. Hydrocarbons from plants; analytical method and observations. *Naturwissenschaften*, 67: 525–533.

CALVIN, M., 1982. Energy agriculture. Presented at the *Symposium on Controversial Topics in Agricultural and Food Chemistry*, American Chemical Society, Kansas City, Mo, U.S.A., 7–12 Sept.: 1–46.

CALVIN, M., 1983. New sources for fuel materials. *Science*, 219: 24–26.

CALVIN, M., NEMETHY, E. K., REDENBAUGH, K. & OTVOS, J. W., 1982. Plants as a direct source of fuel. *Experientia*, 38: 18–22.

CHATTERJEE, A., DAS, B., PASCARD, C. & PRANGE, T., 1981. Crystal structure of a lignan from *Jatropha gossypifolia*. *Phytochemistry* 20: 2047–2048.

CHAWLA, H. M., CHAKRABARTY, K., CHIBBER, S. S., KALIA, A. N. & CHAUDHURY, N. C., 1980. Daphnetin from *Euphorbia dracunculoides*. *Indian Journal of Pharmaceutical Science*, 42: 138–139.

CHEN, Y., TANG, Z.-J., JIANG, F.-X., ZHANG, X.-X. & LAO, AI-NA., 1979. Studies on the active principles of Ze-Qi (*Euphorbia helioscopia* L.), a drug used for chronic bronchitis. *Yao Hsueh Hsueh Pao*, 14(2): 91–95. (via *Chemical Abstracts*; 92: 72680).

CHOPRA, R. N., CHOPRA, I. C. & VARMA, B. S., 1969. *Supplement to Glossary of Indian Medicinal Plants*. New Delhi: Publications & Information Directorate.

CHOPRA, I. C., ABROL, B. K., HANDA, K. L., PARIS, R. & DILLEMAN, G., 1983. *Medicinal Plants of the Arid Zones*. New Delhi: Today and Tomorrow's Printers & Publishers.

CLAUS, E. P., 1961. *Pharmacognosy*, 4th edition. Philadelphia: Lea & Febiger.

CRAVEIRO, A. A., MONTE, F. J., MATOS, F. J. A. & ALENCAR, J. W., 1978. Volatile constituents of *Croton* aff. *agyrophylloids* Muell.-Arg. *Revista latinoamericana de química*, 9: 98–99.

CRAVEIRO, A. A., RODRIGUEZ, A. S., ANDRADE, C. H. S., MATOS, F. J. A., ALENCAR, J. W. & MACHADO, M. I. L., 1981. Volatile constituents of Brazilian Euphorbiaceae. Genus *Croton*. *Journal of Natural Products*, 44: 602–608.

DAUGHERTY, P. M., SINEATH, H. H. & WASTLER, T. A., 1953. Industrial raw material of plant origin, III. Candelilla and candilella wax. *Fette und Seifen*, 55: 113–118.

DUKE, J. A. & AYENSU, E. S., 1985. *Medicinal Plants of China*, 1. Algonac, Michigan: Reference Publications, Inc.

DUTTA, P. K., BANERJEE, D. & DUTTA, N. L., 1972. Euphorbetin: A new bicoumarin from *Euphorbia lathyris* L., *Tetrahedron Letters*, 601–604.

DUTTA, P. K., MAJUMDER, P. C. & DUTTA, N. L., 1975. Synthetic approaches towards bicoumarins. Synthesis of euphorbetin and isoeuphorbetin. *Tetrahedron*, 31: 1167–1170.

EARLE, F. R., McGUIRE, T. A., MALLAN, J., BAGBY, M. O. & WOLFF, I. A., 1960. Search for new industrial oils. II. Oils with high iodine value. *Journal of the American Oil Chemists' Society*, 37: 48–50.

EVANS, F. J. & SOPER, C. J., 1978. The tigliane, daphnane and ingnane diterpenes, their chemistry, distribution and biological activities. A review. *Lloydia*, 41: 193–233.

FALSONE, G. & SPUR, B., 1979. Constituents of Euphorbiaceae. V. Synthesis of calcium 5,5-dimethyl-2-oxo-5,6-dihydro-2H-pyran3,4-dicarboxylate. *Liebigs Annalen die Chemie*, 923–926.

FUERSTENBERGER, G. & HECKER, E., 1977a. New highly irritant *Euphorbia* factors from the latex of *Euphorbia tirucalli* L. *Experientia*, 33: 986–988.

FUERSTENBERGER, G. & HECKER, E., 1977b. The new diterpene 4-deoxyphorbol and its highly unsaturated irritant esters. *Tetrahedron Letters*, 11: 925–928.

GOÑALONS, G. P. & FONTANA, A., 1926. Effect of *Phyllanthus sellowianus* Muell. on blood sugar concentration. *Archives argentinos de enfermedades del aparato digestivo y de la nutrición*, 1: 993–998 (via *Chemical Abstracts*, 22: 988).

GOVARDHAN, CH., REDDY, P. R. & SUNDRARARAMAIAH, T., 1984. 3-Epicycloaudenol and known triterpenes from *Euphorbia caudicifolia*. *Phytochemistry*, 23: 411–413.

GUNASEKERA, S. P., CORDELL, G. A. & FARNSWORTH, N. R., 1980. Constituents of *Nealchornea yapurensis* (Euphorbiaceae). *Journal of Natural Products, 43:* 285–287.

HELLETT, F. P. & PARKS, L. M., 1953. The antispasmodic principle of *Euphorbia pilulifera*. *Journal of the American Pharmaceutical Association 42:* 607–609.

HAMMOUDA, F. M., RIZK, A. M., EL-MISSIRY, M. M. & RADWAN, H. M., 1984. Constituents of the latex of *Euphorbia royleana*. *Fitoterapia, 55:* 245–247.

HART, N. K., JONES, S. R. & LAMBERTON, J. A., 1967. (+)-9-Aza-1-methyl-bicyclo [3,3,1] nonan-3-one, a new alkaloid from *Euphorbia atoto* Forst. *Australian Journal of Chemistry, 20:* 561–563.

HART, N. K., JONES, S. R. & LAMBERTON, J. A., 1969. Hexahydroimidazo-pyrimidines, a new class alkaloids from *Alchornea javanensis*. *Journal of the Chemical Society, D:* 1484–1485.

HART, N. K., JONES, S. R., LAMBERTON, J. A. & WILLIG, R. J., 1970. Alkaloids of *Alchornea javanensis*: the isolation of hexahydroimidazo [1,2a] pyrimidines and guanidines. *Australian Journal of Chemistry, 23:* 1679–1693.

HASSAN, S. Q., AHMED, I., SHERWANI, M. R. K., ANSARI, A. A. & OSMAN, S. M., 1980. Studies on herbaceous seed oils. X. *Fette, Seifen, Anstrichmittel, 82:* 204–205.

HECKER, E., 1968. Cocarcinogenic principles from the seed oil of *Croton tiglium* and from other Euphorbiaceae. *Cancer Research, 28:* 2338–2349.

HEGNAUER, R., 1966. *Chemotaxonomie der Pflanzen*, Bd. 4. Dicotyledoneae: Daphniphyllaceae-Lythraceae, Basel & Stuttgart: Birkhaeuser Verlag.

HILL, A. F., 1952. *Economic Botany—A Textbook of Useful Plants and Plant Products*, 2nd edition. Bombay, New Delhi: Tata McGraw-Hill Publishing Co. Ltd.

HINMAN, C. W., HOFFMAN, J. P., MCLAUGHLIN, S. P. & PEOPLES, T. R., 1980. Hydrocarbon production from arid land plant species. *Proceedings of the 1980 Annual Meeting, American Section of the International Solar Energy Society, Inc.:* 110–114.

HUI, W.-H., LI, M.-M. & NG, K.K., 1975. Terpenoids and steroids from *Macaranga tanarius*. Phytochemistry *14:* 816–817.

HUSAIN, S. R., AHMAD, M. S., AHMAD, S. M., AHMAD, M. & OSMAN, S. M., 1982. Studies on herbaceous seed oils. XIII. *Journal of the Oil Technologists Association of India, 14(2):* 61–63.

ISMAIL, S. I., EL-MISSIRY, M. M., HAMMOUDA, F. M. & RIZK, A. M., 1977. Flavonoids of *Euphorbia geniculata* and *Euphorbia prostrata*. *Pharmazie, 32:* 538.

JEFFERIES, P. R., 1979. Studies on the resin chemistry of some Western Australian plants. *Journal of the Royal Society of Western Australia, 62:* 95–107.

JOHNS, S. R. & LAMBERTON, J. A., 1967. New imidazole alkaloids from a *Glochidion* species (family Euphorbiaceae). *Australian Journal of Chemistry, 20:* 555–560.

KALLIMANIS, G. & PHILIANOS, S., 1980. Constituents chimiques des feuilles d'*Euphorbia acanthothamnos* Heldr. st Start. *Plantes medicinales Phytotherapie, 14:* 233–237.

KAUFMANN, H. P. & KING, B.-W., 1939. Fats from the seeds of *Stillingia sebifera* Trillot. *Fette und Seifen, 46:* 388–390.

KHAFAGY, S. M., MAHMOUD, Z. F. & ABDEL SALAM, N. A., 1979. Coumarins and flavonoids of *Ricinus communis* growing in Egypt. *Planta Medica, 37:* 191.

KINGHORN, A. D., 1979. Carcinogenic irritant Euphorbiaceae. In A. D. Kinghorn, (Ed.), *Toxic Plants;* 137–159. New York: Columbia University Press.

KINGSBURY, J. M., 1964. *Poisonous Plants of the United States and Canada*. New Jersey: Prentice Hall, Inc.

KIRTIKAR, K. R. & BASU, B. D., 1984. *Indian Medicinal Plants, 2:* 2190–2290. Allahabad, India: Lalit Mohan Basu Publisher.

KJAER, A. & FRIIS, P., 1962. Isothiocyanates XLIII. Isothiocyanates from *Putranjiva roxburghii* Wall. including (*S*)-2-methylbutyl isothiocyanate, a new mustard oil of natural derivation. *Acta Chemica Scandinavica, 10:* 936–946.

KOUNO, I., SAISHOJI, T., SUGIYAMA, M. & KAWANO, N., 1983. A xylosyl-glucoside of xanthoxylin from *Sapium sebiferum* root bark. *Phytochemistry, 22:* 790–791.

KUONO, I., SHIGEMATSU, N., IWAGAM, M. & KAWANO, N., 1985. Further phloroglucinol derivatives in the fruits of *Mallotus japonicus*, *Phytochemistry, 24:* 620–621.

KREWSON, C. F. & SCOTT, W. E., 1966. *Euphorbia lagascae* Spreng., an abundant source of epoxyloeic acid; seed extraction and oil composition. *Journal of the American Oil Chemists' Society, 43:* 171–174.

KUPCHAN, S. M., UCHIDA, I., BRANFMAN, A. R., DAILEY, R. G., Jr. & FEI, B. Y., 1976. Antileukemic principles isolated from Euphorbiaceae plants. *Science, 191:* 571–572.

LAW, AH., THRELFALL, D. R. & WHISTANCE, G. R., 1970. Ubiquinones of *Heva brasiliensis*. *Phytochemistry, 9:* 2461.

LEE, Y. C. & NOBLES, W. L., 1959. *Journal of the American Pharmaceutical Association, Scientific Edition, 48:* 162–165.

LEE, K.-H., HAYASHI, N., OKANO, M., HALL, I. H., WU, R.-Y., & MCPHAIL, A., 1982. Lasiodiplodin, a potent antileukemic macrolide from *Euphorbia splendens*. *Phytochemistry, 21:* 1119–1121.

LEWIS, W. H. & ELVIN-LEWIS, M. P. F., 1977. *Medical Botany*. New York, London: John Wiley & Sons.

LI KEN JIE, M. S. F. & SINHA, S., 1981. Fatty acid composition and the characterisation of a novel dioxo C_{18} fatty acid in the latex of *Heva brasiliensis*. *Phytochemistry, 20:* 1863–1866.

MAHMOUD, Z. F. & ABDEL SALAM, N. A., 1979. Coumarins of *Euphorbia terracina* and *Euphorbia paralias* growing in Egypt. *Pharmazie, 34:* 446–447.

MANNERS, G. D. & DAVIS, D. G., 1984. Epicuticular wax constituents of north American and European *Euphorbia esula* biotype. *Phytochemistry, 23:* 1059–1062.

MANSKE, R. H. F., 1968. Alkaloids unclassified and of unknown structure. In R. H. F. Manske (Ed.), *The Alkaloids, Chemistry and Physiology, 10:* 545–595. New York, London: Academic Press.

MATSUNAGO, S. & MORITA, R., 1983. Hopenol-B, a triterpene alcohol from *Euphorbia supina*. *Phytochemistry, 22:* 605–606.

MUELLER, R. & POHL, R., 1970. Die Flavonolglykoside von *Euphorbia amygdaloides* und ihre Quantitative Bestimmung in Verschieden Entwicklungsstadien der Pflanze. 5. Mitteilung ueber die Flavonoide einheimisher Euphorbiaceen. *Planta Medica, 18:* 114–129.

MUELLER, P. & SCHUETTE, H. R., 1968. Biochemistry and physiology of latex. XVIII. 1-Methyl-6-hydroxy-1,2,3,4-tetrahydroxyisoquinoline-3-carboxylic acid in *Euphorbia myrsinites* latex. *Zeitschrift für Naturforschung, B, 23:* 491–493.

NAHRSTEDT, A., KANT, J.-D. & WRAY, V., 1982. Acalyphic, a cyanogenic glucoside from *Acalypha indica*. *Phytochemistry, 21:* 101–105.

NAZIR, M., AHMAD, I., BHATTY, M. K. & KARIMULLAH, 1966. Chemical constituents of *Euphorbia royleana*. *Pakistan Journal of Scientific and Industrial Research, 9:* 38–40.

NEMETHY, E. K., OTVOS, J. W. & CALVIN, M., 1979. Analysis of extractables from one *Euphorbia*. *Journal of the American Oil Chemists' Society, 56:* 957–960.

NEMETHY, E. K., OTVOS, J. W. & CALVIN, M., 1981a. Natural production of high-energy liquid fuels from plants. In D. L. Klass & G. H. Emert (Eds), *Fuels from Biomass and Wastes:* 405–419. Michigan: Ann Arbor Publishers Inc.

NEMETHY, E. K., OTVOS, J. W. & CALVIN, M., 1981b. Hydrocarbons from *Euphorbia lathyris*. *Pure and Applied Chemistry, 53:* 1101–1108.

NOVACK, E. A., CREA, A. E. G. & FALSONE, G., 1980. Inhibition of mitochondrial oxidative phosporylation by 4-deoxyphorbol triesters, a poisonous constituent of the latex of *Euphorbia biglandulosa* Desf. *Toxicon, 18:* 165–174.

OGURI, S. & SHINOHARA, I., 1957. Effects of standing, heating and stirring on mucilage for Japanese paper making. *Kogyo Kagaku Zasshi, 60:* 467–469; (via *Chemical Abstracts, 53:* 7592).

OKUDA, T. & KARU S., 1981. Studies of the tannins and polyphenols of *Mallotus japonicus*. I. *Mallotus japonicus* leaf tannins. *Nippon Kagaku Kaishi:* 671–677. (via *Chemical Abstracts, 95:* 3440).

OKUDA, T. & SENO, K., 1981. Cited in Yoshida *et al.*, 1982.

OKUDA, T., MORI, K. & HATANO, T., 1980a. The distribution of geraniin and mallotusinic acid in the order Geraniales. *Phytochemistry, 19:* 547–551.

OKUDA, T., YOSHIDA, T. & HATANO, T., 1980b. Equilibrated stereo-structures of hydrated geraniin and mallotusinic acid. *Tetrahedron Letters, 21:* 2561–2564.

PARELLO, J. & MUNAVALLI, S., 1965. Phyllantine and phyllantidine, alkaloids of *Phyllanthus discoides*. *Comptes Rendus, 260:* 337–340.

PARIS, R. A., MOYSE-MIGNON, H. & MEN, J. LE., 1955. Alkaloids of *Flueggea virosa*. *Annales pharmaceutiques françaises, 13:* 245–249.

PARTHASARATHY, M. R. & SARADHI, K. P., 1984. A Coumarino-lignan from *Jatropha glandulifera*. *Phytochemistry, 23:* 867–869.

PEOPLES, T. R., ALCORN, S., BLOSS, H. E., CLAY, W. F., FLUG, M., HOFFMAN, J. J., LEE, C. W., LUNA, S., MCLAUGHLIN, S. P. *et al.*, 1981. *Euphorbia lathyris* L.: A future source of extractable liquid fuels. *Biosources Digest 3:* 117–123; (via *Chemical Abstracts, 95:* 222769).

PIATAK, M. D., & REIMANN, K., 1972. Plant investigation. IV. Corollatadiol, a new triterpene from *Euphorbia corollata*. *Tetrahedron Letters, 44:* 4525–4528.

POHL, R., JANISTYN, B. & NAHRSTEDT, A., 1975. Die Flavonolglykoside von *Euphorbia helioscopia*, *E. stricta*, *E. verrucosa* and *E. dulcis*. 9. Mitteilung ueber die Flavonoide eimheimisher Euphorbiaceen. *Planta Medica, 27:* 301–303.

PONSINET, G. & OURISSON, G., 1965. Études chimico-taxonomiques dans la famille des Euphorbiacees—I. Introduction générale et séparation et identification des triterpènes tétracycliques monohydroxyles naturels. *Phytochemistry, 4:* 799–811.

PRAKASH, S., KHAN, M. A., KHAN, H. & ZAMAN, A., 1983. A piperidine alkaloid from *Excoecharia agallocha*. *Phytochemistry, 22:* 1836–1837.

PUNTAMBEKAR, S. V. & KRISHNA, S., 1932. The seeds of *Vateria indica* Linn. as a source of vegetable tallow. *Indian Forester, 58:* 68–74; (via *Chemical Abstracts, 28:* 6005).

RAO, D. N. & ROW, L. R., 1966. Crystalline constituents of Euphorbiaceae. III. The triterpenes of Euphorbia nerifolia. *Current Science, 34:* 432.

RIZK, A. M., 1986. *The Phytochemistry of the Flora of Qatar.* The Scientific and Applied Research Centre, Qatar University. London: Kingprint Press.

RIZK, A. M. & EL-MISSIRY, M. M., 1986. Non-diterpenoid constituents of Euphorbiaceae and Thymelaeaceae. In F. J. Evans (Ed.), *Naturally Occurring Phorbol Esters:* 107–138. Boca Raton: CRC Press.

RIZK, A. M. & RIMPLER, H., 1977. Investigation of *Euphorbia indica*. *Planta Medica, 32:* 177–180.

RIZK, A. M., YOUSSEF, A. M., DIAB, M. A. & SALEM, H. M., 1974. Constituents of Egyptian

Euphorbiaceae. I. Tritrpenoids and related substances of *Euphorbia paralias*. *Zeitschrift für Naturforschung*, 29C: 529–531.
RIZK, A. M., YOUSSEF, A. M., DIAB, M. A. & SALEM, H. M., 1976. Flavonoids of *Euphorbia paralias*. *Pharmazie*, 31: 405.
RIZK, A. M., RIMPLER, H. & ISMAIL, S. I., 1977. Flavonoids and ellagic acid from *Euphorbia hypericifolia* (= *E. indica*). *Fitoterapia*, 48: 99–100.
RIZK, A. M., HAMMOUDA, F. M., SEIF EL-NASR, M. M. & EL-MISSIRY, M. M., 1978. Phytochemical investigation of *Euphorbia geniculata* and *E. prostrata*. *Pharmazie*, 33: 540–541.
RIZK, A. M., AHMED, S. S. & DIAB, M. M., 1979. Further investigation of the flavonoids of *Euphorbia paralias* L. *Planta Medica*, 36: 189–190.
RIZK, A. M., HAMMOUDA, F. M., SEIF EL-NASR, M. M. & ABOU-YOUSSEF, A. A., 1980a. Phytochemical investigation of *Euphorbia peplus*. *Fitoterapia*, 51: 223–227.
RIZK, A. M., HAMMOUDA, F. M., EL-SHAMY, A. M. & EL-MISSIRY, M. M., 1980b. Triterpenoids from latices of *Euphorbia paralias* and *E. geniculata*, *Fitoterapia*, 51: 313–316.
RIZK, A. M., AL-NAGDY, S. A. & EL-MISSIRY, M. M., 1982. Investigation of *Euphorbia granulata*. *Pharmazie*, 37: 737–738.
ROJAS, H. C., 1964. Possibilities of obtaining pulp and paper from some national raw materials. *Tecnologia*, 6: 26–32; (via *Chemical Abstracts, 66:* 66932).
ROSHCHIN, YU. V. & KIR'YALOV, N. P., 1970. Tetracyclic triterpenoid compounds from *Euphorbia condylocarpa*. *Khimiya prirodnykh soedinenij*, 6: 483 (via *Chemical Abstracts, 74:* 10351).
SAHAI, R., DUBE, M. P. & RASTOGI, R. P., 1981. Chemical and pharmacological study of *Euphorbia maddeni*. *Indian Journal of Pharmaceutical Science*, 43: 216–219.
SAINSBURY, M., 1970. Fridelin and epifriedlinol from the bark of *Prunus* and a review of their natural distribution. *Phytochemistry*, 9: 2209–2215.
SARKAR, S. & CHAKRABARTY, M. M., 1956a. Seed fats of the Euphorbiaceae family. I. Chemical examination of the seed fat from *Trewia nudiflora*. *Science and Culture*, 21: 473–474; (via *Chemical Abstracts, 50:* 10427).
SARKAR, S. & CHAKRABARTY, M. M., 1956b. Seed fats of the Euphorbiaceae II. The chemical composition of the seeds fats from *Bischofia javanica* and *Antidesma diandrum*. *Science and Culture*, 22: 336–337; (via *Chemical Abstracts, 51:* 12511).
SCHROEDER, G., ROHMER, M., BECK, J. P. & ANTON, R., 1980. 7-Oxo-, 7 α-hydroxy- and 7 β-hydroxysterols from *Euphorbia fischeriana*, *Phytochemistry*, 19: 2213–2215.
SCHULTE, K. E., REISCH, J. & BORNFLETH, H., 1964. The occurrence of *trans*-dehydromatricaria ester and 1-tridecene-3,5,7,9,11-pentyne in *Ricinus communis*, *Archiv der Pharmazie*, 297: 443.
SEKULA, B. C. & NES, W., 1980. The identification of cholesterol and other sterols in *Euphorbia pulcherrima*. *Phytomistry*, 19: 1509–1512.
SERRA, R. M., 1944. Investigation of quinine in *Phyllanthus nururi* L. *Anales Universidad Santo Domingo*, 8: 295–297; (via *Chemical Abstracts, 41:* 1812).
SHANG, T.-M., WANG, L., LIANG, H.-T., LIN, X.-Y., XIAO, J.-M. & NIU, S.-L., 1979. Studies on the constituents of Mao-Yan-Cao (*Euphorbia lunulata* Bge). *Hua Hsueh Hsueh Pao*, 37: 119–128; (via *Chemical Abstracts, 92:* 3184).
SHIGEMATSU, N., KOUNO, I. & KAWANO, N., 1983. Phloroglucinol derivatives from fruits of *Mallotus japonicus*. *Phytochemistry*, 22: 323–325.
SINGH, A. & SRIVASTAVA, S. N., 1966. Chemical examination of *Euphorbia dracunculoides* Lam. *Indian Journal of Chemistry*, 4: 420.
SOBOLEVA, V. A., 1979. Flavonoids of some *Euphorbia* species. *Khimiya Prirodnykh Soedinenij*, 5: 855–856; (via *Chemical Abstracts, 93:* 3899).
SOBOLEVA, V. A., 1980. Polyphenol compounds of *Ricinus communis* and *Mercurialis perennis*. *Khimiya Prirodnykh Soedinenij*, 123–124; (via *Chemical Abstracts, 92:* 177454).
STARRAT, A. N., 1966. Triterpenoid constituents of *Euphorbia cyparissias*. *Phytochemistry*, 5: 1341–1344.
STARRAT, A. N., 1969. Isolation of Hopenone-B from *Euphorbia cyparissias*. *Phytochemistry*, 8: 1831–1832.
STUART, K. L., CHAMBERS, C. & BYFIELD, D., 1969. Morphinandienone alkaloids from *Croton flavens* L., *Journal of the Chemical Society (C)*, 168–1684.
TAKEMOTO, T. & INGAKI, M., 1958. Constituents of *Euphorbia maculata*. *Yakugaku Zasshi*, 78: 292–294.
TAKEMOTO, T. & ISHIGURO, T., 1966. Constituents of *Euphorbia watanabei*. I. *Yakugaku Zasshi*, 86: 530–533.
TALAPATRA, B., GOSWAMI, S. & TALAPATRA, S. K., 1981. Acalyphamide, a new amide and other constituents of *Acalypha indica* Linn. *Indian Journal of Chemistry*, 20B: 974–977.
TIWARI, K. P., CHOUDHARY, R. N. & PANDY, G. D., 1981. 3-Methoxy-4,6-dihydroxymorphinandien-7-one, an alkaloid from *Croton bonpalandianum*. *Phytochemistry*, 20: 863–864.
TIPATHI, R. D. & TIWARI, K. P., 1980. Geniculatin, a triterpenoid saponin from *Euphorbia geniculata*. *Phytochemistry*, 19: 2163–2166.
TUTIYA, M. & KATO, Y., 1940. Foliaceous woods as raw material from pulp in Formosa. *Journal of the Agricultural Chemical Society of Japan*, 16: 224–226.
UEMURA, D. & HIRATA, Y., 1971. The isolation and structures of two new alkaloids milliamines A and B, obtained from *Euphorbia millii*. *Tetrahedron Letters*, 3673–3676.

VALEN, F., VAN., 1978. Contribution to the knowledge of cyanogenesis in Angiosperms. 10. Communication. Cyanogenesis in Euphorbiaceae. *Planta Medica, 34:* 408–413.

VICKERY, M. L. & VICKERY, B., 1979. *Plant Products of Tropical Africa.* London: MacMillan.

WAGNER, H., HOERHAMMER, L. & KIRALY, I. C., 1970. Flavon-C-Glykoside in *Croton zambezicus. Phytochemistry, 9:* 897.

WARNAAR, F., 1981. Conjugated fatty acids from latex of *Euphorbia lathyris. Phytochemistry: 20:* 89–91.

WATANABE, T. A., ARAKI, T., OGATA, K., GOTO, M. & HOMU, I., 1956. Antibiotic components of plants. II. *Takeda Kenkyusho Nempo, 15:* 129–139. (via *Chemical Abstracts, 51:* 8367).

WATT, J. M. & BREYER-BRANDWIJK, M. G., 1962. *The Medicinal and Poisonous Plants of Southern and Eastern Africa,* 2nd edition. Edinburgh, London: E. & S. Livingstone Ltd.

WOERD, L. A., VAN DER., 1941. The native medicines of the East Indian Archipelago. VII. The diuretic action of the most common herbs used in the Netherland East Indies in native medicine against diseases of the urinary system. *Geneeskundig Tijdschrift voor Nederlandsch-Indië, 81:* 1963–1980. (via *Chemical Abstracts, 36:* 160).

YAMAGUCHI, K., 1970. *Spectral Data of Natural Products, I.* Amsterdam: Elsevier.

YOSHIDA, T., SENO, K., TAKAMO, Y. & OKUDA, T., 1982. Bergenin derivatives from *Mallotus japonicus. Phytochemistry, 21:* 1180–1182.

YOSHITAMA, K., ISHII, K. & YASUDO, H., 1980. A chromatographic survey of anthocyanins in the flora of Japan. *Journal of the Faculty of Science of Shinshu University, 15:* 19–25.